ELEVENTH INTERNATIONAL CONFERENCE ON THE BEARING CAPACITY OF ROADS, RAILWAYS AND AIRFIELDS, VOLUME 1

PROCEEDINGS OF THE ELEVENTH INTERNATIONAL CONFERENCE ON THE BEARING CAPACITY OF ROADS, RAILWAYS AND AIRFIELDS, 8-10 JUNE 2022, TRONDHEIM, NORWAY

Eleventh International Conference on the Bearing Capacity of Roads, Railways and Airfields, Volume 1

Edited by

Inge Hoff
Norwegian University of Science and Technology (NTNU), Trondheim, Norway

Helge Mork
Norwegian University of Science and Technology (NTNU), Trondheim, Norway

Rabbira Saba
Norwegian Public Roads Administration (SVV), Trondheim, Norway

CRC Press is an imprint of the
Taylor & Francis Group, an **informa** business

A BALKEMA BOOK

CRC Press/Balkema is an imprint of the Taylor & Francis Group, an informa business

© 2022 selection and editorial matter, the Editors; individual chapters, the contributors

Typeset by Integra Software Services Pvt. Ltd., Pondicherry, India

The right of the Editors to be identified as the author of the editorial material, and of the authors for their individual chapters, has been asserted in accordance with sections 77 and 78 of the Copyright, Designs and Patents Act 1988.

The Open Access version of this book, available at www.taylorandfrancis.com, has been made available under a Creative Commons Attribution-Non Commercial-No Derivatives 4.0 license.

Although all care is taken to ensure integrity and the quality of this publication and the information herein, no responsibility is assumed by the publishers nor the author for any damage to the property or persons as a result of operation or use of this publication and/or the information contained herein.

Library of Congress Cataloging-in-Publication Data

A catalog record has been requested for this book

Published by: CRC Press/Balkema
               Schipholweg 107C, 2316 XC Leiden, The Netherlands
               e-mail: enquiries@taylorandfrancis.com
               www.routledge.com – www.taylorandfrancis.com

ISBN: 978-1-032-12044-7 (Hbk)
ISBN: 978-1-032-12068-3 (Pbk)
ISBN: 978-1-003-22288-0 (eBook)
DOI: 10.1201/9781003222880

Table of contents

Preface ix

Organizing committee xi

Asphalt materials

Comparing asphalt modified with recycled plastic polymers to conventional polymer modified asphalt 3
G. White & F. Hall

Possibilities of surface protection on road and airport pavements 18
J. Grosek, Z. Nevosad, T. Zavrel, T. Macan & B. Dohnalkova

Developing high performance asphalt mixtures with considerable amounts of recycled asphalt with a new chemical bitumen additive 28
N. Carreño, M. Oeser, M. Zeilinger & O. Fleischel

Study on low-temperature crack resistance of zirconium tungstate modified asphalt mastic and asphalt mixtures 38
Z.S. Pei, J.Y. Yi, D.C. Feng & J.J. Li

Evaluation air void content of drilled porous asphalt mixture cores using non-destructive X-ray computed tomography 48
T. Vieira, J. Lundberg, O. Eriksson, A. Guarin, S. Gong & S. Erlingsson

Experimental investigation on rheological property of bitumen influenced by preheating process 59
Xuemei Zhang & Inge Hoff

Skid resistance properties against RAP content change in surface asphalt mixture 68
M. Pomoni, C. Plati & A. Loizos

Generation and propagation of reflective cracking and thermal cracking of asphalt pavement in cold regions of China 78
W.Y. Zhou, Z.G. Chen, Z.N. Chen, W.J. Qin & J.Y. Yi

Validation of national empirical pavement design approaches for cold recycled asphalt bases 87
M. Winter, K. Mollenhauer, A. Graziani, C. Mignini, H. Bjurström, B. Kalman, P. Hornych, V. Gaudefroy, D. Lo Presti & G. Giancontieri

Ethylene-octene-copolymer as an alternative to styrene-butadiene-styrene bitumen modifier 96
A. Riekstins, V. Haritonovs, R. Merijs-Meri & J. Zicāns

Exploring correlations between the compressive modulus and the flexural modulus of asphalt concrete mixtures 108
A. Jamshidi, D. Alenezi & G. White

Viscoelastic characterization and comparison of norwegian asphalt mixtures using dynamic modulus and IDT tests 118
M. Alamnie, E. Taddesse & I. Hoff

Granular materials

Evaluating the effect of mineralogy and mechanical stability of recycled excavation materials by Los Angeles and micro-Deval test 131
S. Adomako, R.T. Thorstensen, N. Akhtar, M. Norby, C.J. Engelsen, T. Danner & D.M. Barbieri

Freeze-thaw performance of granular roads stabilized with optimized gradation and clay slurry 139
Ziqiang Xue, Jeramy C. Ashlock, Bora Cetin, Sajjad Satvati, Halil Ceylan, Yijun Wu & Cheng Li

Stabilization and Geogrids

Evaluating the cost effectiveness of using various types of stabilized base layers in flexible pavements 153
A. Francois, D. Offenbacker & Y. Mehta

Measurement of modulus of subgrade reaction for geogrid stabilized roadways 162
Prajwol Tamrakar, Mark H. Wayne, Garrett Fountain, David J. White & Pavana Vennapusa

Waste paper ash as an alternative binder to improve the bearing capacity of road subgrades 172
J.J. Cepriá, R. Orejana, R. Miró, A. Martínez, M. Barra, D. Aponte & H. Baloochi

Experiences with the use of stabilisation geogrids in demonstrating an improvement in bearing capacity of recycled materials 182
P. Mazurowski, K. Zamara, C. Gewanlal & J. Kawalec

Airport design

Introducing dynamic asphalt modulus to the design of flexible aircraft pavement structures 195
H. Weisser & G. White

Practical implications for the implementation of the new international airport pavement strength rating system 210
G. White

Analyzing the bearing capacity of materials used in arresting systems as a suitable risk mitigation strategy for runway excursions in landlocked aerodromes 226
M. Ketabdari, E. Toraldo & M. Crispino

Highway design

Impact of introducing longer and heavier vehicles on the bearing capacity of pavement subgrades 239
M.S. Rahman & S. Erlingsson

Post-modern pavement: Theory, Concept, Performance and Challenges 250
A. Jamshidi & G. White

Concrete pavements

Crack behavior of continuously reinforced concrete pavements (CRCP) in Germany   263
M. Moharekpour, S. Hoeller & M. Oeser

Modelling

A model for the permanent deformation behavior of the unbound layers of pavements   277
M.S. Rahman, S. Erlingsson, A. Ahmed & Y. Dinegdae

Development of simplified models to assess pavement structural condition on network level   288
M. Pettinari, S. Baltzer, M. Kalantari & D. Jansen

Predicting asphalt pavement temperatures as an input for a mechanistic pavement design in Central-European climate   299
S. Cho, C. Tóth & P. László

Stress random distributions on railway subgrade surface under train loads considering irregular shapes of ballast   305
J. Xiao, L. Xue, P. Jing & D. Zhang

Modelling and predicting asphalt deflection values with artificial neural networks   315
M. Rahimi Nahoujy & M. Radenberg

Analysis and modelling of the main causes of unsatisfactory quality of transportation infrastructures   325
F.G. Praticò, A. Astolfi & R. Fedele

Influence of acceleration and deceleration of Super Heavy Loads (SHLs) on the service life of pavement structures   335
Ali Morovatdar & Reza S. Ashtiani

Analytic analysis of a grid-reinforced asphalt concrete overlay   346
J. Nielsen, K. Olsen, A. Skar & E. Levenberg

Calibration of fatigue performance model for cement stabilized flexible pavements   355
E. Rodriguez & R. Ashtiani

Case study

Stone mastic asphalt as an ungrooved runway surface: Case study on Emerald airport   367
G. White & H. Dutney

Field investigation of the deterioration of flexible polymer modified pavements: A case study in Northern New England   387
M. Elshaer & C. Decarlo

NDT and APT

Traffic speed deflectometer measurements at the aurora instrumented road test site   399
C.P. Nielsen, P. Varin, T. Saarenketo & P. Kolisoja

Unbound pavement materials' response to varying groundwater table analysed by falling weight deflectometer 409
M. Fladvad & S. Erlingsson

FWD quality assurance in Germany 419
D. Jansen

Identifying weak joints in jointed concrete pavements from TSD measurements by basis pursuit 430
M. Scavone, S.W. Katicha & G.W. Flintsch

Assessing the usefulness of the outer-area method in quantifying the structural capacity of full-scale composite pavement sections subjected to accelerated pavement testing 440
A. Francois, D. Offenbacker & Y. Mehta

Prediction of subgrade soil density using dielectric constant of soils 448
A. Abdelmawla & s. Sonny Kim

Traffic speed deflectometer measurements at copenhagen airport 458
C.P. Nielsen & K. Jensen

Field performance assessment versus asphalt pavement design considerations 466
A. Loizos, K. Gkyrtis, C. Plati & K. Georgouli

Using non-destructive test to validate and calibrate smart sensors for urban pavement monitoring 475
A. Di Graziano, S. Cafiso, A. Severino, F. Praticò, R. Fedele & G. Pellicano

Author index 488

Preface

Bearing capacity is not an easy term to define. It is used to express the traffic load a pavement structure can withstand without showing severe damage. It is not a measure of the ultimate structural strength that will cause a failure state for a single vehicle if exceeded. The ultimate strength of a pavement is normally significantly higher than the bearing capacity. For rehabilitation projects, the bearing capacity is becoming more and more important, and it is a challenging task for an engineer to determine the remaining bearing capacity of a structure and how this capacity could be further utilized.

This conference is the eleventh in the series started in Trondheim 1982 and arranged at four-year intervals under the title Bearing Capacity of Roads and Airfields, BCRA. At the sixth conference in Lisbon, Portugal, railway tracks was added as the third important component. The acronym is since then BCRRA. The conference has been organized at five cities in addition to Trondheim (1982, 1990, 1998, 2005 and 2013): Plymouth, UK (1986), Minneapolis, USA (1994), Lisbon, Portugal (2002), Champaign, USA (2009) and Athens, Greece (2017).

The BCRA/BCRRA conferences have contributed to creating and sharing knowledge in this area through more than 2000 papers and an even higher number of participants in the conferences.

The organizing committee would like to express gratitude to all that have participated in making this conference possible.

Due to the SARS Covid-19 pandemic, the planned 2021 conference had to be postponed one year. However, some of the submitted papers are published in this 2021 Volume 1 of the Proceedings, while the main part will be published in the 2022 Volume 2.

Inge Hoff
Chair of the organizing committee

Organizing committee

Inge Hoff, *Norwegian University of Science and Technology (NTNU), Trondheim, Norway*
Helge Mork, *Norwegian University of Science and Technology (NTNU), Trondheim, Norway*
Rabbira Saba, *Norwegian Public Roads Administration (SVV), Trondheim, Norway*
Leif Bakløkk, *Norwegian Public Roads Administration (SVV), Trondheim, Norway*

Asphalt materials

Comparing asphalt modified with recycled plastic polymers to conventional polymer modified asphalt

G. White & F. Hall
School of Science and Engineering, University of the Sunshine Coast, Sippy Downs, Queensland, Australia

ABSTRACT: As sustainable infrastructure solutions become a focus of societies and governments, the interest in recycling plastic into asphalt mixtures will continue to increase. There are many forms of plastic that can be added to asphalt, but those plastics that partially replace the bituminous binders, as well as improve the mechanical properties of the asphalt mixture, are the most valuable. The effects of 6% (of bitumen mass) of two commercially available recycled plastic products were compared to the effects of 2-6% conventional elastomeric and plastomeric polymers, within a typical dense graded asphalt mixture. The Marshall stability associated with recycled plastic was comparable to that for 2-3% conventional polymer content, while indirect tensile stiffness and deformation resistance were comparable to 4-6% conventional polymer content. In contrast, workability, Marshall flow and moisture damage resistance were not significantly different for recycled plastic and conventional polymers, regardless of the conventional polymer content. It was concluded that the recycled plastic products used in this research should be thought of as sustainable polymers for asphalt mixture modification, rather than as simple extenders of the bituminous phase in asphalt production. However, further work is required to understand the ageing, leaching and fuming of asphalt mixtures modified with recycled plastic, compared to conventional polymers.

Keywords: Asphalt, Modification, Waste, Recycled, Plastic, Polymer

1 INTRODUCTION

The desire to develop sustainable infrastructure, including pavement structures and materials, is ever increasing in recent times (NAPA 2020). Given the diversity of pavement structures, which can include cement concrete, asphalt mixtures, granular crushed rock and natural gravels, the opportunities for sustainable pavement construction are broad and many. When considering sustainability opportunities, it is important to take into account the effect on the durability and expected life of the pavement, as well as the reduction in financial or environmental cost of the more sustainable solution (Jamshidi & White 2019). That is, an initiative that reduces the new pavement's greenhouse gas emissions by 20% is not really sustainable if the pavement only lasts 50% of the life of the conventional solution (Jamshidi & White 2020). It is also important to understand that sustainability initiatives are only viable if the cost to collect, process and reincorporate a recycled material or product is less expensive than the cost of the material or product that it replaces (White 2019). For this reason, the replacement of high cost materials, such as bituminous binder and cement, with recycled or repurposed materials, is of great interest to practitioners and researchers alike.

One opportunity is to replace high cost bituminous binder with low cost recycled plastic in asphalt mixtures (NAPA 2020). However, there are many types and forms of recycled plastic

DOI: 10.1201/9781003222880-1

and only a few of them are well suited to recycling in asphalt mixtures (Brasileiro et al. 2019; Masad et al. 2020). Some binders and asphalt mixtures modified with recycled plastic have been reported to have properties comparable to those generally associated with conventional polymer modification (White & Reid 2020). Consequently, some researchers have recommended a direct comparison to polymer modified binders (PMB) and asphalt mixtures produced with PMB.

This research compares the effects of two commercially available recycled plastic products, to the effects associated with asphalt mixtures modified with the conventional polymers styrene-butadiene-styrene (SBS) and ethyl vinyl acetate (EVA). The aim was to determine the dosage of conventional polymers that gave comparable asphalt performance property values to those associated with the recycled plastics. Common performance-indicating tests were undertaken on otherwise identical asphalt mixtures produced with bituminous binder modified with varying dosages of various conventional polymers and two recycled plastic products.

2 BACKGROUND

2.1 *Approaches to recycling plastic in asphalt mixtures*

As stated above, there are many approaches to recycled plastic in asphalt mixtures. The polymer and form of plastic, the aim of the user and the local logistics all influence the most appropriate approach to make. However, these same factors will also affect the properties of the resulting asphalt mixture and the long-term performance of the road surface. In general, there are four approaches to recycling plastic in asphalt mixtures (White 2020):

- Uncontrolled waste disposal. Uncontrolled incorporation of variable plastics with only minimal processing to reduce the particle size to be comparable to asphalt mixture aggregates. This offers minimal return on the investment and a higher performance risk.
- Aggregate extension. Hard plastic does not melt at typical asphalt production temperatures, but can partially replace or extend the aggregate in the mixture. However, the aggregate is much less expensive than the bituminous binder, so this approach provides a lower return on the investment. Some performance enhancement is often reported due to the reinforcing nature of the plastic, which is often flexible in nature.
- Binder extension. Soft plastics can melt into the bituminous binder to partially replace or extend the binder without necessarily enhancing its properties. This provides a greater return on the investment than aggregate extension, because the binder is significantly more expensive than the aggregate.
- Binder extension and modification. When soft plastic is used to modify, as well as extend, the bituminous binder, the effects can be similar to those associated with traditional polymers for binder and mixture modification.

The use of plastics that can both modify and extend the bituminous binder is the most beneficial approach to recycling plastic is asphalt mixtures. This is because the extended bituminous binder is expensive, but the polymers that would otherwise be used for asphalt mixture performance enhancement are even more expensive.

2.2 *Use of plastic in asphalt on roads*

Many countries have now reported the use of recycled plastic in asphalt mixture production, either as an aggregate extender, a bitumen extender or a binder modifier. For example, Vancouver (Canada) incorporated plastic crate waste as a warm mixed asphalt wax additive in 2012 (Ridden 2012) and Rotterdam (Netherlands) announced a plan to produce recycled plastic segments in a factory for road construction in 2015 (Sani 2015). Also, Jamshedpur (India) reported reducing bitumen usage by 7% by mixing shredded recycled plastic during asphalt production (PTI 2015). More recently, a New Zealand asphalt contractor added shredded engine oil containers to asphalt at Christchurch Airport (Parkes 2018) and an independent asphalt producer includes recycled plastic as a bitumen extender in every tonne of asphalt produced. South Africa commenced recycled plastic trials in 2019, on a road near Jeffrey's Bay (BusinessTech 2019).

In Australia a comparative trial of three recycled plastic extenders and modifiers was constructed in May 2018, which was shortly followed by other trials in Melbourne (IPWEA 2018), Sydney (Tapsfield 2018), Adelaide (Pisani 2018) and Canberra (Roberts 2019). Previous use of recycled plastic in the United Kingdom has recently been expanded to include roads and highways from Cumbria in the north, to Kent in the South, and many counties in between (Doyle 2019). Even the United States has now performed trials in San Diego (Grifin 2018) and more recently in downtown Los Angeles (Houser 2019) as well as a highway near Sacramento (Bousquin 2020).

Many of these trials have simply surfaced or resurfaced pavements with asphalt containing recycled plastic. Some have complimented construction trials with laboratory testing and evaluation of the mixtures. Furthermore, some trials have included otherwise identical asphalt mixtures, but without recycled plastic, to allow side-by-side comparison in the same mixture, in the same environment and subject to the same traffic loading. The side-by-side investigations that included laboratory investigations are particularly useful because they allow the effect of recycled plastic on asphalt mixtures to be objectively measured.

2.3 *Effects of recycled plastic on asphalt mixtures*

There are hundreds of recycled plastic polymers available for use in asphalt mixtures. However, to be practically recycled in asphalt mixtures, a plastic must melt at typical binder blending and asphalt production temperatures, which are typically limited to a maximum of 185°C. They must also be available in appropriate quantities and in a form that makes cleaning and pre-processing economically viable. Furthermore, because reuse as a new plastic product is preferred over recycling into construction materials, selected plastics should be otherwise destined for landfill. When these constraints are considered, polyethylene, polypropylene, polyvinyl chloride and ethyl vinyl acetate become the most viable plastics for recycling in asphalt mixtures (Brasileiro et al. 2019). Because of its high melting point, polyethylene terephthalate, which is commonly known as PET and is abundantly available in the form of plastic drink bottles, is not actually viable (White & Reid 2019). Researchers have concluded that polyethylene is the most viable type of plastic to be recycled as an asphalt mixture modifier (Brasileiro et al. 2019). As a result, many researchers have focused on evaluating the effect of recycled polyethylene in bituminous binder and asphalt mixtures.

Dalhat & Wahhub (2017) shredded and ground low and high density polyethylene, as well as polypropylene, and wet mixed the recycled plastic products into bitumen prior to asphalt manufacture in the laboratory. The viscosity of the binder increased, as did the Performance Grading (PG) of high temperature rating. The asphalt modulus increased and when a typical asphalt pavement was modelled in a pavement management model, the predicted rut depth and top-down longitudinal cracking were both predicted to reduce significantly (Dalhat & Wahhub 2017).

Yin et al. (2020) modified binders and mixtures with 2-3% polyethylene, with and without reactive elastomeric terpolymers to improve storage stability. It was concluded that polyethylene modified mixtures had better deformation resistance and increased stiffness, but reduced moisture damage resistance. It was also concluded that low temperature fracture, intermediate temperature fatigue, block cracking and reflective cracking resistance were all unaffected by the additional of polyethylene (Yin et al. 2020).

Nizamuddin et al. (2020) also assessed bituminous binder modified with polyethylene. The viscosity, softening point and penetration index all increased, and the penetration value decreased, indicating a stiffer binder. The MSCR measured elasticity and deformation resistance were also improved and it was concluded that 3-6% was the optimal dosage, by mass of the bitumen (Nizamuddin et al. 2020).

The products evaluated in this research are commercially available recycled plastics supplied by MacRebur (MacRebur 2020). The type of plastic is not disclosed, but one product is intended to be plastomeric in nature, while the other is intended to be elastomeric. White & Reid (2018) previously investigated these products and found the asphalt mixture modulus increased by 120-250%, wheel track rutting reduced by 0.5-1.8 mm and fracture toughness increased. In related work, White (2019) reported comparable moisture damage resistance and improved fatigue life of asphalt mixtures. The same products were also found to increase the MSCR-based PG of binders by three to four grades (White & Reid 2019) and White & Magee (2019) found similar improvements in deformation resistance and stiffness without adversely affecting the fatigue and moisture damage resistance. Most recently, it was concluded that the effects associated with these products were not significantly different when the plastic was wet-mixed into the bitumen or dry-mixed into the asphalt production plant (White & Hall 2021). It was also concluded that the improvement in binder and mixture properties associated with the two products were comparable to those generally associated with conventional polymers for asphalt production, particularly those associated with the plastomeric EVA (White & Hall 2020).

2.4 *The products*

As stated above, this research compared the effects of two commercially available recycled plastic products on asphalt mixture properties, to those associated with conventional polymers for binder modification and asphalt mixture production. The two products are known as MR6 and MR10 (MacRebur 2020). In 2015, MR6 was first developed to:

- Productively consume a portion of waste plastic otherwise destined for landfill.
- Reduce the cost of new road construction and maintenance.
- Increase the strength and durability of local roads.

MR6 was intended to improve deformation resistance, via an increase in asphalt mixture stiffness. MR10 was then developed to produce a more elastomeric and crack resistant bituminous binder. Both products are manufactured from 100% recycled plastic that was selected for its physical properties, as well as being otherwise not economically recyclable. That is, the plastic used to produce MR6 and MR10 would otherwise be disposed in landfill because there are no viable alternate recycling options available.

3 METHODS, MATERIALS AND RESULTS

Asphalt mixtures were produced with waste plastic modified binder (MR6 and MR10) and with conventional polymer modified binder (SBS and EVA), as well as unmodified penetration grade bitumen. The mixtures were otherwise nominally identical and reflected a common 10 mm sized dense graded and Marshall designed mixture for road surfacing in the United Kingdom. The samples were all tested for properties that are directly or indirectly related to asphalt performance, including deformation resistance, modulus, moisture damage resistance and Marshall properties.

3.1 *Materials*

As stated above, the asphalt mixture was a Marshall-designed, dense graded 10 mm nominal maximum aggregate sized product for local road surfacing. It is commonly used in the United Kingdom and meets the requirements of British BS EN 130108 and PD 6691 for a mixture containing basaltic coarse aggregate (Table 1).

Table 1. Asphalt mixture properties.

Property	Test method	Specification limit	Target value
Binder content (by mass) (%)	BS EN 12697-1	4.5-5.5	5.2
Voids in the aggregate (%)	BS EN 12697-8	18.0-22.0	20.0
Percentage passing (%) standard sieve size (mm)			
Sieve size (mm)	Test method	Specification limit (%)	Target value
10	BS EN 12697-2	90-100	99
6.3		62-68	68
2		25-33	32
1		17-26	22
.063		4-8	7

Control asphalt samples were produced with 50-70 penetration-grade bitumen meeting the requirements of EN 12591. The recycled plastic modified samples all included 6% (by mass of the unmodified bitumen) of the commercial products MR6 or MR10, shown in Figure 1. The SBS and EVA modified binders were produced with 2%, 4% and 6% polymer content, by mass of the unmodified bitumen.

Figure 1. Current (a) MR6 shreddings and (b) MR10 pellets.

All samples were prepared in a laboratory environment. Modified binder samples were first prepared by shearing the recycled plastic or conventional polymer into the 50-70 bitumen binder for 15 minutes using a Silverson laboratory mixer. The binder was then rested at 175°C overnight before being mixed into the pre-graded and pre-heated aggregates at 175°C for 5 minutes. Test specimens were subsequently prepared and tested according to the various methods, which are detailed below.

3.2 *Methods*

Asphalt samples were generally prepared in triplicate and tested using standard United Kingdom methods, which generally reflect European Norm standards, as summarised in Table 2. The triplicate sample results allowed statistical comparisons to be made for most test methods. However, the moisture sensitivity test requires six samples, three conditioned and three not conditioned, to produce one result. Furthermore, only one bulk sample was tested for volumetric composition (aggregate gradation and reference density) and one wheel track result requires two test results, while as many fatigue tests as was necessary were performed to achieve a relationship between initial strain and cycles to failure, with a R^2 value greater than 0.90, for each mixture.

Table 2. Test methods.

Property	Method	Description
Aggregate gradation	BS EN 12697-2	Unwashed sieve analysis after soluble binder extraction
Binder content	BS EN 12697-1	Percentage of binder extracted from a bulk asphalt sample after extraction
Void characteristics	BS EN 12697-8	Voids in the aggregate and air voids in samples compacted by a standard method
Marshall Stability/Flow	BS EN 12697-34	Stability and Flow of samples prepared by 50 blows to each side by a standard Marshall hammer and tested at 60°C
Deformation resistance	BS EN 12697-22	Average deformation following 10,000 passes of a Cooper's wheel tracking wheel at 45°C of samples compacted using a laboratory slab compactor
Moisture sensitivity	BS EN 12697-12	Ratio of indirect tensile strength of conditioned and unconditioned samples, where conditioning includes vacuum saturation followed by 72 hours in a 40°C water bath
Resistance to fatigue	BS EN 12697-24	Indirect tensile fatigue life of samples prepared to 8% air voids content in a gyratory compactor, over a range of initial tensile strain magnitudes to allow a relationship between initial strain and cycles to failure to be determined with a R^2 value greater than 90%

The results were generally analysed by graphical comparison. For asphalt mixture properties measured in triplicate, Student-t tests were performed for the difference of means. For all t-tests, a p-value of 0.05 or less was accepted as an indication that the difference between the two sets of results was statistically significant.

3.3 Results

Table 3 summarises the volumetric bulk sample test results, excluding the aggregate gradations, which are shown in Figure 2. The Marshall property and modulus results are summarised in Table 4 while the wheel tracking and TSR results are in Table 5. Some properties were also tested for the significance of the slope of a regression on the polymer content to determine the significance of the polymer content. This is noted where applicable.

Table 3. Bulk sample volumetric test results.

Mixture	Bulk density (t/m^3)	Binder content (%)
Control	2.322	5.30
SBS-2	2.353	5.31
SBS-4	2.309	5.32
SBS-6	2.324	5.31
EVA-2	2.315	5.31
EVA-4	2.301	5.30
EVA-6	2.345	5.32
MR6	2.298	5.33
MR10	2.338	5.30

Note: the numerical value following SBS and EVA designates the polymer content in the binder.

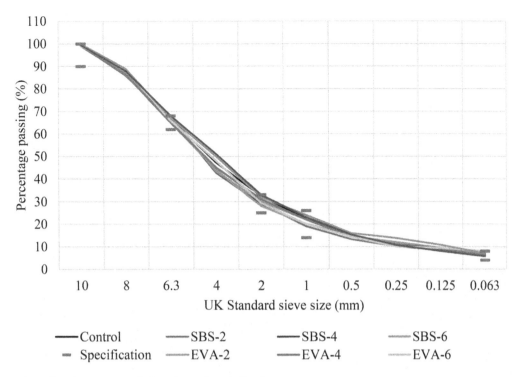

Figure 2. Aggregate gradations for various bulk mixtures.

Table 4. Sample Marshall property and modulus results (triplicate).

Mixture	Air voids (%)	Stability (kN)	Flow (mm)	Modulus (MPa)
Control	8.2	7.1	4.6	4,740
	6.9	8.4	4.2	4,630
	6.5	9.7	2.7	4,760
SBS-2	5.4	8.5	2.9	5030
	6.1	9.5	3.1	5130
	6.2	10.0	3.6	4910
SBS-4	7.6	7.1	3.7	6580
	7.1	7.3	2.7	6150
	8.0	7.1	3.9	5220
SBS-6	6.5	15.0	2.4	5710
	6.4	12.9	2.7	6570
	6.7	13.5	3.5	6230
EVA-2	7.6	6.4	4.4	5030
	6.4	7.7	3.8	5130
	5.6	9.3	3.9	4910
EVA-4	6.5	10.4	3.1	7300
	7.6	10.2	4.4	7180
	7.0	11.7	3.9	8540
EVA-6	6.5	14.4	4.4	11,580
	5.8	16.1	4.1	10,080
	6.3	14.5	5.2	9810
MR6	7.9	9.8	3.5	7150
	8.2	8.8	2.7	7330
	8.3	4.9	3.5	7760
MR10	6.2	11.4	4.3	6120
	7.1	9.3	5.4	6130
	5.8	11.3	3.1	6600

Table 5. Sample wheel tracking and TSR results.

Mixture	Rut depth (mm)	TSR (%)
Control	7.1	82.5
SBS-2	5.0	78.0
SBS-4	3.3	81.1
SBS-6	4.1	84.5
EVA-2	6.9	88.9
EVA-4	3.5	90.1
EVA-6	4.0	88.0
MR6	3.7	81.6
MR10	3.5	81.3

4 DISCUSSIONS

4.1 *Isolation of effects*

Before the effect of polymer content and type can be considered, it is first important to understand that the volumetric composition of the various asphalt mixtures was consistent, so that the effects of the variously modified binders is isolated from any changes in the aggregate gradation or binder content. The aggregate gradations were generally all consistent and within the specification limits, as shown in Figure 2. The binder contents and bulk densities were also consistent (Table 3) with coefficients of variation of 0.2% and 0.8%, respectively, which are both well below common asphalt production variability (White & Jamshidi 2019). As a result, it was determined that the various mixtures were produced with generally similar volumetric composition, meaning that any measured differences in the properties were generally isolated to the difference in the binder modification.

4.2 *Effect on air voids*

The air voids content of the compacted Marshall samples is an indicator of the mixture workability because the Marshall sample preparation method uses 50 blows of a standard hammer on each side of the sample, thereby imparting the same level of compactive energy to all samples. As the air void content increases, the workability of the mixture decreases. However, air voids contents are sensitive to small changes in aggregate gradation. The average air voids content of each mixture is shown in Figure 3. The control mixture had an air voids content of 7.2%. The conventional polymers were associated with variable changes, compared to the control, ranging from 5.9-7.6%. MR10 was associated with 6.4% air voids while MR6 was associated with 8.2% air voids. The plastomeric nature of MR6 likely reduced the workability of the mixture, but the effect was only small and may actually reflect natural variations in aggregate gradation, which affected the conventional SBS and EVA modified mixtures. This was supported by t-test p-values of the comparison to the control mixture results, which ranged from 0.10 to 0.82 for the various conventional polymers, plastics and polymer dosages. Any reduction in workability associated with recycled plastic can only be concluded as being comparable to that associated with conventional plastomeric EVA. More specific workability testing is required to form a more objective conclusion.

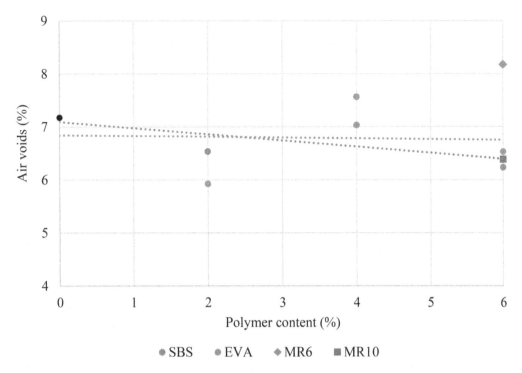

Figure 3. Asphalt mixture air voids contents after Marshall compaction.

4.3 *Effect on Marshall properties*

Marshall properties are well recognised around the world and remain useful measures of the relative contribution of the bituminous binder to asphalt mixture properties, when the mixture composition is otherwise consistent. Marshall stability is generally related to the stiffness of the binder, while the flow is generally related to the ductility or elasticity of the binder.

The Marshall stability generally increased with increasing conventional polymer content (Figure 4) with 6% SBS and 6% EVA associated with 64% and 78% increases in stability, respectively, compared to the unmodified asphalt mixture. These differences were significant (p-values both <0.01). The increase in stability generally reflects the increased binder stiffness and the reduction in binder temperature susceptibility, which are well recognised effects of conventional polymer modification. The effects of recycled plastic were less substantial, with MR6 and MR10 associated with only 5% and 27% increases, respectively. The differences were not significant, with p-values 0.66 and 0.09, for MR6 and MR10, respectively. MR6 and MR10 were associated with results that were comparable to results associated with 2-3% of conventional polymers.

The Marshall flow generally decreased with increasing EVA polymer content, but increased with increasing SBS polymer content (Figure 5). It is likely that the decrease in flow associated with EVA conventional polymer reflects the plastomeric nature of the EVA polymer and the reduced ductility generally exhibited by plastomers. MR6 produced a similar result to 3-4% EVA, while MR10 produced a result that was comparable to 6% SBS. This reflects the intent of MR6 to be a plastomeric modifier and the intent of MR10 to be a more elastomeric modifier. Despite these general comparisons, Marshall flow results are often associated with high variability. Consequently, none of the binders were associated with significant differences to the control mixture, with t-test p-values in the range 0.22-0.99.

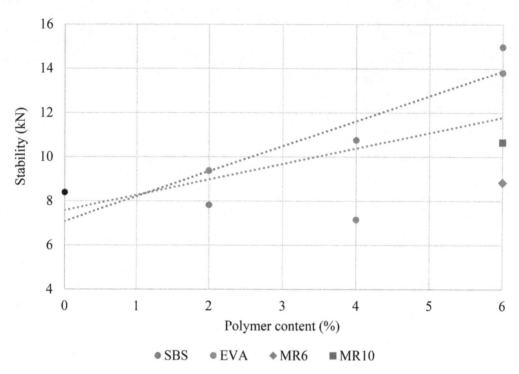

Figure 4. Asphalt mixture Marshall stability.

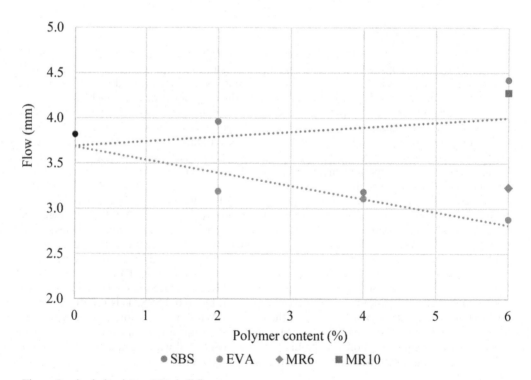

Figure 5. Asphalt mixture Marshall flow.

4.4 *Effect on stiffness*

Stiffness is the only asphalt material property that is directly used in pavement thickness design and/or structural evaluation. Along with the Poisson's ratio, stiffness, also referred to as modulus, is used to characterise asphalt and other materials in layered elastic mechanistic-empirical pavement modelling. The stiffer the asphalt mixture, the lower the pavement thickness needs to be to support the traffic loads over the design life.

Figure 6 shows the increase in modulus associated with increasing conventional polymer content. The effect of EVA polymer content was much greater than the effect of SBS polymer content, with 6% EVA producing a mixture with 123% higher ITSM than the control mixture stiffness, compared to 6% SBS, which produced a mixture with 31% higher ITSM than the control mixture stiffness. It is well established that plastomeric polymers increase asphalt mixture stiffness, while elastomeric polymers are generally associated with minor increases and occasional decreases in modulus, compared to asphalt produced with unmodified bitumen. All conventional PMB ITSM results were significantly higher than the control mixture values, with p-values ranging from <0.01-0.03.

Both recycled plastic products were also associated with significant increases in asphalt mixture ITSM, compared to the control mixture, with p-values <0.01. MR6 was associated with a 57% modulus increase, compared to 33% for MR10. This reflects the intent that MR6 be plastomeric while MR10 is intended to be more elastomeric in nature. The MR6 modulus was equivalent to that associated with approximately 4% EVA, while the MR10 modulus value was higher than all the SBS modulus values.

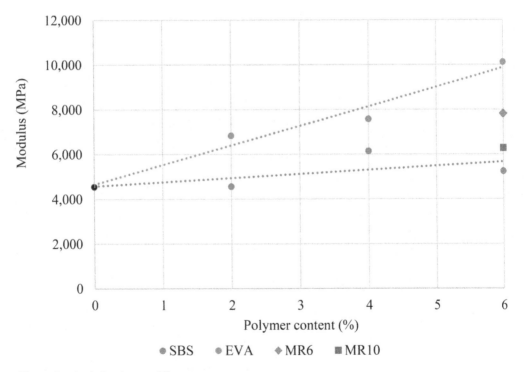

Figure 6. Asphalt mixture stiffness.

4.5 *Effect on wheel track rut depth*

Wheel track rut depth is a universally accepted indicator of deformation resistance of asphalt mixtures (Jamieson & White 2019). Although different jurisdictions use different test machines and protocols, the principle of measuring the depth of a rut after cyclic loading of a small wheel in the laboratory at elevated temperature is established as a sound relative measure of deformation resistance in the field. The rut depth decreased significantly with increasing conventional polymer content (Figure 7). Both SBS and EVA were associated with a 50% reduction for 6% polymer dosage. MR6 and MR10 were associated with similar rut depth values, indicating that these products had a comparable effect to 4-6% of conventional polymer.

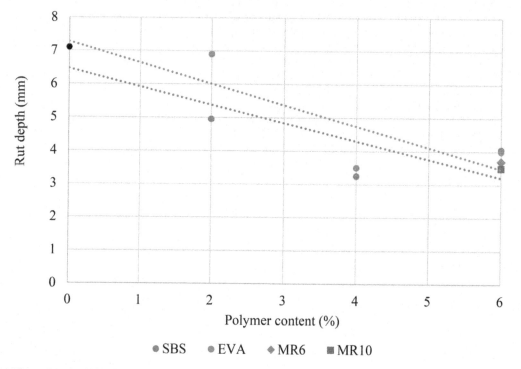

Figure 7. Asphalt mixture wheel tracking depth.

4.6 *Effect on tensile strength ratio*

All but one (2% SBS) asphalt mixture had a TSR that exceeded the 80% value that is generally set as the lower limit for acceptable resistance to moisture damage. Because the control mixture did not have a moisture damage issue (TSR 82.5%) the intent was not to demonstrate any improvement through binder modification. Rather, it was intended to determined whether the various polymers and recycled plastic would have any adverse effect. Overall, SBS was associated with a reduction in TSR while EVA was associated with an increase in TSR (Figure 8). However, regression equation p-values (0.22-0.54) indicated that these effects were not significant. This generally reflects the variable nature of the indirect tensile strength test and the sensitivity of the TSR to small changes in sample composition, resulting in high variability and lower repeatability. Both MR6 and MR10 were associated with TSR results that were comparable to 4% SBS and slightly lower than the control mixtures. Overall, it was determined that recycled plastic had no adverse effect on the moisture damage potential of the asphalt, certainly no different to the generally variable effect of the conventional polymers.

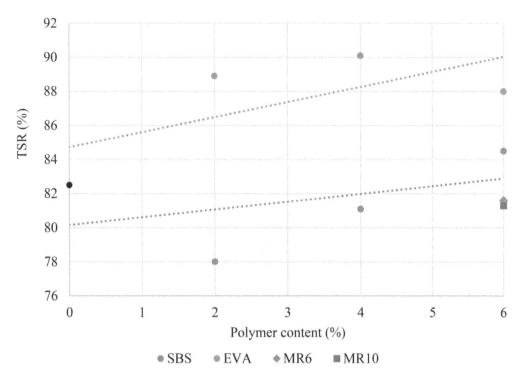

Figure 8. Asphalt mixture TSR.

5 CONCLUSION

Comparative performance-related testing of asphalt mixtures in the laboratory compared the effects of conventional elastomeric (SBS) and plastomeric (EVA) polymers, as well as MR6 and MR10 recycled plastic products, on the mechanical properties of a common dense graded asphalt mixture. It was found the recycled plastic, at a dosage of 6% of the bitumen mass, produced asphalt mixtures with comparable properties to conventional polymers. In particular, the Marshall stability associated with recycled plastic was comparable to that for 2-3% conventional polymer content, while indirect tensile stiffness and deformation resistance were comparable to 4-6% of conventional polymer. In contrast, workability, Marshall flow and moisture damage resistance were not significantly different for recycled plastic and conventional polymers, regardless of the conventional polymer content. It was concluded that the recycled plastic products used in this research should be thought of as sustainable polymers for asphalt mixture modification, rather than as extenders of the bituminous phase in asphalt production. However, further work is still required to understand the ageing of asphalt mixtures modified with recycled plastic, compared to unmodified bitumen and to conventional PMBs. The fuming and leaching potential of recycled plastic modification of asphalt must also be considered.

REFERENCES

Brasileiro, L., Moreno-Navarro, F., Tauste-Martinez, R., Matos, J. & Rubio-Gamez, M. 2019, 'Reclaimed Polymers as Asphalt Binder Modifiers for More Sustainable Roads: A Review', *Sustainability*, vol. 11, pp. no. 646.

Bousquin, J. 2020, 'California DOT repaves highway with recycled plastic bottles', *Construction Dive*, 26 August, accessed 23 September 2020, https://www.constructiondive.com/news/california-dot-repaves-highway-with-recycled-plastic-bottles/584131/.

BusinessTech, 2019, 'Works begins on South Africa's first Plastic Road', BusinessTech, 6 August, accessed 27 September 2020, https://businesstech.co.za/news/motoring/333307/work-begins-on-south-africas-first-plastic-road/

Dalhat, M. A. & Wahhab, H. I., 2017, 'Performance of recycled plastic waste modified asphalt binder in Saudi Arabia', *International Journal of Pavement Engineering*, vol. 18, no. 4, pp. 349–357.

Doyle, S. 2019, 'Plastic road trials expanded by UK Government to stop potholes', *Engineering & Technology*, 31 January, accessed 27 September 2020, https://eandt.theiet.org/content/articles/2019/01/plastic-road-trials-expanded-by-uk-government-to-stop-potholes/.

Grifin, J. 2018, 'On the Road to Solving out Plastic Problem', *UC San Diego News Center*, 25 October, accessed 27 September 2020, https://ucsdnews.ucsd.edu/feature/on-the-road-to-solving-our-plastic-problem.

Houser, K 2019, 'Los Angeles is building a road from recycled plastic bottles', *Futurism*, 25 October, accessed 23 September 2020, https://futurism.com/the-byte/los-angeles-building-recycled-plastic-road.

IPWEA 2018, 'Soft plastics from about 200,000 plastic bags and packaging, and 63,000 glass bottle equivalents will be diverted from landfill to construct a Victorian road', Institute of Public Works Australasia, 29 May, accessed 7 April 2019, http://ipwea.org/blogs/intouch/2018/05/29/this-australian-first-road-will-be-built-from-plas.

Jamieson, S. & White, G. 2019, 'Improvements to the Australian wheel tracking protocol for asphalt deformation resistance measurement', *Australian Geomechanics*, vol. 54, no. 2, pp. 113–121.

Jamshidi, A. & White, G. 2019, 'Use of recycled materials in pavement construction for environmental sustainability', *Eighteenth Annual International Conference on Pavement Engineering, Asphalt Technology and Infrastructure*, Liverpool, England, United Kingdom, 27–28 February.

Jamshidi, A. & White, G. 2020, 'Evaluation of performance of challenges of use of waste materials in pavement construction: a critical review', *Applied Sciences*, vol. 10, no. 226, pp. 1–13.

MacRebur 2020, 'MacRebur Products', Lockerbie, Scotland, United Kingdom, accessed 7 April 2020, http://macrebur.com/pdfs/product/MacReburProductSheet_v1.pdf.

Masad, E., Roja, K. L., Rehman, A. & Abdala, A. 2020, *A Review of Asphalt Modification Using Plastics: A Focus on Polyethylene*, Texas A&M Univeristy, Qatar, Doha.

NAPA, 2019, *Sustainable Asphalt Pavements: A Practical Guide*, National Asphalt Pavement Association, Geenbelt, Maryland, United States of America, November 2019.

Nizamuddin, S., Jamal, M., Gravina, R. & Giustozzi, F. 2020, 'Recycled plastic as bitumen modifier: the role of recycled linear low-density polyethylene in the modification of physical, chemical and rheological properties of bitumen', *Journal of Cleaner Production*, no. 266, pp. 1–12.

Parkes, R. 2018. 'Recycled plastic used in airport asphalt' *Roads & Infrastructure Australia*, 5 April, accessed 7 April 2019, http://roadsonline.com.au/recycled-plastic-used-in-airport-asphalt/.

Pisani, A. 2018, 'SA's first recycled road made from binned plastic and glass', http://news.com.au, 11 December, accessed 7 April 2019, http://l%2522><www.news.com.au/national/south-australia/sas-first-recycled-road-made-from-binned-plastic-and-glass/news-story/e7f03a9b43f797e651256d6bc8a4fc90>.

PTI 2015, 'Jamshedpur's Plastic Roads Initiative Is a Lesson for All Indian Cities!" *India Times*, accessed April 7, 2019, http://www.indiatimes.com/news/india/every-indian-city-needs-to-learn-from-juscos-plastic-roads-in-jamshedpur-232246.html.

Ridden, P. 2012, 'The streets of Vancouver are paved with.... Recycled plastic', *New Atlas*, 1 December, accessed 7 April 2019, http://newatlas.com/vancouver-recycled-plastic-warm-mix-asphalt/25254/.

Roberts, L. 2019, 'Canberra roads to be paved with recycled plastic as Government trials new type of asphalt', *Riotact!*, 5 March, accessed 27 September 2020, https://the-riotact.com/canberra-roads-to-be-paved-with-recycled-plastic-as-government-trials-new-type-of-asphalt/289498.

Sani, S. 2015, 'Forget asphalt: a European city is building a road made entirely out of recycled plastic', Business Insider, 22 July, accessed 7 April 2019, https://www.businessinsider.com.au/a-dutch-city-is-planning-to-build-roads-from-recycled-plastic-2015-7?r=US&IR=T.

Tapsfield, J. 2018, 'Plastic and glass road that could help solve Australia's waste crisis', The Sydney Morning Herald, 2 August, accessed 7 April 2019, www.smh.com.au/environment/sustainability/plastic-and-glass-road-that-could-help-solve-australia-s-waste-crisis-20180802-p4zv10.html.

White, G. 2019, 'Quantifying the impact of reclaimed asphalt pavement on airport asphalt surfaces', *Construction and Building Materials*, no. 197, pp. 757–765.

White, G. 2020, 'A Synthesis on the effects of two commercial recycled plastics on the properties of bitumen and asphalt', *Sustainability*, vol. 12, no. 8594, pp. 1–20.

White, G. & Hall, F. 2020, 'Comparing wet mixed and dry mixed binder modification with recycled waste plastic', *RILEM International Symposium on Bituminous Materials*, Lyon, France, 14-16 December.

White, G. & Hall, F. 2021, 'Laboratory Comparison of Wet-mixing and Dry-mixing of Recycled Waste Plastic for Binder and Asphalt Modification', *100th TRB Annual Meeting: a virtual event*, January.

White, G. & Jamshidi, A. 2019, 'Resetting dense graded airport asphalt production and construction tolerances', *International Journal of Construction Management*, article-in-press.

White, G. & Magee, C. 2019, 'Laboratory evaluation of asphalt containing recycled plastic as a bitumen extender and modifier', *Journal of Traffic and Transportation Engineering*, vol. 7, no. 5, pp. 218–235.

White, G. & Reid, G. 2018, 'Recycled waste plastic for extending and modifying asphalt binders', *8th Symposium on Pavement Surface Characteristics (SURF 2018)*, Brisbane, Queensland, Australia, 2-4 April.

White, G. & Reid, G. 2019, 'Recycled waste plastic modification of bituminous binder', *7th International Conference on Bituminous Mixtures and Pavements*, Thessaloniki, Greece, 12-14 June, pp. 3–12.

White, G. & Reid, G. 2021, 'Recycled plastic as an alternate to conventional polymers for bituminous binder', *7th Eurasphalt and Eurobitume Congress*, Madrid, Spain, 16-18 June.

Yin, F., Moraes, R., Fortunatas, M., Tran, N. Elwardany, M.D. & Planche, J.-P. 2020, Performance Evaluation and Chemical Characterization of Asphalt Binders and Mixtures Containing Recycled Polyethylene, Final Report, Plastic Industry Association, 10 March.

Possibilities of surface protection on road and airport pavements

J. Grosek, Z. Nevosad, T. Zavrel, T. Macan & B. Dohnalkova
Transport Infrastructure Department, Transport Research Centre - CDV, Brno, Czech Republic

ABSTRACT: Concrete silicate structure durability of road pavements is significantly affected by the surrounding environment. It concerns climatic changes, including changes of temperatures, humidity, presence of soluble aggressive agents, vegetation microorganisms, etc. Consequently, expansion reactions in the concrete structure may appear and the distresses reduce its material durability. It is possible to significantly slow down the process of concrete pavement material breakdown by suitable secondary protection that does not form a continuous film on the surface.

In the paper we classified different secondary protection means based on declared chemical bases of the agents. The effect of secondary protection was laboratory tested through a modified method of describing the water absorption and determining resistance against alternate freezing in NaCl environment. We found the material bases of concrete secondary protection products have a crucial effect on the test results. In general, the water absorption results are not related to the results of alternate freezing in NaCl. The applicable mutual conversion cannot be made for this combination of laboratory tests.

Keywords: Durability, concrete, secondary protection, testing

1 INTRODUCTION

The concrete transfers all loading of the environment where the structures are built and must have material durability under static loading in structures. Based on general experience, it is expected that concrete durability may mostly reach 50 to 100 years, based on the structure type.

In the course of the 20th century cement production methods gradually changed and new ingredients and admixtures started to be used. The environment changes due to human activities and over the last decades in some localities there is more and more frequent occurrence of concrete structures whose material durability is significantly reduced. Regarding the Czech Republic, or Central Europe, this particularly concerns partial, discontinuous motorway segments and some airports. Regarding airports, an interesting fact is that fast material concrete destruction occurs regardless on whether the damaged areas (runways, stands, etc.) are loaded by traffic or not.

Core diagnostics by an electron microscope were performed for all cases of concretes with reduced material durability in the Czech Republic. Extremely high presence of expansive crystals and gels was found by chemical discreet analysis under the microscope. They were found exclusively in pores of hydrating cement sealants. In more than 1000 core analyses no sample of expansive or disintegrating aggregates was found in the structure. In all cases, concurrent occurrence of crystals (gels) of sulphate, alkali-silicate and calcareous expansion was identified (Figure 1).

All contractual qualitative parameters (with some rare exceptions) were maintained for the concrete production based on all available producers' and investors' documents and materials. Investors neither found, nor complained about any faults during construction.

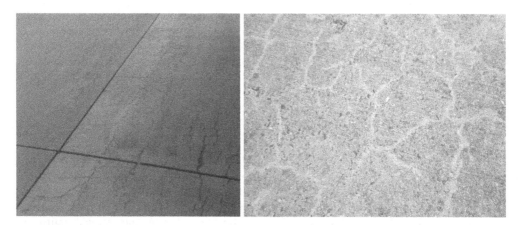

Figure 1. Examples of different types of material destruction of concrete structure.

2 TECHNICAL ANALYSIS

Nowadays, concrete road pavements fail to meet the designed durability. Apparently, the problem is not based on construction indiscipline or keeping the right technology for construction, therefore, it is necessary to discover what happens. We need to start with hydration.

Cement binder hydration is a complex natural process. It has been described according to many different models. Each one of the models has its validity, nevertheless none of them deals with the issue of reduced material durability. If we view hydration as a natural process of gradual dissolving of cement (clinker minerals) and subsequent gradual recrystallisation from gels and solutions, Mendeleev-Clapeyron equation can be used as an example (Nevosad, 1980). In an adjusted form it can be derived that the consumed energy A_h [J] during hydration from the time of mixing binder with water until the time t_h [s] may be in the form of (Feynman 1980 and Nevosad 2015):

$$A_h = \int_0^{t_h} \left(\frac{RTm}{\mu} - \frac{S_{pc}^4 m^5}{t^2 V^2} \right) dt. \qquad (1)$$

from this it can be derived that theoretical hydration time t_h [s] is

$$t_h = \frac{m^2 S_{pc}^2}{V} \sqrt{\frac{\mu}{RT}}, \qquad (2)$$

where **R** is the universal gas constant 8 314.3 [J°K^{-1}mol^{-1}], **T** temperature [°K], **m** binder mass [kg], **μ** mass of mean gram molecule [kgmol^{-1}], **S$_{pc}$** cement specific surface [m^{2}kg^{-1}], **t** time [s], **V** volume of hydration system [m^3].

Example: If we put in the relationship (2) average values for volume V = 1 m^3, constant temperature T = 293°K, m = 350kg, S$_{pc}$ = 300 m^{2}kg^{-1}, μ = 230 kgmol^{-1}, theoretical time of hydration will be around 14 years. Since there are no ideal conditions in concretes for free courses of reactions according to (1), the real hydration time must be considerably longer in comparison with the theoretical one.

The relationships (1) and (2) define that the reaction takes place even with the presence of every potential reagent. The process is not only entered by clinker minerals, but also by foreign components. They are either discharged from binders, additives, and admixtures from other concrete mixture components, or they enter the concrete structure from the environment. If this occurs, the further hydration course is irreversibly affected.

In case any secondary protection of concrete surface is applied, it will have a certain effect in terms of further hydration continuation. The benefits of secondary protection should be significantly higher than its potential risks. Therefore, it is very important what criteria and where and in which structure they will be selected for the diagnostics of the secondary protection choice.

3 LABORATORY TESTS

If we follow the relationships (1) and (2), the hydration as a process is not defined in any relationship with strengths in concrete. This can be understood in such way that mechanical properties are just side properties of forming concrete. It is reflected in the fact that fresh concrete can have measurable strength immediately after compaction. This is known and confirmed, e.g. with small prefabricated products produced by the technology of vibropress. At the same time hydration as a process may theoretically run in such way that even when hydration is completed, the structures fail to reach the required strength. In extreme cases it may fail to reach even measurable strengths. This is also known.

The relationships (1) and (2) are conditioned by the presence of cement and certain amount of free water. Furthermore, the process is based on temperature, time and basic properties of the binder. The material durability will be the longest if the process is affected by as few changes as possible. This is related to an ability to limit the size of moisture gradients in the structure as well as the number of moisture cycles. These are phenomena related to the introduction of foreign substances in concrete and the amount of released hydrating products. The concrete sensitivity to changes in the environment humidity (rain, dry spells) may be defined by a suitable water absorption test. The concrete sensitivity to the surface breakdown by the introduced aggressive agents is usually defined by the test of combined effect of freezing and real aggressive agents.

3.1 *Modified water absorption test*

We modified the water absorption test so that the relationship (1) was accepted in the maximum extent. In regular intervals we measured the course and magnitude of concrete relative water absorption within a given time period. We experimentally verified with aerated and non-aerated concrete that 120 minutes of immersion is sufficient. The test on sufficiently matured specimens is performed in such way that the specimens are at first dried to constant mass. Thus, the lowest mass of the specimen is determined. Then the same specimens are immersed in water for 7 days and the highest specimen mass is determined. The difference in mass gives the maximum possible mass of water which the concrete specimen may absorb in common conditions (Grosek et al. 2019 and Hodakova et al. 2017).

Subsequently, the specimen is left to get dry in laboratory out of the storage chamber and without any protection and treatment until it reaches constant mass (usually approx. 14 days). The specimens keep the so-called balanced moisture, which is given by temperature and relative humidity of the laboratory environment. This makes the specimens ready for the test. Only in this phase the secondary protection is applied. It is applied over the whole surface, i.e. along all six sides in case of cubes. The water absorption measurement is performed after perfect drying/hardening of all compared secondary protection means.

The specimens are weighed before immersion and then fully immersed in water. They are lifted out from water in 15-min intervals, weighed and then immersed in water again. The result is converted into relative percentage of the degree of water saturation of every specimen for every time interval. The outcome is a time curve of the course of concrete water saturation. The kinetics of saturation of untreated concretes is compared to the kinetics of saturation of concretes with applied secondary protection.

3.2 *Combined effect of alternate freezing and aggressive agents*

The test is to closely simulate the conditions negating the hydration course description according to relationships (1) and (2). It is meant to be the most destructive combination of destructive environmental effects on concrete. We recommend to take into account long-term experience, which may differ in different regions. The condition is that when selecting this test, at least two combined destructive effects must be applied. The combined effect has an impact on concrete and its structure more destructively than the effect of the sum of two identical, individually testing effects.

In conditions of a landlock country, such as the Czech Republic, it is unnecessary to take into account the effects of the sea, blizzards, tornados, etc. However, very frequent changes in temperature above and below the freezing point occur in winter. Therefore, the combined effect of alternate freezing in the road salt solution is the most troublesome. Road salt is relatively frequently used for winter road maintenance. Czech standards conventionally replaced the road salt with 3% NaCl solution and specified in detail the temperature changes and delay of freezing cycles (ČSN 73 1326). The temperature difference ranges between -20°C and +20°C and one cycle duration is 2 hours. The mass of spontaneously broken off concrete particles is converted into units of g/m^2 in regular intervals, usually after 25 cycles. It is an extremely severe test for concretes, even when compared to other tests of the same type.

3.3 *Laboratory test results for specimens of 100/100/100 mm*

Based on the available means of concrete secondary protection, the selected products were those which

- are available in the Middle Europe,
- are applicable with universal low-pressure and high-pressure sprayers,
- are designed for application in the exterior,
- do not contain organic solvents (i.e. just emulsions, dispersions and solutions),
- penetrate the concrete top surface layer,
- do not form a continuous visible film on the treated concrete.

Twelve different products met these requirements. For their more detailed classification we only used the information specified by manufacturers, i.e. from material or technical sheets of the products. They are the basic available information materials. Users should get enough information from those to make a decision whether and which concrete secondary protection means to use.

We performed more detailed classification of the selected products based on the basic material bases of the secondary protection products. The first group are silicate-based products (silanes, siloxanes, water glass, and their combinations). The second one is a group of oil-based products (flax varnish, linseed oil, tung oil, etc.). The last group includes products without manufacturer's specification of the basic material basis. Those are products described as balanced substance mixtures (with no further information) or products which, without specified basic material bases, define the final effect (product crystallisation, waterproof surface, etc.).

Laboratory tests were performed on series of aerated and non-aerated concrete. The composition of concrete mixtures of aerated and non-aerated concretes differed just with the presence or absence of aerating agents.

3.3.1 *Modified water absorption test*

This modified water absorption test showed that in the course of the first 120 minutes after immersion the untreated concrete is continuously saturated by water in a non-linearly increasing function. The untreated concrete reached nearly 70% of its maximum saturation. The course of the relative water saturation is reduced for all 12 tested different secondary protection means The course of the relative water saturation is reduced for all 12 tested different secondary protection means in comparison to untreated concrete. The effect of material base of tested impregnations is significant for this test, differences are obvious in Figures 2 to 4.

Regarding silicates, Figure 2, there are two significant branches. One of them forms nearly water-impermeable treatment on the surface (values between 10 and 17%), the other one just slightly slows down the water absorption. Regarding oils, Figure 3, the most balanced and at the same time significant values of water absorption slowdown were reached within the scope of tested products. Regarding the group of secondary protection products with unspecified material bases, Figure 4, the overall water absorption slowdown is the best and the most balanced. Reached values are close to water-impermeability of the treated concrete surfaces (values between 10 and 18%).

The modified water absorption test reached roughly comparable results of secondary protection effects, both with aerated and non-aerated concretes. The absolute values of water mass that entered the concrete were higher with non-aerated concretes.

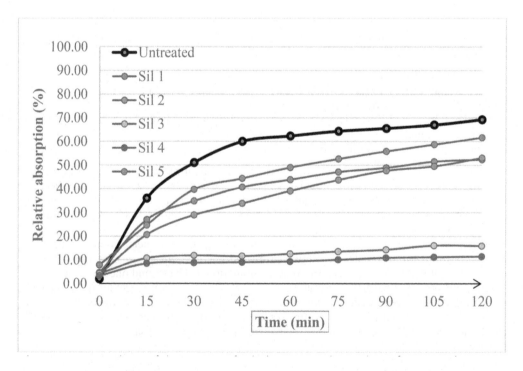

Figure 2. Comparison of the course of relative water absorption of untreated concrete and treated concrete through silicate-based secondary protection.

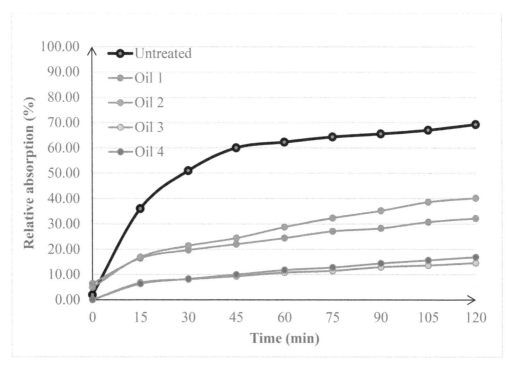

Figure 3. Comparison of the course of relative water absorption of untreated concrete and treated concrete through oil-based secondary protection.

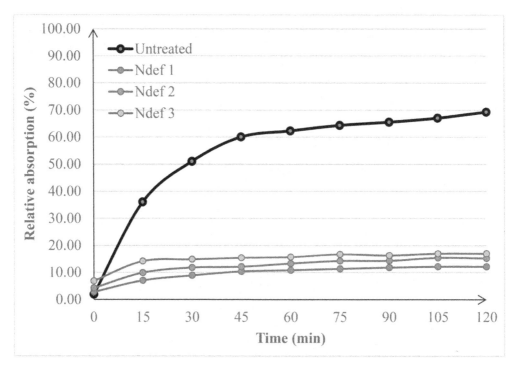

Figure 4. Comparison of the course of relative water absorption of untreated concrete and treated concrete through secondary protection with unspecified bases.

3.3.2 Combined effect of alternate freezing in NaCl environment

The effect of material bases of tested impregnations is significant for this test as well. The waste mass reduction obvious in Figures 5 to 7 has different dependencies in comparison to the previous modified absorption tests. Some products of silicate-based group and the group of unspecified bases are, after 100 freezing cycles, even worse than the compared untreated concrete.

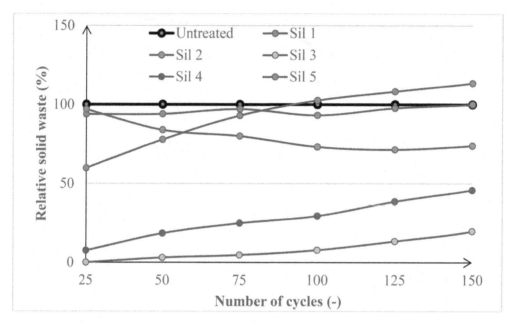

Figure 5. Comparison of breakdown of untreated concrete and treated concrete with silicate-based secondary protection during freezing in NaCl environment.

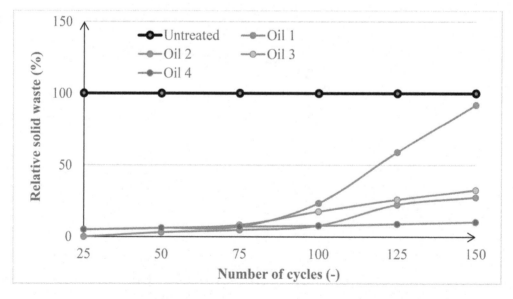

Figure 6. Comparison of breakdown of untreated concrete and treated concrete with oil-based secondary protection during freezing in NaCl environment.

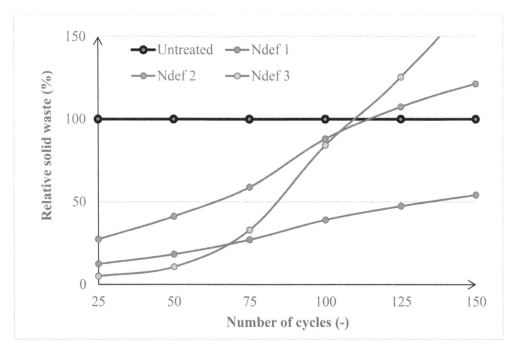

Figure 7. Comparison of breakdown of untreated concrete and treated concrete with secondary protection without specified bases during freezing in NaCl environment.

The test was performed with the same specimens after the end of the modified water absorption tests. We obtained the direct relationship of results for each individual specimen with the aim to generally prove or refute the inter-relationship of these tests. The difference between the behaviour of aerated and non-aerated concrete is crucial for this test.

The testing continued further with aerated concretes and was interrupted after reaching 300 freezing cycles with NaCl solution (Czech standard). The damage of specimens was so minimal that it was still possible to continue. In the same freezing procedure in NaCl environment the non-aerated concretes completely disintegrated after 50 or 75 cycles. In the Czech Republic the criteria for demonstration tests for monolithic aerated concretes of road pavements are set to 150, and for control tests in constructions to 100 freezing cycles in NaCl environment.

4 CONCLUSION

Nowadays, some exposed exterior concrete structures, particularly motorways, airports, bridges, etc., face unusual problems. Despite the fact that all standards and regulations are met, quality materials are used, and the construction is performed without serious problems, the material durability of concretes is sometimes reduced. The situation is all the more serious that it occurs in similar extent for loaded and unloaded (not overrun) road pavement structures. The question is whether the reduced durability is caused by the effects that are sufficiently discussed by the existing projects.

If not, the current attitude to concrete needs to be corrected. We introduce one of the potential approaches. Here, the most important parameter is considered the concrete material durability instead of traditional concrete mechanical properties. This change allowed to view hydration as a process which is fully defined according to natural laws. There may be more mathematical notations, but we work with a physically mathematic relationship (1).

Is there any direct relationship between the changed priorities of the new concrete durability and the existing concrete strengths? There is not, since strength is not the direct outcome of the process description (1). There are just indirect relationships between material durability and concrete strengths.

It is not surprising since in most countries mass production of concrete with strengths even over 15 MPa immediately after compaction is very common. At that moment hydration does not even start. Specifically, this concerns small prefabricated products by vibropresses and others. On the other hand, some cases are known when a mixture comprising cement, fillers and water fails to reach typical, and in extreme cases actually measurable strengths, even after months. This is in accordance with the relationship (1).

Another problem is how to test material durability of concretes in laboratory as well as in construction sites. Every environment is different. The individual differences must be taken into account for the diagnostics and testing. Therefore, we believe that there should not be a universal test for the parameter of concrete material durability, as it is common with concrete strength determination.

The mentioned assessment of agents is useful for applications of additional secondary protection means which do not leave a continuous film on the concrete surface and enter into the concrete top surface. Regarding the applications to concrete road pavements, we selected our modified test of relative water absorption of concrete that lasts 120 minutes and where solid waste is determined after alternate freezing in NaCl environment.

The relative water absorption test simulates the slowdown effect of the surface protection against water absorption and evaporation from the concrete surface, Figures 2, 3 and 4. The waste mass after alternate freezing simulates the effect of surface protection against the silicate structure breakdown. The measured results were converted into relative values for the comparative untreated concrete, Figures 5, 6 and 7. This simplifies the idea about their effect.

The results in Figure 2 to Figure 7 show that the majority of secondary protection means that have very high relative effectiveness against water absorption and evaporation have low to very low relative protection effectiveness against the silicate structure breakdown in the environment of alternate freezing in NaCl. At the same time, the secondary protection means that have the highest relative effectiveness against the silicate structure breakdown in the environment of alternate freezing in NaCl have the lowest relative effectiveness against water absorption and evaporation.

The material basis of the tested secondary protection means appears to be very important as well. If resistance against alternate freezing in NaCl is required, the oil-based products (the most common in the Czech Republic) ranked the best. If resistance against concrete water absorption and evaporation is required, the best are the combined agents of so-called unspecified bases. Regarding the results, the silicate-based secondary protection products move along a wide range of results, most commonly in between the previously mentioned groups.

The tests for concrete material durability determination need to be carefully selected and assessed. As obvious from our results, it is unsubstantiated to derive concrete durability from the absorption results in relation to joint effects of frost and salts and vice-versa.

ACKNOWLEDGEMENTS

This article was produced with the financial support of the Ministry of Transport within the programme of long-term conceptual development of research institutions.

REFERENCES

Feynman, R.P, Leighton, R.B., Sands, M., 1980. *Feynman's lectures in physics.* Bratislava, Slovakia.
Nevosad, Z., 1980. *Methodology of the concrete composition design and calculation of strength depending on temperature and time.* Dissertation Thesis, Brno, Czechoslovakia.

Nevosad, Z., 2015. *Main Problems of Cement Concrete* Proceedings of the Protecting Your Concrete Infrastructure with Dual Crystal Technology, National Concrete Pavement Tech Center, Des Moines, USA.

Grosek, J., Nevosad, Z., Stryk, J., 2019. *Can we reduce the degradation process of concrete pavement?* Proceedings of the XXIV. Seminar Ivana Poliacka: Mitigating the Effects of Climate Change on Roads, Jasna, Slovakia.

Hodakova, D., Zuzulova, A., Capayova, S. and Schlosser, T., 2017. *Impact of Climate Characteristics on Cement Concrete Pavements*. Proceedings of the 17th International Multidisciplinary Scientific Geo-Conference (SGEM), Varna, Bulgaria.

ČSN 73 1326, 1984. *Resistance of cement concrete surface to water and defrosting chemicals*. Czech Standard, UNMZ, Prague, Czech Republic.

Frybort, A., Vsiansky, D., Stulirova, J., Stryk J., Gregerova M., 2018. *Variations in the composition and relations between alkali-silica gels and calcium silicate hydrates in highway concrete*. Materials Characterization, Czech Republic, vol. 137, pp 91–108.

Developing high performance asphalt mixtures with considerable amounts of recycled asphalt with a new chemical bitumen additive

N. Carreño & M. Oeser
Institute of Highway Engineering, RWTH Aachen University, Aachen, NRW, Germany

M. Zeilinger & O. Fleischel
BASF SE, Ludwigshafen, RLP, Germany

ABSTRACT: Due to the increase of heavy traffic worldwide, many roads are being repaved to be able to support this demand. This generates an ever-growing amount of reclaimed material which is increasingly challenging to recycle due to bitumen aging. A new type of reactive chemical additive (B2Last®), tested at the laboratory of the Institute of Highway Engineering of the RWTH Aachen University, has shown to increase the asphalt performance by creating an elastic network within the bitumen. The additive reacts with the functional groups within the neat and oxidized bitumen producing a good performing bitumen in mixtures with or without reclaimed asphalt. The additive has also shown to increase the workability of the asphalt mixtures, as well as the performance against rutting and fatigue, without disrupting the low temperature behavior. By using softer base bitumen, like a neat 70/100 bitumen, the modified bitumen is able to perform well at very low temperatures, and also at very high ones. This is seen on laboratory bitumen and asphalt tests, as well as on mixture samples taken from the construction of a real-life track close to Munich. There, a 200-meter-long binder layer was built with 50% reclaimed asphalt in June 2019. With a small amount of additive mixed with a neat 70/100 bitumen, the workability on site of the mixture was improved, and the performance was on the same level as the reference track, where an SBS modified bitumen was employed.

Keywords: Modified bitumen, performance testing, reclaimed asphalt, asphalt recycling

1 INTRODUCTION

The increase in traffic and the changing climatic conditions are bringing flexible pavements to its limits. Many roads are being partially or completely repaved to be able to withstand this increase in demand. As a result, more and more reclaimed asphalt material is generated, which is increasingly challenging to recycle due to bitumen oxidative aging. Mixtures with high ratios of recycled asphalt are usually avoided, since they are difficult to re-process (due to high viscosity) and can compromise asphalt performance, specially the low temperature behavior (Krishna Swamy et al. 2011). This creates some challenges for the industry, since the production of reclaimed asphalt increases year by year.

The purpose of this paper is to present the findings on the use of a novel chemically modified bitumen which has been employed during the construction of real test track. The additive is an isocyanate-based modifier developed by BASF. This additive reacts chemically with the functional groups of the bitumen, creating a cross-linked polymeric network between the polar fractions of the bitumen (mainly asphaltenes, resins and even aromatics), corresponding mainly to the n-heptane insoluble fraction of bitumen (see Figure 1a). Specifically, the reactive groups contained in the polymeric chains of the polyurethane-precursor react with the nucleophilic groups within the bitumen, creating a flexible network made of urethane functionalities, but not limited to those ones. This network can increase the elastic behavior of the bitumen, impacting the high temperature performance and limiting its thermal susceptibility.

Oxidative aging of bitumen during the service life of a pavement generates more functional groups in the binder (Lesueur 2009). Therefore, when using this isocyanate-based additive in mixtures with reclaimed asphalt, the additive has more anchoring points to react with, as it can be seen in Figure 1b. This is believed to improve the homogenization and the compatibilization of both phases into the asphalt mix with high amounts of reclaimed asphalt and to ease the workability of the asphalt mixture during construction. Upon addition of the additive to neat bitumen and subsequent mixing with aggregates, a quick reaction leads to a full immobilization of the additive whereas not all isocyanate functionalities are consumed. During the mixing process with aggregates, the additive reacts with anchoring points of reclaimed asphalt and further ones which are generated in situ by oxidation. Once the crosslinking reaction is completed, which is believed to happen at the last phase of the pavement construction (generation of further anchoring points at laydown), the modified bitumen's viscosity increases and has a similar behavior to polymer modified bitumen. This type of modification does not require any swelling time nor high blending energy (e.g. high shear mixer) prior to the asphalt mixture production. The results of the preliminary studies can be found in (Carreño et al. 2020).

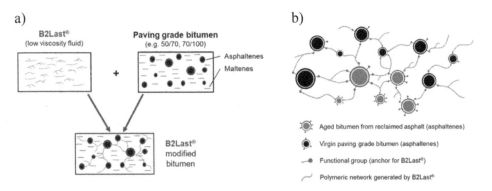

Figure 1. Working principle of chemical modification with B2Last® (www.b2last.com).

Isocyanate-based bitumen modification is not completely new. State of the art on similar use of this modification type has been well described in literature. (Cuadri et al. 2013; Cuadri et al. 2015; Navarro et al. 2007) have all used reactive chemical addition but in a different manner, due to the versatility this type of chemical allows. Prior to the blending, different organic polyols have been used to previously deactivate the isocyanate to form a polyurethane network within the bitumen. Enough free -NCO groups are left unreacted to react during a second step with the functional groups of the bitumen.

2 TEST SECTION

In this publication, the results of a test track construction with a total length of 200 meters are presented. The construction took place in June 2019 in the outskirts of Munich (A-96), where the binder layer was paved using a neat 70/100 bitumen, modified with 1.7 % of the isocyanate-based additive by weight referred to the total bitumen content (neat and reclaimed). The test track was built in a motorway service area (Figure 2), using Hot Mix Asphalt (HMA). For comparison, a control track on the same station was chosen since it is supposed to be submitted to similar traffic load and temperatures. Asphalt mixture samples were taken from both test and control track. The base layer of the control track was made from a 25/55-55 RC bitumen, which is an elastomeric polymer modified bitumen.

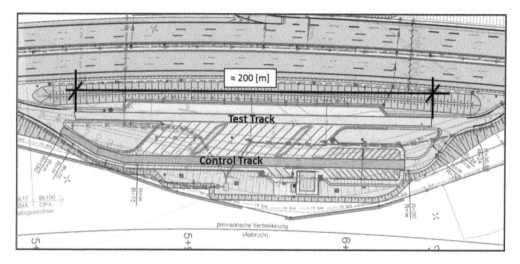

Figure 2. Sketch of test track and control track.

In this investigation, the similar methodology that has been applied throughout the joint research project was applied. First, the modification was analyzed through conventional and rheological tests. Afterwards, the relevant asphalt tests were performed on asphalt samples produced in the laboratory with mixture taken at the construction site. Additionally, longitudinal and cross section profile measurements were planned for both test and control tracks with a S.T.I.E.R. measuring vehicle. The first measurement took place in June 2020, whereas the second one will be investigated 6 months later. Since a comparison is not yet possible, these results haven't been included in this publication.

3 MATERIALS

3.1 *Isocyanate-based additive*

The additive used in the investigation is an isocyanate-based additive, called B2Last®, specifically designed for bitumen modification, developed by BASF. It is a black liquid with a low viscosity (210 mPas at 25°C). As mentioned previously, this additive reacts chemically with the polar fractions of the bitumen creating a cross-linked network among the asphaltenes, resins and even aromatics fractions. This reaction occurs as soon as the additive comes into contact with the bitumen. Knowing that the chemical value of aged bitumen binder from reclaimed asphalt pavement is higher than neat bitumen, meaning the polar phase bears more reactive groups, such additive will react even in a better way with aged bitumen. For the construction, and in this investigation, a 70/100 penetration grade bitumen was used as base bitumen to be modified with the novel additive.

3.2 Asphalt mixture and laboratory samples for bitumen and asphalt tests

The focus of this investigation was set on the binder layer. To contrast the results, samples of mixture were taken from a reference track that is expected to be submitted to the same amount of traffic load. The reference (or control) track was paved with a 25/55-55 RC bitumen. This is an elastomer modified bitumen with a needle penetration between 25 and 55 dmm and a softening point above 55 °C. The RC at the end usually stands for a slightly higher polymer content, but no information on the effective amount of SBS was available. All bitumen tests were performed on laboratory extracted material.

The binder layer of both test and control tracks was an AC 22 B S (dense graded asphalt concrete mixture with a maximum aggregate size of 22 mm), both with 50% of reclaimed asphalt. The mixture properties of the mixtures can be seen in Table 1. These values are from asphalt mixtures gathered at the construction site during paving operations.

Table 1. Main Properties of both asphalt mixtures.

Properties		B2Last®	Reference
Minerals		Limestone filler Crushed Gravel 50% Reclaimed Asphalt (diverse Aggregates)	Limestone filler Crushed Gravel 50% Reclaimed Asphalt (diverse Aggregates)
Bitumen		70/100 + 1,7% B2Last	25/55-55 RC
Bitumen Content	[wt. %]	4,4	4,2
Air Void Content	[vol.%]	5,0	4,7

As required in Germany, as soon as a new element is incorporated to a mixture, a suitability test must be performed. As starting point, the standard recipe for the PmB was taken. Only small adaptations were made to obtain a satisfactory air void content, so both granulometric curves were very similar.

For bitumen tests, bitumen was extracted using an infraTest Asphalt Analyzer following the corresponding German standards. For asphalt testing, samples were sawed and drilled from roller segment compactor plates, depending on the geometries required for the different tests following the German standards.

4 METHODS

4.1 Bitumen modification at asphalt mixing plant

In order to produce isocyanate-based modified asphalt mixture at the mixing plant, two small adjustments had to be operated at the plant: i) a pump including piping elements was installed to dose the additive into the bitumen weighing chamber; ii) a propeller stirrer was installed in the weighing chamber to shortly homogenize the additive into bitumen before the whole liquid phase is pumped to the mixer. The modification process didn´t affect the normal production operations or temperatures (approximately 170 °C during production and 165 °C during construction) of the mixing plant, allowing it to operate effectively. One part of the reaction occurs when the additive comes in contact with the bitumen and subsequently with aggregates during the mixing process (additive immobilization). The other part takes place during transport and road construction (completing the crosslinking reaction). Due to the process setup, the only way to study the modification is through bitumen extraction after asphalt mix production.

4.2 Bitumen testing

To study the effect of the isocyanate-based additive on bitumen, different tests were performed. Aside from the standard tests such as softening point and elastic recovery tests, different rheological tests were investigated. To characterize the high temperature behavior of the binder, the Multiple Stress Creep Recovery test (MSCR) and the Bitumen-Typisierung-Schnell-Verfahren (BTSV, in English: Binder-Fast-Characterization-Test) were done with the Dynamic Shear Rheometer (Anton Paar MCR 702 Multidrive). The MSCR has proven to correlate well with the rutting performance of asphalt mixtures (D'Angelo 2009; Zhang et al. 2015) and can indicate the behavior of a binder at high temperatures. This test is made with the 25 mm plate-plate system at 60 degrees (1 mm gap). The sample is loaded with a constant shear stress of 3,2 kPa for one second and then left to relax for nine seconds. The latter is repeated ten times. The recovery and permanent deformation after each cycle is recorded and used to obtain the average percent recovery (R) and the non-recoverable creep compliance (J_{nr}).

The BTSV method, on the other hand, is a novel method that allows to determine the upper limit of the elastoplastic range via a rheology method (Alisov et al. 2020). This test is performed by applying an oscillatory shear stress of 500 Pa at a frequency of 1,59 Hz, while the temperature is raised from 20 °C to 90 °C at a rate of 1,2 K/min. Using the 25 mm plate-plate system, the complex shear modulus and phase angle are obtained for the entire temperature range. From this test, two key parameters are obtained: T_{BTSV} (Temperature at a complex shear modulus of 15 kPa) and the δ_{BTSV} (Phase angle at T_{BTSV}). The first parameter is meant to correct the softening point for polymer modified bitumen, since the softening point is able to characterize only neat bitumen, whereas the phase angle is meant to help quantify the modification level of bitumen. This test is planned to replace the softening point in Germany in a near future.

Additionally, the bending beam rheometer was used to perform three-point bending tests at two different temperatures in order to determine the higher temperature between the one at which the m-value reaches 0,300 (iso-m-value temperature) and the one at which the stiffness reaches 300 MPa (iso-modulus temperature). The low temperature continuous grade was determined by using the ASTM D7643 standard. Following the Superpave binder specification and (Bahia and Anderson 1995), the values measured with the BBR represent properties at a pavement design temperature 10 °C lower.

4.3 Asphalt testing

In this investigation, the rutting and low temperature behavior were the main performances that were evaluated. The main attributes of the binder layer are to stay stable against permanent deformation, and have a good low temperature behavior, to not crack in the winter. The rutting behavior was studied with two tests: the wheel tracking test and the uniaxial cyclic compression test. Both of them are standards tests used in Germany to study the rutting susceptibility of asphalt layers, since they are very practical and correlate well with the rutting performance of roads (Karcher 2012; Radhakrishnan et al. 2019). Both tests were performed following the respective German standard. The wheel tracking test was performed by loading simultaneously two 80 mm thick roller compactor plates (320 mm x 260 mm) for 20.000 cycles using a rubber wheel (700 N) at 60 °C. The uniaxial cyclic compression test consisted in loading cylindrical 60 mm thick specimens drilled from roller compactor plates for 10.000 cycles at 10 Hz with a Haversine load (load impulse 0.2 s long with afterwards a break 1.5 s long) and at 50 °C.

To study the low temperature behavior, two tests were carried out: the thermal stress restrained specimen test (TSRST) and the uniaxial tensile strength test (UTST). Both tests were performed following the respective German standard as well. The first test consists on holding the length of a prismatic asphalt beam constant while lowering the temperature from 20 °C to -40 °C at a rate of 10 K/hr. Since the asphalt specimen are not able to contract, cryogenic stresses build up within the sample, thus producing eventually a temperature crack. In the UTST, prismatic specimens are pulled apart at a strain rate of 0.625 %/min and at different temperatures. In this case, due to availability of asphalt mixture, this test was performed only at -10 °C for both materials.

5 RESULTS AND DISCUSSION

5.1 *Bitumen results*

All bitumen test results can be found in Table 2. On the one hand, one can see that both bitumen have a similar softening point. Due to the different nature of the modification with the isocyanate-based modifier compared to a classical polymer modification (e.g. SBS), the elastic recovery test is not applicable. This test is thought only to identify elastomeric polymers in polymer modified bitumen and does not represent in any way bitumen performance. The deformations to which the material is submitted do not represent real pavement deformations. Many federal agencies in Europe still require this test for all polymer modified bitumen which makes the use of other additives rather difficult. The cross-linked network formed by the polyurethane-precursor-based additive enhances the elasticity of bitumen at small deformations, similar to the ones that a flexible pavement is submitted to, but not at large displacements as is intended with the elastic recovery test.

In the BTSV test, a clear difference between softening point and the T_{BTSV} can be seen, although the difference between both materials remains fairly the same. The δ_{BTSV}, shows that the polymer modified reference material is more elastic and thus slightly higher modified than the isocyanate-based modified bitumen. With regards to the MSCR test, the results go in hand with what was observed in the results of the BTSV test. Due to the nature of the modification, the percent recovery of the reference materials is higher than the B2Last®-modified bitumen, while the non-recoverable creep compliance is slightly lower. With regards to the high temperature performance it can be stated that both materials are on a very similar level.

Table 2. Bitumen test results.

Bitumen Tests		Unit	70/100+1.7% B2Last®	25/55-55 RC
Softening Point		[°C]	65.6	67.0
Elastic Recovery		[%]	not determined	47
BTSV	T_{BTSV}	[°C]	60.3	62.0
	δ_{BTSV}	[°]	72.0	69.8
MSCR	Recovery	[%]	16.5	35.9
	Jnr	[1/kPa]	0.530	0.284
BBR	T° (m-value = 0.300 or S = 300 MPa)	[°C]	-23.5	-22.4

With respect to the low-temperature behavior, it can be concluded that both materials showed a very similar behavior as well. The three-point bending beam results reflected that the highest temperature between the iso-m-value and the iso-modulus temperatures of both materials is 1.1 K apart from each other, being the isocyanate-based modified bitumen the material with the better low-temperature value. This is to be expected since, depending on the base (neat) bitumen and the modification method, the influence of the B2Last® additive should be negligible. Therefore, by modifying a soft binder, it is possible to have a good low temperature behavior, and improve the high temperature behavior, thus extending the useful temperature interval of the bitumen. Additionally, by using a softer binder, the workability can be improved, since the viscosity is then lower. In summary, the bitumen tests results demonstrate that the novel additive can deliver a comparable performance.

5.2 Asphalt results

5.2.1 Rutting stability

The results of the wheel tracking test (Figure 3a) show a very similar rutting performance for both modifications. None of the curves have an inflection point, and the average rut depth at the end of the 20.000 cycles was 3.35 mm for the B2Last® modified base layer, while the polymer modified base layer had a higher average rut depth of 3.67 mm. This correlates with what was seen on a bitumen level and again indicates that both materials perform similarly with regard to stability against permanent deformation.

The UCCT (Figure 3b) demonstrates a similar rutting behavior as well. In this case, the polymer modified variant has a slightly higher performance than the isocyanate-based modified variant. None of the samples presented an inflection point and the main difference lies in the strain and strain-rate at the end of the 10.000 cycles. The reference material had a strain on average of 36,5 ‰, while the novel additive had a strain on average of 40,8 ‰. The strain rate after the 10.000 cycles are $7,2\ ‰*10^{-4}/n$ and $12,7\ ‰*10^{-4}/n$, respectively. In summary, bitumen and asphalt test results show subtle differences in the properties of both materials. It can be concluded that they show a very similar and comparable performance.

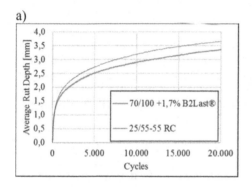
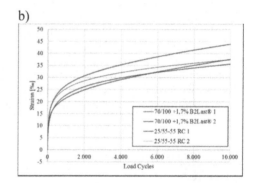

Figure 3. Results of wheel tracking and uniaxial cyclic compression tests.

5.2.2 Low-temperature behavior

In the case of the low temperature behavior, small differences between both materials could be identified. The uniaxial tensile strength test at -10 °C (Table 3) shows that the reference material has a higher tensile strength, although a comparable elongation at break. The elastic network created by the isocyanate-based modified bitumen is mostly effective in the high temperature range. Depending on the base bitumen (mainly on the ratio of asphaltenes to maltenes), it can have a positive or negligible influence on the low temperature behavior. Laboratory experiments (not included in this publication) have shown that higher amounts of the additive are required when using softer binders like a 70/100.

The thermal stress restrained test (Figure 4) shows and inverted picture in comparison to the UTST. In this case, cryogenic stresses appear to build up at lower temperatures for the isocyanate-modified asphalt, which means that the material is less susceptible to thermal stresses. This effect is common for chemically modified bitumen. Nevertheless, the cracking temperatures are very similar. The chemical additive had a break temperature on average at -28 °C, while the reference material on average at -27.1 °C. In summary, the low temperature behavior of the B2Last® based material depends highly on the neat base binder used. In this case, as in the case of the rutting performance, it can be stated that both materials perform at the same level in the low temperature range.

Table 3. Results of uniaxial tensile strength test at -10°C.

Sample	Tensile Strength [MPa]	Elongation at Break [‰]
25/55-55 RC	4,459	4,464
70/100 +1,7% B2Last®	2,675	5,923

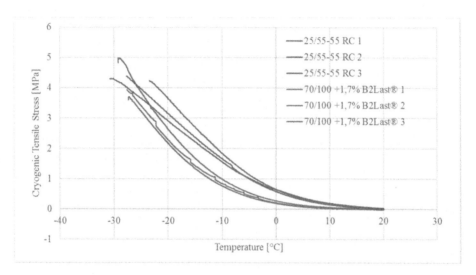

Figure 4. Thermal stress restrained specimen test results.

5.3 *Additional results*

In addition to the bitumen and asphalt tests, the construction of the test track with HMA provided key information about the possibilities that the B2Last® additive offers. During the construction of the test track, the material showed very good workability, considering it had 50% of reclaimed asphalt content. The compaction window was longer, and the odor was different to the SBS-modified asphalt. With these experience in mind, laboratory tests were performed to explore if the novel additive could allow asphalt production at lower temperatures, like for Warm Mix Asphalt processes, following the German Recommendations for Temperature Reduction of Asphalt (FGSV 2011). The results of this investigation lie outside the scope of this publication, since the results were submitted elsewhere, but the results showed that a temperature reduction was possible.

Since the construction of the test track presented here, two other paving trials have been successfully performed at on average 30 °C lower, still with high amounts of reclaimed asphalt and with the performance to the respective reference material (in both cases SBS modified asphalt mixtures). In both cases, the amount of the isocyanate-based additive was higher, between 2.0 % and 2.5 % by weight, referred to the total bitumen content (neat and reclaimed).

6 CONCLUSIONS

According to the analysis presented here, the following statements can be made:

- The bitumen tests show that the polyurethane-precursor-modified bitumen has a similar rutting and low-temperature behavior compared to the PmB RC reference bitumen. By using a softer bitumen (like a neat PEN 70/100 bitumen), it is possible to cover a wide range of temperatures. Since the working principle of the novel additive is very different compared to classical elastomeric polymer modification, the elastomer-specific elastic recovery test is not applicable.
- The asphalt tests showed a comparable rutting and low-temperature behavior between both asphalt mixtures as well, confirming what was seen on a bitumen level. Additionally, the novel additive showed a lowering of the thermal susceptibility.
- The additive could be successfully blended in the mixing plant, without affecting production times nor temperatures. It allows an on-demand dosage and doesn't require any swelling time since it reacts quickly at addition to the bitumen and during mixing with aggregates (additive immobilization). During the paving process, the crosslinking of bitumen components by the additive is finalized and thus the modified asphalt achieves the elastic behavior it requires to endure traffic.
- The isocyanate-based additive showed an increase in the workability of the asphalt mixture during the first paving trial. This finding allowed to focus research in this direction and discover the ability of the modifier to produce high performing modified asphalt mixture at lower production and paving temperatures.
- The longitudinal and cross-section profiles of both reference and test track will be controlled every 6 months. Further analysis will be made with the data retrieved from these measurements.

The novel additive has the potential to be very helpful to the asphalt industry. It enables the production of modified asphalt mixtures on demand, without great technical requirements. Due to its nature, it works well with high amounts recycled materials, and, additionally, it allows the production of high performing warm-mix-asphalt mixtures.

REFERENCES

Alisov, Alexander; Riccardi, Chiara; Schrader, Johannes; Cannone Falchetto, Augusto; Wistuba, Michael P. (2020): A novel method to characterise asphalt binder at high temperature. In *Road Materials and Pavement Design* 21 (1), pp. 143–155. DOI: 10.1080/14680629.2018.1483258.

Bahia, H. U.; Anderson, D. A. (1995): The Development of the Bending Beam Rheometer; Basics and Critical Evaluation of the Rheometer. In J. C. Hardin (Ed.): Physical Properties of Asphalt Cement Binders. 100 Barr Harbor Drive, PO Box C700, West Conshohocken, PA 19428–2959: ASTM International, 28–28–23.

Carreño, Nicolás; Renken, Lukas; Schatz, Waldemar; Zeilinger, Michael; Bokern, Stefan; Fleischel, Olivier; Oeser, Markus (Eds.) (2020): New type of chemical modification of asphalt binders to enhance the performance of flexible. pavements Proceedings of the 7th Eurasphalt & Eurobitume Congress v1.0. 7th Eurasphalt & Eurobitume Congress. Madrid.

Cuadri, A. A.; García-Morales, M.; Navarro, F. J.; Partal, P. (2013): Isocyanate-functionalized castor oil as a novel bitumen modifier. In *Chemical Engineering Science* 97, pp. 320–327. DOI: 10.1016/j.ces.2013.04.045.

Cuadri, A. A.; Partal, P.; Ahmad, N.; Grenfell, J.; Airey, G. (2015): Chemically modified bitumens with enhanced rheology and adhesion properties to siliceous aggregates. In *Construction and Building Materials* 93, pp. 766–774. DOI: 10.1016/j.conbuildmat.2015.05.098.

D'Angelo, John A. (2009): The Relationship of the MSCR Test to Rutting. In *Road Materials and Pavement Design* 10 (sup1), pp. 61–80. DOI: 10.1080/14680629.2009.9690236.

FGSV (2011): Merkblatt für Temperaturabsenkung von Asphalt. M TA. Ausg. 2011. Köln (FGSV R2 - Regelwerke, FGSV 766).

Karcher, Carsten (Ed.) (2012): Evaluation of the Deformation Resistance of Asphalt Mixes By Cyclic Compression Tests. E&E 2012 Conference Paper.

Krishna Swamy, Aravind; Mitchell, Luke F.; Hall, Steven J.; Sias Daniel, Jo (2011): Impact of RAP on the Volumetric, Stiffness, Strength, and Low-Temperature Properties of HMA. In *J. Mater. Civ. Eng.* 23 (11), pp. 1490–1497. DOI: 10.1061/(ASCE)MT.1943-5533.0000245.

Lesueur, Didier (2009): The colloidal structure of bitumen: consequences on the rheology and on the mechanisms of bitumen modification. In *Advances in colloid and interface science* 145 (1-2), pp. 42–82. DOI: 10.1016/j.cis.2008.08.011.

Navarro, F. J.; Partal, P.; Martínez-Boza, F.; Gallegos, C.; Bordado, J. C. M.; Diogo, A. C. (2007): Rheology and microstructure of MDI–PEG reactive prepolymer-modified bitumen. In *Mech Time-Depend Mater* 10 (4), pp. 347–359. DOI: 10.1007/s11043-007-9029-2.

Radhakrishnan, Vishnu; Chowdari, G. Surendra; Reddy, K. Sudhakar; Chattaraj, Rajib (2019): Evaluation of wheel tracking and field rutting susceptibility of dense bituminous mixes. In *Road Materials and Pavement Design* 20 (1), pp. 90–109. DOI: 10.1080/14680629.2017.1374998.

Zhang, Jun; Walubita, Lubinda F.; Faruk, Abu N.M.; Karki, Pravat; Simate, Geoffrey S. (2015): Use of the MSCR test to characterize the asphalt binder properties relative to HMA rutting performance – A laboratory study. In *Construction and Building Materials* 94, pp. 218–227. DOI: 10.1016/j.conbuildmat.2015.06.044.

Study on low-temperature crack resistance of zirconium tungstate modified asphalt mastic and asphalt mixtures

Z.S. Pei, J.Y. Yi* & D.C. Feng
School of Transportation Science and Engineering, Harbin Institute of Technology, Harbin, China

J.J. Li
China Railway Eryuan Engineering Group CO.LTD, Chengdu, China

ABSTRACT: During the service time of asphalt pavement, low-temperature cracking is one of the main deteriorations which could influence the performance of pavement and shorten the service life of the pavement. One of the main causes of low-temperature cracking is that the thermal stress inside the asphalt mixture cannot well relax along with time and then accumulate continuously until exceeding the tensile strength when the temperature drops suddenly in winter. To improve the thermal cracking resistance of asphalt mixture, zirconium tungstate (ZT) with a negative expansion property was introduced to partly replace the mineral powder in asphalt mortar and asphalt mixtures in this study. From the thermal contraction test, it was conducted that the ZT could reduce the low-temperature contraction coefficient of asphalt mortar as well as asphalt mixture. Besides, the contraction coefficient of mortar had a good positive correlation with the contraction coefficient of mixtures under different ZT content conditions. Results of the Thermal Stress Restrained Specimen Test showed that the cracking temperature and transition temperature of asphalt mixtures reduced with the increase of ZT contents. The research of this subject can provide a new idea for solving the low-temperature contraction cracking and design of asphalt pavement.

Keywords: Asphalt mortar, asphalt mixtures, zirconium tungstate, low-temperature cracking

1 INTRODUCTION

In the course of pavement service, low-temperature cracking is one of the main failure modes, which has appeared in many countries in the world (Das et al. 2013, Son et al. 2019). Because most of the materials have the characteristics of thermal expansion and cold contraction, it is difficult to avoid temperature shrinkage cracks in the pavement under extreme temperature (Park et al. 2016, Judycki, 2016). For example, in China, the technical index of asphalt pavement design specification for the high-grade highway is that no more than three low-temperature cracks can be allowed within 100 meters after completion acceptance. However, low-temperature cracking not only affects the driving comfort of the road surface but also has a greater impact on the continuity and safety of the road surface, which will increase the fuel cost and time cost of driving (Aflaki and Hajikarimi 2012, Sun et al. 2018). Besides, water and impurities could corrode the pavement structure through

---

*Corresponding author

DOI: 10.1201/9781003222880-4

cracks, and the water remaining in the pores will cause further damage to the pavement structure when it is frozen at low temperatures (Ren and Sun 2017, Ahmad and Khawaja 2018).

To improve the ability of pavement structure to resist low-temperature cracking, scholars have carried out a lot of related research. Results indicate that the asphalt binder has a significant influence on the low-temperature performance of asphalt mixture (Judycki, 2014). However, asphalt binders show "hard" and "brittle" mechanical properties in the low-temperature environment. Besides, with the aging process, the asphalt becomes more "hard" and "brittle", which brings adverse impact on the low-temperature performance of asphalt mixture (Wu et al. 2012, Wang et al. 2013). Therefore, many researches focus on improving the performance of asphalt binders to improve the ability of asphalt mixture to resist low temperature cracking. The modification of asphalt has aroused great interest of researchers. The results show that styrene-butadiene-styrene (SBS) or rubber powder can improve the low-temperature performance of asphalt binder and reduce the low-temperature cracking of the asphalt mixture (Nam and Bahia 2009, Hajikarimi et al. 2013). Besides, the addition of 0.2 ~ 0.4% lignin fiber can also improve the low-temperature performance of asphalt mixture (Luo et al. 2019, Yue et al. 2019). The thermal conductivity of the asphalt mixture can be reduced by adding expanded polypropylene (EPP) microspheres to the asphalt mixture, to change its thermodynamic performance (Shi et al. 2019). Besides, the use of lightweight aggregate can reduce the thermal conductivity of asphalt mixture, to improve its mechanical properties at low temperatures (Khan and Mrawira 2008). Open-graded friction course (OGFC) has a lower thermal shrinkage coefficient and is more potential to resist low temperature cracking than asphalt concrete (AC) (Islam et al. 2018). Although researchers have carried out a lot of research on asphalt binder, aggregate, and mixture gradation to improve the low-temperature cracking performance of asphalt mixture, the effect is still not particularly obvious. The main reason is that the contraction stress of the asphalt mixture in the low-temperature environment is difficult to avoid. If the thermal stress in the low-temperature environment can be greatly reduced, the low-temperature cracking resistance of asphalt mixture will be greatly improved. Negative thermal expansion materials can be tried to achieve this function.

Zirconium tungstate (ZT) was successfully prepared in 1996 by Mary et al (Mary et al. 1996). They found that the negative thermal expansion temperature range of zirconium tungstate is - 273 ℃ ~ 777 ℃, and the isotropic properties of ZT was good. This discovery makes it possible for the practical application of negative thermal expansion materials. ZT has been widely used in the preparation of composite materials due to its wide range of reverse expansion temperature range and isotropy. ZT was combined with aluminum alloy to make the alloy. When the volume of ZT was three times that of aluminum alloy, the contraction coefficient of the alloy could reach 0 (Matsumoto et al. 2003). ZT and zirconia were used to produce composite materials, and the expansion coefficient of the composites was close to 0 when the mass of zirconia was twice that of ZT (Niwa et al. 2004). Also, ZT was used to improve the performance of cement concrete, and the results show that the contraction coefficient of cement concrete was close to 0 when the mass fraction of ZT reaches 60% (Kofteros et al. 2001). If the thermal contraction coefficient of asphalt mixture can be improved by using ZT, it will greatly change the anti-cracking ability of asphalt mixture at low temperatures.

Based on the above analysis, ZT with a negative expansion property was introduced to partly replace the mineral powder in asphalt mastic and asphalt mixtures in this study. Through the performance changes of asphalt mortar and asphalt mixture, the influence of zirconium tungstate on the low-temperature crack resistance of asphalt mixture was explored and evaluated. The research of this subject can provide a new idea for solving the low-temperature contraction cracking and design of asphalt pavement.

2 MATERIALS

2.1 Asphalt

The 70 base asphalt selected in this project is the asphalt binder commonly used in northern China, and its technical indicators are shown in Table 1. The test method refers to "Standard test methods of bitumen and bituminous mixtures for highway engineering" (JTG E20-2011).

Table 1. Technical index of 70 base asphalt.

Technical indexes	Test result	Criteria	Testing method
Penetration (25 °C, 100 g, 5 s)/0.1 mm	69	60-80	JTG T0604-2011
Penetration index	-0.15	$-1.5 \sim +1.0$	
Extension (10 °C)/(cm)	31	≥ 15	JTG T0605-2011
Extension (15 °C)/(cm)	>100	≥ 100	
Softening point (R&B)/°C	51	≥ 45	JTG T0606-2011
Density (15°C)/kg/m^3	1.003	-	JTG T0603-2011
Residue after RTFOT			
Quality change/%	0.53	± 0.8	JTG T0610-2011
Residual penetration ratio (25°C)/%	63	≥ 61	JTG T0604-2011
Extension (10 °C)/(cm)	11	≥ 6	JTG T0605-2011

2.2 Fillers

ZT is introduced to partly replace the mineral powder in asphalt mastic and asphalt mixtures. The appearance of ZT used in this study is a bright green powder, and the purity is above 99.5%. The mineral powder is made of limestone, and its technical indicators are shown in Table 2, and the test method refers to "Test methods of aggregate for highway engineering" (JTG E42-2005). In this study, the fillers are divided into five kinds, and the mass ratio of ZT to the total filler is 0%, 20%, 40%, 60%, and 80%, respectively.

Table 2. Technical index of mineral powder.

Technical indexes	Test result	Criteria	Testing method
Density/kg/m^3	2.63	≥ 2.50	JTG E42-T0352-2005
Moisture content/%	0.61	≤ 1	JTG E42-T0103-2005
Hydrophilic coefficient	0.6	< 1	JTG E42-T0353-2005
Appearance	No agglomeration	No agglomeration	-

2.3 Asphalt concerete

The asphalt concrete (AC-13) is used as the test object, and the aggregate and gradation are carried out according to the requirements of "Technical specification for construction of highway asphalt pavement" (JTG F40-2004). The optimum asphalt content is determined to be 4.9% through the mixture proportion design. Besides, the mixing ratio of asphalt mortar is selected as the mass ratio of asphalt and filler in AC-13.

3 TEST METHODS

3.1 *Thermal contraction test of asphalt mortar*

The coefficient of thermal contraction of asphalt mortar refers to the volume change value with the change of unit temperature, which could well characterize the low-temperature performance of asphalt mortar (Islam and Tarefder 2015). The low-temperature performance of asphalt mortar deteriorates with the increase of temperature contraction coefficient.

To measure the volume deformation of the mortar more accurately, the contraction coefficient test device developed by the research group is used to carry out the test, as shown in Figure 1. During the test, the prepared asphalt mortar specimen is put into the metal chamber of the equipment, and then the top cover is covered. The fixed bolts are tightened, and the base is adjusted to make the measuring instrument in a horizontal state. Then ethanol is injected into the glass calibration tube, and the device is placed in the environment of different temperatures. Finally, the change value of the mortar volume is calculated according to the change of ethanol liquid scale.

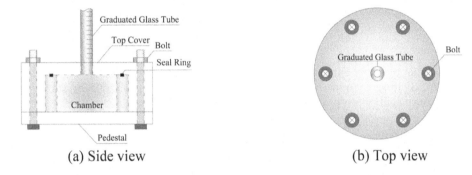

Figure 1. Device of asphalt mortar contraction coefficient.

During the test, constant temperature equipment is used for temperature control. Firstly, the initial temperature is set, and the test device is put into the temperature control equipment for 4 hours to read the initial value of the calibration tube. Then the temperature is adjusted to the next setting value and the value of the calibration tube is read after 4 hours of heat preservation. In this way, the volume changes of asphalt mortar at -20 °C, -10 °C, and 0 °C are measured. In the process of temperature change, the volume of the test device and ethanol will change as well, resulting in system errors. Therefore, the metal specimen with a known volume contraction coefficient and the same volume as the standard asphalt mortar specimen is used to calibrate the system errors. The calibration process is also carried out at -20 °C, -10 °C, and 0 °C. Finally, the thermal contraction coefficient of asphalt mortar is calculated according to formula (1).

$$\beta_i = \frac{(L_{i+1} - L_i) - \Delta V_{system-errors}}{(T_i - T_{i+1}) \times V_{specimen}} \quad (1)$$

where β_i is the coefficient of thermal contraction of asphalt mortar, L_i is the value of calibration tube at time T_i, $\Delta V_{system-errors}$ is the system errors calibrated by the metal specimen, $\Delta V_{specimen}$ is the volume of asphalt mortar specimen.

3.2 Thermal contraction test of asphalt mixture

The contraction capacity of asphalt mixture can well reflect its low-temperature performance. Many parameters affect the thermal contraction coefficient of asphalt mixture, including mineral aggregate gradation of asphalt mixture, asphalt properties, and asphalt dosage. The thermal contraction coefficient of the asphalt mixture is about $2.21 \times 10^{-5} \sim 3.33 \times 10^{-5}$ (/ °C) in several studies (Zeng and Shields 1999, Akentuna et al. 2018). To verify the low-temperature performance of ZT modified asphalt mixture, a test on the thermal contraction coefficient of the asphalt mixture is carried out.

The test system includes an incubator for temperature control, a test chamber for holding test pieces, a displacement sensor system, and a temperature sensor system. The asphalt concrete specimens are formed by the method of the wheel rolling and cut into 200 mm × 35 mm × 30 mm prisms.

The initial temperature was set at 20 °C with a cooling rate of 5 °C/h. The temperature range is set at - 20 °C ~ 20 °C. Because the measurement system will shrink due to the change of temperature, the instrument without the test piece is calibrated, and the calibration temperature range is - 20 °C ~ 20 °C. Finally, the thermal contraction coefficient of the asphalt mixture is calculated according to formula (2).

$$\alpha = \frac{(L_{i+1} - L_i) - L_{system-errors}}{L_{specimen} \times (T_{m+1} - T_m)} \qquad (2)$$

where α is the thermal contraction coefficient of asphalt mixture, L_i is the value of displacement sensor at time T_i (mm), $L_{system-errors}$ is the system errors calibrated by standard test, $L_{specimen}$ is the height of test piece (mm).

3.3 The thermal stress restrained specimen test

The device of thermal stress restrained specimen test is composed of a temperature control system, loading system, data acquisition system, and measurement system, as shown in Figure 2. The device can accurately control the length variation of the specimen with an accuracy of 3 μm. The maximum allowable deformation of asphalt mixture is 4 cm. The temperature sensor is used to measure the test temperature, and the accuracy is 0.25 °C. The accuracy of the stress measuring device is 10N.

(a) Device of test   (b) Specimen   **(c) Test process**

Figure 2. Test of thermal stress restrained specimen.

4 RESULTS AND DISCUSSIONS

4.1 Coefficient of thermal contraction asphalt mortar

The contraction coefficient of asphalt mortar containing 0%, 20%, 40%, 60%, and 80% ZT was tested. At least 3 parallel tests were conducted for each group. Finally, the contraction coefficient of asphalt mortar is obtained, as shown in Table 3.

It can be seen from Table 3 that the thermal contraction coefficient of asphalt mortar is not completely the same in different temperature ranges. However, with the addition of ZT, the thermal contraction coefficient of asphalt mortar decreases gradually. The results show that the negative expansion of zirconium tungstate can reduce the thermal contraction coefficient of asphalt mortar. The average contraction coefficient of 80% group is 60.55% lower than that of the 0% group. The addition of zirconium tungstate can effectively slow down the contraction trend of asphalt mortar in the low-temperature environment.

Table 3. Test results of volume contraction of asphalt mortar.

Content of ZT in filler	T_i (°C)	Contraction coefficient (/ °C×10^{-4})			Average contraction coefficient (/ °C×10^{-4})
0%	0	7.24	8.44	7.52	6.92
	-10	5.89	5.65	6.75	
	-20				
20%	0	6.69	5.94	6.95	6.12
	-10	6.03	5.62	5.46	
	-20				
40%	0	6.77	5.96	5.68	5.32
	-10	4.36	4.87	4.26	
	-20				
60%	0	3.48	4.1	3.67	4.33
	-10	5.2	4.44	5.06	
	-20				
80%	0	2.29	1.98	2.19	2.73
	-10	3.18	3.2	3.54	
	-20				

4.2 Coefficient of thermal contraction asphalt mixture

The displacement variation under unit temperature is taken as the index of the contraction coefficient of the asphalt mixture. In the experiment, the cooling rate was 5 °C/h, and the data were collected every 1 hour. The contraction of asphalt mixture with different content of ZT is obtained, as shown in Figure 3. According to formula (2), the thermal contraction coefficient of the asphalt mixture is calculated, as shown in Table 4.

Table 4. Thermal contraction coefficient of asphalt mixture.

Content of ZT in filler	Contraction coefficient (×10^{-5}/°C)	Correlation coefficient
0%	3.34	0.9884
20%	3.08	0.9817
40%	2.83	0.9645
60%	2.78	0.9598
80%	2.51	0.9489

The contraction coefficient of asphalt mixture (0% group) without ZT is 3.34 × 10-5 (/ °C) in the range from - 20 °C to 20 °C, which is very close to the contraction coefficient of asphalt mixture measured by other scholars (Dong et al. 2012, Akentuna et al. 2018). It can be seen from Table 4 that the contraction coefficient of the asphalt mixture gradually decreases with the increase of ZT content. The contraction coefficient of 80% group was 24.85% lower than that of the 0% group. Compared with the contraction coefficient test of asphalt mortar, the decrease range of the contraction coefficient of the asphalt mixture with ZT is smaller. The main reason is that the stone accounts for the majority of the asphalt mixture, and the contraction coefficient of the mixture is affected by the properties of the stone. On the whole, the low-temperature contraction trend of asphalt mixture is weakened by adding ZT.

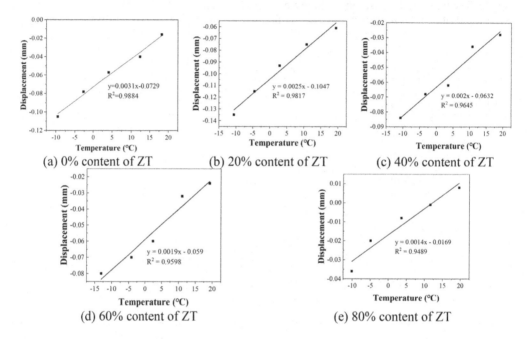

Figure 3. Fitting results of contraction test of asphalt mixture.

Asphalt mixture is composed of asphalt mortar and stone, and there is a close connection on low-temperature performance between asphalt mortar and asphalt mixture. The correlation of contraction coefficient between ZT modified asphalt mortar and asphalt mixture was analyzed. The contraction coefficient of ZT modified asphalt mixture decreases with the decrease of contraction coefficient of the asphalt mortar, and the correlation coefficient reaches to 0.9365, as shown in Figure 4. It shows that the contraction coefficient of the mixture has a good positive correlation with the contraction coefficient of mortar.

4.3 Rusults of the thermal stress restrained specimen test

The thermal stress restrained specimen test can simulate the stress process of pavement cracking at low temperatures. Among many evaluation indexes, the cracking temperature and the transition temperature have good consistency in evaluating the low-temperature performance of the asphalt mixture. In this study, the low-temperature performance of ZT modified asphalt mixture is evaluated by cracking temperature and transition temperature. The results of the thermal stress restrained specimen test are shown in Figure 5.

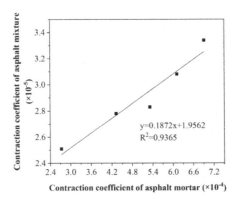

Figure 4. Fitting results of contraction coefficient of asphalt mortar and asphalt mixture.

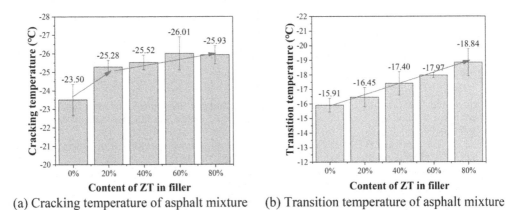

(a) Cracking temperature of asphalt mixture (b) Transition temperature of asphalt mixture

Figure 5. Results of thermal stress restrained specimen test.

It can be seen from Figure 5 that the cracking temperature and transition temperature of asphalt mixture gradually decrease with the increase of ZT content, which shows that the low-temperature performance of asphalt mixture could be improved by adding ZT. And the cracking temperature of the test group added ZT is at least 1.8 °C lower than that of the ordinary asphalt mixture. Besides, the cracking temperature of the 80% group was 10.34% lower than that of the 0% group, and the transition temperature of the 80% group was 15.27% lower than that of the 0% group. It shows that the low-temperature performance of the ZT-modified asphalt mixture is better than that of ordinary asphalt mixture. Moreover, the cracking temperature and transition temperature of the asphalt mixture have a good consistency.

Although the cracking temperature of asphalt mixture only decreases by 2-3 °C after adding ZT, the thermal stress restrained specimen test is also related to the size effect of the specimen, which can not fully represent the real service condition of asphalt pavement. Therefore, the test results mainly show the influence trend of ZT addition on the low-temperature performance of asphalt mixture and provide some new ideas to improve the low-temperature performance.

5 CONCLUSION

(1) The negative expansion of ZT can reduce the thermal contraction coefficient of asphalt mortar. And the average contraction coefficient of asphalt mortar of 80% group is 60.55% lower than that of 0% group.
(2) The contraction coefficient of asphalt mixture gradually decreases with the increase of ZT content. And the contraction coefficient of the 80% group was 24.85% lower than that of the 0% group. Besides, the contraction coefficient of the asphalt mixture has a good positive correlation with the contraction coefficient of the asphalt mortar.
(3) The low-temperature performance of ZT modified asphalt mixture is better than that of ordinary asphalt mixture. The cracking temperature of the asphalt mixture of 80% group is 10.34% lower than that of 0% group, and the transition temperature of the asphalt mixture of 80% group is 15.27% lower than that of 0% group.
(4) The addition of ZT could improve the cracking resistance of asphalt mixture at low temperatures. Although the cost of ZT is high at present, the unit price of material has been greatly reduced with the improvement of preparation technology in recent years. In the following study, other mechanical properties of asphalt mixtures mixed with ZT will be fully verified. We hope that this research can provide a new idea for solving the low-temperature shrinkage cracking and design of asphalt pavement.

REFERENCES

Aflaki, S. and Hajikarimi, P., 2012. *Construction and Building Materials*. Implementing viscoelastic rheological methods to evaluate low temperature performance of modified asphalt binders, 36, 110–118.

Ahmad, T. and Khawaja, H., 2018. *International Journal of Multiphysics*. Review of Low-Temperature Crack (LTC) Developments in Asphalt Pavements, 12 (2), 169–187.

Akentuna, M., Kim, S. S., Nazzal, M., et al., 2017. *Journal of Materials in Civil Engineering*. Asphalt Mixture CTE Measurement and the Determination of Factors Affecting CTE, 29 (6), 04017010.

Das, P. K., Jelagin, D., Birgisson, B., 2013. *Construction and Building Materials*. Evaluation of the low temperature cracking performance of asphalt mixtures utilizing HMA fracture mechanics, 47, 594–600.

Dong, D., Huang, X., Li, X., et al., 2012. *Construction and Building Materials*. Swelling process of rubber in asphalt and its effect on the structure and properties of rubber and asphalt, 29, 316–322.

Hajikarimi, P., Aflaki, S., Hoseini, A. S., 2013. *Construction and Building Materials*. Implementing fractional viscoelastic model to evaluate low temperature characteristics of crumb rubber and gilsonite modified asphalt binders, 49, 682–687.

Islam, M. R., Rahman, A. S. M. A., Tarefder, R. A., 2018. *Journal of Cold Regions Engineering*. Open Graded Friction Course in Resisting Low-Temperature Transverse Cracking in Asphalt Pavement, 32 (2), 04018006.

Islam, M. R. and Tarefder, R. A., 2015. *Journal of Materials in Civil Engineering*. Coefficients of Thermal Contraction and Expansion of Asphalt Concrete in the Laboratory, 27 (11), 04015020.

Judycki, J., 2016. *International Journal of Pavement Engineering*. A new viscoelastic method of calculation of low-temperature thermal stresses in asphalt layers of pavements, 19 (1), 24–36.

Judycki, J., 2014. *Construction and Building Materials*. Influence of low-temperature physical hardening on stiffness and tensile strength of asphalt concrete and stone mastic asphalt, 61, 191–199.

Khan, A. and Mrawira, D., 2008. *Journal of Testing and Evaluation*. Influence of Selected Mix Design Factors on the Thermal Behavior of Lightweight Aggregate Asphalt Mixes, 36 (6), 492–499.

Kofteros, M., Rodriguez, S., Tandon, V., et al., 2001. *Scripta Materialia*. A preliminary study of thermal expansion compensation in cement by ZrW2O8 additions, 45 (4), 369–374.

Luo, D., Khater, A., Yue, Y. C., et al., 2019. *Construction and Building Materials*. The performance of asphalt mixtures modified with lignin fiber and glass fiber: A review, 209, 377–387.

Mary, T. A., Evans, J. S. O., Vogt, T., et al., 1996. *SCIENCE*. Negative thermal expansion from 0.3 to 1050 Kelvin in ZrW2O8, 272 (5258), 90–92.

Matsumoto, A., Kobayashi, K., Nishio, T., et al., 2003. *Materials Science Forum*. Fabrication and Thermal Expansion of Al-ZrW2O8 Composites by Pulse Current Sintering Process, 426-432 (3), 2279–2284.

Nam, K. and Bahia, H. U., 2009. *Journal of Materials in Civil Engineering*. Effect of Modification on Fracture Failure and Thermal-Volumetric Properties of Asphalt Binders, 21 (5), 198–209.

Niwa, E., Wakamiko, S., Ichikawa, T., et al., 2004. *Journal of the Ceramic Society of Japan*. Preparation of Dense ZrO_2/ZrW_2O_8 Cosintered Ceramics with Controlled Thermal Expansion Coefficients, 112 (1305), 271–275.

Park J. W., et al., 2016. *Polymer Testing*. Characteristic shrinkage evaluation of photocurable materials, 56, 344–353.

Ren, J. L. and Sun, L., 2017. *Engineering Fracture Mechanics*. Characterizing Air Void Effect on Fracture of Asphalt Concrete at Low-temperature using Discrete Element Method, 170, 23–43.

Shi, X. J., Rew, Y., Ivers, E., et al., 2019. *International Journal of Pavement Engineering*. Effects of thermally modified asphalt concrete on pavement temperature, 20 (6), 669–681.

Son S., Said I. M., Al-Qadi, I. L., 2019. *Construction and Building Materials*. Fracture properties of asphalt concrete under various displacement conditions and temperatures, 222, 332–341.

Sun, Z. Q., Xu, H. N., Tan, Y. Q., et al., 2019. *Construction and Building Materials*. Low-temperature performance of asphalt mixture based on statistical analysis of winter temperature extremes: A case study of Harbin China, 208, 258–268.

Wang, D., Linbing, W., Christian, D., et al., 2013. *Journal of Materials in Civil Engineering*. Fatigue Properties of Asphalt Materials at Low In-Service Temperatures, 25 (9), 1220–1227.

Wu, S. P., Zhu, G. J., Pang, L., et al., 2008. *Key Engineering Materials*. Influences of Aging History on Low Temperature Performance of Asphalt Concrete, 385–387, 493–496.

Yue, Y. C., Abdelsalam, M., Luo, D., et al., 2019. *Materials*. Evaluation of the Properties of Asphalt Mixes Modified with Diatomite and Lignin Fiber: A Review, 12 (3), 400.

Zeng, M. L. and Shields, D. H., 1999. *Canadian Journal of Civil Engineering*. Nonlinear thermal expansion and contraction of asphalt concrete, 26 (1), 26–34.

Evaluation air void content of drilled porous asphalt mixture cores using non-destructive X-ray computed tomography

T. Vieira
Swedish National Road and Transport Research Institute, Linköping, Region Östergötland, Sweden

J. Lundberg
Division of Transport and Roads, Department of Technology and Society, Lund University, Lund, Sweden

O. Eriksson
Swedish National Road and Transport Research Institute, Linköping, Region Östergötland, Sweden

A. Guarin & S. Gong
Division of Building Materials, Department of Civil and Architectural Engineering, KTH – Royal Institute of Technology, Stockholm, Sweden

S. Erlingsson
Swedish National Road and Transport Research Institute, Linköping, Region Östergötland, Sweden
Division of Building Materials, Department of Civil and Architectural Engineering, KTH – Royal Institute of Technology, Stockholm, Sweden
Faculty of Civil and Environmental Engineering, University of Iceland, Reykjavik, Iceland

ABSTRACT: Tomography technology is not usual when investigating drilled asphalt cores properties. Currently, there is no internationally recognised standard tomography method for asphalt pavements. Tomography provides, however, a non-destructive alternative to traditional, usually destructive, testing of drilled cores. Furthermore, tomography offers possibilities, which traditional laboratory analyses do not. It is not straightforward to distinguish mastic and aggregates in tomography results while air void content is less difficult to assess. To have a more reliable assessment of drilled cores properties found by tomography, the method must be carefully planned, executed, and the results compared to those of traditional laboratory methods. In this work, analysis was carried out using six in-situ drilled cores from a double layered porous asphalt pavement. This allowed a comparison of tomography and a standardised conventional laboratory air void measurement on the exact same samples. Comparisons of the air voids found by tomography in all three directions were also carried out, estimating how anisotropic and heterogeneous the samples are, which is not possible using traditional laboratory tests. As few as four tomography slices can give enough precision in the determination of air void content for the porous layers. No more than eight slices per sample were needed in the suggested tomography method. The statistical results did not indicate that the air void content determined by tomography is different from the laboratory results.

Keywords: Tomography, non-destructive test, destructive test, air void content, porous asphalt drilled cores

DOI: 10.1201/9781003222880-5

1 INTRODUCTION

Porous pavements were developed as a method to increase traffic safety by draining run-off water from the road surface and reduce splash and spray (Dawson et al., 2009). A more recent application of porous pavements has been to reduce tyre/road noise at its source, especially due to its high air void content (Sandberg and Ejsmont, 2002, Vieira et al., 2019, Vieira, 2020).Another potential application is to reduce non-exhaust particle emissions from the road surface/tyre interaction (Norman and Johansson, 2017, Lundberg et al., 2019, Lundberg, 2020). The noise reduction is relevant considering that excessive exposure to noise leads to health impacts, including increased blood pressure, sleep disturbance, among others (World Health Organization, 2011). Exposure to particles are also related to health impacts, including, for instance, respiratory diseases, cardiovascular diseases and cancer, among others (World Health Organization, 2013, World Health Organization, 2016).

Porous pavements, compared to traditional dense pavements, have certain requirements when it comes to the construction process. Under its lifecycle, the porous layers are subjected to clogging due to particles and road dust (Sandberg and Ejsmont, 2002, Norman and Johansson, 2017, Lundberg et al., 2020); this, in turn, decreases its acoustical benefits. Another important aspect is the decreased durability, especially where studded winter tyres are used (Sandberg and Ejsmont, 2002); this affects the maintenance requirements and activities, which also affects the resulting lifecycle costs and benefits.

In order to achieve the desired properties, it is important to have an appropriate quality control method. Presently, quality control is performed with traditional laboratory methods that are typically destructive and do not allow a detailed quantification of the internal structure, including the air void content. One relatively novel method is the use of X-ray computed tomography (CT-scan) to obtain a detailed three-dimensional description of the internal structure of drilled cores, however, there is currently no standard nor commonly accepted method to use tomography in order to evaluate air void content. It is also needed to explore how the air void content obtained by tomography compares to the results obtained by traditional laboratory methods.

To contribute to a better understanding of the tomography method, its applicability to evaluate air void content and validate it against traditional laboratory methods, six drilled cores were collected in Linköping, Sweden, from a double layered porous asphalt concrete (DLPAC). The investigation summarized here is intended to be a steppingstone to the development of a method to evaluate air void content using CT-scan, and, in continuation, other material properties, such as binder content, as well as degradation processes, *e.g.* aggregate cracking and clogging of air voids. This paper is based on the master thesis by Gong (2020) in which more details can be found.

The overall aim of this study is to develop a methodology to determine the air void content from drilled cores from porous asphalt pavements using CT-scanning; the results were validated by comparison to traditional laboratory methods. Furthermore, another specific objective was to determine different kinds of air void content, including total and interconnected air voids for each asphalt layer and drilled core.

2 METHODS

2.1 *Traditional laboratory determination of air void content and layer thickness*

The air void content is currently most commonly determined using traditional laboratory methods. In this study, the air void content was determined using the European standard EN 12697-8 (Swedish Standards Institute, 2019). In short, the air void content is determined by measuring the bulk and maximum density of the specimen. All traditional laboratory tests were carried out after the CT-scans, as destructive laboratory measurements were also performed, including binder determination. Layer thickness was measured using a measurement tape at four equidistant locations around the lateral surface of the drilled cores.

2.2 X-ray computed tomography (CT-scanning)

X-ray CT scanning is a method which allows a three-dimensional determination of the internal structures of a specimen, in this case a drilled core of asphalt concrete (Guarin et al., 2012) in this case a drilled core of asphalt concrete. The X-ray CT-scanning device used in this study was the NSI X5000 located at KTH, laboratory of the Department of Civil and Architectural Engineering in Stockholm, Sweden; it is a seven-axis X-ray imaging system with 225 kV and 450 kV X-ray tubes (focal spot sizes of 5 μm and 400 μm, respectively). It has software for scan calibration, real time visualization and post-processing. A schematic view of the CT-scan components is displayed in Figure 1a. The CT-scan provides a series of grey-scale images which are stitched together to form a three-dimensional model of the specimen. This greyscale represents the density of the material where darker tones translates to lower densities. Figure 1b shows the different cutting planes used.

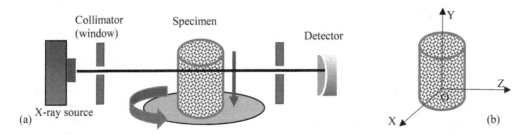

Figure 1. (a) Schematic of the X-ray CT-scan procedure. Figure based on Masad et al. (2002). (b) The different cutting planes defined by their normal direction, X, Y, and Z.

2.3 CT-scan process and slice acquisition

The general procedure includes sample preparation, calibration, data acquisition and reconstruction; further information can be found in Onifade (2013). In a few words, the specimen is scanned on a rotational table in the CT when geometrical configuration between x-ray source, sample, and detector panel (Figure 1b), as well as appropriate energy levels are verified. When the scan is completed, a series of greyscale 2D images is generated; the greyscale is related to density of the material, where darker tones mean lower densities. When the 2D images are put together, the three-dimensional object is generated

After scanning of the specimen, CT images are obtained using a reconstruction process which converts sinograms into two-dimensional slice images. With the used settings, each CT-scan produced around 1500 slices in 16-bit scale per sample per direction.

2.4 Image processing

Image processing was required to differentiate between aggregates, mastic (asphalt binder and aggregate particles), and air voids, as well as, to determine the total and interconnected air void content. This was done using an open-source image processing software named ImageJ, v 1.0 from National Institutes of Health (Schneider et al., 2012). A flowchart depicting the process is presented in Figure 2 and it is described in more detail than here in Gong (2020). After the slice import, the image extension is changed to an 8-bit image format to allow further analysis. Cropping is applied to obtain the area of interest. The minimum auto threshold function is then applied to adjust the grey levels to more clearly identify the air voids in relation to the other material phases. A binary transformation was applied to the image thresholding results. Using this image, interconnected air voids were identified and classified according to the Lefebvre air void classification (Lefebvre, 1993). The interconnected air voids are, in this case, air voids connected to the surface as illustrated in Figure 3. The software measures the interconnected air voids area in each image and calculates the interconnected air voids content in the entire drill core.

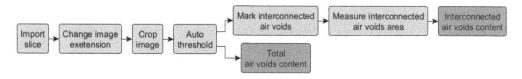

Figure 2. Flowchart describing the image analysis processing using the ImageJ software. Green colours mark output results.

Figure 3. A binary transformation slice where the red areas are classified as interconnected air voids, black areas as non-interconnected air voids and white areas as mastic and aggregates. Figure source: Gong (2020).

2.5 *Qualitative analysis method*

Different slices can be selected in all three directions by adjusting the cutting plane. After reviewing all slices, representative slices for each drilled core were selected based on the following parameters: (i) interface between different layers (ii) air void content (relatively large or small), (iii) aggregate sizes, (iv) aggregate segregation, (v) aggregates' characteristics, e.g. cracking. Although criteria (iii), (iv), and (v) are not within the scope of this paper, they were used for the qualitative characterization of each drilled core.

2.6 *Quantitative analysis method*

2.6.1 *Layer thickness*
Layer thickness was measured both in the laboratory, and by using CT-scan slices. For the CT-scan, slices in the X-direction was used. The layer thickness was measured on random points in the interface between layers in each CT-slice. This was repeated for five randomly chosen slices for each drilled core, and the average was calculated and used as the CT-scan layer thickness.

2.6.2 Air void content

The quantitative analysis was done by investigating the effect of number of asphalt layers, number of slices from the CT-scan data, different analysis directions and combinations of them, on the calculated air voids content. Regression analysis was performed for each combination regarding air void content between CT-scan (e_{CT}) and traditional laboratory methods (e_{lab}) where traditional laboratory was used as explanatory variable and CT-scan as response variable:

$$E[e_{CT}] = \beta_0 + \beta_1 \cdot e_{lab} \quad (1)$$

where $E[e_{ct}]$ is the expected value of e_{ct} for the given value of e_{lab}, β_1 is the slope and β_0 is the intercept. For each combination three separate hypotheses H were tested:

$$H_1 : \beta_0 = 0 \quad (2)$$

$$H_2 : \beta_1 = 1 \quad (3)$$

$$H_3 : \beta_0 = 0 \text{ and } \beta_1 = 1 \quad (4)$$

Hypothesis H_1 and H_2 were tested using two-tailed t-test. H_3 were tested using partial F-test. For all test, the significance level of 5 % was used.

The Evaluation Index (EI) was used to evaluate the accuracy using the slope and intercept of the regression analyses:

$$EI = \left(\hat{\beta}_1 - 1\right)^2 + \left(\hat{\beta}_0 - 0\right)^2 + \left(R^2 - 1\right)^2 \quad (5)$$

where $\hat{\beta}_1$ is the estimated slope, $\hat{\beta}_0$ is the estimated intercept and R^2 is the coefficient of determination. However, this evaluation has the limitation of not being scale invariant.

3 SITE AND DRILLED CORES DESCRIPTION

3.1 Site

The investigated DLPAC was located at Industrigatan in northeast Linköping, Sweden (Figure 4). It was constructed in two layers during august 2018. The layers were designed with an intended air voids content of 23.1 %. The top layer and bottom layer were designed to have a thickness of 25 mm and 55 mm respectively and a maximum aggregate size of 11 mm and 16 mm respectively. The DLPAC was built on top of part of the old dense pavement at the site. The road section is about 600 meters long with an average annual daily traffic of about 14,700 vehicles of which about 7 % are heavy vehicles. In Sweden, winter tyres (non-studded and studded) are mandatory during December through March given winter conditions, and studded tyres are common on light vehicles. The studded tyre share at Industrigatan was measured to between 51 % and 66 % for the light vehicles.

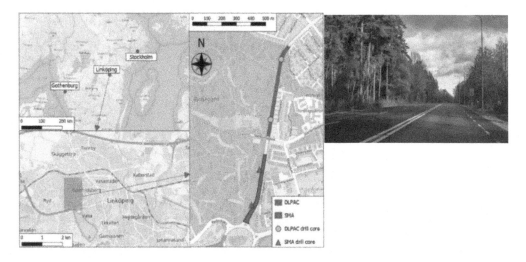

Figure 4. Left: Location of the double layered porous asphalt section in Linköping, Sweden. Adapted from Lundberg et al. (2020). Right: photo of Industrigatan in the northern driving direction at the double layered porous asphalt.

3.2 *Drilled cores*

Six drilled cores were taken for this study. Three drilled cores were taken each at the north and the south site of the DLPAC in the northbound direction (see Figure 4). At each site, one drilled core was taken per wheel track and one between wheel tracks. The drilled cores were taken in early November 2018, approximately two months after construction.

4 RESULTS AND DISCUSSION

4.1 *Traditional laboratory determination of air void content*

The results of the laboratory assessment of the drilled cores are presented in Table 1; it can be seen that the top layer is closer to the intended design with air void content between 18.6 – 23.4 % compared to the intended 23.1 %. The bottom layer, however, is far from the intended design with an air void content between 6.1 – 10.8 % compared to the intended 23.1 %. This is visible from the CT-scans, as illustrated in Figure 5. Also, the layer thickness deviates from the intended design, with larger variations for the bottom layer. The binder content is instead close to the intended design.

Table 1. Comparison between the intended design values and the laboratory determination for air void content, layer thickness and binder content.

Layer	Air void content [%]		Layer thickness [mm]		Binder content [%]	
	Design value	Drilled core analysis	Design value	Drilled core analysis	Design value	Drilled core analysis
Top	23.1	18.6 – 23.4	25	20 – 30	6.6	6.0 – 6.3
Bottom	23.1	6.1 – 10.8	55	35 – 60	6.4	6.4 – 6.8

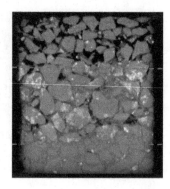

Figure 5. A CT-scan of one of the drilled cores. The top and middle layers are the DLPAC layers, while the bottom layer is the old dense pavement. The different grey-tones represents the material density where darker tones represent lower densities.

4.2 *Qualitative analysis*

The results from the qualitative analysis is summarised in Table 2. It was possible to detect different levels of segregation (for example Figure 6a) in the different drilled cores and the different layers within each core. It was also possible to visually estimate the different amounts of air void content within each core and layer (Figure 6a and b). Different interconnectivity levels were observed as well (Figure 6b). In general, both air void content and levels of interconnectivity are higher in the top layer compared to the bottom layer. Comparing the two sites, no clear difference is seen for the top layer regarding the air void content. However, at the northern site a more prominent densification in the bottom layer is visible. This is also true regarding the levels of interconnectivity. For segregation, a similar pattern is seen, although with less prominence. This indicates inhomogeneity of this test section, possibly due to the construction process or compaction by traffic.

Table 2. A qualitative comparison of similarities between the different drilled cores. RW = right wheel track, BW = between wheel track and LW = left wheel track.

Drilled core	Air void content		Interconnectivity		Segregation
	Top	Bottom	Top	Bottom	Both layers
North RW	4	4	4	5	7
North BW	5	3	5	2	3
North LW	7	4	7	4	4
South RW	6	6	7	7	5
South BW	4	5	5	7	4
South LW	5	4	4	4	3

Note: Lower values indicate less observed features and higher values indicate more observed features.

Figure 6. (a) Example on segregation and (b) interconnectivity. Figure source: Gong (2020).

4.3 *Quantitative analysis*

4.3.1 *Layer thickness*

The quantitative analysis allowed calculation of the layer thickness for all drilled cores. In average, the difference compared to the laboratory measurements was 8.9 %. However, the difference was found to have a large variation, between 0.1 % and 30.6 %. This variation was due to the possibility to analyse the internal variation in layer thickness using the CT-scan method, which is not possible using traditional laboratory methods.

4.3.2 *Significance testing – difference between laboratory method and CT-scan*

Linear regression was performed on all combinations of layers and directions with number of slices varying between four, six and eight slices. An example of the results is presented in Figure 7, where R2, the intercept and the slope for the regression is presented. The statistical analysis of this regression analysis resulted in an intercept and a slope with p-values of 0.952 and 0.918 when testing hypotheses 1 and 2 respectively. Hypothesis 3 was tested with partial F-test but was not significant ($p > 0.999$). The results, when considering different number of slices, different combination of layers and different combinations of directions are presented in Table 3 with the regression coefficients, *i.e.* the estimated slope, $\hat{\beta}_1$, the estimated intercept, $\hat{\beta}_0$, the coefficient of determination, R^2 and the evaluation index, *EI*. The t-tests and partial F-tests did not indicate that there is any difference between tomography and traditional laboratory results, and that four slices are enough to determine the air void content. However, the EI and R^2 increase slightly when the number of slices is increased from 4 to 6, and then to 8.

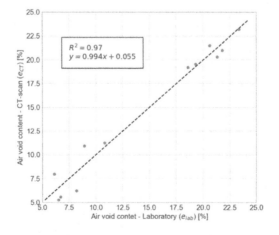

Figure 7. Air void content from the traditional laboratory method and from tomography (CT). Each point is the average air void content for the X, Y and Z directions for six slices for each porous layer of the drilled cores.

Table 3. Accuracy comparison of different combination. The optimal combination is highlighted by a *, being the X+Z with 8 slices.

Directions	Number of slices	$\hat{\beta}_1$	$\hat{\beta}_0$	R^2	EI
X	4	0.9516	0.0022	0.9637	0.00366
	8	0.9628	0.0008	0.9667	0.00249
Y	4	1.0171	-0.0048	0.9651	0.00153
	8	0.981	0.001	0.9732	0.00108
Z	4	0.9755	0.0016	0.9681	0.00162
	8	0.9949	0.0005	0.9681	0.00104
X+Z	4	0.9636	0.0019	0.9681	0.00235
	8	0.9788	0.0007	0.9687	0.00143
Y+Z	4	0.9963	-0.0016	0.9684	0.00101
	8	0.9879	0.0008	0.9714	0.00097*
X+Y	4	0.9843	-0.0013	0.9672	0.00132
	8	0.9719	0.0009	0.9705	0.00166
X+Y+Z	4	0.9814	0.0003	0.9687	0.00133
	8	0.9796	0.0008	0.9704	0.00129

The air void content for each layer, as well as the interconnected air void content and the overall air void content are presented in Table 4 for each drilled core. These results are comparable to the ones obtained by traditional laboratory methods. This was previously shown in Figure 7.

Table 4. Results of average air void content [%] using the directions indicated in parenthesis.

Air void content	North RW	North BW	North LW	South RW	South BW	South LW
Top layer (Average Y+Z)	20.80	20.25	23.26	19.31	19.57	20.97
Bottom layer (Average Y+Z)	6.18	5.37	5.80	10.94	10.67	7.95
Interconnected (Average X+Z)	3.35	3.01	4.40	3.23	2.40	2.84
Overall (Average X+Z)	11.96	10.88	11.86	13.15	11.94	12.20

5 CONCLUSIONS AND RECOMMENDATIONS

5.1 Conclusions

This study has developed a methodology for using CT-scanning for the determination of air void content of porous asphalt concrete drilled cores. Six drilled cores samples from Linköping, Sweden, were used for this purpose, in total 12 samples. From the results of this study, it was concluded that CT-scanning, in combination with image analysis were found to be suitable tools to analyse the air void content of drilled cores of porous pavements.

From the qualitative analysis it was noted that all drilled cores showed some common features: air void content in top layers were larger than bottom layers; interconnectivity between air voids were lower in the bottom layers than the top layers; and aggregate segregation was observed in both top and bottom layers. Sample characteristics that differed between samples were found to be: aggregate cracking was observed for some drilled cores in the bottommost layer; and in some samples air voids were observed in the interface between porous layers and porous and dense layers.

From the quantitative analysis it was shown that different kinds of air void content were found to be quantifiable using CT-scanning and the proposed methodology. Four slices were found to be sufficient to determine an accurate air void content compared to traditional laboratory results.

5.2 Recommendations and research needs

Based on the study of drilled cores using CT scan, some recommendations and research needs are presented here. A method should be developed to determine not only air void content but also more asphalt mixture properties (e.g. aggregate size distributions, binder content, mastic content, aggregate content etc.), either individually or in combination with the air void structure. The relationship between different kinds of air void content and noise absorption, particle emissions should be evaluated, and a numerical relationship or model should be developed. The process of air void structure investigation using CT scanning should be developed to become automatized. The results presented here should be generalized by repeating the study on more drilled core samples, preferably using a mix of different types of porous pavements with different degrees of densification as well. The methodology developed in this study should also be applied to aggregate characterization or air void characterization of cement concrete pavements as well.

ACKNOWLEDGEMENTS AND FUNDING

The authors wish to acknowledge Leif Söderberg (NCC industry AB) who was hired to take the drilled cores samples. Also acknowledged are Ramudden AB which were hired to keep us safe during the acquisition of the drilled cores. Andreas Waldemarson (VTI) is acknowledged for performing the traditional laboratory measurements. This study was carried out as a master thesis project at KTH. The drilled cores acquisition was funded by Linköping.

REFERENCES

Dawson, A., Kringos, N., Scarpas, T. & Pavsic, P. 2009. Chapter 5: Water in the Pavement Surfacing. *In*: Dawson, A. (ed.) *Water in Road Structures - Movement, drainage effects*. The Netherlands: Springer.

Gong, S. 2020. *Investigation of Air Void Structure in Double Layer Porous asphalt based on X-ray Computed Tomography*. Maste of Science, KTH Royal Institute of Technology: Stockholm, Sweden.

Lefebvre, G. 1993 *Porous asphalt*, Permanent International Association of Road, Congresses. Paris.

Lundberg, J. 2020. *Road Surface and Tyre Interaction: Functional Properties affecting Road Dust Load Dynamics and Storage*. KTH Royal Institute of Technology: Stockholm, Sweden.

Lundberg, J., Gustafsson, M., JanhÄll, S., Eriksson, O., Blomqvist, G. & Erlingsson, S. 2020. Temporal variation of road dust load and its size distribution - a comparative study of a porous and a dense pavement *Water, Air, & Soil Pollution (submitted)*.

Lundberg, J., Vieira, T., Blomqvist, G., Gustafsson, M., JanhÄll, S., Genell, A. & Erlingsson, S. 2019 PM10 Emissions from a Porous Pavement, and Implication for Noise Reduction, Initial Results. *European Aerosol Conference*, 2019 of Conference Gothenburg, Sweden.

Masad, E., Jandhyala, V. K., Dasgupta, N., Somadevan, N. & Shashidhar, N. 2002. Characterization of Air Void Distribution in Asphalt Mixes using X-ray Computed Tomography. *Journal of Materials in Civil Engineering*, 14, pp. 122–129. doi: 10.1061/(ASCE)0899-1561(2002)14:2(122).

Norman, M. & Johansson, C. 2017 Emission of PM10 and coarse particles from "silent" asphalt. *22nd International Transport and Air Pollution Conference* 2017 of Conference Zürich, Switzerland.

Sandberg, U. & Ejsmont, J. A. 2002. *Tyre/road noise reference book*, Kisa, Sweden, Informex.

Swedish Standards Institute 2019, SS-EN 12697-8: Bituminous mixtures – Test methods – Part 8: Determination of void characteristics of bituminous specimens. Stockholm, Sweden: Swedish Standards Institute.

Vieira, T. 2020. *Tyre/road interaction: A holistic approach to the functional requirements of road surfaces regarding noise and rolling resistance*. 2013 Doctoral thesis, comprehensive summary, KTH Royal Institute of Technology: Stockholm.

Vieira, T., Lundberg, J., Genell, A., Sandberg, U., Blomqvist, G., Gustafsson, M., JanhÄll, S. & Erlingsson, S. 2019 Porous pavement for reduced tyre/road noise and improved air quality – Initial results from a case study. *In*: Montreal, Icsv26 Local Committee In (ed.) *International Congress on Sound and Vibration*, 2019 of Conference Montreal, Canada. Canadian Acoustical Association, p. pp. 8.

World Health Organization. 2011 *Burden of disease from environmental noise: quantification of healthy life years lost in Europe*. Copenhagen, Available at: http://www.euro.who.int/__data/assets/pdf_file/0008/136466/e94888.pdf.

World Health Organization. 2013 *WHO 2013. Review of evidence on health aspects of air pollution – REVIHAAP project: Final technical report*. Copenhagen, Denmark.

World Health Organization. 2016 Ambient air pollution: A global assessment of exposure and burden of disease. Geneva, Switzerland.

Experimental investigation on rheological property of bitumen influenced by preheating process

Xuemei Zhang & Inge Hoff
Department of Civil and Environmental Engineering, Norwegian University of Science and Technology, Trondheim, Norway

ABSTRACT: Bitumen samples for laboratory tests are normally preheated at moderate temperature. However, the preheating process is suspect of influencing the aging process and bitumen property. This research aims to investigate the rheological property of bitumen influenced by preheating process. For this purpose, the matrix bitumen and bitumen aged by Thin-Film Oven Test (TFOT) were preheated for seven cycles, respectively. Dynamic Shear Rheometer (DSR) test was conducted to explore the influence of preheating process on rheological properties of bitumen. The results showed that preheating process shortened the differences in rheological properties between matrix and aged bitumen over preheating cycle. The effect of preheating process on bitumen could be divided into three stages: rapid change stage, stable stage, and aged stage. The rapid change stage, in which bitumen volatilizes rapidly resulting in apparent changes in rheological properties, typically occurs in the first two preheating cycles. The stable stage in which bitumen remains few changes with preheating cycle lasts for a longer period (3-5 preheating cycles). The aged stage, in which excessive heating time is applied on bitumen resulting in the aging of bitumen, usually starts from the sixth preheating cycle. The consequences of this study provide practitioners with proposals to reduce experimental error for bituminous tests by avoiding preheating bitumen more than six times.

Keywords: Bitumen, preheating process, rheological property, three stages

1 INTRODUCTION

Bitumen as a viscoelastic material plays a significant role in determining the service life of asphalt pavement, although it only accounts for 4-6 % of the gross mass of asphalt concrete (Li et al., 2019; Morova, 2013; Serin et al., 2012). The viscoelastic behaviour of bitumen could be comprehensively characterized by the rheological properties of bitumen. Rheological properties of bitumen primarily include deformation resistance, elastic/viscous response, rutting resistance, and resistance to fatigue cracking, which are of great importance in predicting asphalt pavement distresses (Remisova et al., 2016; Tao et al., 2019).

For laboratory tests, bitumen is inevitably preheated several times for subsequent tests. To be specific, when a barrel of bitumen is obtained from an asphalt factory, a barrel of bitumen should be preheated (up to 100 °C) to fluid and taken out to make subsequent test samples according to NS-EN 12594:2014. Thus, bitumen might be heated several times until the bitumen bucket is empty. However, several preheating cycles are suspected of aging bitumen i.e., excessive heating time would promote the volatilization or oxidation of bitumen (Ye et al., 2019).

DOI: 10.1201/9781003222880-6

In addition, microwave heating and induction heating as the major healing techniques can also be regarded as a heating action to bitumen (Li et al., 2020; Liu et al., 2020; Trigos et al., 2020). Previous studies showed that the temperature of asphalt concrete after microwave heating and induction heating could reach up to 91 °C and 83 °C, respectively (Gonzalez et al., 2018; Sun et al., 2020). These temperatures are beyond the softening point of bitumen. Therefore, the healing process performed on asphalt concrete could be regarded as the preheating process.

Therefore, the influence of the preheating process on the rheological properties of bitumen should be paid more attention. For this purpose, two kinds of bitumen, matrix bitumen and aged bitumen by Thin Film Oven Test (TFOT) aging, were preheated for seven cycles, respectively. The rheological properties of both matrix bitumen and aged bitumen were investigated by Dynamic Shear Rheometer (DSR) test. The conclusions obtained from the research could give some suggestions on reducing experimental error for bituminous tests.

2 MATERIALS AND METHODS

2.1 Materials

The matrix bitumen applied in this research was obtained from company Veidekke in Norway. Aged bitumen was aged by thin film oven test (TFOT) according to NS-EN 12607-2:2014. The basic properties of bitumen are shown in Table 1. The penetration of bitumen is tested according to NS-EN 1426:2015 standard. The mass loss caused by TFOT aging is calculated according to formula 1.

$$\text{Mass loss} = \frac{m_A - m_U}{m_U} \times 100\% \qquad (1)$$

Where m_U is the mass of matrix bitumen, m_A is the mass of TFOT aged bitumen.

Table 1. Basic properties of bitumen.

The number of preheating cycles	Penetration [mm]		Mass loss [%]
	Matrix bitumen	TFOT aged bitumen	
1	7.71	5.44	-0.05
2	7.66	5.72	-0.03
3	7.41	5.79	-0.02
4	7.52	5.85	0
5	6.99	5.40	-0.02
6	7.01	5.52	-0.02
7	7.05	5.75	-0.01

2.2 Methodology

2.2.1 Thin film oven test aging
Thin-film oven test aging (TFOT) test was conducted to simulate the short-term aging of bitumen during storage and transportation. The test parameters are shown in Table 2.

2.2.2 Dynamic shear rheometer
In this research, rheological behaviours of bitumen were measured using Dynamic Shear Rheometer (DSR). The test parameters are presented in Table 3. The temperature is chosen as 10 °C and 20 °C, which is reasonable and practical temperature condition for bitumen in reality.

Table 2. Test parameters.

Test parameters	Value	Test parameters	Value
Diameter of metallic pan	140 ± 1 mm	Oven temperature	163 °C
The mass of each sample	50 g	Aging/heating time	5 h
Thickness of bitumen film	3.2 mm	Cooling time	2 h

Table 3. Parameters value.

Setting	Value for rheological property
Angular frequency	10 rad/s
Amplitude gamma	0.5 %
Temperature range	10 °C, 20 °C
Diameter of the plate	8 mm
Sample thickness	2 mm

2.2.3 Preheating procedure

In terms of the preheating procedure, a barrel of bitumen is heated at 90 °C for 1.5 hours in accordance with the NS-EN 12594:2014 standard so that bitumen was liquid enough for conducting subsequent tests. This procedure is called the preheating of bitumen.

The flow diagram of the preheating cycle for matrix bitumen and aged bitumen is presented in Figure 1. Firstly, a bucket of bitumen is preheated n times and then placed in a metallic plate for TFOT test. After the preheating cycles or TFOT aging, the rheological properties of both matrix bitumen and aged bitumen are studied by DSR tests. Meanwhile, U-n indicates matrix bitumen after n preheating cycles, A-n indicates TFOT aged bitumen after n preheating cycles. Regarding the number of preheating cycles, both matrix bitumen and TFOT aged bitumen are preheated for 1, 2, 3, 4, 5, 6, and 7 cycles, respectively.

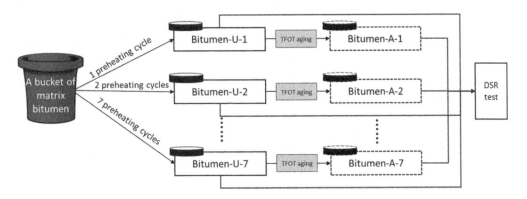

Figure 1. The flowchart of preheating cycle for matrix bitumen and aged bitumen.

3 RESULTS AND DISCUSSIONS

3.1 *Complex modulus*

The complex modulus (G*) is defined as the total resistance to deformation of bitumen, and higher complex modulus indicates better resistance to deformation (Farias et al., 2016). Figure 2 shows the complex modulus of matrix and aged bitumen over preheating cycle at 10 °C and 20 °C. Regarding matrix bitumen, U-3 and U-4 performed the highest values of complex modulus, and the complex modulus of U-1 was the smallest. These phenomena show little correlation in complex modulus between matrix bitumen and preheating cycle. However, in terms of aged bitumen, A-2 and A-1 have the biggest complex modulus, and A-7 had the smallest complex modulus. Comparing matrix bitumen and aged bitumen, the preheating process changed the order of complex modulus of bitumen samples.

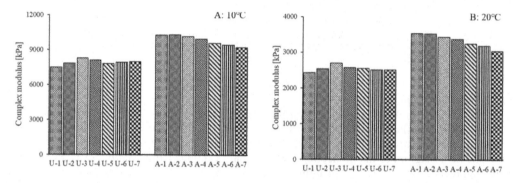

Figure 2. The complex modulus of samples versus preheating cycle.

In order to evaluate the complex modulus of bitumen influenced by the preheating cycle, the change of complex modulus between aged bitumen and matrix bitumen under the same preheating cycle is calculated and shown in Figure 3. The change of complex modulus decreased with the increasing preheating cycle. This phenomenon means that the preheating process causes a significant impact on bituminous resistance to deformation, and more preheating cycles would result in smaller differences in complex modulus between matrix bitumen and aged bitumen. Bitumen samples after 1 and 2 preheating cycles showed relatively higher values of complex modulus than other samples, which means the deformation resistance of bitumen is vulnerable to preheating procedure at the beginning (1-2 preheating cycles). Bitumen samples after 3, 4, 5, and 6 preheating cycles showed similar values of the change of complex modulus, which indicates these samples are relatively stable. Besides, bitumen after 7 preheating cycles had the smallest change of complex modulus than middle four samples. This conclusion demonstrates that the matrix bitumen after 7 preheating cycles is slightly aged so that the difference in complex modulus between matrix and aged bitumen is relatively small.

3.2 *Phase angle*

Phase angle (δ) is defined as the ratio of the loss to the storage components of the complex modulus, the phase angle of bitumen is between 0° to 90°, a higher δ indicates better viscous response and worse elastic response of bitumen (He et al., 2019). Figure 4A and Figure 4B show the results of the phase angle of samples at 10 °C and 20 °C after different preheating cycles, respectively. U-6 and U-4 have relatively bigger phase angles among matrix samples. U-1 has the smallest δ among matrix samples. Regarding aged bitumen, A-6 and A-7 have the

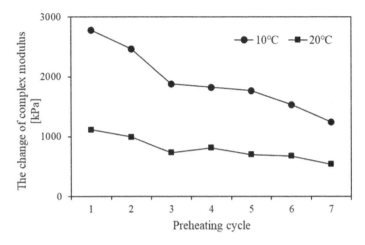

Figure 3. The changes in complex modulus between matrix and aged bitumen versus preheating cycle.

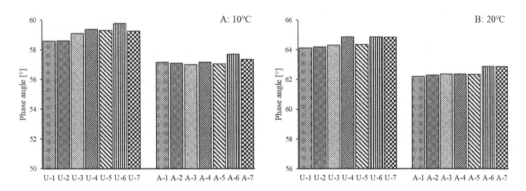

Figure 4. The phase angle of samples versus preheating cycle.

biggest phase angle, and the phase angle of A-3 and A-5 is relatively small. These results indicate that the preheating process has limited influence on the viscoelastic response of matrix bitumen but affects the short-term aging on bitumen.

The contrast of matrix bitumen and aged bitumen is estimated, which results in the difference of phase angle between matrix bitumen and aged bitumen over preheating cycle shown in Figure 5. The changed value of phase angle of all samples is below 3°. Phase angle showed a different changing trend over the preheating cycle compared to complex modulus. The phase angle of bitumen decreased before 4 preheating cycles and remained few changes afterward. Meanwhile, the phase angles of bitumen samples after 1 and 2 preheating cycles were the biggest among all samples. These facts indicate that bitumen is intensely affected at the beginning (1-2 preheating cycles) and then keeps stable after 4 preheating cycles.

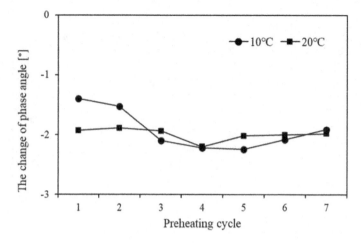

Figure 5. The changes in phase angle between matrix and aged bitumen versus preheating cycle.

3.3 *Rutting factor*

Rutting factor is defined as the ratio of complex modulus to sin δ. It reflects the rutting resistance of bitumen. A higher rutting factor indicates better performance to resist rutting/permanent deformation (Sedaghat et al., 2020). Figure 6A and Figure 6B show the result of the rutting factor of samples at 10 °C and 20 °C after different preheating cycles, respectively. U-3 and U-4 had higher rutting factors, and U-1 and U-5 had smaller rutting factors among all matrix bitumen samples. These results indicate little correlation between rutting factor of matrix bitumen and preheating cycle. For aged bitumen, A-2 and A-1 had the higher rutting factors, and A-7 and A-6 had the smaller rutting factors. These data indicate that matrix bitumen and aged bitumen perform differently to resist rutting under different preheating cycles. Based on the different orders of samples' rutting factor between matrix and aged bitumen, preheating cycles might affect the degree of TFOT aging on bitumen.

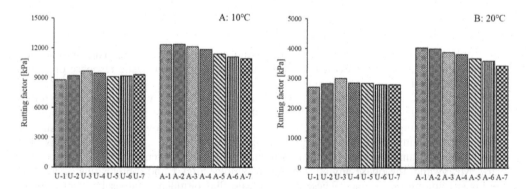

Figure 6. The rutting factor of samples versus preheating cycle.

In order to study the effect of preheating cycle on the rutting factor of bitumen, the changes in rutting factor between matrix and aged bitumen after different preheating cycles are calculated and shown in Figure 7. The changes in rutting factor of bitumen samples after 1 and 2 preheating cycles were relatively noticeable; samples after 3, 4, or 5 preheating cycles showed similar differences in rutting factor; sample after 7 preheating cycles had the smallest change in rutting factor. These facts show that the change in rutting resistance of bitumen decreased with the increase of preheating cycles. Besides, bitumen is more stable when the number of preheating cycles is in the range of 3 to 5. Due to the instability of samples, the difference of rutting factor of sample-1 and sample-2 is apparent. While after 6 or 7 preheating cycles, the bitumen is already aged to some extent so that the difference in rutting factor between matrix and aged bitumen is relatively small.

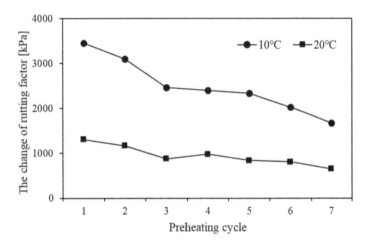

Figure 7. The change in rutting factor between matrix and aged bitumen versus preheating cycle.

3.3 *Fatigue factor*

Fatigue factor is defined as the product of complex modulus and sin δ, it is used to evaluate the performance against fatigue distress of bitumen. A higher fatigue factor would contribute to a worse ability to resist fatigue cracking (Ameri et al., 2018). Figure 8 shows the fatigue factor of samples at 10 °C and 20 °C after different preheating cycles. Regarding the matrix bitumen, U-1 has the lowest fatigue factor, U-3 has the highest fatigue factor. However, in terms of aged bitumen, A-2 has the biggest fatigue factor, followed by A-1, and A-7 has the smallest fatigue factor. These facts indicate that the preheating procedure has little impact on the ability to resist fatigue cracking of matrix bitumen, while it could influence the evolution of bitumen aging leading to the changed order in fatigue factor of aged bitumen.

How preheating procedure influences the fatigue factor is estimated by comparing the fatigue factor between matrix and aged bitumen after the same preheating cycle, which is presented in Figure 9. The change in fatigue factor of samples decreased over preheating cycle, which indicates that more preheating cycles would shorten the differences in fatigue factor between matrix bitumen and aged bitumen. As seen from Figure 9, bitumen samples after 1 and 2 preheating cycles showed higher values in the change of fatigue factor than other samples, which indicates that bitumen is easily influenced with a severe degree after one or two

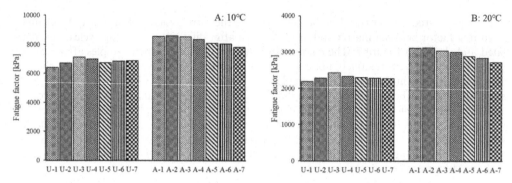

Figure 8. The fatigue factor of samples versus preheating cycle.

preheating cycles. However, after 3, 4, 5, and 6 preheating cycles, the samples showed similar changes in fatigue factor, which indicates that the sample after 3, 4, 5, and 6 preheating cycles have a relatively stable state. Finally, the bitumen with seven preheating cycles had the smallest difference in fatigue factor, which shows that bitumen is already aged to a certain extent so that the difference of matrix and aged bitumen is previously smaller than that after fewer preheating cycles.

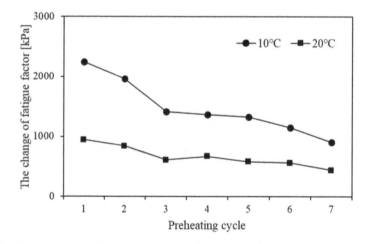

Figure 9. The change in fatigue factor between matrix and aged bitumen versus preheating cycle.

4 CONCLUSIONS

For laboratory tests, bitumen is generally preheated at moderate temperature to fluid for the following tests. While the preheating process would influence bitumen rheological properties and the outcome of the following tests. Therefore, this research studied the effect of preheating process on the rheological properties of matrix bitumen and TFOT aged bitumen. The obtained conclusions were summarized as follows.

The preheating process had a limited impact on the rheological properties of matrix bitumen, whereas the preheating process has a significant influence on the short-term aging on bitumen. The difference in rheological properties between matrix bitumen and aged bitumen decreased over preheating cycle. There are three stages during preheating process, i.e., rapid change stage, stable stage, and aged stage. The rapid change stage normally occurs in the first two preheating cycles since bitumen is the most unstable and easily volatilized at the beginning resulting in apparent differences in rheological properties. The stable stage in which the bitumen remains few changes in rheological properties lasts for a longer period (3 - 5 preheating cycles). The aged stage, in which bitumen is slightly aged resulting in small changes in rheological parameters, normally starts from the sixth preheating cycle.

The conclusions could provide engineers with proposals on how to reduce experimental error for bituminous tests. For example, it is better not to preheat the bitumen sample more than six times to avoid bitumen aging, which could affect the subsequent test results.

DECLARATION OF INTERESTS

The authors declare that they have no known competing financial interests or personal relationships that could have appeared to influence the work reported in this paper.

REFERENCES

Ameri, M., Mansourkhaki, A., Daryaee, D., 2018. Evaluation of fatigue behavior of high reclaimed asphalt binder mixes modified with rejuvenator and softer bitumen. Constr Build Mater 191, 702–712.

Farias, L.G.A.T., Leitinho, J.L., Amoni, B.D.C., Bastos, J.B.S., Soares, J.B., Soares, S.D.A., de Sant'Ana, H.B., 2016. Effects of nanoclay and nanocomposites on bitumen rheological properties. Constr Build Mater 125, 873–883.

Gonzalez, A., Norambuena-Contreras, J., Storey, L., Schlangen, E., 2018. Self-healing properties of recycled asphalt mixtures containing metal waste: An approach through microwave radiation heating. J Environ Manage 214, 242–251.

He, R., Zheng, S.N., Chen, H.X., Kuang, D.L., 2019. Investigation of the physical and rheological properties of Trinidad lake asphalt modified bitumen. Constr Build Mater 203, 734–739.

Li, H.C., Yu, J.Y., Wu, S.P., Liu, Q.T., Wu, Y.Q., Xu, H.Q., Li, Y.Y., 2020. Effect of moisture conditioning on mechanical and healing properties of inductive asphalt concrete. Constr Build Mater 241.

Li, Y.Y., Wu, S.P., Liu, Q.T., Nie, S., Li, H.C., Dai, Y., Pang, L., Li, C.M., Zhang, A.M., 2019. Field evaluation of LDHs effect on the aging resistance of asphalt concrete after four years of road service. Constr Build Mater 208, 192–203.

Liu, K., Dai, D.L., Fu, C.L., Li, W.H., Li, S.Q., 2020. Induction heating of asphalt mixtures with waste steel shavings. Constr Build Mater 234.

Morova, N., 2013. Investigation of usability of basalt fibers in hot mix asphalt concrete. Constr Build Mater 47, 175–180.

Remisova, E., Zatkalikova, V., Schlosser, F., 2016. Study of Rheological Properties of Bituminous Binders in Middle and High Temperatures. Civ Environ Eng 12(1), 13–20.

Sedaghat, E., Taherrian, R., Hosseini, S.A., Mousavi, S.M., 2020. Rheological properties of bitumen containing nanoclay and organic warm-mix asphalt additives. Constr Build Mater 243.

Serin, S., Morova, N., Saltan, M., Terzi, S., 2012. Investigation of usability of steel fibers in asphalt concrete mixtures. Constr Build Mater 36, 238–244.

Sun, G.Q., Hu, M.J., Sun, D.Q., Deng, Y., Ma, J.M., Lu, T., 2020. Temperature induced self-healing capability transition phenomenon of bitumens. Fuel 263.

Tao, G.Y., Xiao, Y., Yang, L.F., Cui, P.D., Kong, D.Z., Xue, Y.J., 2019. Characteristics of steel slag filler and its influence on rheological properties of asphalt mortar. Constr Build Mater 201, 439–446.

Trigos, L., Gallego, J., Escavy, J.I., 2020. Heating potential of aggregates in asphalt mixtures exposed to microwaves radiation. Constr Build Mater 230.

Ye, W.L., Jiang, W., Li, P.F., Yuan, D.D., Shan, J.H., Xiao, J.J., 2019. Analysis of mechanism and time-temperature equivalent effects of asphalt binder in short-term aging. Constr Build Mater 215, 823–838.

Skid resistance properties against RAP content change in surface asphalt mixture

M. Pomoni, C. Plati & A. Loizos
Laboratory of Pavement Engineering, National Technical University of Athens (NTUA), Greece

ABSTRACT: Many countries use the Reclaimed Asphalt Pavement (RAP) into traditional asphalt mixtures for road surface layers, at a content of approximately 10–30%. So far, many studies have investigated the performance of these mixes dealing mainly with existing fundamental issues of the mix design process. The present study focuses on their functional performance and most specifically, on their skid resistance properties when used at surface layers containing RAP. To that purpose, a laboratory process is designed for the fabrication of both RAP-content asphalt mixtures and traditional Hot-Mix-Asphalt (HMA) mixtures in order to assess their skid resistance performance. HMA mixes are used as reference considering that these mixtures are most often used for road construction. The HMA was taken from an in-situ layering process. The emphasis was put on the impact of weather changes on materials' performance. In an attempt to reflect field conditions, a laboratory simulation was performed to consider some weather changes (i.e. temperature, raining effect and contamination). Skid resistance measurements were performed on the fabricated specimens using the British Pendulum Tester. The analysis results showed promising aspects for the use of RAP in wearing courses for all the followed simulations. Overall, it is argued that the use of RAP did not appear to weaken skid resistance levels, providing as such an additional merit in its use apart from its low carbon footprint, in opposition to the production of new HMA mixes.

Keywords: RAP, skid resistance, weather changes, laboratory simulation, British Pendulum Tester

1 BACKGROUND & OBJECTIVE

Recycling of pavement materials is a common practice nowadays in an effort to limit, as much as possible, the use of virgin materials. This procedure started from the early 70s due to the high cost of crude oil during the Arab oil embargo (Federal Highway Administration, 2011). Therefore, there have been numerous applications in a worldwide scale, not only for pavement rehabilitation projects but also for new road construction activities.

The most popular recycled pavement material is the Reclaimed Asphalt Pavement (RAP), which is extensively used in asphalt mixtures during the last decades. However, besides its sustainable identity, there are concerns amongst the agencies regarding its long-term performance and the long-term cost-effectiveness that can be achieved over the pavement service life (Sotoodeh-Nia et al. 2019). The main concerns refer to the rheological properties of the RAP material that progressively change. Basically, the different properties between the fresh Hot-Mix Asphalt concrete (HMA) and the RAP have produced evidence regarding the susceptibility of the final mixture to low temperatures and fatigue cracking (Sotoodeh-Nia et al. 2019, Al-Qadi et al. 2007). Hence, several approaches have been investigated for the incorporation

DOI: 10.1201/9781003222880-7

of RAP material in HMA, especially during pavement rehabilitation, where the upper asphalt layer is usually replaced. Focusing on the surface layer and according to FHWA (2011), it seems that the majority of highway agencies in the US tend to use RAP materials at a percentage of 10% – 30%. However, the usage of contents 10%-20% is more usual (Figure 1).

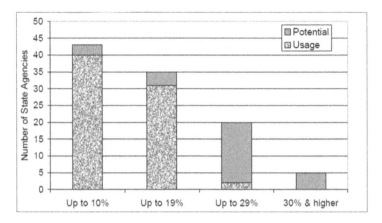

Figure 1. Usage and potential of RAP material (%) in the surface layer (FHWA, 2011).

Knowledge on the usage of RAP on the surface layer is mainly concentrated around the mechanical properties of the asphalt mixture. However, considering that the surface layer comes in direct contact with the tyres of the moving vehicles, the provided tyre-pavement contact is also a matter of concern (Plati et al. 2020a). The pavement wearing course should have appropriate characteristics for providing adequate skid resistance with regards to road safety requirements under various weather conditions. Adequate skid resistance levels enable vehicles to speed up, brake, slow down and move safely around curvatures (Flintch et al. 2012).

With regards to the above, the present study aims to provide some preliminary findings on the skid resistance performance of HMA wearing courses containing RAP. This study is part of an ongoing research that aims at a more thorough assessment of the skid resistance performance of pavement surfaces consisting of both conventional and sustainable materials. To meet the study goals, asphalt mixtures with variable RAP contents were fabricated in the laboratory in order to meet the specifications of wearing courses, while a laboratory procedure to simulate weather changes was designed and followed.

2 SKID RESISTANCE & WEATHER CHANGES

The two basic components that are critical to the skid resistance development are: macrotexture and microtexture. They are both interrelated to the properties of the materials used for pavement surface construction (Plati et al. 2019, Flintch et al. 2012). In particular, aggregate gradation in the asphalt mixture, air void content and binder properties mainly influence the pavement macrotexture, while aggregates mineralogy and their shape mainly affect microtexture levels (Kane and Edmonsson 2020, White et al. 2019, Plati et al. 2017). Towards this, it is obvious that the material properties used for the wearing course are determinant for the provided skid resistance.

Besides material properties, tyre-pavement interaction is susceptible to weather changes as well. This means that temperature changes, raining events and the presence of dust or snow on the pavement surface can affect the provided skid resistance (Plati et al. 2020b). On the one hand, this sort of weather changes affects the tyre-pavement contact, as is the case of water presence beneath the tyre tread. On the other hand, the asphalt mixture characteristics influence the tyre-pavement contact during various weather conditions. For instance, the aggregate asperities (level of microtexture) can penetrate the film of water during a rainfall and enable the tyre tread to directly contact the pavement surface (Plati et al. 2020b, Do et al. 2013).

In general, skid resistance level is subjected to variations throughout a year due to the seasonal weather variations (Pomoni et al. 2020a, Pomoni et al. 2020b, Do et al. 2013). Typically, increases in temperature decrease the skid resistance level. The same happens when dust is present on the surface. In addition, although the raining effect is dangerous for hydroplaning, it can also clean the pavement surface from the remaining dust and vehicle oils (contaminants) and provide a clean surface with increased skid resistance levels (Figure 2) (Plati et al. 2020b, Pomoni et al. 2020b).

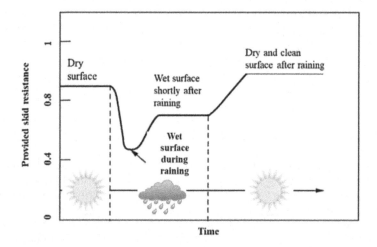

Figure 2. Typical changes in skid resistance levels due to weather changes.

Considering the above, it should be acknowledged that the use of conventional HMA in wearing courses as a standard practice, has been investigated in several researches regarding the provided skid resistance. The introduction though of new or recycled materials into pavement engineering pinpoints the necessity to assess the contribution of these materials to the provided skid resistance levels and cover the related knowledge gaps. Hence, with regards to the above concerns, at this stage of research the focus was put on the investigation of indicative weather changes effect on the provided skid resistance.

3 FABRICATION OF HMA & RAP MIXES

For the accomplishment of the laboratory-based study, asphalt mixture slabs were fabricated according to the following categories:

3.1 *HMA slab (0% RAP addition): Slab A*

Loose hot-mix asphalt (HMA) concrete was collected before being used for layering a wearing course/antiskid surface during a road construction project. The specific HMA was an open-graded mixture with the O-5 mix designation according to the ASTM D3515-01 standard (2001). Steel slag was used into the mix design as the aggregate type and a polymer-modified bitumen of 25-55/70 grade penetration was added. The mixture contained a modified binder at a percentage of 5% by mass of the mixture.

The mixture was compacted according to the described procedure in the CEN - EN 12697-33 standard (2019). The machinery utilized for the compaction process was a rolling compactor, which is a laboratory device that simulates field compaction of asphalt materials contained in specified moulds of certain dimensions (300 mm*300 mm*40 mm height). The amount of the asphalt mixture that was placed in the moulds depended on the maximum density of the asphalt mixture and the desired level of the air voids. After the static load of the compactor on the mixture was accomplished, a specific air void content was reached. Herein, the accomplished air voids were measured and were found around 16% using the related standard (ASTM D3203, 2017).

3.2 *HMA & RAP slab (30% RAP addition): Slab B*

At this type of asphalt mixture, 30% of RAP was used combined with the loose HMA as described above, in a percentage of 70%. The utilized RAP material had similar properties with the HMA, but it aged more than 5 years until the time of the road rehabilitation activities when it was moved out and collected for the research purpose.

The mixture was heated in an oven and a low amount of asphalt bitumen was added (1 – 2 % approximately). Both original HMA and RAP were mixed properly to produce a homogeneous mixture, which was thereafter compacted with the rolling compactor. The achieved air void content was 9% (ASTM D3203, 2017).

3.3 *HMA & RAP slab (10% RAP addition in the form of fine aggregates): Slab C*

The specific type of asphalt mixture included 10% of RAP combined with the loose HMA in a percentage of 90%. However, differently from the previous mixture, the RAP material was finer, as only the fine aggregates of the RAP were added in the mixture. More specifically, only the material with aggregate size lower than 4mm was added in order to assess how this kind of variation could affect the provided skid resistance. By following the same process as previously described for mixing and compacting the HMA and RAP material, the achieved air-void content was found to be 7% (ASTM D3203, 2017).

4 EXPERIMENTAL PROCESS

4.1 *Weather conditions simulation*

Regarding the weather changes that affect skid resistance levels, it was attempted to develop a procedure in order to simulate some basic weather conditions that can occur in the field. As an overview, both dry and wet surfaces were examined under the conditions described in Figure 3.

The first part of the investigation was dedicated to the assessment of temperature changes. Hence, the temperatures of 10, 20 and 30°C were simulated in order to investigate skid resistance variations against these temperature changes. The desired temperature was each time achieved through proper air-conditioning of the area surrounding the slabs as well as through a temperature-controlled chamber where the slabs were stored. Prior to measurements, a small amount of water was rinsed on the surface of the slabs as described in the British Pendulum Tester BPT standard (ASTM E303- 93, 2018).

In the second part of the investigation, the presence of water and dust on pavement surfaces was simulated at three stages. The temperature was kept constant at 20 °C, as this temperature can be in general observed annually in southern European countries. At first, the slabs were totally dry and clean and then, a small amount of water was rinsed on the fabricated surfaces to simulate the condition after a short rainfall event on a clean road surface (Figure 4a).

Thereafter, the water was naturally removed from the slab and a new mixture of both dust (limestone-based) and water was placed on the surface to simulate the presence of contamination. This type of contamination can be formed after a long dry period when an immediate

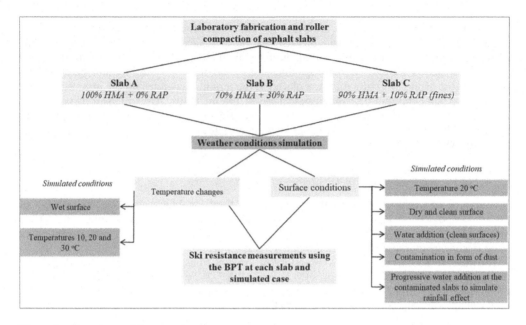

Figure 3. Experimental framework.

Figure 4. (a) Wet and clean surface (b) Contaminated surface (c) Contaminated surface after the rainfall simulation.

short rainfall occurs producing a slippery and contaminated surface (Figure 4b). Finally, a continuous rainfall event was simulated by rinsing a certain amount of water for a couple of minutes on the contaminated surfaces (Figure 4c).

4.2 *Skid resistance measurements*

The BPT was utilized to measure the provided skid resistance under the afore-mentioned conditions. The BPT system is a portable friction device developed in the 1960s in the United Kingdom. It is a static device and its use is standardized (ASTM E303-93, 2018). A smooth rubber slider mounted on the pendulum arm is released from the horizontal position. The slider reaches a certain speed (~ 10 km/h, by gravity) when the rubber touches the road surface. The output of each test is the British Pendulum Number (BPN). Zero BPN value corresponds to a totally smooth surface while 150 BPN is the upper limit, representing practically an abnormal rough surface.

It has to be also mentioned that in many studies, the BPN index has been connected with the microtexture of a surface rather than the provided skid resistance (Pomoni et al. 2020, Pratico and Astolfi 2017). However, under the framework of the particular study where new surfaces were tested, microtexture is more critical (Plati and Pomoni 2019, Vaiana et al. 2012).

Thus, by measuring the microtexture component, a good indication for the skid resistance level can be provided, as well.

At each of the simulated conditions described in section 3.1, BPN measurements were conducted. The average of four measurements per each condition corresponded to the characteristic BPN values that are presented in the following section.

5 RESULTS

At first, the impact of temperature on the provided skid resistance levels of the three slabs is assessed (Figure 5).

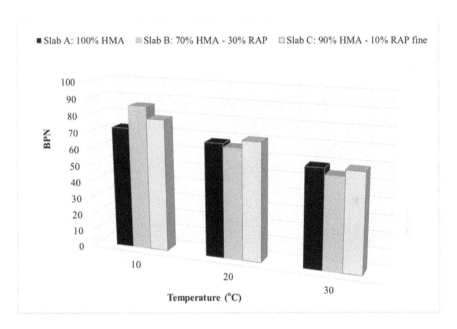

Figure 5. BPN and temperature changes.

It can be seen that a temperature increase decreases skid resistance, something which is in accordance with previously existing knowledge (Kogbara et al. 2016, Srirangam et al. 2015). This remark is valid for all the RAP contents under consideration. In particular, the rate of decrease appeared to be more intense for the slab B that consists of RAP at a percentage of 30%. Also, the addition of RAP seems to be more influencing for the BPN levels at the temperature of 10 °C. This result may be connected with the differences in air void content of the mixtures which might be critical for the low temperatures.

Regarding the second part of the investigated simulated weather changes, the BPN values are presented for each slab at Figures 6-8. At each figure, the horizontal axis represents the condition of the slab surface, as water mass was progressively added during the experiments.

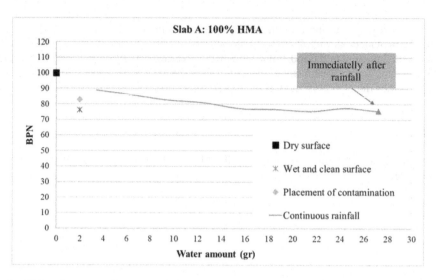

Figure 6. Skid resistance performance of slab A (100% HMA and 0% RAP) against weather changes.

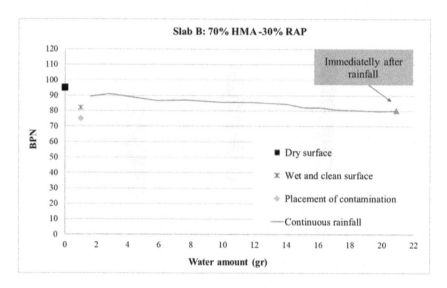

Figure 7. Skid resistance performance of slab B (70% HMA and 30% RAP) against weather changes.

For all the fabricated slabs, skid resistance is higher when the surfaces are dry. The HMA surface presents the greater values, something which is expected due to the increased air void content that provides a rougher surface than the other slabs with lower air void contents.

Moreover, the addition of water tends to reduce the skid resistance level. However, the slabs performance against contamination is not similar. An opposite trend can be observed in the relationship between the wet, clean and contaminated surface in slab A (Figure 6) as opposed to slabs B and C (Figures 7 and 8). The placement of contamination in the surface totally made by HMA increases the BPN levels and this is probably connected with the highest air void content (17%) of slab A. Due to the high air voids, the mix of water and dust inserts into them, so it does not drastically intervene with the BPT slider-pavement surface contact. On the contrary, this process is not facilitated at the RAP-based slabs. Due to the lower air voids, the mix of water and dust cannot

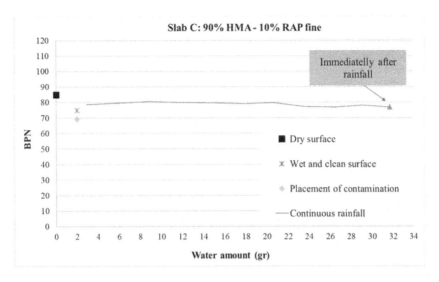

Figure 8. Skid resistance performance of slab C (90% HMA and 10% RAP) against weather changes.

penetrate into the slabs, and as a consequence remains between the BPT slider-pavement contact.

Afterwards, as rainfall starts cleaning the contaminated surfaces, a better skid resistance level is observed for each slab. However, it has to be pinpointed that the BPN in slabs B and C does not significantly vary during the simulated continuous rainfall event, probably due to the fact that air voids are blocked by the trapped contamination. This phenomenon is more profound for the case of slab C. In particular, the slope of the curve in slab C is practically near zero, since both the presence of RAP fines as well as its lower air void content did not enable the contamination runoff. On the other hand, the HMA surface present greater changes in the BPN values shortly after the initiation of the rainfall, because part of the contamination may run off through the air voids faster and more effectively, in comparison with the other two slabs.

A final remark is that at the end of the rainfall simulation, a lower BPN value was measured in slab A than slabs B and C. This observation may produce evidence that RAP-based mixtures could exhibit an equally satisfactory behavior in terms of their skid resistance performance. However, further research is needed to make reliable remarks.

6 CONCLUSION

In the present study, it was proved that the addition of RAP did not provide negative effect on the skid resistance levels for the simulated conditions. This seems as an additional merit in its use apart from the low carbon footprint, in opposition to the new HMA mixes. Also, the air void content seemed to be critical both for the temperature changes as well as the simulated surface conditions (rainfall etc.) The high air void content provided a rougher surface with higher skid resistance values as it facilitated the water runoff during the simulated rainfall events.

In respect to the argument that the usage of the BPT device has been connected more with the provided microtexture rather than the skid resistance performance, it can be stated that the effect of macrotexture was not totally ignored within this study. Since the air void content that is absolutely connected with the provided macrotexture, appeared to affect the provided skid resistance at the investigated slabs, it can be argued that macrotexture and its influence were indirectly taken into consideration as well.

Overall, the study highlighted the need to assess skid resistance performance of the asphalt mixtures containing recycled materials. Besides their structural soundness, roads are expected to be serviceable and safe for road users depending among others, on the provided skid resistance. Future research steps include the combined consideration of both the traffic and weather conditions effect on the provided skid resistance levels of pavement surfaces with recycled materials. Since accelerated pavement testing is not a feasible solution for preliminary testing, the potential of using these materials at wearing courses could be reliably assessed through extensive laboratory investigations that can act as a basis for future applications in the field.

ACKNOWLEDGEMENT

The research was conducted at the premises of the Laboratory of Pavement Engineering of the National Technical University of Athens (NTUA). This research is co-financed by Greece and the European Union (European Social Fund- ESF) through the Operational Programme 'Human Resources Development, Education and Lifelong Learning' in the context of the project "Strengthening Human Resources Research Potential via Doctorate Research" (MIS-5000432), implemented by the State Scholarships Foundation (IKY).

REFERENCES

Al-Qadi, I.L., Elseifi, M. and Carpenter, S.H., 2007. *Reclaimed asphalt pavement—a literature review*. Report no. FHWA-ICT-07-001. Rantoul, IL: Illinois Center for Transportation.
ASTM D3203/D3203M-17, 2017. *Standard Test Method for Percent Air Voids in Compacted Asphalt Mixtures*. ASTM International, West Conshohocken, PA.
ASTM D3515–01, 2001. *Standard Specification for Hot-Mixed, Hot-Laid Bituminous Paving Mixtures*. ASTM International, West Conshohocken, PA.
ASTM E303–93, 2018. *Standard Test Method for Measuring Surface Frictional Properties Using the British Pendulum Tester*. ASTM International, West Conshohocken, PA.
CEN - EN 12697–33, 2019. *Bituminous mixtures - Test method - Part 33: Specimen prepared by roller compactor*, p. 22
Do, M.-T., Cerezo, V., Beautru, Y. and Kane, M., 2013. *Modeling of the connection road surface microtexture/water depth/friction*. Wear, 302(1-2): 1426–1435.
FHWA, 2011. *Reclaimed Asphalt Pavement in Asphalt Mixtures: State of the Practice*. Federal Highway Administration, Publication No. FHWA –HRT-11-021, April 2011.
Flintsch, G.W., de Leon, E., McGhee, K.K. and Najafi, S., 2012. *The Little Book of Tire Pavement Friction*. Pavement Surface Properties Consortium, 2012, Version 1.
Kane M. and Edmondson, V., 2020. *Long-term skid resistance of asphalt surfacings and aggregates' mineralogical composition: Generalisation to pavements made of different aggregate types*. Wear, https://doi.org/10.1016/j.wear.2020.203339.
Kogbara, R.B., Masad, E.A., Kassem, E., Scarpas A.T. and Anupam, K., 2016. *A state-of-the-art review of parameters influencing measurements and modeling of skid resistance of asphalt pavements*. Construction and Building Materials, 114, 602–617.
Plati, C. and Pomoni, M., 2019. *Impact of traffic volume on pavement macrotexture and skid resistance long-term performance*. Transportation Research Record: Journal of Transportation Research Board, 2673(2): 314–322.
Plati, C., Pomoni, M., Loizos, A. and Yannis, G., 2020a. *Stochastic prediction of short-term friction loss of asphalt pavements: a traffic dependent approach*. In the Proceedings of the 9th International Conference on Maintenance and Rehabilitation of Pavements (Mairepav9), Raab C. (eds), Lecture Notes in Civil Engineering, vol 76. Springer, Cham. https://doi.org/10.1007/978-3-030-48679-2_86
Plati, C., Pomoni, M. and Georgouli, K., 2020b. *Quantification of skid resistance seasonal variation in asphalt pavements*. Journal of Traffic and Transportation Engineering, 7(2): 237–248.
Plati, C., Pomoni, M. and Stergiou, T., 2017. *Development of Mean Profile Depth to Mean Texture Depth Shift Factor for Asphalt Pavements*. Transportation Research Record: Journal of Transportation Research Board, 2641(1): 156–163.
Plati, C., Pomoni, M. and Stergiou, T., 2019. *From Mean Texture Depth to Mean Profile Depth: Exploring possibilities*. In the Proceedings of 7th International Conference on Bituminous Mixtures and

Pavements (7ICONFBMP), Nikolaides & Manthos (eds), Taylor & Francis, pp. 639–644, June 12–14, 2019, Thessaloniki, Greece, http://dx.doi.org/10.1201/9781351063265-86.

Pomoni, M., Plati, C., Loizos, A. and Yannis, G., 2020a. *Investigation of pavement skid resistance and macrotexture on a long-term basis*. International Journal of Pavement Engineering, https://doi.org/10.1080/10298436.2020.1788029.

Pomoni, M., Plati, C. and Loizos, A., 2020b. *How Can Sustainable Materials in Road Construction Contribute to Vehicles' Braking?* Vehicles, 2(1), 55–77, https://doi.org/10.3390/vehicles2010004.

Pratico, F.G. and Astolfi, A., 2017. *A new and simplified approach to assess the pavement surface micro- and macrotexture*. Construction and Building Materials, 148: 476–483.

Sotoodeh-Nia, Z., Manke, N., Williams, R.C., Cochran, E.W., Porot, L., Chailleux, E., Lo Presti, D., Carrión, A. and Blanc, J., 2019. *Effect of two novel bio-based rejuvenators on the performance of 50% RAP mixes – a statistical study on the complex modulus of asphalt binders and asphalt mixtures*. Road Materials and Pavement Design, https://doi.org/10.1080/14680629.2019.1661276.

Srirangam, S.K., Anupam, K., Scarpas, A. and Kasbergen, C., 2015. *Development of a thermomechanical tyre-pavement interaction model*. International Journal of Pavement Engineering, 16(8): 721–729.

Vaiana, R., Capiluppi, G.F., Gallelli, V., Iuele, T. and Minani, V., 2012. *Pavement Surface Performances Evolution: An Experimental Application*. Procedia - Social and Behavioral Sciences, 53: 1149–1160.

White, G., Ward, C. and Jamieson, S., 2019. *Field evaluation of a handheld laser meter for pavement surface macro texture measurement*. International Journal of Pavement Engineering, https://doi.org/10.1080/10298436.2019.1654103.

Generation and propagation of reflective cracking and thermal cracking of asphalt pavement in cold regions of China

W.Y. Zhou
School of Transportation on Science and Engineering, Harbin Institute of Technology, Heilongjiang Harbin, China

Z.G. Chen
Jilin Transportation Research Institute, Changchun, Jilin, China

Z.N. Chen
School of Transportation on Science and Engineering, Harbin Institute of Technology, Heilongjiang Harbin, China

W.J. Qin
Jilin Transportation Research Institute, Changchun, Jilin, China

J.Y. Yi*
School of Transportation on Science and Engineering, Harbin Institute of Technology, Heilongjiang Harbin, China

ABSTRACT: In order to resist heavy traffic loads and reduce the risk of surface damage, rutting and settlements over time, cement-treated materials are often utilized in pavement construction in China. However, these materials exhibit drying and thermal shrinkage, resulting in cracking in the cement-treated materials. These cracking may further result in reflective cracking of the asphalt material during service. Thermal cracking is another type of cracking in the cold region of China. All these cracking generate with the micro-cracks inside the materials under tensile stress and develop with the effects of traffic and environmental loadings. As the initiated micro-cracks are invisible, it is rather difficult to predict where and when the macro-cracks will appear. In this study, the distributed optical fibre sensors were separately embedded in the cement-treated base layer, the bottom layer, and the middle layer. Then the strain distribution along with the distributed optical fibre sensors was measured every 6 months since the winter of 2018. We have found something useful for further interpretation and analysis about how the reflective cracking and thermal cracking in asphalt pavement generate and propagate.

Keywords: Asphalt pavement, reflective cracking, thermal cracking

1 INTRODUCTION

Due to its high performance(Thøgersen et al. 2004; Yeo 2008), semi-rigid base asphalt pavement has been the main form of pavement structure of high-class highways in China. However, we are still unable to thoroughly solve the cracking problems of it. Once the cracking

*Corresponding author

DOI: 10.1201/9781003222880-8

generate and propagate, the overall strength of the pavement will be descended. And with the water intrusion, the performance of the pavement will be further destroyed. Normally, cement-treated materials are widely employed as the base materials for semi-rigid base asphalt pavement. Factors such as drying and thermal shrinkage of cement-treated materials can generate reflective cracking in asphalt pavement. Moreover, thermal cracking as another typical cracking usually occurs in the cold region of China. To boost the cracking resistant research, more fundamental studies are needed to better clarify how the reflective cracking and thermal cracking in asphalt pavement generate and propagate.

Many researchers were looking for research methods to evaluate reflective cracking and thermal cracking. Applications of alternative methods, such as laboratory experiment(Ji et al. 2019; Gonzalez-Torre et al. 2015), fracture mechanics(Wagoner et al. 2005; Oshone et al. 2019; Xie and Wang 2020), finite element method(Wang and Zhong 2019; Wang et al. 2018; Dave et al. 2007; Kim and Buttlar 2015) and so on, have been investigated. However, laboratory experiment has deviation comparing with the conditions of the actual pavement due to procedures and different sizes. Fracture mechanics and finite element method are both the ideal state, and some parameters are derived from the laboratory experiment. In recent years, the technology of fiber-optic cables has been introduced into civil engineering research. As a usable long-term health monitoring method, this application can provide relatively correct records, and avoid full-scale loading experiment that is expensive and service-disruptive. Skar et al. employed high-resolution fiber-optic distributed strain sensing to conclude the fundamental expressions concerning the relationship between the location of distinct strain points and the variation of soil parameters(Skar et al. 2019). Levenberg et al. studied soil support characterization in slab-on-grade constructions with fiber-optic distributed strain sensing (Levenberg et al. 2020). Furthermore, this research method has been gradually employed in pavement engineering to monitor the cracking. Xiang et al. monitored asphalt pavement based on optical fibre sensors(Xiang and Wang 2017). Their works contained the design of the optical fibre sensor, experimental investigations, and in-situ of urban asphalt pavement. Chapeleau et al. installed the optical fibers near the bottom of the asphalt layer to assess cracking based on accelerated pavement testing facility(Chapeleau et al. 2017).

Related studies indicated that the micro-cracks occurred in the initial stage of cracking formation, and coalesced into a macro-crack with the effects of traffic and environmental loadings(Ling et al. 2020). This study attempted to collect useful information to indicate the generation and propagation of macro-cracking based on Chang-Ji Expressway in the cold region of China. The technology of optical fibre sensors was employed to evaluate reflective cracking and thermal cracking. Different from the above researches, the strain of cement-treated base layer was also concerned.

2 OPTICAL FIBRE SENSORS EMBEDDED DETAILS

Based on the structure design of Chang-Ji Expressway, the layout of optical fibre sensors is shown in Figure1. The layout of optical fibre sensors in every layer is the same except for the length. More detailed information follows Table 1.

In preparation, we found the external of the optical fibre sensors was not well resistant to high temperatures. As a result of this, the optical fibre sensors were wrapped by heat-insulating materials to prevent the high temperature generated during the construction of the pavement, which would cause the melting of the external of the optical fibre sensors.

The installation process of optical fibre sensors in the cement-treated base layer was shown in Figure2. After curing, the cement-treated base layer was cut by the joint cutter following the layout of optical fibre sensors. And the cement-treated base layer in the seam was smashed and transferred manually to avoid sharp gravel generating shear force to cut the optical fibre sensors, eventually leaving a groove with a depth of 5 cm and width of 5 cm. The straightened optical fibre sensors were put into the groove and finally buried and compacted with the mixed cement mortar. Preliminery tests showed that the deformation between the cement-treated base layer and the optical fibre sensors was concerted.

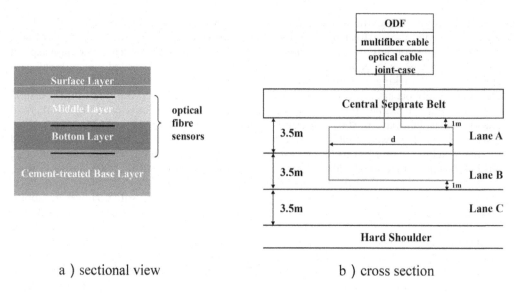

a) sectional view   b) cross section

Figure 1. Optical fibre sensors layout.

Table 1. Information of the layout of optical fibre sensors.

Layer	Bid-Section	d/m
Cement-treated base layer	K403-900	210
Bottom layer	K404-300	130
Middle layer	K404-300	110

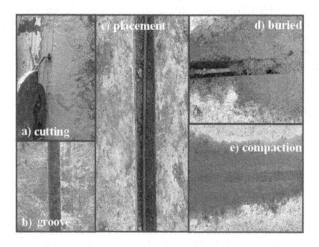

Figure 2. The embedded process of optical fibre sensors in the cement-treated base layer.

The embedded process of optical fibre sensors in the bottom layer and the middle layer was similar to these procedures, as shown in Figure 3. After measurement and marking, the no compaction asphalt mixture layer was smashed into large blocks of mixed stone and transferred manually. Similarly, a depth of 5 cm and width of 5 cm groove was formed. As soon as the straightened optical fibre sensors were put into the groove, the groove would be buried manually with the remaining asphalt mixture to avoid destruction of the optical fibre sensors due to the vibration of the road roller.

Figure 3. The embedded process of optical fibre sensors in asphalt mixture layer.

And then, the survival rate of the optical fibre sensors was inspected using a red light pen. The light pen was normally at work, which indicated the optical fibre sensors had no breakage, Figure 4 depicts. This also illustrated from the side that the manual compaction had little influence on the optical fibre sensors. What's more, the joints of the optical fibre sensors were numbered and done waterproof treatment.

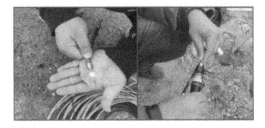

Figure 4. The inspection of the optical fibre sensors.

Finally, the multifiber cables were introduced into the storage box through the embedded PVC pipe. And the ODF shelf was installed. After that, every optical fibre sensor was welded to the corresponding multifiber cable through the optical cable joint case. Moreover, a steel container was set up to establish the monitor room. These procedures were described in Figure 5.

3 MONITORING EQUIPMENT

Long-distance Distributed Brillouin Optical Fibre Analyzer RP1020 made by Realphotonics was employed to collect data. More information about RP1020 is described in Table 2.

Figure 5. The procedures of optical fibre sensors assembly process.

Table 2. Information about RP1020.

Sensed distance	Spatial resolution	Sampling resolution	Precision	Swept frequency range
20 km	20 cm	5 cm	1°C/20 $\mu\varepsilon^3$	10-13 GHz

4 RESULTS

The frequency shift values were obtained through RP1020, and the strain values were calculated by Formula(1). Positive numbers represent tensile strain, and negative numbers represent compressive strain.

$$Strain = \frac{test\ frequency\ shift - reference\ frequency\ shift}{strain\ coefficient} - \frac{temperature\ change \times temperature\ coefficientt}{strain\ coefficient} \tag{1}$$

Illustrate: temperature coefficient—1.12MHZ/°C;
strain coefficient—0.0482MHZ/μ_ε;
the measured value is GHZ, converted to MHZ, 1GHZ=1000MHZ.

The optical fibre sensors in the cement-treated base layer were embedded in July 2018, and the measured values were used as the reference frequency shift. And the values measured in October 2018, March 2019, and November 2019 were taken as the test frequency shift. The temperature of the cement-treated base layer at different measurement times is displayed in Table 3. The strain values of the cement-treated base layer were calculated as shown in Figure 6, and the locations of the peak and trough strain values were marked in Figure 7. At regular intervals, the weak positions would occur the tensile strain, which was consistent with the cracking characteristic of the cement-treated base layer.

Table 3. The temperature of the cement-treated base layer.

Date	July 2018	October 2018	March 2019	November 2019
Temperature °C	30.1	12.6	11	6.8
Reference temperature °C	30.1	30.1	30.1	30.1
Temperature difference °C	0	-17.5	-19.1	-23.3

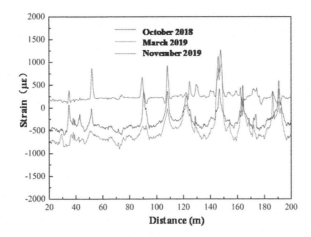

Figure 6. The strain values of the cement-treated base layer.

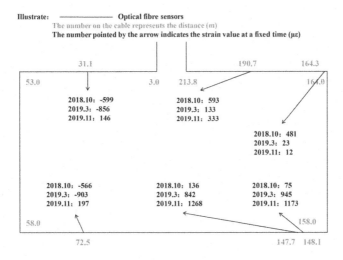

Figure 7. The location of peak and trough strain values of the cement-treated base layer.

The optical fibre sensors in the bottom layer were embedded in October 2018, and the measured values were employed as the reference frequency shift. Moreover, the values measured in March 2019, and November 2019 were regarded as the test frequency shift. However, the values measured in November 2019 were missing due to the destruction of the optical fibre sensors. As a result, the strain values and the locations of the peak and trough strain values were described in Figure 8 and Figure 9.

The optical fibre sensors in the middle layer were embedded in October 2018, and the measured values were employed as the reference frequency shift. Additionally, the values measured in March 2019, and November 2019 were taken as the test frequency shift. Similarly, the strain values and the locations of the peak and trough strain values were described in Figure 10 and Figure 11.

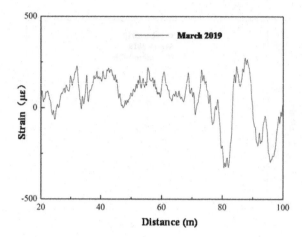

Figure 8.　The strain values of the bottom layer.

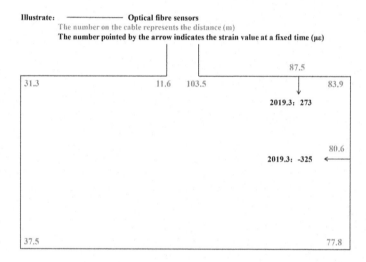

Figure 9.　The location of peak and trough strain values of the bottom layer.

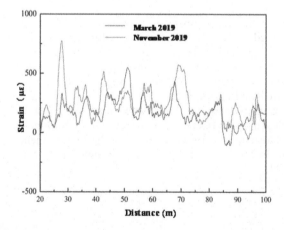

Figure 10.　The strain values of the middle layer.

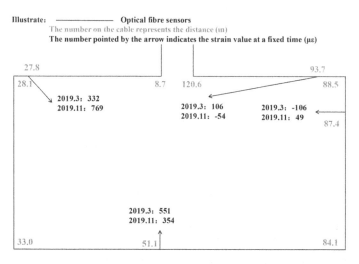

Figure 11. The location of peak and trough strain values of the middle layer.

5 CONCLUSION

It was interesting to find that some high tensile strain appeared in the first several months along the traffic direction for the cement-treated base layer. In the spring of 2019, the strain of the cement-treated base layer was mostly compressive, and a few points showed as the tensile strain. In the winter of 2019, the strain of the cement-treated base layer was displayed as the tensile strain.

The strain of the asphalt mixture layer generally showed as the tensile strain. However, the compressive strain appeared at some certain points. Comparing the strain values of the middle layer in the spring with the ones in the winter of 2019, it can be seen that the changing trend of the strain values in the spring resembled the one in the winter. Moreover, the strain in the winter was higher than the ones in the spring.

Besides, the high tensile strain of the cement-treated base layer was significantly higher than that of the asphalt mixture layer. The high tensile strain indicates that the micro-cracks may develop into macro-cracks. Combined with the fact that there was no obvious cracking on the pavement, especially thermal cracking, the cement-treated base layer was more prone to micro-cracks than the asphalt mixture layer, which gradually developed into reflection cracking under the action of traffic and environmental loadings. The monitoring mission of Chang-Ji Expressway is still in progress, and the measurement will be conducted once half a year or a year, and the follow-up situation will continue to be analyzed.

ACKNOWLEDGMENTS

Thanks for the financial support of Jilin Transportation Science and Technology Projects (2017ZDGC-2-1), and the National Natural Science Foundation of China (No.51878229).

REFERENCES

Chapeleau, X., J. Blanc, P. Hornych, J. Gautier, And J. Carroget. 2017. Assessment Of Cracks Detection In Pavement By A Distributed Fiber Optic Sensing Technology. *Journal Of Civil Structural Health Monitoring* 7 (4):459–470.

Dave, E. V., S. H. Song, W. G. Buttlar, And G. H. Paulino. 2007. Reflective And Thermal Cracking Modeling Of Asphalt Concrete Overlays. *Advanced Testing And Characterization Of Bituminous Materials* 2:1241–1252.

Gonzalez-Torre, I., M. A. Calzada-Perez, A. Vega-Zamanillo, And D. Castro-Fresno. 2015. Evaluation Of Reflective Cracking In Pavements Using A New Procedure That Combine Loads With Different Frequencies. *Construction And Building Materials* 75:368–374.

Ji, R. Y., T. Mandal, And H. Yin. 2019. Laboratory Characterization Of Temperature Induced Reflection Cracks. *Journal Of Traffic And Transportation Engineering* 7 (5).

Kim, J., And W. G. Buttlar. 2015. Analysis Of Reflective Crack Control System Involving Reinforcing Grid Over Base-Isolating Interlayer Mixture. *Journal Of Transportation Engineering* 128 (4):375–384.

Levenberg, E., A. A. Klar, And A. Skar. 2020. Soil Support Characterization In Slab-On-Grade Constructions With Fiber-Optic Distributed Strain Sensing. Paper Read At Geo-Congress 2020.

Ling, M., X. Luo, Y. Chen, F. Gu, And R. L. Lytton. 2020. Mechanistic-Empirical Models For Top-Down Cracking Initiation Of Asphalt Pavements. *The International Journal Of Pavement Engineering* 21 (4):464–473.

Oshone, M., E. V. Dave, And J. E. Sias. 2019. Asphalt Mix Fracture Energy Based Reflective Cracking Performance Criteria For Overlay Mix Selection And Design For Pavements In Cold Climates. *Construction And Building Materials* 211 (Jun.30):1025–1033.

Skar, A., A. Klar, And E. Levenberg. 2019. Load-Independent Characterization Of Plate Foundation Support Using High-Resolution Distributed Fiber-Optic Sensing. *Sensors (Basel, Switzerland)* 19 (16):3518.

Thøgersen, F., C. Busch, And A. Henrichsen. 2004. Mechanistic Design Of Semi-Rigid Pavements. *Danish Road Institute, Report* 138.

Wagoner, M., W. Buttlar, G. Paulino, And P. Blankenship. 2005. Investigation Of The Fracture Resistance Of Hot-Mix Asphalt Concrete Using A Disk-Shaped Compact Tension Test. *Transportation Research Record: Journal Of The Transportation Research Board* 1929:183–192.

Wang, X., K. Li, Y. Zhong, Q. Xu, And C. Li. 2018. Xfem Simulation Of Reflective Crack In Asphalt Pavement Structure Under Cyclic Temperature. *Construction And Building Materials* 189 (Nov.20):1035–1044.

Wang, X., And Y. Zhong. 2019. Influence Of Tack Coat On Reflective Cracking Propagation In Semi-Rigid Base Asphalt Pavement. *Engineering Fracture Mechanics* 213:172–181.

Xiang, P., And H. Wang. 2017. Optical Fibre-Based Sensors For Distributed Strain Monitoring Of Asphalt Pavements. *International Journal Of Pavement Engineering* 19 (9):842–850.

Xie, P., And H. Wang. 2020. Analysis Of Temperature Variation And Thermally-Induced Reflective Cracking Potential In Composite Pavements. *Transportation Research Record Journal Of The Transportation Research Board* 2674 (10):862865176.

Yeo, R. 2008. *Fatigue Performance Of Cemented Materials Under Accelerated Loading: Influence Of Vertical Loading On The Performance Of Unbound And Cemented Materials*.

Validation of national empirical pavement design approaches for cold recycled asphalt bases

M. Winter & K. Mollenhauer
Universität Kasse, Germany

A. Graziani & C. Mignini
Università Politecnica delle Marche, Italy

H. Bjurström & B. Kalman
Statens väg- & transportforskningsinstitut (VTI), Sweden

P. Hornych & V. Gaudefroy
Université Gustave Eiffel, France

D. Lo Presti
Nottingham Transportation Engineering Centre (NTEC), UK
Università degli Studi di Palermo, Italy

G. Giancontieri
Università degli Studi di Palermo, Italy

ABSTRACT: The applied pavement design procedures differ from country to country within Europe. Especially the use of cold recycling materials which is handled similar to hot-mix asphalt (HMA) in the most available guidelines isn't sufficient researched. To verify developed procedures, practical trial sections are indispensable. Within this study cold recycled asphalt base layers (CRAB) were assessed according to the German procedure for monitoring and evaluation of road surface conditions. It can be concluded, that the general applied empirical design approach, in which the CRAB layer thickness is increased by 50 % compared to a HMA base will result in a feasible pavement structure. In general, the re-designing of the sections according to various national empirical pavement design approaches partly results in high differences for same layer types.

Keywords: Validation, Pavement Design, Cold Recycling, Reclaimed Asphalt

1 INTRODUCTION

Because of the European political goals regarding CO_2-reduction gain more and more importance, each industry sector has to look for possibilities to slow down the greenhouse effect. Due to the resulting restrictions as well as the increasing maintenance work within the road network alternative construction methods are required. A solution to handle this circumstances could be the usage of cold recycled materials (CRM).

In comparison to hot mix asphalt (HMA) the recycling of reclaimed asphalt (RA) is more simple for cold recycling (CR). The procedure provides recycling rates of usually ≥ 75 %

DOI: 10.1201/9781003222880-9

(Mollenhauer & Simnofske, 2016) and is applied at ambient temperature. This results in a reduction of around 10 % for greenhouse gas emissions associated with a complete road structure (Giani et al., 2015).

Despite if there is practical experience with cold recycled materials and additionally well-researched mechanical properties of CRM in laboratory-based projects, the pavement design procedure often leads to higher structural thickness compared to HMA. This can be justified because of the missing fatigue failure criteria for those materials. Therefore, the pavement design procedures in practice are different and may result in the same thickness estimations compared to hot bituminous materials (e.g. in Switzerland, (VSS Schweiz, 2011)) up to thickness-increase factors of e. g. 50 % (Bocci et al., 2010) compared to standard HMA. Especially the time-dependent change of material strength during curing, which results in increasing bearing capacity of Cold Recycled Asphalt Bases (CRAB) structures over months or even years (Godenzoni et al., 2018), is one reason for this.

These thickness differences result in varying costs for the environmental-friendly technology and therefore in extremely different application rates within the various European countries (EAPA, 2018). This contribution analyses the various pavement structures with CRAB and tries to validate the designs regarding achieved pavements performance.

2 METHODOLOGY

As there are several road sections in Europe which contain cold recycled or cold bounded material as base layer, an assessment of the structural performance for existing pavements in four selected countries is conducted. To validate the national empirical pavement design procedures from Germany, Italy and UK, the observations regarding structural design, loading parameters and pavement conditions for the existing CRAB pavement sections are assessed. The validation is split in 3 phases.

Firstly, the individually applied traffic load parameter is calculated for all pavement structures. These are the number of equivalent standard axle loads, which are based on the composition and volume of heavy traffic as well as the aimed service lifetime.

In a second step, each empirical design procedure is applied for each of the CRAB pavement sections. This will allow the direct comparison of the three design procedures.

At last, the structural performance is estimated by taking the actual service lifetime into account. By these means, a theoretical residual service life parameter is estimated for each section which can be compared with the actually observed pavement condition.

3 PERFORMANCE AND CONDITION PARAMETER OF EXISTING STRUCTURES

3.1 *Studied sections*

From each selected country (GE, ITA, SWE, UK) 2-4 road sections were selected, where cold recycling material has been included in the base layer. In Figure 1 the pavement design of all studied sections are shown. Structural designs for the road sections are presented from the surface layer down to the bitumen or cement bound bottom layer. This is the CRAB layer for all road sections, except for two of the UK sections and one Italian section where cement bound foundation is below the CRAB layer. The structures are demonstrating a large variation in traffic loading and climatic conditions (location) as well as design approach. (Bjurström et al., 2020).

3.2 *Structural performance*

The internationally applied procedures for monitoring differ considerably. For allowing a comparison of the structural design success, a common procedure for the assessment of the pavement condition is required. Therefore, the German procedure for monitoring and

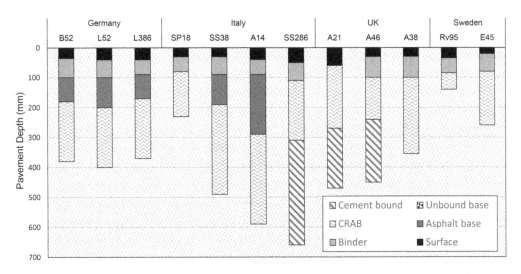

Figure 1. Layer thicknesses for the studied road sections illustrated from the surface down to the deepest bound layer (Bjurström et al., 2020).

evaluation of road surface conditions of federal highways (FGSV, 2001) was applied to all pavements. The measured surface condition is transferred to condition quality categories as followed:

< 1,5: very good/new road structure
1,5 – 2,49: good
2,5 – 3,49: satisfactory
3,5 – 4,49: sufficient, but maintenance methods should be prepared
≥ 4,5: defective, maintenance is required shortly

The transfer functions from the measured condition value to the quality categories is calculated linearly for the transfer parameters according to Table 1. The permanent deformation in terms of rut depth measurement in millimeters was available for the pavements in Germany, Sweden and UK. The proportion of cracking in the road surface was available for pavements from Germany only. In order to include this assessment for the other pavements, the proportion of cracked surfaces was estimated by analyzing photographs taken from the road pavement. Here, Google Street View photos (https://www.google.com/maps) were partially used, which allowed the rough assessment of cracks. As the sections were of various lengths, and the categories varied within each section, the arithmetic mean category value was calculated as well as the standard deviation. From these, the 95 % quantile category value was calculated in order to assess the variability of the pavement condition.

Table 1. Surface condition transfer parameters according to (FGSV, 2001).

Surface condition quality category	1,5	2,5	3,5	4,5
Rut depth [mm]	4	7	10	20
Proportion of cracked surface [%]	1	7	15	25

The results of the surface condition assessment are listed in Table 2. For each section the year of the implementation of the CRA pavement is given (year of rehabilitation). Further, the newest result for the surface monitoring assessment is given as well as the average daily traffic of heavy vehicles. For the rutting depth and the cracking, the mean quality values are given (in bold) as well as the 95% quantile (in italic).

In general, the mean surface conditions of most of the assessed sections is very good or good. The Italian section SP18 is the only one which results in satisfactory condition with a mean quality higher than 2,5.

Table 2. Pavement condition according to the German standard (FGSV, 2001) of each section since rehabilitation.

Country	Section	Year of Rehabilitation	Last year of Measurement	Daily heavy traffic	quality value* rutting	cracking
GER	B52	2009	2015	3900	**1,88** *2,18*	**1,04** *1,20*
GER	L52	2011	2017	60	**1,28** *2,00*	**1,00** *1,00*
GER	L386	2007	2017	365	**1,50** *1,79*	**1,36** *3,45*
SWE	Rv95	2014	2019	380	**2,10** *2,25*	**1,35** *2,75*
SWE	E45	2012	2019	333	**1,74** *1,81*	**1,54** *2,79*
ITA	SS38	2007	2019	1850		**1,00** *1,00*
ITA	SS268	2016	2017	2115		**1,00** *1,00*
ITA	SP18	2008	2019	250		**2,74** *4,50*
ITA	A14	2007	2019	11000		**1,18** *1,73*
UK	A46	2006	2018	3664	**1,64** *1,84*	**1,36** *1,82*
UK	A21	2002	2018	11700	**1,56** *2,01*	**1,00** *1,00*

* status value written in bold and a 95-% quantile written in italic

However, when considering the variation of the surface condition, additional sections can be identified with problematic performance on a part of the total length. This is indicated by higher 95%-quantile values on sections L386 (GER), Rv95 and E45 (SWE).

4 VALIDATION OF NATIONAL EMPIRICAL DESIGN APPROACHES

4.1 *Comparative evaluation of the traffic loads*

A common design parameter is the traffic load. In all design procedures, the average daily traffic of heavy vehicles and the aimed service lifetime is used for calculating the number of equivalent standard axle loads (ESAL). Here, the three design procedures apply different load values for the standard axle load (ITA, UK: 8,2 t; GER, SWE: 10 t). Additionally, different aimed service lifetimes are considered: GER, ITA, SWE: 30 a; UK: 40 a.

In order to base the different pavement design on the same traffic load estimations, the individual number of equivalent standard axles of the considered country, where the structure is located is used as an input value. The other two traffic volume parameters are then calculated, by application of the "power 4-law". By these means, the different approaches for traffic

Table 3. Traffic load parameters of CRAB validation structures (Millions of equivalent standard axle loads (ESAL: GER, SWE, ITA; msa: UK)).

	Section	ADT [veh/d]	Proportion of heavy veh. [%]	Service life [a]	GER/SWE ESAL10, 30 a	ITA ESAL8,2, 30 a	UK msa (8,2 t; 40 a)
GER	B52	26000	15	11	**47,5**	105,1	131,4
	L52	1500	4	9	**0,5**	1,0	1,3
	L386	7000	5	13	**2,8**	6,3	7,8
ITA	SP18	5000	5	12	2,7	**6,0**	7,5
	SS38	30000	6	13	13,6	**30,0**	37,5
	A14	44000	25	13	67,8	**150,0**	187,5
	SS268	19661	11	4	10,9	**24,0**	30,0
UK	A21	47714	25	18	8,7	19,2	**25,6**
	A46	19192	19	14	5,8	12,8	**17,0**
	A38	37000	10	14	11,9	26,3	**35,0**
SWE	RV95	3136	12	6	**10,2**	22,6	29,3
	E45	1233	27	8	**9,2**	20,3	26,5

assessment and estimation of the equivalent load parameters are biased. The resulting numbers of ESAL are summarized in Table 3. The basic traffic parameter is printed in bold, from which the other two parameters were calculated.

4.2 *Re-design of the existing pavements*

In this section, the existing pavements were re-designed according to the national standards of Germany (FGSV, 2005, 2012), UK (Merrill et al., 2004) and Italy (Autonome Provinz Bozen - Südtirol, 2016; Consiglio Nazionale Delle Ricerche, 1995). For the design procedures of Germany and Italy, the deformation modulus E_{V2} is the parameter for considering subbase bearing capacity. For the UK, the design depends on the foundation class. For deformation moduli between two foundation classes, the layer thickness according to UK design was linearly interpolated. The design procedures define different bearing capacity classes. For the catalogue systems (GER, ITA), this results in different modulus classes applied.

Table 4 shows an example for the re-design of the Italian pavement of SS268. The actual pavement structure has a total thickness of bound layers of 31 cm. The re-design results in higher thickness values: GER: 36 cm, UK: 35,5 cm and IT: 30 cm (with additional stabilized subbase). Obviously, the actual thickness of the pavement is lower than the designed values according to German and UK pavement design.

The surface condition values (mean/95%-quantile) for this section are 2,74/4,5 (compare Table 2). The observed cracking in the pavement correlates to the undersized design.

The resulting total thickness values of the bituminous layers for all studied road sections are summarized in Table 5. For the re-designed thickness results, the thickness-deviation of the actual pavements are given and highlighted by different font types:

• Bold: actual thickness is higher compared to re-designed thickness
• Standard: actual thickness is the same (± 2 cm) as the re-designed thickness
• Italic: actual thickness is lower than the re-designed thickness.

The three assessed empirical design procedures identify over- and under-designed pavement structures in the assessed roads similarly. Despite of differences in re-design thickness of up to 7 cm (section E45), the German and UK design procedure are capable to identify the same sections as under- (red) or overdesigned. However, the pavements designed according to the UK specifications are generally thinner compared to the German design thickness, with B52 as an exception.

Table 4. Results for redesigned pavement of the Italian section S268.

SS268	Current state	Redesigned pavement		
		ESAL10	msa8.2	ESAL8.2
ESAL/msa	0,8	10,9[1]	30,0	24,0
deformation modulus subbase	160[2]	150	160	160
surface	5	4	4	3
binder	6	0	6	7
base	0	8	0	0
CRAB	20	24	25,5	20
stabilized base layer	35			30
Total asphalt thickness	31	36	35,5	30

1) Extrapolation within the design procedure; 2) Estimation because of missing data

For the Italian re-design results, the thickness of all structures is lower compared to English or German designs. However, these CRAB structures are based on top of a cement stabilized layer, which gives higher bearing capacity. Generally, the same differences between the actual pavement structure thickness and the designed ones are identified also with the Italian method with the B52 as an exception. In three cases (SP18, SS268 and A46) the difference between pavement thickness and re-designed thickness is lower than ± 2 cm.

Four of the pavements (L52, L386, SS38 and A14) seem to be over-designed according to all design procedures applied. Three of these structures are identified by a very low crack condition value and nearly show no crack damage at all (see Table 2). The exception for this conclusion is section L286, where cracking could be observed in some specific spots and which originally showed transversal cracking shortly after construction and which was maintained.

Of the six structures that are identified as under-designed: SP18, SS286, A21, A46, Rv95 and E45, three actually show some (Rv95, E45) or considerable (SP18) crack damages.

4.3 Calculation of structural condition parameter

As can be observed from Table 5, there is only a low correlation between the thickness differences between design thickness and actual thickness and the observed cracking in the pavements. One reason is, that the design life is considerably higher for all of the assessed structures than the actual service lifetime (compare Table 3). In order to consider the actual received traffic loading on the assessed pavements, a theoretical structural condition parameter is calculated according to a German guideline document (FGSV, 2019).

Therefore, the required thickness ($DI_{erf.}$) of the bound layers are calculated by considering the traffic load class and the bearing capacity of the unbound subbase. This thickness is compared to the "active" thickness ($DI_{vorh.}$) of the assessed pavement structure, where the actual thicknesses of the structural layers are reduced according to the age of the layer.

The active thickness of the existing pavement DI_{vorh} is given in the following equation:

$$DI_{vorh.} = \sum_{i}(D_i * Aq_{it}) \tag{1}$$

where:
D_i = Thickness of the bound layer i [cm]
$Aq_{i\ t}$ = age-related factor of thickness equivalation of the layer i (see Table 6)
$Aq_{i\ t}$ = MIN ($Aq_{i\ max}$; MAX ($Aq_{i\ min}$; Equation according to Table 6))

Table 5. Real and design thickness according to different empirical design approaches.

Section	Actual thickness [cm]	Thickness according to national standards* [cm]			Crack condition value: (mean 95%-quantile)
		GER	UK	ITA	
B52	38	38	42	35	1,04
Δ [cm]		0	4	-3	1,2
L52	40	**28**	**28,5**	21	1,0
Δ [cm]		**-12**	**-12**	-19	1,0
L386	37	32	27,5	21	1,36
Δ [cm]		-5	-10	-16	3,45
SP18	23	32	37	21	2,74
Δ [cm]		9	14	-2	4,5
SS38	49	40	34,5	36	1,0
Δ [cm]		-9	-15	-13	1,0
A14	59	38	44	35	1,18
Δ [cm]		-21	-15	-24	1,73
SS268	31	36	35,5	30	1,0
Δ [cm]		5	5	-1	1,0
A21	27	34	32	31	1,0
Δ [cm]		7	5	4	1,0
A46	24	34	30	25	1,36
Δ [cm]		10	6	1	1,82
A38	35,5	36	34	36	-
Δ [cm]		0,5	-1,5	0,5	-
Rv95	14	40	33,5	31	1,35
Δ [cm]		26	20	17	2,75
E45	26	40	33	31	1,54
Δ [cm]		14	7	5	2,79

* standard: same thickness (± 2 cm), **bold**: actual thickness is higher, *italic*: actual thickness is lower

Table 6. Age-related factor of thickness equivalation for different types of layers.

Layer type	Age-related factor of thickness equivalency $Aq_{i,t}$ with t = age of layer [a]
Surface asphalt	$0,35 < 1,0392 - t * 0,0392 < 1$
Mastic asphalt	$0,4 < 1,0192 - t * 0,0192 < 1$
Asphalt Binder	$0,4 < 1,0400 - t * 0,0200 < 1$
Asphalt Base	$0,5 < 1,0200 - t * 0,0100 < 1$
Hydraulic bound	$0,33 < 0,5540 - t * 0,0070 < 0,54$

The equations for calculation $DI_{erf.}$ are based on the German design catalogue structures and have input parameters linked to the number of equivalent axle loads (10 t, 30 a) and the subbase bearing capacity in terms of deformation modulus E_{V2}. The resulting loading class (Bk) refer to the upper limit of million 10-t-ESAL for each design catalogue column and regression equation.

The ratio of these values (called "thickness comparison number DVZ") is used as an indicator for the structural thickness reserve, which is already reduced by traffic loading (eq. 1).

$$DVZ = \frac{DI_{vorh.}}{DI_{erf.}} \qquad (2)$$

This number is the basis for assessment of a structural condition indicator, where a DVZ of 0,9 (thickness reserve is reduced by 10%) is considered as "very good" and a DVZ of 50%

(structural thickness is reduced to half of its initial value) is considered as defective. By this estimation, a structural condition parameter SW is obtained, which can be directly compared with surface condition parameters. The determination of the parameter SW is based on a standardization function, which takes into account the thickness comparison number DVZ.

This comparison is shown in Figure 2, where the observed surface parameter (maximum of the 95 %-quantile of cracking or rutting) is plotted against the structural condition parameter SW_B. Considering the different design approaches, there is a relatively good correlation between these two parameters, representing all assessed CRAB pavements. This further indicates, that the observed performance is linked to the structural design of these pavements and can be used as a parameter to evaluate the success of an applied pavement design.

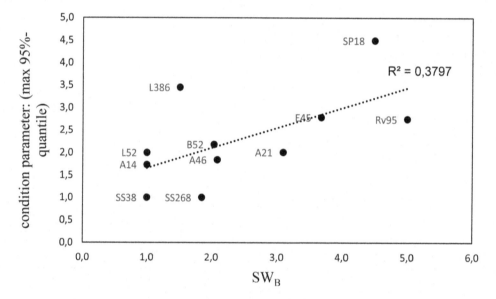

Figure 2. Comparison of structural condition parameter SW_B with surface cracking conditions for the assessed CRAB pavements.

5 CONCLUSIONS

The structural performance of existing CRAB pavement structures was assessed in detail. Regarding the surface condition performance, eight of these 12 structures showed none or only very little rutting or cracking, two showed some cracking which can be considered as moderate and only two sections showed defective surface cracking on some parts of the pavement area. This observed surface performance could be correlated against the difference between actually applied structural thickness and the required thickness (according to German, Italian or UK design) only roughly. However, when the actual service life is considered within this comparison, a close correlation between the surface condition and the theoretical design life could be identified. In this way it could be observed, that the structures of the Swedish pavements as well as an Italian pavement, have a shorter remaining expected service life. It must be emphasized, that a German procedure was applied for estimating the theoretical service life, which is based on German pavement design thickness results.

However, the good surface conditions of most assessed sections clearly show, that the construction of durable pavements with CRAB is possible. For three of the four sections with only moderate or even defective condition, the thickness of the pavement structures could be identified as not sufficient for the traffic loading conditions. Therefore, it can be concluded, that the generally applied empirical design approach, in which the CRAB layer thickness is increased by 50 % compared to a HMA base will result in a feasible pavement structure.

For the assessment of cracking phenomena in pavements with CRAB layers, additional research is required in order to identify the origin of the cracks. It has to be identified if the crack result from traditional bottom-up cracks in the CRAB layer or rather in the asphalt surfacing as a result of low thickness and low bearing capacity of the CRAB layer. This may be related to the CRAB composition (e.g. cement content) but could not be analyzed within this contribution.

ACKNOWLEDGEMENT

The work reported here is part of the project "CRABforOERE – Cold recycled asphalt bases for optimized energy & resource efficient pavements" funded by the CEDR Transnational Road Research Programme Call 2017: Materials funded by Austria, Belgium-Flanders, Denmark, Germany, Netherlands, Norway, Slovenia, Sweden and the United Kingdom

REFERENCES

Autonome Provinz Bozen - Südtirol. (2016). *Verzeichnis der Bezugsbauweisen zur Dimensionierung von Asphaltstrassen.*
Bjurström, H., Kalman, B., Mollenhauer, K., Winter, M., Graziani, A., Mignini, C., Lo Prest, D., Giancontieri, G., & Gaudefroy, V. (2020). Experiences from cold recycled materials used in asphalt bases: a comparison between five European countries.
Bocci, M [M.], Canestrari, F., Grilli, A., Pasquini, E., & Lioi, D. (2010). Recycled Techniques and Environmental Issues Relating to the Widening of an High Traffic Volume Italian Motorway. *International Journal of Pavement Research and Technology*(3), 171–177.
Consiglio Nazionale Delle Ricerche (1995). *Catalogo delle pavimentazioni stradali. Bollettino Ufficiale CNR*. Rom. Consiglio Nazionale Delle Ricerche.
EAPA. (2018). *Asphalt in Figures*. Belgium. European Asphalt Pavement Association.
FGSV (2001). *FGSV-Arbeitspapier Nr.9/A 1.2 zur ZEB.* (Arbeitspapier, 9). Köln. FGSV-Verlag.
FGSV. (2005). *Merkblatt für Kaltrecycling in situ im Straßenbau - M KRC*. FGSV-Verlag.
FGSV. (2012). *Richtlinien für die Standardisierung des Oberbaus von Verkehrsflächen: RStO 12*. FGSV-Verlag.
FGSV (2019). *Richtlinien zur Bewertung der strukturellen Substanz des Oberbaus von Verkehrsflächen in Asphaltbauweisen*. (Richtlinie). Köln. FGSV-Verlag.
Giani, M. I., Dotelli, G., Brandini, N., & Zampori, L. (2015). Comparative life cycle assessment of asphalt pavements using reclaimed asphalt, warm mix technology and cold in-place recycling. *Resources, Conservation and Recycling, 104*, 224–238.
Godenzoni, C., Graziani, A., Bocci, E., & Bocci, M [Maurizio] (2018). The evolution of the mechanical behaviour of cold recycled mixtures stabilised with cement and bitumen: field and laboratory study. *Road Materials and Pavement Design, 19*(4), 856–877.
Merrill, D., Nunn, M., & Carswell, I. (2004). *A guide to the use and specification of cold recycled materials for the maintenance of road pavements: TRL Report TRL 611* (TRL 611). London. Department of Transport.
Mollenhauer, K., & Simnofske, D. (2016). Tolerances for inhomogeneity of pavement structure for in-situ cold recycling. In K. Suchý (Ed.), *6th Eurasphalt & Eurobitume Congress: 2016* (1st ed.). CTU Prague.
VSS Schweiz (2011). *Dimensionierung des Strassenaufbaus; Unterbau und Oberbau*. (SN 640324). Zürich. Verband der Strassen- und Verkehrsfachleute

Ethylene-Octene-Copolymer as an alternative to Styrene-Butadiene-Styrene bitumen modifier

A. Riekstins
Faculty of Civil Engineering, Department of Roads and Bridges, Riga Technical University, Riga, Latvia
SJSC Latvian State Roads, Riga, Latvia

V. Haritonovs
Faculty of Civil Engineering, Department of Roads and Bridges, Riga Technical University, Riga, Latvia

R. Merijs-Meri & J. Zicāns
Faculty of Materials Science and Applied Chemistry, Institute of Polymer Materials, Riga Technical University, Riga, Latvia

ABSTRACT: This paper investigates the perspective of the use of an unfamiliar bitumen modifier - Ethylene-Octene-Copolymer (EOC) - for the production of polymer modified bitumen (PMB). For the study, bitumen 70/100 was modified at a laboratory with EOC and SBS modifiers. From the industrial bitumen manufacturer, SBS modified bitumen 45/80-55 (SBS fact) was obtained. The rheological properties of the laboratory-produced and the industrially-produced bitumen were compared. In addition to the industrial binder, two laboratory-produced PMB compositions with the lowest viscosity on the one hand (70/100+2%EOC) and the highest viscosity on the other hand (70/100+4%SBS) were used to design asphalt mixtures. The following performance tests were used to evaluate physical properties of the obtained asphalt mixtures: wheel tracking test (WTT), thermal strain restricted specimen test (TSRST) and four-point bending test (4PB). In comparison to the laboratory-produced SBS modified bituminous binder containing asphalt mixture the EOC modified counterpart demonstrates a better low-temperature cracking behaviour, approaching the values characteristic of the SBS fact bound asphalt mixture. In its turn, rutting and fatigue resistance of the SBS lab modified bitumen containing asphalt mixtures are higher in comparison to EOC lab modified systems. Even higher fatigue resistance of SBS fact containing systems is evidently connected with a better mixing quality due to technical aspects ensured in bitumen production plant.

Keywords: polymer modified bitumen, ethylene-octene-copolymer, styrene-butadiene-styrene, asphalt mixture, performance tests

1 INTRODUCTION

Increasing traffic intensity, environmental concerns and limited budget keep industry and academicians searching for more sustainable asphalt mixtures. Bitumen as the most expensive constituent of asphalt directly affects all performance properties of the layer. Through bitumen modification, it is possible to improve the properties of asphalt binder (Singh, Chopra, Jain, Kaur, & Kamotra, 2019). As modifiers are not cheap, they noticeably impact the price of the end product and therefore they are mostly used on high traffic intensity roads

DOI: 10.1201/9781003222880-10

(Brasileiro, Moreno-Navarro, Tauste-Martínez, Matos, & Rubio-Gámez, 2019) and just in the top layer. Currently, the most used modifier on the market is styrene-butadiene-styrene (SBS) although it has several drawbacks such as poor storage stability and low ageing resistance (Ren, Liu, Fan, Wang, & Erkens, 2019; Zhu, Birgisson, & Kringos, 2014). Because of the huge demand, the price of SBS elastomer has significantly increased in recent years and the forecast is that it will continue to grow thus affecting the price of asphalt. (Market Watch, 2020).

One way to reduce the costs is to use thinner asphalt layer(s) so that less bitumen and also a modifier are needed. In Latvia, the thickness of the top layer is usually designed to be 35-40 mm. The main aim of the wearing course is to provide a smooth surface and load transfer. The thickness of the top layer could be optimized if better quality aggregates and modified bitumen are used (Kragh et al., 2011) in case the reduction of the wearing course is not compensated by another thicker structural layer. Very thin asphalt concrete (BBTM) is a promising alternative to conventional asphalt mixtures (A. Riekstins, Haritonovs, Abolins, Straupe, & Tihonovs, 2019; Arturs Riekstins, Haritonovs, & Straupe, 2020). BBTM could be used on high-intensity roads because of the durable mineral carcass. Typically, the thickness of the wearing course is 20 to 30 mm. By using BBTM instead of conventional asphalt mixtures such as asphalt concrete (AC) or stone mastic asphalt (SMA), it is possible to reduce the consumption of aggregates and binder up to 50%. Such reduction is particularly important for countries and regions that import aggregates and/or bitumen. However, the use of BBTM mostly increases the use of PMB. Thus, it is not possible to avoid the use of polymers, but it is possible to find an alternative to SBS.

A cheaper and more promising polymer might be ethylene-octene copolymer (EOC) which has not been used for commercial modification yet. In the authors' previous studies, EOC showed improved bitumen high-temperature properties as well as ageing resistance, suggesting that this thermoplastic elastomer has the potential to be used as a bitumen modifier. (Merijs-Meri, Abele, Zicans, & Haritonovs, 2019; Zicans, Ivanova, Merijs-Meri, Berzina, & Haritonovs, 2019).

Bitumen modification parameters and technologies have a great influence on the performance of asphalt binder. It is well known that bitumen modification in the factory is done by colloidal mixers, while preparation of bituminous compositions in the laboratory is usually done by high shear mixers. A direct comparison can give false information about the efficiency of a modifier. Consequently, in order to develop an efficient alternative asphalt binder, it is relevant but at the same time, scientists should be careful to directly compare the properties of the laboratory-produced binders with industrially manufactured commercial counterparts.

2 OBJECTIVE

In this research, the performance of the EOC modified bitumen binder in comparison to the traditional SBS, manufactured both at laboratory and industrial scales was evaluated. The comparison was done on two levels – (1) polymer modified bitumen (PMB) binder testing, (2) asphalt mixture testing. For this purpose, a dynamic shear rheometer (DSR), a wheel tracking test (WTT), a thermal stress restrained specimen test (TSRST) and a four-point bending test (4PB) were used.

3 MATERIALS AND METHODS

A flowchart of the experimental program is demonstrated in Figure 1. Initially, two types of PMB binder were developed at the laboratory. In comparison to traditionally used SBS, EOC was used as a bitumen modifier. The content of either of the polymer modifiers was fixed at 2 wt. % and 4 wt.%. As a reference, industrially produced PMB 45/80-55 was used, in which SBS content could be estimated in the range of 3-4 wt.% independence on the used test procedure (Ratajczak & Wilmański, 2020). At first, rheological properties were tested and evaluated for all the obtained bitumens. Based on the rheological properties for further investigations bituminous composition with 4 wt.% of SBS (in general, the highest viscosity and shear modulus at the lowest phase angle value over the temperature range investigated) and 2 wt. % of EOC (in general, the lowest viscosity and shear modulus at the highest phase angle value over the temperature range investigated) were chosen to characterize the influence of the laboratory developed modified bitumens on the performance of asphalt mixtures. This allowed evaluating the effect of the addition of both the lowest viscosity and the highest viscosity PMB on the performance of the asphalt mixtures. Volumetric tests of the designed asphalt mixtures were performed to make sure that air voids are within the planned range. Subsequently, mixtures that met the requirements were tested for the following performance-based tests: WTT, TSRST, 4PB.

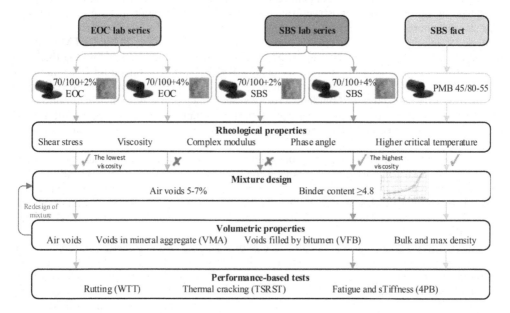

Figure 1. Flowchart of the experimental program.

3.1 Materials

3.1.1 Bitumen
For the needs of the research, two different binders were obtained - bitumen 70/100 and PMB 45/80-55. Bitumen 70/100 was used for modification purposes and PMB for comparison purposes.

3.1.2 Modifiers
Two different elastomers – linear SBS and EOC with 1-octene group content of 38% - were used as polymeric modifiers. SBS is the most popular bitumen modifier on the market. On the other hand, EOC is not currently used to modify bitumen commercially. Both SBS and EOC are elastomers, showing low glass transition temperatures and high ultimate elongations; EOC, however, is less viscous. Images of SBS and EOC modifiers are shown in Table 1.

3.1.3 Aggregates
For the production of asphalt mixtures crushed quartz-diorite fractions of 8/11 mm, 5/8 mm and 0/5 mm were used. Dolomite flour was used as a filler. All materials meet the highest level (S-1 class) of the Latvian Road specification requirements (VAS "Latvijas Valsts ceļi", 2019).

Table 1. Comparison of SBS and EOC modifiers.

Type	EOC	SBS
Structure of the repeating unit	-[CH-CH$_2$]$_n$-[CH$_2$-CH$_2$-]$_m$- | CH$_2$-CH$_2$-CH$_2$-CH$_2$-CH$_2$-CH$_3$	-[CH$_2$-CH]$_x$-[CH$_2$-CH$_2$-]$_y$-[CH$_2$-CH]$_z$-
Photo		
Density, kg/m^3	870	940
Melt flow rate, g/10 min.	5(190°C/2.16kg)	1(200°C/5kg)
Amount used to modify bitumen, wt. %	2 and 4	2 and 4

3.2 Methods

3.2.1 Preparation of PMB binders

PMB binders with 2 wt.% and 4 wt.% of either EOC or SBS were prepared as follows: initially, the bitumen was heated up to 140°C in a laboratory oven, then the vessel with the melted bitumen was placed in a silicone oil bath and, when the temperature reached 170°C, the elastomer was gradually added and mixing was continued for 3 hours at 185-188°C at the speed of 6000-7000 rpm by using Silverson L5M-A high shear laboratory mixer equipped with a square hole screen. Thus, the prepared modified bitumen binders, were poured in a separate vessel, and kept at sub-zero temperatures until measurements were taken to avoid settling. The process of obtaining the industry and the laboratory-produced bituminous binders is schematically depicted in Figure 2 for SBS fact and Figure 3 for SBS lab and EOC lab.

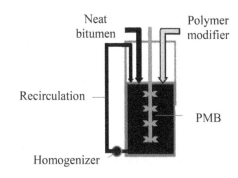

Figure 2. PMB made by plant equipment.

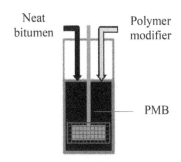

Figure 3. PMB made by high shear laboratory mixer.

3.2.2 Mixture design

Granulometric curves of mixtures are demonstrated in Figure 4. According to the experimental program, depicted in Figure 1, three mixtures were compared differing by the polymer binder used and its content – 70/100+2wt.%EOC (EOC lab), 70/100+2wt.%SBS (SBS lab) and 45/80-55 (SBS fact). Based on the previous research results (A. Riekstins et al., 2019) two most perspective granulometric compositions were selected (solid lines in Figure 4). The same granulometric distribution was selected for EOC lab and SBS fact. A slightly different granulometric distribution was selected for laboratory-made SBS PMB. The target air voids content for all the mixtures was within 5% and 7%. Thus, it was obtained that the optimal bitumen content for EOC lab and SBS fact mixtures is 5.2% and for SBS lab mixture - 4.8%.

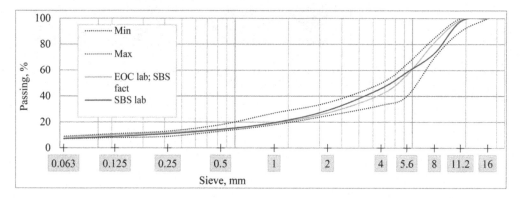

Figure 4. Granulometric distribution of asphalt mixtures. The dotted lines show granulometric distribution limits.

3.2.3 Rheological properties of bituminous binders

Rheological properties of commercial bitumen and the laboratory developed bituminous compositions with EOC and SBS were tested according to AASHTO T315 by using a 25 mm plate-plate configuration with a 1 mm gap. In brief, the tests were performed at 12% shear strain by increasing temperatures with a step of 6PPP°PPPC until the critical temperature at which the value of $|G^*|/\sin(\delta)$ decreased below 1 kPa.

3.2.4 Volumetric properties of asphalt mixtures

All Marshall samples with a diameter of 100 mm were made according to EN 12697-30. The impact compactor was used, and 50 blows were applied to each side. Samples were compacted at 150°C. Bulk density, maximum density, air voids, voids in mineral aggregate (VMA), voids filled by bitumen (VFB) and binder content were measured.

3.2.5 Rutting resistance

WTT was used to evaluate the rutting resistance of the developed asphalt mixtures. The test was performed according to EN 12697-22. The specimens were tested according to procedure B at 60°C degrees in the air by a small device. The design thickness of the wearing course was 25 mm. Referring to the standard, the nominal thickness of the samples with a maximum particle size of 11 mm should be at least 40 mm. Therefore, a two-layer specimen system was used to evaluate the ruts (see Figure 5). Firstly, the base course with the thickness of 60 mm was compacted. A tack coat layer was applied on the top of the base layer. Emulsion with a residual bitumen content of 60% was used. Finally, BBTM layer was compacted on top of the base layer. The samples were tested under 700 N moving wheel cyclic load at the rate of 26.5 cycles per minute. For each sample, 20000 passes (10000 cycles) were made. Proportional rut depth (PRD_{AIR}) and wheel tracking slope (WTS) were recorded.

Figure 5. Two-layer specimen system for WTT.

3.2.6 *Low-temperature resistance*

TSRST was used to detect the ability of mixtures to resist thermal stresses when ambient temperature is decreased by a constant rate. The test was performed according to EN 12697-46. The starting temperature of the test was 20°C and it was lowered with the rate of 10°C/h. For each mixture, a slab (320×260×90) was compacted by the roller compactor according to EN 12697-33. Two prismatic samples were cut out with the dimensions 40×40×160. Failure stress and failure temperature were recorded.

3.2.7 *Fatigue and stiffness*

4PB test was used to evaluate fatigue resistance and stiffness of asphalt mixtures. The fatigue resistance test was performed according to EN 12697-24 and the stiffness test - according to EN 12697-26. Both of the tests were done at +10°C. For each mixture, one slab with the dimensions 420×320×90 mm was compacted by roller compactor according to EN 12697-33. Four beams were cut out from each slab with the dimensions 50×50×420 mm. First, the stiffness was tested for all beams. For the stiffness test, 1000 cycles were applied at the constant deformation amplitude of 50 μm/m (microstrain). Small deformation amplitude and a low number of cycles were assumed not to damage the investigated samples; therefore, the same beams were tested to fatigue resistance. Relative deformation for each mixture was individual and was based on the first beam test result (relative deformation for the samples ranged from 200 to 450 μm/m). Fatigue resistance and stiffness were compared.

4 RESULTS AND DISCUSSION

4.1 *Bitumen rheological properties*

Rheological properties of 70/100, EOC and SBS modified virgin bitumen compositions, as well as virgin SBS fact, measured at various temperatures within a linear viscoelasticity region (LVER), are summarized in Table 2. The effects of UV irradiation and temperature of the properties of laboratory modified bituminous binders have been reported elsewhere (Zicans et al, 2019), where it was concluded that using these ageing protocols allowed to estimate the ageing of bitumen by obtaining comparable results to standard RTFOT tests and real aged bitumen. In the current manuscript, only the results of rheological properties of virgin PMB binders are reported as the research on the presented systems continues.

Table 2. Bitumen rheological properties.

Param. T, PPP°PPPC	Bitumen composition	Shear stress, Pa	Viscosity, Pa s	Shear modulus, kPa	Phase shift angle, deg
64	70/100	167	139	1.386	84
	70/100 + 2% EOC	319	265	2.647	82
	70/100 + 2% SBS	340	281	2.811	79
	70/100 + 4% EOC	554	460	4.598	78
	70/100 + 4% SBS	533	442	4.420	69
	PMB 45/80-55	633	525	5.251	65
70	70/100	81	67	0.673	86
	70/100 + 2% EOC	157	130	1.302	84
	70/100 + 2% SBS	165	137	1.369	81
	70/100 + 4% EOC	284	236	2.355	81
	70/100 + 4% SBS	292	242	2.420	71
	PMB 45/80-55	355	295	2.945	66
76	70/100 + 2% EOC	79	66	0.658	86
	70/100 + 2% SBS	84	69	0.694	83
	70/100 + 4% EOC	148	123	1.229	83
	70/100 + 4% SBS	167	138	1.381	73
	PMB 45/80-55	206	171	1.710	67
82	70/100 + 4% EOC	80	66	0.622	85
	70/100 + 4% SBS	99	82	0.822	73
	PMB 45/80-55	122	101	1.013	69
88	PMB 45/80-55	74	61	0.609	71

As it is demonstrated all the polymer-modified bituminous compositions show a higher shear stress, viscosity and shear modulus values along with lower phase shift angle values, denoting a more structured system (see Figure 6). In comparison, the introduction of SBS ensured larger increments of viscosity and shear modulus and correspondingly larger decrements of phase shift angle values than EOC modified compositions. As a result, the rheological properties of SBS based systems were closer to the commercial polymer-modified bitumen. This is easily explained by viscosity values of neat polymer modifiers, the viscosity of EOC being considerably lower than that of SBS. Another reason for this is differences in macromolecular structures of the used polymers, the structure of EOC being more linear. Consequently, the addition of EOC to bitumen in general leaves a smaller impact on the flowability of the composition. It is also worth mentioning that the rheological behaviour of the investigated bituminous binders was also influenced by the fact that the tests were performed over the temperature range, below the glass transition of styrene moieties in SBS and at or above melting of crystalline phase of EOC (ca 40-70°C).

The temperature susceptibility of the investigated bituminous systems also improved upon the addition of polymer modifier, especially SBS. The temperature susceptibility of the laboratory modified 70/100+4%SBS system was rather close to that of commercial polymer-modified bitumen.

Higher critical temperatures of the investigated bituminous systems can be derived from |G*|/sin(delta) plots. As one can see, the critical temperature is considerably increased (by 16PPP°PPPC) upon the introduction of a polymer modifier within the system. As already expected, the higher increment was ensured by the addition of the SBS modifier. Nevertheless,

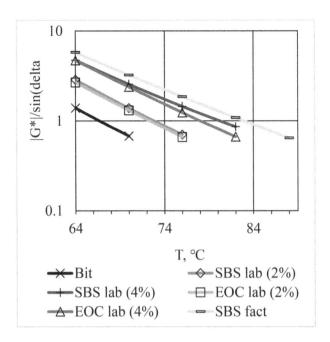

Figure 6. Higher critical temperature of neat bitumen 70/100 (Bit), SBS fact, as well as SBS lab and EOC lab-based systems. In the brackets mass percentage of the polymer is given.

it is clearly shown that also EOC modifier ensures a certain increment in rutting resistance of the 70/100 grade bitumen.

4.2 *Physical properties*

Conventional volumetric test results are summarized in Table 3. Similar bulk and maximum densities were observed for all the mixtures. Mixtures were designed so that air voids are within the limit of 5 to 7%. The obtained SBS fact obtained slightly lower air void content by 0.6% than mixture EOC lab that was with the same bitumen content and granulometric composition. The highest air void content of 6.9% was obtained by the mixture SBS lab. Relatively low air void content was chosen for BBTM mixtures because of climatic conditions. In Latvia, many freeze-thaw cycles can be observed around a year which could affect the stiffness and service life of the mixture with high air void content (Teltayev, Rossi, Izmailova, & Amirbayev, 2019). Another drawback is the wide usage of studded tyres (Kingdom, 2014; Kragh et al., 2011). Both factors negatively affect the properties of high porosity mixtures.

Table 3. Volumetric properties of asphalt mixtures.

Parameter	EOC lab	SBS lab	SBS fact
Bulk density, Mg/mPPP^3	2.373	2.373	2.393
Max density, Mg/mPPP^3	2.539	2.548	2.542
Air voids, %	6.5	6.9	5.9
VMA, %	18.9	18.1	18.2
VFB, %	65.8	61.9	67.3
Binder content, %	5.2	4.8	5.2

4.3 Rutting performance

Rutting resistance results are shown in Figure 7. All the mixtures showed a similar tendency of rut development. Mixture SBS fact with a rut depth of 3.8 mm and WTS with the value of 0.11 mm/1000 cycles demonstrated the best rutting resistance among the investigated asphalt mixtures. SBS lab showed a slightly weaker result with the rut depth of 4.3 mm and the WTS value of 0.13 mm/1000 cycles. EOC lab showed the lowest rutting resistance with the rut depth of 4.9 mm and the WTS value of 0.15 mm/1000 cycles. Overall, huge differences between mixtures were not observed.

Currently in Latvia BBTM wearing course is not specified but in the Road Specifications, two requirements are set for conventional used wearing courses (AC and SMA): WTS value, which is specific for different road classes, and PRD_{AIR}. All the mixtures correspond to the second-highest road class according to Latvian State Roads (LSR) Road Specifications (the number of heavy trucks in 24 hours is up to 2000) (VAS "Latvijas Valsts ceļi", 2019), at the same time, only SBS fact meets PRD_{AIR} requirement of 16% (4 mm rut depth for a 25 mm thick sample). The obtained results confirm findings in the literature that BBTM wearing course could be used on high traffic intensity roads (A. Riekstins et al., 2019; Sol-Sánchez, García-Travé, Ayar, Moreno-Navarro, & Rubio-Gámez, 2017). It should also be reminded that usually the evaluation of rutting resistance is not done for a two-layer specimen system, therefore the results of the WTS, in this case, reflect the ability of both the top layer and the bottom layer to withstand permanent deformations.

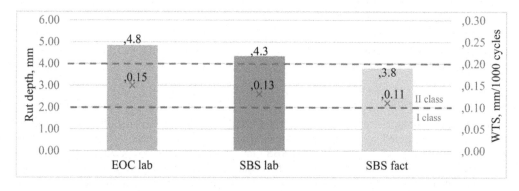

Figure 7. Rutting resistance results of asphalt mixtures. The grey dashed line shows the PRD requirement of wearing course. The red dashed line shows the compliance of asphalt mixtures into classes.

4.4 Thermal cracking

Results from the TSRST test are depicted in Figure 9. All the mixtures showed very similar results with the critical cracking temperature between -26.0°C and -27.1°C and with a slightly larger difference in critical strength between 5.000 MPa and 6.028 MPa. The results show that samples of SBS fact bound mixture broke at the lowest temperature of -27.1°C and at the highest stress of 6.028 MPa in comparison to other mixtures obtained. Both laboratory-made mixtures demonstrated very similar results even if different modifiers (SBS and EOC) and concentrations (4% and 2%) were used. SBS lab demonstrated the weakest result most likely because of lower bitumen content and different granulometric composition. From different studies, it is known that bitumen content and granulometric composition directly influence the low-temperature performance of a mixture (Isacsson & Zeng, 1998; Mollenhauer & Tušar, 2016).

Overall, all **BBTM** mixtures showed very good resistance to thermal cracking. In Latvia in the future, it is planned to implement TSRST as a quality control test for asphalt mixtures. Based on local weather conditions it is planned to use a requirement of -22.5°C for all types of the wearing courses. Convincingly, all **BBTM** mixtures surpass the proposed requirement.

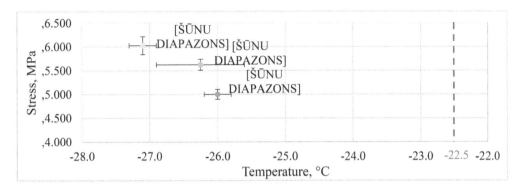

Figure 8. TSRST cracking temperatures for all PMB asphalt mixtures. The red dashed line shows the requirement of minimal cracking temperature.

4.5 *Stiffness and fatigue*

Stiffness and fatigue test results are illustrated in Figure 9 and Figure 10. Contrary to the results of the previous physical properties (rutting and thermal cracking) difference between asphalt mixtures produced by laboratory and industrially manufactured bitumen are much higher. SBS fact shows significantly higher fatigue strength then SBS lab and EOC lab (see Figure 10). What's more SBS fact also showed a better ability to withstand higher strain amplitudes. The main reason for that might be a more elastic and less stiff binder. SBS fact in comparison to EOC lab and SBS lab is less stiff (9762 MPa versus 10212 MPa and 11043 MPa). Other important factors that may influence fatigue behaviour are 1) possible concentration differences of polymer modifier within bitumen matrix, 2) presence of specific additives in the commercially produced bitumen 3) as well as the grade of the neat bitumen (before modification). It is known that the content of light fractions (maltenes) can significantly impact the properties of asphalt (Hofko et al., 2016). As it can be seen in Figure 2 and Figure 3 there exist important differences between the equipment used for the production of bituminous binders at industrial and laboratory scales, which results in a lower mixing degree in the case of laboratory-produced bitumen formulations.

Due to smaller differences between density values of SBS modifier and virgin bitumen reduced stratification is expected within the system in comparison to EOC modified bituminous composition, yielding better long-term behaviour of SBS modified bitumen, represented by fatigue tests. Consequently, SBS lab containing systems demonstrate better fatigue resistance than the EOC lab ones even at lower bitumen content and higher mixture stiffness. In line with the obtained results, the bitumen modification with SBS has been more successful. At the same time, the bitumen mixture with the EOC modifier also showed competitive results at lower strain amplitudes (up to 300 μm/m).

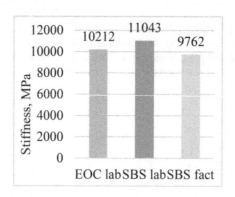

Figure 9. Stiffness modulus for all PMB asphalt mixtures.

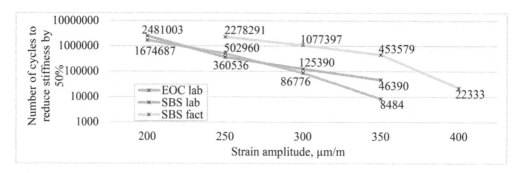

Figure 10. Fatigue strength as a function of relative deformation.

5 CONCLUSIONS

Test results demonstrate that in comparison to virgin bitumen the developed PMB compositions are characterized by a higher shear stress, viscosity and shear modulus values along with lower phase shift angle values, denoting a more structured system. Besides, temperature susceptibility and rutting resistance of the developed bituminous systems are improved upon the addition of the polymer modifier. Although rheological properties of both SBS based systems (SBS lab and *SBS* fact) denote better bituminous binder behavior within asphalt mixtures, EOC modified systems also demonstrate certain improvement in comparison to virgin bitumen. The difference between the rheological properties of SBS laband SBS fact is most probably explained by dissimilar mixing protocols, as well as unknown composition of SBS fact formulation, possibly containing specific admixtures for improvement of processability as well as compatibility within the system. Better dispersion of SBS within bitumen matrix in comparison to EOC containing systems is most probably attributed to closer density values of SBS and bitumen leading to better fatigue resistance as well as other functional properties of these systems. It is worth of mentioning that asphalt mixture with EOC modified bitumen also demonstrates good performance properties which mostly comply with the highest requirements of Latvian road specifications for wearing course.

However, further research is necessary to make EOC containing bituminous binders more competitive with SBS modified systems, particularly by addressing compatibility issues. Improvement on laboratory mixing procedure as such should be also improved (e.g. customization of the mixing head, the introduction of bitumen recirculation) to make a prediction of the properties of a factory modified bitumens more realistic.

ACKNOWLEDGEMENTS

This study was supported by the Latvian Council of Science on the basis of the Cabinet of Ministers of the Republic of Latvia Regulation N° 725 "Evaluation of Fundamental and Applied Research Projects and Administration of Finance" within the project: N°LZP-2018/ 1-0237 "Innovative use of industrial by-products for sustainable asphalt pavement mixtures".

The authors gratefully acknowledge SLLC "Latvian State Roads" for cooporation in obtaining laboratory results.

REFERENCES

Brasileiro, L., Moreno-Navarro, F., Tauste-Martínez, R., Matos, J., & Rubio-Gámez, M. del C. (2019). Reclaimed polymers as asphalt binder modifiers for more sustainable roads: A review. *Sustainability (Switzerland)*, *11*(3), 1–20. https://doi.org/10.3390/su11030646

Hofko, B., Eberhardsteiner, L., Füssl, J., Grothe, H., Handle, F., Hospodka, M., ... Scarpas, A. (2016). Impact of maltene and asphaltene fraction on mechanical behavior and microstructure of bitumen. *Materials and Structures/Materiaux et Constructions*, *49*(3), 829–841. https://doi.org/10.1617/s11527-015-0541-6

Isacsson, U., & Zeng, H. (1998). Low-temperature cracking of polymer-modified asphalt. *Materials and Structures/Materiaux et Constructions*, *31*(1), 58–63. https://doi.org/10.1007/bf02486415

Kingdom, U. (2014). *CEDR Transnational Road Research Programme Call 2012: Noise: Integrating strategic noise management into the operation and maintenance of national road networks*. (D).

Kragh, J., Nielsen, E., Olesen, E., Goubert, L., Vansteenkiste, S., & De Visscher, J. (2011). OPTHINAL Optimization of Thin Asphalt Layers. *Era-Net Road*, (March), 82. Retrieved from http://www.brrc.be/pdf/15/doc/OPTHINAL_FinalReport.pdf

Market Watch. (2020). Styrene Butadiene Styrene Market 2020 Size,Share Global Future Trend, Segmentation, Business Growth, Top Key Players Analysis Industry, Opportunities and Forecast to 2023. Retrieved from https://www.marketwatch.com/press-release/styrene-butadiene-styrene-market-2020-sizeshare-global-future-trend-segmentation-business-growth-top-key-players-analysis-industry-opportunities-and-forecast-to-2023-2020-08-18

Merijs-Meri, R., Abele, A., Zicans, J., & Haritonovs, V. (2019). Development of polyolefine elastomer modified bitumen and characterization of its rheological and structural properties. *Bituminous Mixtures and Pavements VII*, 52–59. https://doi.org/10.1201/9781351063265-9

Mollenhauer, K., & Tušar, M. (2016). *Functional Durability-related Bitumen Specification (FunDBitS) Identified correlations between bitumen and asphalt properties.*

Ratajczak, M., & Wilmański, A. (2020). Evaluation of laboratory methods of determination of sbs content in polymer-modified bitumens. *Materials*, *13*(22), 1–16. https://doi.org/10.3390/ma13225237

Ren, S., Liu, X., Fan, W., Wang, H., & Erkens, S. (2019). Rheological properties, compatibility, and storage stability of SBS latex-modified asphalt. *Materials*, *12*(22). https://doi.org/10.3390/ma12223683

Riekstins, A., Haritonovs, V., Abolins, V., Straupe, V., & Tihonovs, J. (2019). Life cycle cost analysis of BBTM and traditional asphalt concretes in Latvia. *Engineering for Rural Development*, *18*. https://doi.org/10.22616/ERDev2019.18.N400

Riekstins, Arturs, Haritonovs, V., & Straupe, V. (2020). *Life cycle cost analysis and life cycle assessment for road pavement materials and reconstruction*. *5*, 118–135.

Singh, H., Chopra, T., Jain, S., Kaur, A., & Kamotra, S. (2019). Effect of aggregate type and polymer modification on the performance of bituminous concrete mixes. *International Journal of Applied Science and Engineering*, *16*(1), 1–13. https://doi.org/10.6703/IJASE.201906_16(1).001

Sol-Sánchez, M., García-Travé, G., Ayar, P., Moreno-Navarro, F., & Rubio-Gámez, M. C. (2017). Evaluating the mechanical performance of Very Thin Asphalt Overlay (VTAO) as a sustainable rehabilitation strategy in urban pavements. *Materiales de Construccion*, *67*(327), 1–15.

Teltayev, B. B., Rossi, C. O., Izmailova, G. G., & Amirbayev, E. D. (2019). Effect of freeze-thaw cycles on mechanical characteristics of bitumens and stone mastic asphalts. *Applied Sciences (Switzerland)*, *9*(3), 1–18. https://doi.org/10.3390/app9030458

VAS "Latvijas Valsts ceļi". (2019). *Ceļu specifikācijas 2019*.

Zhu, J., Birgisson, B., & Kringos, N. (2014). Polymer modification of bitumen: Advances and challenges. *European Polymer Journal*, *54*(1), 18–38. https://doi.org/10.1016/j.eurpolymj.2014.02.005

Zicans, J., Ivanova, T., Merijs-Meri, R., Berzina, R., & Haritonovs, V. (2019). Aging behavior of bitumen and elastomer modified bitumen. *Bituminous Mixtures and Pavements VII*, 47–51. https://doi.org/10.1201/9781351063265-8

Exploring correlations between the compressive modulus and the flexural modulus of asphalt concrete mixtures

A. Jamshidi, D. Alenezi & G. White
School of Science, Technology and Engineering, University of the Sunshine Coast, Queensland, Australia

ABSTRACT: Structural performance of the asphalt pavement can be evaluated through the compressive and flexural moduli. Various predictive moduli models have different independent variables as inputs. In this study, two distinct methodologies were used to calculate the compressive modulus (E*) based on the flexural modulus, temperature, and frequency. The first model involved an analytical approach based on matrix analysis while the second was based on ridge regression. The effects of binder type can be characterized via a component's transformation matrix. The results of statistical modeling showed a linear relationship between E* and the independent variables. Comparing outputs of statistical modeling and the matrix analysis showed a good relationship with the experimental results of E*. Such methodologies are proposed to improve the current models.

Keywords: Dynamic modulus, Flexural modulus, Fatigue, Rutting

1 INTRODUCTION

The structural design of asphalt pavement has been changing due to the development of mechanistic approaches. The empirical methods based on the California Bearing Ratio (CBR) were replaced with advanced analytical methods through the application of finite element and layer elastic theories, implemented in empirical-mechanical softwares. Furthermore, new criteria and structural parameters are being developed to more realistically characterize the structural performance of asphalt mix, in comparison to those which are offered by previous methods. One of the structural parameters is the compressive modulus (E*) of asphalt concrete over a range of temperature and speed combinations, which is often referred to as the dynamic modulus. Witczak proposed a model to estimate E* based on the volumetric properties of asphalt mix and the rheological properties of the binder. Later, researchers modified the model to include different engineering properties. For example, Witczak 1-37A (1999) produced a model of E* based on eight input parameters that characterize aggregate particles, asphalt binder rheological properties, binder-aggregate interactions, and loading conditions, as shown in Equation (1).

$$\mathrm{Log}E^* = -1.249937 + 0.029232 P_{200} - 0.001767(P_{200})^2 - 0.002841 P_4 - 0.058097 V_a$$

$$- 0.802208 \frac{V_{A_{ef}}}{(V_{A_{ef}} + V_a)} + \frac{3.871977 - 0.0021 P_4 + 0.003958(P_{38})^2 + 0.005470 P_{34}}{1 + e^{(-0.603313 - 0.313551 \log f - 0.393532 \log \mu)}}$$

(1)

where
E*: Compressive modulus of asphalt (in 10^5 psi)

DOI: 10.1201/9781003222880-11

P_{200}: Percentage of passing the No.200 sieve, by total aggregate weight
P_4: Percentage retains on the No.4 sieve, by total aggregate weight (cumulative)
Va: Percentage of air voids in the mix, by volume
V_{Aef}: Percentage of effective asphalt binder content, by volume
P_{38}: Percentage retains on the 3/8 inch sieve, by total aggregate weight (cumulative)
P_{34}: Percentage retains on the 3/4 inch sieve, by total aggregate weight (cumulative)
F: Frequency of loading (Hz)
μ: Asphalt binder's viscosity

To consider the effect of the frequency of loading on the binder, the shear dynamic modulus and the phase angle of the asphalt binder were included, and this is referred to as the Witczak 1-40D model (Equation 2). With the adapted model, the dynamic response of the binder and binder-aggregate interaction is better explained. In other words, the researchers tried to propose integrated models that encompassed volumetric properties of the mixture and rheological properties of the binder in the temperature range at which the binder behaved as a viscoelastic. The use of the viscoelastic properties of the binder was a breakthrough in the development of accurate predictive models because sensitivity analyses carried out by Yousefdoost et al. (2013) showed that intrinsic characteristics related to asphalt binders play a key role in the accuracy of the predicted compressive modulus.

$$\text{Log}E^* = -0.349 + 0.754 G_A^{*-0.0052}$$
$$\times \left(6.65 - 0.032P_{200} + 0.0027P_{200}^2 + 0.011P_4 - 0.0001P_4^2 + 0.006P_{38}\right.$$
$$\left. -0.00014P_{38}^2 - 0.08V_a - 1.06\frac{V_{Aef}}{(V_{Aef}+V_a)}\right)$$
$$+ \frac{2.56 + 0.03V_a + 0.71\frac{V_{Aef}}{(V_{Aef}+V_a)} + 0.012P_{38} - 0.0001P_{38}^2 - 0.01P_{34}}{1+e^{(-0.7814-0.578585\log G_A^* - 0.8834\log \delta_A)}} \quad (2)$$

where
E^*: Compressive modulus of asphalt (psi)
P_{200}: Percentage of passing the No.200 sieve, by total aggregate weight
P_4: Percentage retains on the No. 4 sieve, by total aggregate weight (cumulative)
Va: Percentage of air voids in the mix, by volume
V_{Aef}: Percentage of effective asphalt binder content, by volume
P_{38}: Percentage retains on the 3/8 inch sieve, by total aggregate weight (cumulative)
P_{34}: Percentage retains on the 3/4 inch sieve, by total aggregate weight (cumulative)
G_A^*: Dynamic shear modulus of the asphalt binder (psi)
δ_A: Asphalt binder's phase angle (°)

Also, Christensen et al. (2003) proposed the Hirsch model (Equation 3). This was based on the volumetric properties of the mix and the shear modulus of the binder. In addition to the complex shear modulus, Al-Khateeb et al. (2006) added a complex shear modulus at the glass point of the binder as a rheological characteristic of the binder (Equation 4). All these predictive models are useful in the analysis and design of asphalt pavements. However, the models cannot be used for all the utility conditions. For example, Sing et al. found that the Hirsch and Al-Khateeb predictive models are reliable at low temperature ranges, while the Witczak models yield accurate outputs at high temperatures. Also, the models show bias at low temperatures and in high air voids. Thus, further investigation is required to improve the accuracy of the predictive models.

$$E^* = P_C \times \left[4,200,000\left(1 - \frac{VMA}{100}\right) + 3G_A^*\frac{VFA \times VMA}{10,000}\right] + \frac{(1-K)}{\left[\frac{1-\frac{VMA}{100}}{4,200,000} + \frac{VMA}{3VFAG_A^*}\right]} \quad (3)$$

$$K = \frac{\left(20 + \frac{\text{VFA} \times 3G_A^*}{\text{VMA}}\right)^{0.58}}{650 + \left(\frac{\text{VFA} \times 3G_A^*}{\text{VFA}}\right)^{0.58}}$$

where
E*: Compressive modulus of asphalt (psi)
VMA: Percentage of voids in mineral aggregates in compacted mix, by volume
VFA: Percentage of voids filled with asphalt binder in compacted mix, by volume
G_A^*: Dynamic shear modulus of the asphalt binder (psi)

$$E^* = 3 \times \left(\frac{100 - \text{VMA}}{100}\right) + \left(\frac{\left(90 + 1.45\frac{G_A^*}{\text{VMA}}\right)^{0.66}}{1100 + \left(0.13\frac{G_A^*}{\text{VMA}}\right)^{0.66}}\right) \times G_g^* \quad (4)$$

where
E*: Compressive modulus of asphalt (psi)
G_g^*: Complex shear modulus of asphalt binder in glassy state in Pa, which is assumed to be GPa

In addition, the other researchers tried to correlate E* and the other engineering properties of mixtures. As an example, Apeyagi (2001) found that E* and gradation could be considered as potential rutting specification indicators for quality control and quality insurance purposes in the field. Wang et al. (2011) characterized the effects of binder migration on E*. The results indicated that E* decreased as binder migrated. The reason is that the migration changed the binder content in the mix and therefore the viscoelasticity changed. In the top part of the mix, E* was somewhat higher than in the lower part. In other words, the elastic property of the aggregate was dominant, while the viscous property of the binder controlled the structural performance of the mix in the bottom. As the temperature increased, more binder migrated, which resulted in segregation of the aggregate particles on the upper surface. In parallel with this problem, the structural performance of mix under the bottom transited from interlocking aggregate particles into hydrostatic pressure of the binder. As a consequence, E* drastically reduced as temperature increased or the loading frequency decreased in the samples with the asphalt migration phenomenon.

Although E* is a key parameter for structural design of asphalt pavement using Mechanistic Pavement Design Guide (MEPDG), more structural parameters exist. Among the main criteria is the flexural modulus (F_M) of the mixture that governs its fatigue characteristics. Similar to E*, F_M depends on binder type, additive type, content, temperature, frequency of loading, and aggregate gradation (White 2020, Bekele et al. 2019, Moreno et al. 2013, Fu & Harvey, 2007). Therefore, an estimation of the correlation between E* and F_M can be useful in the development of future performance-related predictive models. Such models can be used for the structural analysis and design of pavements. In addition, the models can be used to calculate E* based on laboratory or field results of the F_M. In this research, the E* and F_M of asphalt mixes prepared using different binder types were evaluated. Then, the correlation between compressive and flexural moduli of the asphalt mix was estimated based on the laboratory results using a proposed statistical and analytical model.

2 MATERIALS AND METHODS

Three asphalt binders were used in this study. Table 1 shows the properties of the binders. C320 is a common unmodified bitumen for asphalt production (AS 2008-2013) while M1000 is an acid-modified multigrade bitumen (White 2016). Jetbind™ is a heavily modified proprietary binder for heavy duty pavement surfaces (White and Embleton 2015). Figure 1 shows the aggregate gradation of mixture, which was typical dense graded asphalt for road and airfield pavement surfacing.

The samples were tested by using a Universal Testing Machine (UTM) and conditioned in its environmental chamber. The conditioning temperatures were 10 °C, 20 °C, and 30 °C. The frequency of loading for cylinder and beam samples for the compressive and fatigue tests were 0.10 Hz, 1 Hz, 10 Hz, and 20 Hz.

Table 1. Engineering properties of the asphalt binders.

Property	C320	M1000	JetBind™
Softening point (°C)	50	60	69
Viscosity at 135°C (Pa.s)	0.55	1.03	—
Elastic Recovery at 0.1 kPa	3.8	42.4	64.40
Elastic Recovery at 3.2 kPa	1	23.70	52.70
Cumulative creep at 0.10 kPa	3.87	0.59	0.19
Creep compliance at 3.20 kPa	4.15	0.87	0.34

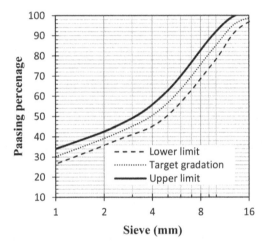

Figure 1. Aggregate size distribution of asphalt mixture.

The compressive modulus of the specific asphalt mix was determined in accordance with the American Association of State Highway and Transportation Officials (AASHTO) T 342-11. To determine the flexural modulus, a four-point beam fatigue test was implemented in accordance with AGPT-T233-06. Three replicate specimens were prepared for each test and mean value at each temperature and frequency was chosen as final result.

3 RESULTS AND DISCUSSION

Figure 2 shows E* and F_M at various temperatures. It can be seen that curve trends were comparable. Due to the use of various types of binder, the E* and F_M of the samples were different. For example, the F_M of the sample containing C320 which was tested at 10 °C and 0.10 Hz was 6.40 (Figure 2d), while corresponding values of samples prepared with M1000 and Jetbind™ were 7.44 GPa and 7.73 GPa, respectively.

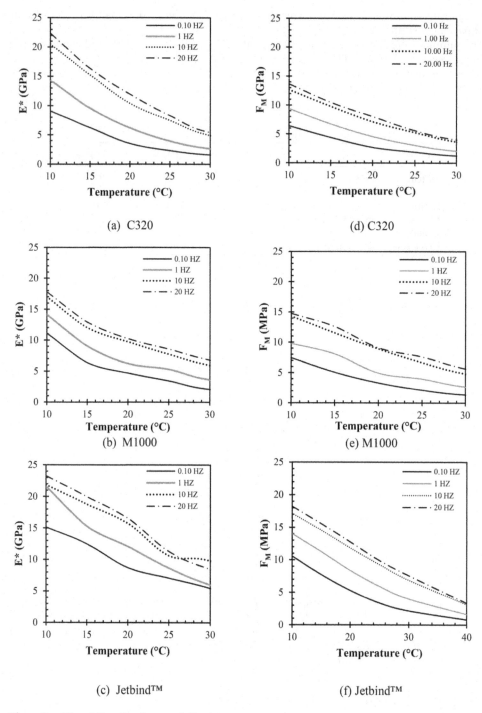

Figure 2. E^* and F_M of various asphalt mixtures.

In addition, the E* of the C320 mixture was 9.07 GPa at 10 °C, which decreased by 61% at 20 °C. The reductions were 52% and 42% for the M1000 and Jetbind™ samples. The amount of reduction in F_M was higher than E* for all the binders. The potential reason is that the construction materials were generally more resistant to compressive strain/stress compared to flexural strain/stress. Since the compressive modulus test is a type of compressive test, the values of E* were higher than the F_M of samples for all the binders and temperatures.

The reason behind the similar trend of E* and F_M at various temperatures can be attributed to the Poisson effect of materials in which the compressive and shear stress/strain in one direction can result in strain/stress in the other. Therefore, any change in E* may influence the other performance-related properties such as F_M. Estimating such correlations can be useful in finding multi-variable predictive models based on performance-related properties, rheological characteristics of the binder, and aggregate size.

It is easy to calculate the F_M based on the E* via transformational matrices built using the temperature and frequency of testing. In such an analysis, the matrix of E* is assumed as the input matrix (Figure 3) and the output matrix is F_M. Since the temperature and frequency are identical for both tests, the rows and columns of the matrix are arranged by the results of E* and F_M.

The transformational matrix converts the F_M value to equivalent E* value. Since the values of F_M are lower than those of E*, the component of the transition matrix is greater than 1. Clearly, the components of the transition matrix change as the binder type changes. Thus, comparing the transition matrices of different binders shows the effects of rheological characteristics of the asphalt mixtures.

$$\begin{bmatrix} F_{Mij} & \cdots & F_{Min} \\ \vdots & \ddots & \vdots \\ F_{Mmu} & \cdots & F_{Mnu} \end{bmatrix} \cdot (\Psi) = \begin{bmatrix} E_{ij} & \cdots & E_{in} \\ \vdots & \ddots & \vdots \\ E_{mu} & \cdots & E_{nu} \end{bmatrix}$$

Figure 3. Matrix of E* and F_M for the mixture.
where n is number of rows, m is number of columns, E_{ij}, E_{nu}, E_{mu}, and E_{in} are components of the matrix of compressive modulus, and F_{Miu}, F_{Mnu}, F_{Mmu}, and F_{Mju} are components of the flexural modulus matrix.

In other words, E* transforms to F_M via a function of Ψ. Mathematically speaking, Ψ is a transition for two different moduli from the compressive modulus to the flexural modulus. Therefore, Ψ can be a constant value, linear operator, or multi-variable function, depending on the number and trend of parameters included in the matrices. In this study, the input matrix contained F_M at various temperatures and frequencies and the output matrix showed E* at the same temperature and frequency. In other words, there was one output per input.

The main advantage of the proposed matrix analysis is that the transformational matrix shows the trend of each component in terms of the independent variables, such as the frequency and test temperature. In addition, there is no limitation to the number of components in the input and output matrix. However, it is recommended that a square matrix be built to find the transformational matrix based on a linear operator. The large matrices require mathematical software to find the operator. Figure 4 shows the input and output matrices of the mixture produced, using C320 binder in a square matrix (4 × 4). For example, the operator E* of the mix (9.07 GPa) at 10°C and 0.10 Hz transform to F_M of 6.39 GPa at the same temperature and frequency via the operator. Note that this methodology is more efficient for minimum 3×3 matrix (or 9 components in the matrix).

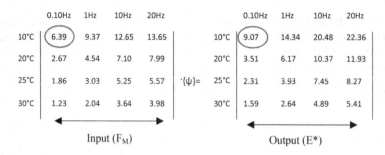

Figure 4. Input and output of C320 mixture.

Figure 5 shows the transformational matrices of the C320, M1000, and Jetbind™ mixtures. The operator transformed values of one engineering property to another. It should be noted that the component of each matrix was different in that it indicated the effect of the binder type. Further, the components varied, which showed the trend of changes at a temperature and frequency sweep. It is obvious that more components in the matrix provide more data for characterization of the trend of components of the transformational matrix. By analyzing the coefficients of the transformational matrices, it is found that E* was between 10% and 160% higher than F_M. On average E* was 41%, 30% and 50% higher than F_M, for the C320, M1000 and Jetbind™ mixtures, respectively.

$$\{\psi\} = \begin{vmatrix} 1.41 & 1.53 & 1.62 & 1.63 \\ 1.31 & 1.35 & 1.48 & 1.49 \\ 1.24 & 1.29 & 1.41 & 1.48 \\ 1.29 & 1.29 & 1.34 & 1.40 \end{vmatrix} \qquad \{\psi\} = \begin{vmatrix} 1.49 & 1.44 & 1.17 & 1.21 \\ 1.56 & 1.26 & 1.10 & 1.13 \\ 1.63 & 1.25 & 1.16 & 1.12 \\ 1.50 & 1.38 & 1.25 & 1.21 \end{vmatrix}$$

(a) C320   (b) M1000

$$\{\psi\} = \begin{vmatrix} 1.46 & 1.58 & 1.25 & 1.27 \\ 1.64 & 1.45 & 1.50 & 1.30 \\ 2.10 & 1.47 & 1.14 & 1.14 \\ 2.60 & 1.56 & 1.42 & 1.12 \end{vmatrix}$$

(c) Jetbind™

Figure 5. Transformation of matrix of mixtures.

Equation 5 shows predictive models proposed using the Ridge Regression method based on the F_M, temperature, and frequency (independent variable) of mixtures. Table 2 shows the results of the ridge regression modeling adopted in this study. It can be seen that there was a nearly linear correlation between the E* and the independent variables. The effects of the binder types can be seen in terms of coefficients of the equation. For example, the coefficient of the flexural modulus in the C320 mixture was 0.99, which was two times greater than the corresponding coefficient in the M1000 and Jetbind™. Further, the constant value of the Jetbind™ mixture was 14.68, which was almost two times as much as the values in the other mixtures. This difference can be attributed to the high E* in the Jetbind™ (Figure 2).

$$E* = \begin{cases} 0.99F_M - 0.18T + 0.12f_r + 6.67 & R^2 = 0.90 \quad C320 \\ 0.42F_M - 0.14T + 0.10f_r + 7.75 & R^2 = 0.88 \quad M1000 \\ 0.58F_M - 0.33T + 0.16f_r + 14.68 & R^2 = 0.88 \quad Jetbind^{TM} \end{cases} \quad (5)$$

E* and M_F are in GPa; T is temperature (°C) and fr is frequency (Hz)

To compare the outputs of the predictive models and matrix analysis, the average of the components was considered as an operator in each mix. Figure 6 shows the laboratory results

Table 2. Summary of Results of the ridge regression modelling.

Independent variables	Mix	Standard error	t value (scaled)	Pr>t	Significance
Intercept	C320	NA	NA	NA	NA
Temperature		1.30	11.97	$<2e^{-16}$	Significant
frequency		1.33	4.85	$1.22\,e^{-6}$	Significant
F_M		1.30	26.41	$<2e^{-16}$	Significant
Intercept	M1000	NA	NA	NA	NA
Temperature		1.26	9.76	$<2e-16$	Significant
frequency		1.32	4.01	5.88 e-6	Significant
F_M		1.25	12.32	$<2e-16$	Significant
Intercept	Jetbind™	NA	NA	NA	NA
Temperature		2.90	9.63	$<2e-16$	Significant
frequency		2.74	3.09	0.00	Significant
F_M		2.95	10.25	$<2e-16$	Significant

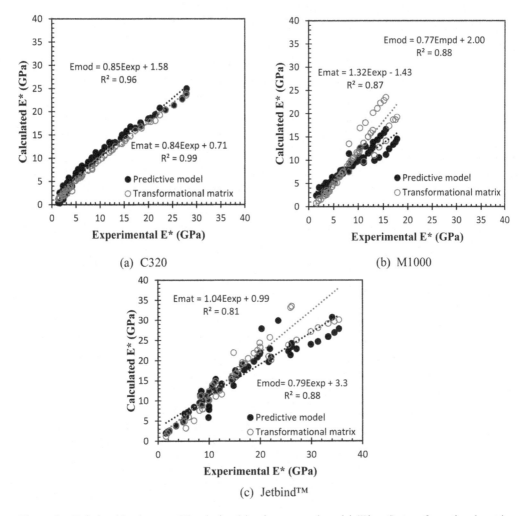

(a) C320

(b) M1000

(c) Jetbind™

Figure 6. Relationships between E* calculated by the proposed model (E*mod), transformational matrix (E*mat), and experiments.

of E* in comparison to the outputs of the predictive models and the matrices. It can be seen that there was a good correlation between the two methods and the experimental results. According to data represented in Figure 5, the predictive models show the higher correlation coefficients compared to the transformational matrixes. However, both methods can provide comparable outputs with the experimental results of E*.

4 CONCLUSION

Compressive and flexural moduli are different key parameters in the analysis of the structural performance of asphalt mixtures. However, their trends are comparable at temperature and frequency sweeps, which can be correlated together. The correlations between the compressive and flexural moduli of mixtures containing different binder types were evaluated. In this regard, two different methods were used; matrix analysis and statistical modeling. For each mixture, a transformational matrix was proposed to convert the flexural modulus to the compressive modulus. The results showed that the components of the matrix as the operator depend on the binder type, test temperature, and frequency. Also, the results of ridge regression showed that there is a linear relationship between the compressive modulus and the flexural modulus. It was found that there is a satisfying relationship between the outputs of the proposed approaches, the predictive model and the matrix analysis, with the experimental results. The proposed methods can be used for mechanistic analysis of the design of asphalt pavements. However, the flexural modulus cannot simply be measured as a substitute for the compressive modulus, due to the complex interactions between the aggregate skeleton and the viscoelastic behavior of the binders. It is recommended that more variables such as aggregate gradation, binder type, rheological properties of the binder, and the other engineering properties of mixtures are considered in the proposed methods.

REFERENCES

AASHTO, T. 2011. 342-11. *Standard method of test for determining dynamic modulus of hot mix asphalt (HMA)*. American Association of State Highway & Transportation Officials, Washington DC.

AGPT-T233-06. 2006. *Fatigue life of compacted bituminous mixes subject to repeated flexural bending*, Sydney, Australia.

Al-Khateeb, G., Shenoy, A., Gibson, N., & Harman, T., 2006. *A new simplistic model for dynamic modulus predictions of asphalt paving mixtures*. Journal of the Association of Asphalt Paving Technologists, 75.

Apeagyei, A. K., 2011. *Rutting as a function of dynamic modulus and gradation*. Journal of Materials in Civil Engineering, 23(9), 1302–1310.

Bekele, A., Ryden, N., Gudmarsson, A., & Birgisson, B., 2019. *Effect of Cyclic low temperature conditioning on stiffness modulus of asphalt concrete based on non-contact nResonance testing method*. Construction and Building Materials, 225, 502–509.

Christensen Jr, D. W., Pellinen, T., & Bonaquist, R. F., 2003. *Hirsch model for estimating the modulus of asphalt concrete*. Journal of the Association of Asphalt Paving Technologists, 72.

Fu, P., & Harvey, J. T., 2007. *Temperature sensitivity of foamed asphalt mix stiffness: Field and lab study*. International Journal of Pavement Engineering, 8(2), 137–145.

Moreno, F., Sol, M., Martín, J., Pérez, M., & Rubio, M. C. 2013., *The effect of crumb rubber modifier on the resistance of asphalt mixes to plastic deformation*. Materials & Design, 47, 274–280.

Singh, D., Zaman, M., & Commuri, S. 2011., *Evaluation of predictive models for estimating dynamic modulus of hot-mix asphalt in Oklahoma*. Transportation research record, 2210(1), 57–72.

Wang, H., Zhan, S., & Liu, G. 2019., *The effects of asphalt migration on the dynamic modulus of asphalt mixture*. Applied Sciences, 9(13), 2747.

White, G & Embleton, K., 2015. *Next generation binder for airport asphalt*. 16th AAPA International Flexible Pavement Conference, Gold Coast, Queensland, Australia, 13-16 September.

White, G., 2016, *Inter-batch and Inter-feedstock variability of an acid modified bitumen*, Road Materials and Pavement Design, vol. 17, no. 3, pp. 658–677.

White, G., 2020, *Incorporating binder type into fatigue life characterization of airport pavement surfaces*, International Journal of Pavement Research and Technology, no. 13, pp. 40–47.

Yousefdoost, S., Vuong, B., Rickards, I., Armstrong, P., & Sullivan, B., 2013. *Evaluation of dynamic modulus predictive models for typical Australian asphalt mixes.* In Proceedings of the 15th AAPA International Flexible Pavements Conference.

Viscoelastic characterization and comparison of Norwegian asphalt mixtures using dynamic modulus and IDT tests

M. Alamnie & E. Taddesse
Department of Civil and Structural Engineering, University of Agder (UiA), Grimstad, Norway

I. Hoff
Department of Civil and Transport Engineering, Norwegian University of Science and Technology, Trondheim, Norway

ABSTRACT: This paper presents the comparison of stiffnesses of two types of asphalt concrete mixes (Ab11 and Agb11) fabricated in the laboratory using two commonly used stiffness tests. (1) uniaxial compression dynamic modulus (E*) and (2) Indirect tensile resilient modulus (Mr) tests are conducted at various temperatures, frequencies, and loading times. As expected, both E* and Mr values decrease as temperature increases and frequency decreases. It is observed that the Mr is generally less than the E*. However, this phenomenon is dependent on frequency, temperature, and material type. The viscoelastic property of the two mixes is characterized using the time-temperature superposition principle. Master curves are developed using the mathematical sigmoid and the rheological 2S2P1D models. It has been observed that the rheological model has shown difficulty to converge optimization error to fit Mr master curve while sigmoid function accurately fitted the measured data. Mr is compared with E* both in frequency and time domains using master curves. A closer comparison is observed with the storage modulus((E')and time-domain relaxation modulus E(t) master curves. Considering the inherent time-domain property of Mr, it is recommended to compare with E' or E(t). From the master curves and data analysis and by considering inputs such as energy loss, modular ratio, frequency, temperature, and other volumetric properties, a rigorous Mr vertical shifting function can be developed. Finally, a material database is obtained for Norwegian asphalt mixes.

Keywords: Viscoelastic, Dynamic Modulus, Resilient Modulus, 2S2P1D Model

1 INTRODUCTION

Asphalt concrete (mixture) is a composite material constituted of aggregate particles embedded in a bitumen/binder. The asphalt binder firmly adheres to the aggregate particles and binds them to form asphalt concrete (a.k.a. bituminous mixture). Asphalt concrete is a viscoelastic material at small strain or stress levels, whereas it behaves in a nonlinear viscoelastoplastic manner at high temperatures and large deformations. Since the pioneering of the mechanistic-based design method, enormous research efforts are put forth towards developing more accurate and complicated testing methods. Various field and laboratory test methods are developed to characterize asphalt concrete in linear, nonlinear, viscoelastic, or viscoplastic states. A wide range of material data is needed to calibrate advanced pavement design models, which required extensive testing using advanced test equipment. However, equipment cost,

DOI: 10.1201/9781003222880-12

complexity equipment/test method, and requirements skilled labor/time are imminent challenges.

For this reason, asphalt material characterization still relies on simple performance indicators such as Indirect tensile, Marshall stability, and other rheological tests. The Indirect Tensile (IDT) resilient modulus test is a standard test used as a performance indicator to assess stiffness, permanent deformation, and fatigue cracking of asphalt concrete (Bennert et al., 2018). The small strain dynamic modulus test is another well-known test for linear viscoelastic characterization of asphalt concrete in the frequency domain. In essence, the IDT strength is a relative indicator of the resistance of the asphalt mixture to tensile loads. Efforts have been made to correlate the IDT resilient modulus and dynamic modulus values (Lacroix et al., 2007, Ping and Xiao, 2008, Loulizi et al., 2006).

This study aims to characterize and compare dynamic modulus and IDT resilient modulus tests using two laboratory fabricated asphalt concrete mixtures used in Norway. For the test campaign, the small strain (viscoelastic) range is considered for both test methods. The asphalt mixtures are selected based on binder stiffness (i.e., soft and relatively harder polymer-modified binder mixes). Moreover, the study investigates the correlation between E* and Mr, and Mr with storage dynamic modulus (E′) using the superposition principle and through the use of master curves. In addition, a material database is obtained for the Norwegian asphalt concrete mixes.

2 THEORETICAL BACKGROUND

The performance of asphalt concrete is dependent on the fundamental thermodynamic variables (temperature and pressure) and loading frequency. The effect of temperature has been studied with more emphasis than pressure. The superposition principle of Time and Temperature (TTSP) is used to develop viscoelastic models. Several researchers have validated the applicability of TTSP for asphalt concrete with and without damage (i.e., in linear or small strain, and nonlinear or large strain states) (Schwartz et al., 2002, Zhao and Richard Kim, 2003, Yun et al., 2010, Nguyen et al., 2013). The TTPS is used to shift experimental data (at various temperatures, stress, frequency, strain rate, etc.) to a specified reference parameter (such as temperature, stress, or strain rate). A material that satisfies the superposition principle is called a thermo-rheologically simple material. Using TTSP, the same material response can be obtained either at low temperatures and longer loading times (low frequency) or at high temperatures but short loading times (high frequency). The relationship can be described as follows (Equation 1).

$$|E*(T,f)| = |E*(f_R)| \tag{1}$$

Where, $f_R = f \times aT$ = reduced frequency; f = frequency; aT = shift factor for temperature (T).

The horizontal shifting behavior allows for the processed data to form a single curve but with a wide frequency beyond the experimental window. The curve generated by shifting experimental data is called the master curve. The Time-temperature shift factor (amount of horizontal shift to a reference temperature) and the master curve enable us to predict the linear viscoelastic (LVE) behavior over a wide range of time, frequency, stress, strain, etc. Different shift factor models are available in the literature (Yusoff et al., 2011). The Williams, Landel, and Ferry (WLF) function (Williams et al., 1955) is the commonly used model (Equation 2).

$$log(a_T) = -\frac{C_1(T - T_r)}{(T - T_r) + C_2} \tag{2}$$

Where C_1, C_2 – WLF constants

3 EXPERIMENTAL PROGRAM

3.1 Materials and sample production

In this study, two widely used asphalt mixtures in Norway (identified as Ab11, Agb11) are used for investigation. The mixes are collected from a production plant supplied by an asphalt producer. Both mixes have a nominal maximum aggregate size (NMAS) of 11 mm, as shown in the granulometric distribution (Table 1). The Ab11 is a polymer-modified mix (70/100 binder grade and 5.6% content), and it is mainly used in high-traffic pavement. On the other hand, the Agb11 is a soft mixture mainly applied for low to medium-traffic roads and contains 5.83 % content of a 160/220 grade type binder. The sampled mix is reheated at 150°C for 4 hours before compaction in a gyratory compactor. The 150 mm diameter and 180 mm height cylindrical samples are cored and cut to get the final specimens for dynamic modulus (100 mm × 150 mm height) and resilient modulus (100 mm × 45 to 50 mm thickness).

Table 1. Aggregate gradation (Ab11 and Agb11).

Sieve Size (mm)	% Passing	
	Agb11	Ab11
16	100	100
11.2	95	95
8	77	70
4	56	48
2	41.5	36
1	31.5	27.5
0.25	15	15.5
0.063	7.5	10

3.2 Mechanical characterization of asphalt concrete

In the dynamic modulus (E*) test, a uniaxial compressive sinusoidal (harmonic) load is applied on a specimen without a rest period (no delayed elastic rebound) at different frequency sweeps (Figure 1 a). The steady-state stress is applied so that it maintains an axial strain amplitude (ε_o) between 50 to 100 μm/m and smaller (Gayte et al., 2016, Schwartz et al., 2002). A strain amplitude up to 150 μm/m is recommended in compression (Levenberg and Uzan, 2004). However, the idealized steady-state (harmonic) condition is deviated due to transient behaviors, such as non-uniform loading, nonlinearity, or any other phenomenon (Gayte et al., 2016). A complex number (E*) defines the stress-strain relationship under continuous sinusoidal loading. The theoretical sinusoidal evolutions of the three measured parameters with time (strain, stress, and dynamic modulus) are expressed in Equations 3 to 5.

$$\varepsilon(t) = \varepsilon_o \sin(\omega t) \qquad (3)$$

$$\sigma(t) = \sigma_o \sin(\omega t + \varphi) \qquad (4)$$

$$E^* = \frac{\sigma_o}{\varepsilon_o} e^{j\varphi} = |E^*|(\cos\varphi + i\sin\varphi) = |E^*|e^{j\varphi} \qquad (5)$$

Where, $|E^*| = E' + iE''$, $E' = |E*|\cos\varphi$ – storage modulus and $E'' = |E^*|\sin\varphi$ – loss modulus; ε_o, σ_o – the axial strain and stress amplitudes, respectively; i – complex number ($i^2 = -1$); φ – phase angle [0, 90°] which is calculated using time lag (t_l) in strain signal and loading period (t_p) of the stress signal (i.e., $\varphi = 360° \times \frac{t_l}{t_p}$).

On the other hand, in the resilient modulus (Mr) test, a haversine compressive load is usually applied with a rest period along the vertical diametral plane of a cylindrical asphalt concrete specimen (Figure 1 b). The load should be selected to avoid damage to the sample at different pulse and rest periods. A minimum of 200N axial force and 50 microstrains target deformation are selected. The Mr is calculated using the measured recoverable strain (deformation) during the unloading and the rest period with an assumed constant Poisson ratio, μ=0.35 (EN 12697-26). It should be noted that the haversine load with a rest period can be decomposed into a cycle of true sinusoidal load and a creep load.

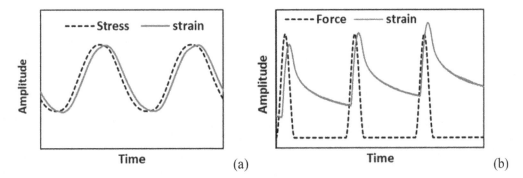

Figure 1. Stress-strain signal responses (not to scale) (a) Dynamic Modulus (b) Resilient Modulus.

Three sets of linear variable differential transducers (LVDTs) are mounted at 120 degrees apart for the dynamic modulus test, while two flat end LVDTs are used for the resilient modulus test. The instrumented specimen is conditioned at target temperature (at least two hours) in a separate chamber and then installed for testing in a UTM-130 machine with a temperature chamber. A dummy sample is used to control the surface and core temperature of a sample.

4 TEST RESULTS

The dynamic modulus test is conducted at three target temperatures (5, 21, 40 °C) and seven frequencies (10, 5, 2, 1, 0.5, 0.2, 0.1 Hz) on two asphalt mixtures (Ab11 and Agb11). Similarly, the resilient modulus test is conducted at five temperatures (5, 10, 21, 30, and 40 °C) and different loading times. The loading times are selected to match the frequency used for the dynamic modulus test. The dynamic and resilient moduli test results are shown in Figure 2 and Figure 3. As expected, both dynamic and resilient moduli decrease as temperature increases and frequency reduced. It is important to note that the Mr value is highly dependent on the magnitude of the applied load and selected loading time. Hence, the pulse load is selected based on the estimated stiffness at target temperature and frequency. In this study, ten conditioning cycles are applied at each frequency, and the last five load pulses are used to estimate the Mr value. A tuning test is conducted to estimate the initial modulus for dynamic modulus at each test temperature. The coefficient of variance (%CV) for the last five loading pulses is generally less than 10% in all measurements at different loading times and temperatures.

To compare E* and the Mr, the loading time or frequency should be equivalent in both tests (i.e., $\omega = 1/t$ or $f = 1/2\pi t$). The ratio of E* to Mr is computed (*i.e., $E^* = \kappa \times Mr$ where κ is a ratio*) at the same frequency and temperature. The average κ values are 1.27 and 1.12 for Ab11 and Agb11 asphalt concrete mixtures, respectively. As shown in Figure 4, the ratio (κ) is dependent on temperature, frequency, and asphalt type. It is observed that κ increases as temperature and frequency increase.

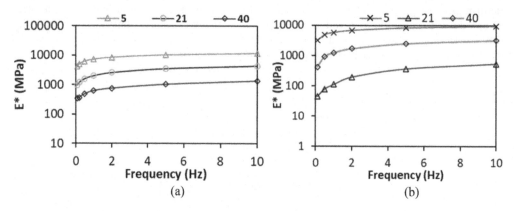

Figure 2. Dynamic Modulus (E*) test results (a) Ab11 (b) Agb11.

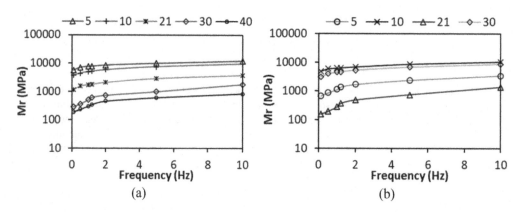

Figure 3. Resilient Modulus (Mr) test results (a) Ab11 (b) Agb11.

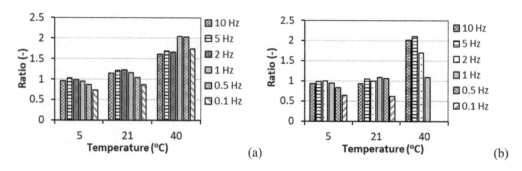

Figure 4. Dynamic to Resilient Modulus Ratio at same temperature and frequency (a) Ab11 (b) Agb11.

5 COMPARISON BETWEEN DYNAMIC AND RESILIENT MODULI

5.1 Master curve construction for E* and Mr

The test data are shifted horizontally (along the logarithmic time or frequency axis) to construct a master curve at a reference temperature. In this paper, two different models are used to develop a master curve for dynamic and resilient moduli. The first is the mathematical Sigmoid function (Equation 6) (Pellinen et al., 2002) and the second type is a rheological model referred to as the 2S2P1D model (Equation 7). The 2S2P1D model was developed at the DGCB laboratory of the ENTPE (Olard and Di Benedetto, 2003, Tiouajni et al., 2011). The model consists of 2 Springs, 2 Parabolic creep elements, and 1 Dashpot in the continuous spectrum.

$$\log|E^*| = \delta + \frac{(\alpha - \delta)}{1 + \exp(\eta - \gamma \log f_R)} \quad (6)$$

where $|E^*|$ - the dynamic modulus; δ, α – the minimum and maximum logarithm dynamic modulus, respectively; η, γ – shape factors; $f_R = f \cdot a_T$ – reduced frequency; a_T – time-temperature shift factor.

$$E^*_{2S2P1D}(\omega) = E_{00} + \frac{E_0 - E_{00}}{1 + \delta(i\omega\tau)^{-k} + (i\omega\tau)^{-h} + (i\beta\omega\tau)^{-1}} \quad (7)$$

Where ω – the angular frequency ($\omega = 2\pi f$); δ, k, h – constants such that $0 < k < h < 1$;
E_{00} – static modulus ($\omega \to 0$); E_0 – the glassy modulus ($\omega \to +\infty$); η – the Newtonian viscosity of the dashpot [$\eta = (E_0 - E_{00})\beta\tau$]; β – a constant that depends on the viscosity of the dashpot; τ – the characteristic time [$\tau = aT*\tau_0$] where τ_0 is determined at a reference temperature.

The shift factors and master curves presented in this paper are all at 21°C reference temperature, and modeling constants are determined by nonlinear optimization using the solver function in Microsoft Excel. As shown in Figure 5 and Figure 6, both the sigmoid and the 2S2P1D models accurately fit a smooth dynamic modulus master curve for both mixtures (Ab11 and Agb11). However, the rheological 2S2P1D model has difficulty producing a smooth resilient modulus master curve. On the other hand, the sigmoid function has shown a good fit for the measured Mr data. The dynamic modulus master curve is generally bigger

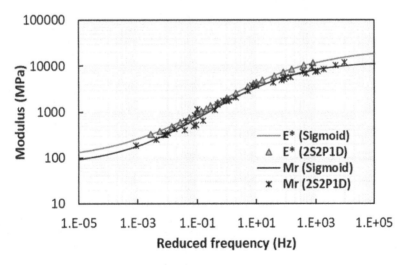

Figure 5. Dynamic modulus (E*) and Resilient Modulus (Mr) Master Curves (Ab11).

than the resilient modulus at the same temperature and frequency due to the rest period in the resilient modulus test (i.e., recovered strain is larger than strain amplitude in dynamic modulus) (Loulizi et al., 2006). However, this conclusion is not always the case. As shown in Figure 6, the Mr curve is above the E* at low frequency (less 1Hz) and below the E* curve at high frequencies for the Agb11 asphalt concrete. The corresponding time-temperature shift factors are shown in Figure 7.

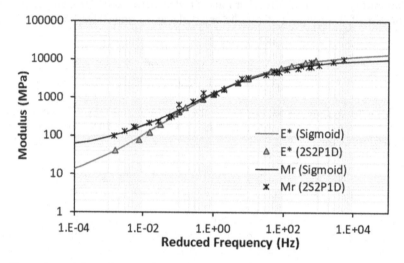

Figure 6. Dynamic modulus (E*) and Resilient Modulus (Mr) Master Curves (Agb11).

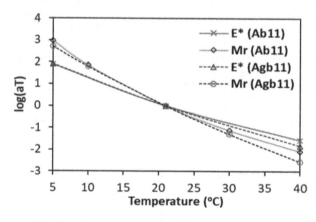

Figure 7. Temperature shift factors.

5.2 *Storage dynamic modulus (E') vs. resilient modulus (Mr)*

During the deformation of a viscoelastic body, part of the total work of deformation is dissipated as heat through viscous losses, but the remainder of the deformational energy is stored elastically (Tschoegl, 1989). The dissipated energy per unit volume (J/m^3 per cycle) due to viscous effects is the loss modulus. The ratio of loss ($\Delta W = \pi \sigma_o \varepsilon_o \sin\varphi$) to maximum stored ($W = \frac{\sigma_o \varepsilon_o}{2}$) energy per cycle (also called specific loss) is given in Equation 8.

$$\frac{\Delta W}{W} = \frac{\pi \sigma_0 \varepsilon_0 \sin\varphi}{\frac{\sigma_0 \varepsilon_0}{2}} = 2\pi \sin\varphi \qquad (8)$$

Where σ_o, ε_o, φ – stress amplitude, strain amplitude, and phase angle, respectively. Note that the specific loss is independent of the stress and strain amplitude.

The storage modulus (E') is the relevant and crucial quantity used to compute the time-domain viscoelastic properties (compliance and relaxation modulus) of asphalt concrete. As Mr is a time-domain test, it is imperative to compare the two moduli at similar conditions. The average ratios ($\kappa = E'/Mr$) are found to be 1.12 and 0.84 for Ab11 and Agb11, respectively. Figure 8 (a and b) shows the E' and Mr master curves for both specimens, and *Figure 9* presents the comparison. One can observe the changes in Figure 5 and Figure 6, and Figure 8. For the Ab11 mixture, E' is very close to Mr at intermediate frequencies (0.1 to 10 Hz or the experimental window) while it barely changes from E* at low and high reduced frequencies. For Agb11, the reduced frequency where E* is equal to Mr (about 1.2 Hz) has shifted to the right (about 10 Hz) for E' and Mr curves.

The loss ratio of Ab11 and Agb11 is presented in Figure 10. A mixture with a softer binder (Agb11) has more energy loss than the stiffer mixture (Ab11), as expected. The maximum energy loss has occurred around the intermediate frequency. Furthermore, the bell-shaped energy loss curve indicates the thermo-rheological simplicity of the tested materials.

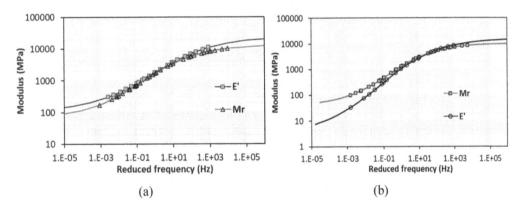

Figure 8. Master curves for E' and Mr using Sigmoid function (a) Ab11 (b) Agb11.

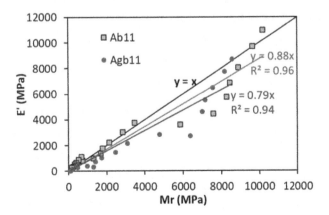

Figure 9. E' versus Mr for Ab11 and Agb11.

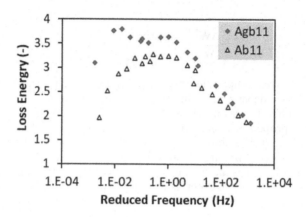

Figure 10. Energy loss for Ab11 and Agb11 mixtures.

5.3 *Relaxation modulus (E(t)) vs. resilient modulus (Mr)*

Although the frequency domain dynamic modulus data can give sufficient information about asphalt concrete's viscoelastic properties, the time domain relaxation modulus and creep compliance are usually used in performance prediction. The generalized Maxwell model (GM) in parallel is used to obtain relaxation modulus, E(t). The one-dimensional E(t) a takes the following Prony form.

$$E(t) = E_\infty + \sum_{m=1}^{M} E_m \left(e^{(-t/\rho_m)} \right) \qquad (9)$$

Where E_∞ - Long-term (equilibrium) modulus; E_m - components of the relaxation modulus; ρ_m – components of relaxation time; and M - the total number of the Maxwell elements (one Maxwell element is composed of one elastic spring and one viscous dashpot connected in series).

The Prony coefficients are determined by the least-squares method from dynamic storage modulus (E'). A good fit is obtained by optimizing eight Prony coefficients (E_m and ρ_m). As can be seen from Figure 5, Figure 8, and Figure 11, the Mr curve is less than the E' and E(t) master curves. The E' vs. Mr and E(t) vs. Mr curves have the same goodness of fit except the x-axis domain is frequency and time, respectively.

Finally, although IDT is more of a quality control test, vital physical properties can be extracted. The virtues of IDT (such as simplicity, economy, repeatability, versatility) are the advantages for industrial applications over the companion dynamic modulus test. Mr alone cannot explain the viscoelastic property of asphalt concrete. Therefore, it is envisaged that advanced analyses should be undertaken by correlating with E* to deduce the fundamental properties (such as relaxation and compliance moduli) from already collected and stored IDT test data.

6 CONCLUSION

This paper presents comparative analyses between dynamic modulus (E*) and Indirect Tensile Resilient modulus (Mr) in linear viscoelastic range. Two different asphalt mixes are used for the investigation. The time-temperature superposition principle is used to construct E* and Mr master curves using sigmoid and 2S2P1D models. Based on the analyses and observations presented in this study, the following conclusions can be drawn.

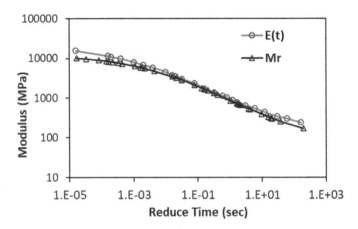

Figure 11. Master curves for $E(t)$ and Mr for Ab11.

- The thermo-rheological simplicity of two asphalt concrete mixtures (Ab11 and Agb11) is investigated using the time-temperature superposition principle.
- An S-shaped master curve is developed using the sigmoid and the 2S2P1D functions for both resilient and dynamic moduli test data. Unlike the sigmoid function, the rheological 2S2P1D model showed some waviness to reflect a smooth Mr master curve. This could be related to the model's inherent sinusoidal nature and is suitable for the harmonic stress-strain evolution.
- Comparison between the three quantities (i.e., E^*, E', and Mr) is performed at the same temperature and frequency. The average ratios between E^* and Mr are found to be 1.27 for Ab11 and 1.12 for Agb11 mixes. Similarly, reduced average ratios of 1.12 and 0.84 are found between E' and Mr due to the deduction loss modulus. It is recognized that the ratios of E^* and Mr is dependent on several factors like mixe type, temperature, frequency, etc.
- The time-domain relaxation modulus (E(t)) master curve is constructed using the Prony method. Almost identical goodness of fit (96%) is found like the E' master curve. Generally, the comparison of time-domain quantities of E^* with Mr is reasonable and practical.
- In addition, a material database is obtained for Ab11 and Agb11 mixtures at wide frequencies and temperatures that can be utilized for the Norwegian pavement management system.
- This work is a preliminary study comparing Mr with E^* for asphalt concrete by developing master curves in both the time and frequency domain. More advance vertical shifting models are needed with more datasets.

REFERENCES

Bennert, T., Haas, E. & Wass, E. 2018. *Indirect Tensile Test (IDT) to Determine Asphalt Mixture Performance Indicators During Quality Control Testing In New Jersey*. Transportation Research Record: Journal Of The Transportation Research Board, 2672, 394–403.

Gayte, P., Di Benedetto, H., Sauzéat, C. & Nguyen, Q. T. 2016. *Influence Of Transient Effects For Analysis Of Complex Modulus Tests On Bituminous Mixtures*. Road Materials And Pavement Design, 17, 271–289.

Lacroix, A., Khandan, A. A. M. & Kim, Y. R. 2007. *Predicting The Resilient Modulus Of Asphalt Concrete From The Dynamic Modulus*. Transportation Research Record: Journal Of The Transportation Research Board, 2001, 132–140.

Levenberg, E. & Uzan, J. 2004. *Triaxial Small-Strain Viscoelastic-Viscoplastic Modeling Of Asphalt Aggregate Mixes*. Mechanics Of Time-Dependent Materials, 8, 365–384.

Loulizi, A., Flintsch, G. W., Al-Qadi, I. L. & Mokarem, D. 2006. *Comparing Resilient Modulus And Dynamic Modulus Of Hot-Mix Asphalt As Material Properties For Flexible Pavement Design*. Transportation Research Record: Journal Of The Transportation Research Board, 1970, 161–170.

Nguyen, M. L., Sauzéat, C., Di Benedetto, H. & Tapsoba, N. 2013. *Validation Of The Time–Temperature Superposition Principle For Crack Propagation In Bituminous Mixtures*. Materials And Structures, 46, 1075–1087.

Olard, F. & Di Benedetto, H. 2003. *General "2S2P1D" Model And Relation Between The Linear Viscoelastic Behaviours Of Bituminous Binders And Mixes*. Road Materials And Pavement Design, 4, 185–224.

Pellinen, T. K., Witczak, M. W. & Bonaquist, R. F. *Asphalt Mix Master Curve Construction Using Sigmoidal Fitting Function With Non-Linear Least Squares Optimization*. Recent Advances In Materials Characterization And Modeling Of Pavement Systems, 2002. 83–101.

Ping, W. V. & Xiao, Y. 2008. *Empirical Correlation Of Indirect Tension Resilient Modulus And Complex Modulus Test Results For Asphalt Concrete Mixtures*. Road Materials And Pavement Design, 9, 177–200.

Schwartz, C. W., Gibson, N. & Schapery, R. A. 2002. *Time-Temperature Superposition For Asphalt Concrete At Large Compressive Strains*. Transportation Research Record: Journal Of The Transportation Research Board, 1789, 101–112.

Tiouajni, S., Di Benedetto, H., Sauzéat, C. & Pouget, S. 2011. *Approximation Of Linear Viscoelastic Model In The 3 Dimensional Case With Mechanical Analogues Of Finite Size*. Road Materials And Pavement Design, 12, 897–930.

Tschoegl, N. W. 1989. *Energy Storage And Dissipation In A Linear Viscoelastic Material. In:* TSCHOEGL, N. W. (Ed.) The Phenomenological Theory Of Linear Viscoelastic Behavior: An Introduction. Berlin, Heidelberg: Springer Berlin Heidelberg.

Williams, M. L., Landel, R. F. & Ferry, J. D. 1955. *The Temperature Dependence Of Relaxation Mechanisms In Amorphous Polymers And Other Glass-Forming Liquids*. Journal Of The American Chemical Society, 77, 3701–3707.

Yun, T., Underwood, B. S. & Kim, Y. R. 2010. *Time-Temperature Superposition For HMA With Growing Damage And Permanent Strain In Confined Tension* And *Compression*. Journal Of Materials In Civil Engineering, 22, 415–422.

Yusoff, N. I. M., Chailleux, E. & Airey, G. D. 2011. *A Comparative Study Of The Influence Of Shift Factor Equations On Master Curve Construction*. International Journal Of Pavement Research And Technology, 4, P. 324–336.

Zhao, Y. & Richard Kim, Y. 2003. *Time–Temperature Superposition For Asphalt Mixtures With Growing Damage And Permanent Deformation In Compression*. Transportation Research Record: Journal Of The Transportation Research Board, 1832, 161–172.

Granular materials

Evaluating the effect of mineralogy and mechanical stability of recycled excavation materials by Los Angeles and micro-Deval test

S. Adomako, R.T. Thorstensen, N. Akhtar & M. Norby
Department of Engineering and Science, University of Agder, Grimstad, Norway

C.J. Engelsen
Department of Building and Infrastructure, SINTEF Community, Oslo, Norway

T. Danner
Department of Architecture, Materials and Structures, SINTEF Community, Oslo, Norway

D.M. Barbieri
Department of Civil and Environmental Engineering, Norwegian University of Science and Technology, Trondheim, Norway

ABSTRACT: This paper presents the findings of the mechanical performance of recycled excavation materials (REM). The aim was to document the stable performance of REM and to suggest its suitability for unbound applications. Test operations were performed at different schedules with the Los Angeles (LA) and micro-Deval (MD) test machine. X-ray diffraction (XRD) analysis was performed to examine the effect of mineralogy on the performance of REM. The results showed that REM showed consistent LA performance, but significant variations were observed in MD values. XRD results showed that up to 20% of chlorite and mica minerals had no influence on the LA performance of REM.

Keywords: Recycled excavation materials, Los Angeles test, micro-Deval test, Mica, Chlorite

1 INTRODUCTION

Recycled excavation materials (REM) present many opportunities in the construction industry. The substitution of conventional and non-renewable construction materials with stabilized REM is gaining global interest due to the environmental, economic, and technical performance. Regarding material substitutions, significant factors such as the type of application, geological properties, conformity with technical requirements (Barbieri et al. 2019; Arulrajah et al. 2012; Sasanipour and Aslani 2020), and location, are mainly considered.

The transformation process of natural ores including weathering is considered a significant parameter in determining the relationship between geological properties and mechanical performance. Considering mineralogy, distinct variations exist in the mineral properties of the same nomenclature of aggregates such as limestones, dolomites and quartzites (Gehringer et al. 2012). REM in general is characterised by a wide range of compositional and geological properties. Hence, uniformity of mechanical performance within the same source or from different sources is not always achieved. In addition, the literature does not provide comprehensive documentation of Los Angeles (LA) and micro-Deval (MD) values of REM to

DOI: 10.1201/9781003222880-13

demonstrate the variations or stabilization of strength properties given the possibility that soft rocks i.e., mudstone, sandstone, conglomerate rock etc. could be a composition of REM. Soft rocks are characterized by low engineering properties and certain geological characteristics e.g., poor cementation, weak intergranular boundaries, clayey minerals etc.

Regarding the effect of mineral composition on strength properties of rocks, the literature present unique findings based on the type and formation of mineral group. Considering primary minerals, it was reported in a review by Adomako et al. (2021), that quartz and feldspar largely contributed to the LA and MD performance of rocks, and generally, about 20% of soft minerals (mica) had no significant effect on the performance of rocks. Furthermore, the spatial distribution of grain size and shape were important factors considered for overall assessment of the performance. Ajagbe et al. (2015) mentioned that increased content of quartz improved the LA resistance of metamorphic rocks. The study by Pang et al. (2010) and Afolagboye et al. (2016) showed that increased content of quartz or feldspar, or a proportional ratio of both resulted in high resistance to LA. The study by Ademila (2019) also showed the relationship between LA of granite gneiss and granite rocks and the content of quartz and feldspar. Regarding quartzite rocks, the same study reported low LA performance although quartzite had high amount of quartz with insignificant amount of feldspar.

In other mechanical test, Hemmati et al. (2020) studied the effects of mineralogy and textural characteristics on compressive and tensile strengths of crystalline igneous rocks. Their study found that the quartz to K-feldspar size ratio was the major indicator that influenced the performance. The differences in connection to mineralogy on each strength property showed that K-feldspar influenced the tensile strength and plagioclase, and quartz influenced the compressive strength. In the case of granitic rocks, Sousa (2013) mentioned that large amount of feldspar increases strength property while quartz resulted in cracking due to its low capacity to withstand stress. Conversely, Tuğrul and Zarif (1999) reported that quartz had low or no cleavage planes and therefore it had the tendency to improve strength while feldspar reduced the strength property of aggregates.

Mica minerals (biotite and muscovite) represent a group of sheet silicates with basal cleavage characteristics and are mainly formed by activities of chemical weathering. There is a strong relationship between the content of mica minerals and mechanical performance (Tandon and Gupta 2013). Given the amount of mica, the study by Anastasio et al. (2016) and Fortes et al. (2016) reported that about 20% of mica does not compromise the strength performance of meta-greywacke. In addition, both studies reported that sparkle traces of mica found in amphibolite did not affect the performance. Nålsund (2010) also showed that 20% of mica content could not be classified to have a negative influence on the performance of rocks. According to Amuda et al. (2014) the probable cause to high resistance to LA in porphyritic granite was the low content of mica. Similarly, the study by (Solomon et al. 2021, manuscript in preparation) found that a positive relationship between mica and the LA of porphyritic granite. Another important consideration given to the effect of mica is the distribution and structure of the grain boundaries (Fortes et al. 2016).

This study was aimed at evaluating the stability of LA and MD performance of REM and to investigate the effect of soft minerals on the performance. The REM samples varied interms of composition. Natural aggregates (NA) were included in the experimental study in order to compare the results. Test operations were performed at different schedules. The results of the study may increase the confidence to apply REM in unbound construction.

2 METHODOLOGY

2.1 Sample preparation

Velde Pukk operates an innovative and sustainable recycling plant (CDE Aggmax ™ system) for the production of REM. Table 1 below describes the REM and NA used in the study. In total, 25 batches of REM and NA were collected from the operational plant within the period of 09/10/2019 to 05/12/2019. 18 samples represented REM and 7 consisted of 100% NA.

Batched samples weighed 22-50 kg and were tested following the dates 09/10/2019 - 17/10/2019 (Batch 1), 29/10/19 -06/11/19 (Batch 2) and 19/11/19 - 05/12/19 (Batch 3). Both REM and NA showed no visible coatings.

Table 1. Description of REM and NA samples used in the experimental study.

Sample name	Type of material	Processing	Source	Particle-Size (d/D) mm
REM	Excavation material	Separation, washing & fractionation	Velde Pukk (Sandnes)	4/16
NA	Crushed	Two-three stage crushing	Velde Pukk (Sandnes)	8/16

2.2 Samples used for X-ray diffraction (XRD) analysis

Table 2 below shows the composition of fractions obtained for XRD analysis after LA tests. Three samples of REM and two NA were selected. The XRD analysis was performed on pulverized fraction i.e., passing the 1.6 mm sieve after LA. This approach helped to establish concrete relationships between the disintegration effect of LA on samples composed of soft minerals.

Table 2. Composition of samples obtained for XRD analysis.

Sample name	Fraction for XRD after LA	Composition					
		Granite/ Gneiss/ Quartzite/ Feldspathic rock	Phyllite	Dark rock	Asphalt	Mortar	Light-weight particles
NA-4	<1.6 mm	100					
NA-5	<1.6 mm	100					
REM-5	<1.6 mm	71	25	2	2		
REM-6	<1.6 mm	91	6		1	1	1
REM-13	<1.6 mm	98	2	< 1			

2.3 Physical and mechanical test

The particle-size distribution (PSD) was based on the procedure described in NS-EN 933-1.

The LA test measures the fragmentation resistance of construction materials. The CEN-1097-2 specifies particle size of 10/14 mm for the test. A test mass of 5000 ± 5 g was poured into cylindrical drum and eleven steel balls were gently added to the test mass. One complete test cycle was 500 revolutions of the steel drum at 31-33 revolutions per minute. Afterwards, the tested mass was washed on a 1.6 mm sieve, oven dried to a temperature of (110 ± 5) °C and the mass loss (%) was determined.

The MD test was performed following the description in CEN-1097-1. The particle size for MD test was 10/14 mm and mass of the test sample was 500 g. Spherical steel balls of 5000 g in weight was added to the test mass and 2.5 ± 0.5 L of water was gently poured into each drum containing the specimen. Each test cycle was up to 2000 revolutions and after the test, the tested sample was washed on 1.6 mm sieve, dried in the oven and the average loss of mass (%) for two test specimens was calculated as the micro-Deval coefficient.

3 RESULTS AND DISCUSSION

3.1 *Physical and mechanical performance*

In Figure 1, the gradation curves shown represent the materials used for the XRD analysis. Attempt was made to evaluate the dense composition of the samples after LA, and which sample had highly pulverized particles. Comparing the PSD before and after LA, it was observed that pulverized test portions after LA showed that REM-6 had about 9% of < 4mm fraction while all other samples had about 5% below 4 mm. In addition, the pulverization degree followed a systematic pattern. This was confirmed by the mass retained on < 1.6 mm sieves measured for LA coefficient.

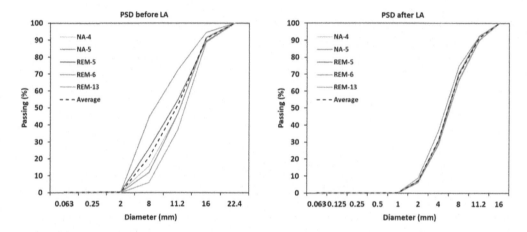

Figure 1. PSD of REM and NA samples before and after LA.

Regarding the LA and MD of REM and NA, Figure 2, Figure 3 and Figure 4 shows the performance of the samples in each batch. For the MD, simultaneous test operations were carried out at Velde following the same protocol of batch number and test dates (Velde AS, 2020). The LA and MD threshold applied on base and subbase materials in Norway were used as reference (NPRA, 2014).

In Figure 2, consistent LA values of 23% and 6% for MD were observed in both NA-1 and NA-2. The LA values of REM were very close. REM-1 recorded 28%, REM-2 and REM-4 showed the same LA performance of 27% and REM-3 had 26%.

Considering the MD, REM-1 recorded the highest at 20%, followed by REM-2 at 15%. Both REM-3 and REM-4 had 9% and 7%, respectively, showing a high resistance to abrasion. While the LA limit requirement was met by all REM samples, REM-1 did not meet the MD threshold.

Figure 3 shows the LA and MD performance for samples in batch 2. The LA values for NA-4 and NA-5 varied at 27% and 22% respectively. However, the MD did not show any significant variation between them; NA-4 had MD of 8% and NA-5 showed 6% MD value.

Regarding REM, REM-5, and REM-8 had the same LA of 25%. REM-6 recorded the highest LA at 28% amongst the REM received in batch 2. REM-7 and REM-9 also had LA of 24% in each case and REM-10 had LA of 26%. REM in this batch met the LA criteria.

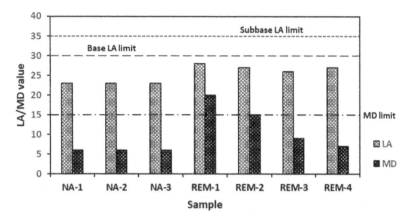

Figure 2. LA and MD performance of batch 1 samples received on 09/10/2019 - 17/10/2019.

For MD values of REM, REM-5 had MD of 17, being the highest amongst the studied REM in this batch. REM-6, REM-7 and REM-9 had the same MD of 10%. REM-8 and REM-10 had MD of 9 and 7, respectively. Only REM-5 exceeded the MD threshold.

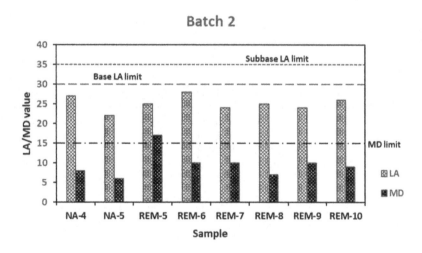

Figure 3. LA and MD performance of batch 2 samples received on 29/10/19 - 06/11/19.

Figure 4 shows the performance of batch 3 samples. NA-6 and NA-7 had LA of 29% and 24% respectively. MD test on NA-6 was not performed. However, NA-7 had a good MD of 6%. Accordingly, it could be assumed that the MD performance of NA-6 would not significantly vary from NA-7.

For REM samples, close LA values was observed. REM-11, REM-12 and REM-16 had 26% LA value. REM-13 and REM-15 also had 24% of LA, and 25% of LA was recorded for both REM-14 and REM-17. The LA of REM-18 was 27%.

Regarding the MD of REM, both REM-11 and REM-13 showed the same resistance to abrasion with MD of 10%. REM-12 and REM-14 also showed MD of 11%. Both REM-15 and REM-17 had the same MD of 8%. The MD for REM-16 was 9% and REM-18 produced MD of 12%. In batch 3, the LA and MD performance met the threshold.

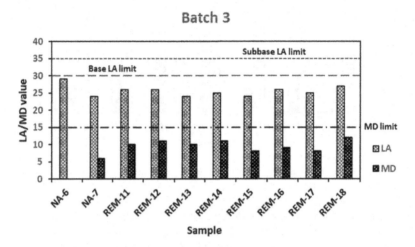

Figure 4. LA and LA and MD performance of batch 3 samples received on 19/11/19 - 05/12/19.

3.2 *Relationship between LA and mineralogy of REM and NA*

In Figure 5, the mineral composition of pulverized samples of REM and NA obtained from < 1.6 mm sieves after LA test is shown. The XRD results showed high peaks of quartz present in all the samples. In addition, significant peaks of mica and chlorite was found in REM-5 and REM-13. The amount of quartz present in all samples ranged from 34-38%. Other minerals such as albite, microcline and anorthite of feldspar group were also present but not displayed in Figure 5. The connection between the low resistance to abrasion by MD in REM-5 could be explained by the content of both mica and chlorite. In Table 2, 25% phyllite materials formed a component of REM-5 and 2% in REM-13.

According to Razouki and Ibrahim (2019) the degree of abrasion of rock samples is connected to the hardness value measured by Mohs hardness scale. Mica and chlorite are phyllosilicates with low hardness value ranging from 2-4. Interms of LA, the difference between REM-5 and REM-13 was not significant. Previous findings (Anastasio et al. 2016, Fortes et al. 2016 and Nålsund 2010) have suggested that 20% of mica content does not significantly reduce the strength performance i.e., LA. The results of this study was consistent with the findings.

4 CONCLUSIONS

This study investigated the stability of mechanical performance of 18 recycled excavation materials (REM) and compared the performance to 7 natural aggregates (NA) samples. In addition, the mineralogy of REM and NA were analyzed to draw final conclusions on the relationship between the strength and mineralogy. The Los Angeles (LA) and micro-Deval (MD) test methods were used to evaluate the strength properties and X-ray diffraction analysis was performed on pulverized samples <1.6 mm obtained after LA test.

Generally, the findings showed consistent LA performance of REM, but wide variations of MD performance was observed. The LA values of REM met the limit criteria for base and

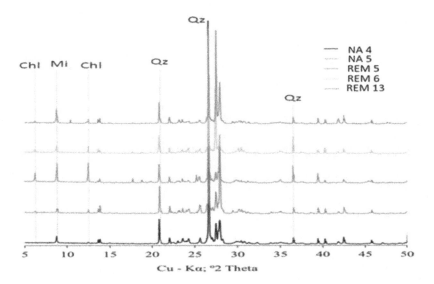

Figure 5. XRD results showing mineral peaks of <1.6 mm fraction of REM and NA. Chl, Chlorites; Mi, Mica; Qz; Quartz are the dominant minerals shown.

subbase layers according to Norwegian Public Roads Administration Handbook (N200). Regarding the MD, REM-1 and REM-5 samples exceeded the threshold.

The results for mineralogy showed that REM-5 contained about 20% of chlorite and mica, however, the LA performance was not influenced. Perhaps, the low MD resistance of REM-5 could be attributed to the content of these weak minerals. This study contributes to establish the stable performance of REM and to raise the confidence of end users about the technical benefits.

ACKNOWLEDGEMENTS

The work presented in this paper is part of the ongoing project MEERC (More Efficient and Environmentally friendly Road Construction), partly funded by the Research Council of Norway (NFR) [project number 273700] and Sorlandets Kompetansefond. In addition, thanks to Velde Pukk for the MD test values.

REFERENCES

Barbieri, D.M., Hoff, I. and Mørk, M.B.E., 2019. *Innovative stabilization techniques for weak crushed rocks used in road unbound layers: A laboratory investigation.* Transportation Geotechnics, 18, pp.132 – 141. https://doi.org/10.1016/j.trgeo.2018.12.002

Arulrajah, A., Ali, M.M.Y., Piratheepan, J. and Bo, M.W., 2012. *Geotechnical properties of waste excavation rock in pavement subbase applications.* Journal of materials in civil engineering, 24(7), pp.924 – 932.

Sasanipour, H. and Aslani, F., 2020. *Durability properties evaluation of self-compacting concrete prepared with waste fine and coarse recycled concrete aggregates.* Construction and Building Materials, 236, p.117540. https://doi.org/10.1016/j.conbuildmat.2019.117540

Gehringer, E., Read, D. and Tutumluer, E., 2012. *Characterization of ballast performance in heavy axle loading (HAL).* In Proceedings of AREMA 2012 Annual Conference.

Adomako, S., Engelsen C.J. and Thorstensen R.T, Barbieri D.M. *Review of the relationship between aggregates geology and Los Angeles and micro-Deval tests.* Bulletin of Engineering Geology and the Environment. 2021 Jan 12:1 – 8.

Ajagbe, W.O., Tijani, M.A. and Oyediran, I.A., 2015. *Engineering and geological evaluation of rocks for Concrete Production.* LAUTECH Journal of Engineering and Technology, 9(2),pp.67–79. Retrieved from https://laujet.com/index.php/laujet/article/view/85

Pang, L., Wu, S., Zhu. J., Wan. L., 2010. *Relationship between retrographical and physical properties of aggregates.* Journal of Wuhan University of Technology-Mater. Sci. Ed. 25(4): 678–681. https://doi.org/10.1007/s11595-010-0069-0

Afolagboye, L.O, Talabi, A.O, Akinola, O.O., 2016. *Evaluation of selected basement complex rocks from Ado-Ekiti, SW Nigeria, as source of road construction aggregates.* Bulletin of Engineering Geology and the Environment 75(2): 853–865. http://doi.org/10.1007/s10064-015-0766-1

Ademila, O., 2019. *Engineering geological evaluation of some rocks from Akure, Southwestern Nigeria as aggregates for concrete and pavement construction.* Geology Geophysics and Environment 45(1): 31 – 43. http://doi.org/10.7494/geol.2019.45.1.31

Hemmati, A., Ghafoori, M., Moomivand, H. and Lashkaripour, G.R., 2020. *The effect of mineralogy and textural characteristics on the strength of crystalline igneous rocks using image-based textural quantification.* Engineering Geology, 266, p.105467.

Sousa, L.M., 2013. *The influence of the characteristics of quartz and mineral deterioration on the strength of granitic dimensional stones.* Environmental earth sciences, 69(4), pp.1333–1346. https://doi.org/10.1007/s12665-012-2036-x

Tuğrul, A. and Zarif, I.H., 1999. *Correlation of mineralogical and textural characteristics with engineering properties of selected granitic rocks from Turkey.* Engineering geology, 51(4), pp.303–317. https://doi.org/10.1016/S0013-7952(98)00071-4

Tandon, R.S. and Gupta, V., 2013. *The control of mineral constituents and textural characteristics on the petrophysical & mechanical (PM) properties of different rocks of the Himalaya.* Engineering Geology, 153, pp.125–143. https://doi.org/10.1016/j.enggeo.2012.11.005

Anastasio, S, Fortes A.P.P., Kuznetsova, E., Danielsen, S.W., 2016. *Relevant Petrological Properties and their Repercussions on the Final use of Aggregates.* Energy Procedia 97: 546–553. https://doi.org/10.1016/j.egypro.2016.10.073

Fortes, A.P.P, Anastasio, S, Kuznetsova, E, Danielsen, S.W., 2016. *Behaviour of crushed rock aggregates used in asphalt surface layer exposed to cold climate conditions.* Environmental Earth Sciences 75(21): 1414. https://doi.org/10.1007/s12665-016-6191-3

Nålsund, R., 2010. *Effect of grading on degradation of crushed-rock railway ballast and on permanent axial deformation.* Transportation Research Record 2154(1): 149–155. https://doi.org/10.3141/2154-15

Amuda, A.G. Uche, O.A.U. and Amuda A.K., 2014. *Physicomechanical Characterization of Basement Rocks for Construction Aggregate: A Case Study of Kajuru Area, Kaduna, Nigeria.* IOSR Journal of Mechanical and Civil Engineering 11(6): 46–51.

NS-EN 933-1: 2012 Tests for geometrical properties of aggregates - *Part 1: Determination of particle size distribution - Sieving method,* CEN_Standard_Norge, 201.

NS-EN 1097-2: 2010 Tests for mechanical and physical properties of aggregates. *Part 2: Methods for the determination of resistance to fragmentation,* CEN_Standard_Norge.

NS-EN 1097-1: 2011 Tests for mechanical and physical properties of aggregates *Part 1: Determination of the resistance to wear (micro-Deval),* CEN_Standard_Norge.

Velde AS, "Resultat LA og MDE," utg., 2020.

NPRA. Håndbok N200 Vegbygging. Norway: Vegdirektoratet;2014.

Razouki, S.S. and Ibrahim, A.N., 2019. *Improving the resilient modulus of a gypsum sand roadbed soil by increased compaction.* International journal of pavement Engineering, 20(4), pp.432–438. https://doi.org/10.1080/10298436.2017.1309190

Freeze-thaw performance of granular roads stabilized with optimized gradation and clay slurry

Ziqiang Xue
Graduate Research Assistant, Department of Civil, Construction and Environmental Engineering, Iowa State University, Ames, IA, USA

Jeramy C. Ashlock
Richard L. Handy Associate Professor, \epartment of Civil, Construction and Environmental Engineering, Iowa State University, Ames, IA, USA

Bora Cetin
Associate Professor, Department of Civil and Environmental Engineering, Michigan State University, East Lansing, MI, USA

Sajjad Satvati
Graduate Research Assistant, Department of Civil, Construction and Environmental Engineering, Iowa State University, Ames, IA, USA

Halil Ceylan
Professor, Department of Civil, Construction and Environmental Engineering, Iowa State University, Ames, IA, USA

Yijun Wu
Graduate Research Assistant, Department of Civil, Construction and Environmental Engineering, Iowa State University, Ames, IA, USA

Cheng Li
Assistant Professor, School of Highway, Changan University, Xi'an, Shanxi Province, China

ABSTRACT: A recently-developed granular road stabilization method was applied to four test sections in different Iowa counties. This new method uses an optimized target gradation achieved by blending existing surface materials with two to three fresh aggregate types to achieve the tightest possible particle packing. A clay slurry is also incorporated into the top two inches during mixing of the surface aggregates to increase the plasticity and binding characteristics to reduce material loss. The performance of the test sections was measured using field dynamic cone penetrometer (DCP) tests, falling weight deflectometer (FWD) tests, and nuclear density gauge tests. The test results are compared to those of adjacent control sections as well as test sections stabilized using Portland cement and steel slag. Among the stabilization methods applied in the test sections, the Optimized Gradation with Clay Slurry stabilization method and the cement-treated 4" surface course was the most costeffective.

Keywords: Stabilization, granular roadway, elastic modulus, falling weight deflectometer, shear strength, dynamic cone penetrometer

DOI: 10.1201/9781003222880-14

1 INTRODUCTION

Granular-surfaced roads commonly provide access between rural and urban areas for hauling agricultural products to market, but high traffic loads and large numbers of freeze-thaw cycles in seasonally cold regions can damage the road surface, especially during springtime. Water present in the near-surface layers experiences phase changes due to the freezing and thawing [1]–[3], resulting in aggregate deterioration and various distresses such as loss of crown, washboarding, potholes, rutting, and raveling. Moreover, the mechanical properties of the surface layers, including gradation, shear strength, and stiffness, are altered due to the material property changes. Regular maintenance is therefore required to maintain road serviceability, which can be costly for DOTs or secondary roads agencies.

Stiffness and strength, the main factors in road design, depend on the quality, shape, aggregate size, thickness, and density of the surface layer [4]–[6]. Environmental conditions and traffic loads alter the surface elastic modulus and shear strength, and knowledge of such mechanistic factors can help in evaluating road performance. Previous studies have shown the significant effects of freeze-thaw cycles on degrading the mechanical properties of granular layers [7]–[11]. Stabilization of granular roadways with quarry fines byproducts has been proven to improve these properties by increasing plasticity and providing tighter bonding between aggregates [12]–[14]. Quarry fines are typically found in one of three distinct formats: screenings, pond fines, and baghouse fines, all of which are waste products of rock-crushing processes [15].

This study investigates the effects of utilizing a clay slurry quarry fines byproduct to provide improved binding between surface aggregates in four granular-surface road test sections. FWD and DCP tests were conducted on test sections and control sections to monitor the elastic modulus and shear strength of the surface aggregate layers subjected to seasonal freezing and thawing cycles. Nuclear gauge density tests were also conducted to compare the density and moisture content of the granular surface layers. The performance of sections stabilized with the clay slurry and an optimized gradation was compared with that of adjacent test sections stabilized with Portland cement and steel slag, as well as control sections.

2 MATERIALS

To cover a wide range of the state's climate conditions, subgrade types, and aggregate sources, the test sections were constructed in Cherokee, Hamilton, Howard, and Washington counties distributed across Iowa (Figure 1). The test sections were constructed by stabilizing the top 4" of the granular surface layers. For the Optimized Gradation with Clay Slurry (OGCS) sections constructed in all four counties, the top 4" of surface course materials were blended in optimized proportions, after which clay slurry was sprayed and mixed into the top 2". The 4"-thick steel slag sections were constructed in Howard and Cherokee counties using conventional granular roadway construction methods, and a 4" thick cement-treated surface course was constructed in Washington County using the county's wheel-loader mounted milling machine.

For the OGCS test sections, an optimized gradation using existing surface materials and locally available fresh aggregates was first determined using the spreadsheet tool published along with the TR-685 project report [25]. A predetermined depth of existing aggregates was first ripped using a motor grader, after which fresh aggregates were spread and blade-mixed in several passes to achieve a uniform gradation. To improve upon a previous study that used powdered bentonite for plasticity, a newly available clay-slurry quarry product was applied to the optimized gradation mixture during construction of the OGCS test sections in all four counties. The clay slurry was sprayed onto the aggregate mixtures using a self-unloading tanker truck, then blade-mixed into the top 2" of aggregates using numerous motor grader passes until the material had somewhat dried out. The surface was then shaped, and a few compaction passes were applied using rubber-tire rollers to accommodate the wet mixture that tended to stick to smooth-drum vibratory rollers. To achieve an immediately drivable surface, a light covering of fresh surface aggregates (approximately two 13-ton truckloads for each 500-ft long test section) were spread over the top and pressed in with the rubber-tired roller. While the resulting surfaces were very wet immediately following construction, they were passable by traffic and typically dried up within 24-48 hours.

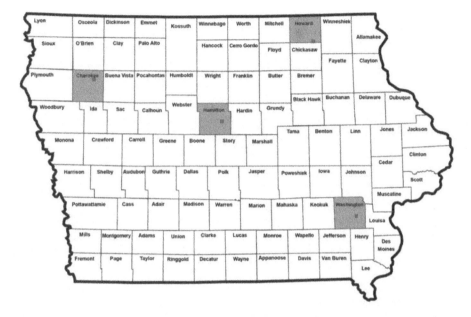

Figure 1. Locations of the test sections in four Iowa counties (Red dots show locations of the actual test sections).

2.1 *Steel slag surface*

Steel slag (SS) is an industrial byproduct commonly used as a stabilizer or amendment in roadway systems. Under compaction and road traffic, SS usually evolves into a high strength material with the capability for self-stabilization. In this study, the SS material was collected from Phoenix Services, LLC, in Wilton, IA. The full depth of the existing surface aggregate was first ripped down to the subgrade and hauled away and the slag then dumped in thin lifts sprayed with water and bladed. After shaping the surface, at least four passes of a rubber-tire roller were made, followed by a few passes of a vibratory smooth-drum roller.

2.2 *Cement treated aggregate surface course*

In this study, 7% Portland cement by dry weight was blended into a 4-in. (101.6-mm) thick surface aggregate course in Washington County without treating the subgrade. The cement was first spread over the test section, then mixed full-depth using a 60-in. wide RoadHog milling machine which was mounted on a Caterpillar 938M Wheel Loader and attached to a water feeder truck by a hose system. After mixing, additional water was sprayed with a water truck and the surface was shaped with a grader. The surface was compacted with four passes of a rubber tire roller immediately followed by a smooth-drum vibratory roller to reduce the compaction delay time.

2.3 *Laboratory testing of construction materials*

After construction, laboratory tests were performed on samples from both the control sections and the stabilized sections. Sieve and hydrometer analyses were performed according to ASTM Methods D6913 and D7928 respectively, along with Atterberg limits tests (ASTM D4318) with classification by the AASHTO and USCS systems (ASTM D3282 & D2487). The index properties and USCS classifications of the materials sampled from the field sections summarized in Table 1 show that, despite the addition of clay-sized particles, the OGCS sections all had lower fines contents than their nearby control sections that contained aggregates abraded and broken down by traffic loading. The control section materials in Cherokee county were non-plastic, i.e., sandy river gravel was the primary source of surfacing material there, while in the other counties crushed limestone having fines with some plasticity was typically used. The slag sections also had very little fines contents from the slag source discussed in this paper. All sections were constructed on a relatively stiff subgrade layer.

3 TESTING METHODOLOGY AND RESULTS

To determine freezing and thawing effects, field tests were performed at five points on each section during winter 2018 and spring 2019. FWD and DCP tests were conducted to evaluate the elastic modulus and shear strength, respectively, of the granular surface and subgrade layers, and nuclear-density gauge tests were performed to determine moisture content and density of each test section. The test methods and results are described in the following sections.

3.1 *Dynamic Cone Penetration (DCP) tests*

DCP tests were conducted to track relative changes in shear-strength profiles in terms of correlated California bearing-ratio (CBR) values. The road system profile was treated as a two-layered system consisting of an aggregate surface layer over a subgrade layer. The thicknesses of the surface layers were estimated based on the DCP data by determining the point at which the plot of cumulative depth versus the number of cumulative blows exhibited a sudden change in slope [1]. The correlated CBR values (denoted DCP-CBR) were then averaged over the depth of the aggregate surface layer for both winter 2018 and spring 2019 field tests.

As shown in Figure 2, the DCP-CBR values for both the OGCS and control sections in Washington County increased from winter to spring, while the values for the 4" cement-treated section were erratic but mostly decreased. In Howard County, the OGCS, control, and 4" slag sections all experienced decreases in DCP-CBR values from winter to spring, with the strength loss generally being smallest in the OGCS section and greatest in the control section. In Hamilton County, the OGCS section exhibited higher DCP-CBR values in the spring, while those of the control section did not change much. In Cherokee County, the OGCS section exhibited a significant decrease in DCP-CBR values (but still greater than the control section's), while both the control and slag sections exhibited little change.

The primary difference setting Cherokee apart from the other three counties is that Cherokee have a continuous thick layer of sandy gravel grading to sand and finally transitioning to silt at a depth of a few feet, while the other three counties have a stiff clay subgrade directly below the 4" surface aggregate layers. Also, 13% of the optimized gradation in Cherokee

Table 1. Soil index properties of the surface materials sampled from control sections and after-constructed stabilized sections.

County	Hamilton		Washington			Howard			Cherokee		
Stabilization method	OGCS	Control	OGCS	4" Cement	Control	OGCS	4" Slag	Control	OGCS	4" Slag	Control
Particle-size Distribution Results (ASTM D6913/D6913M-17/D4318-17)											
Gravel content (%)	60	20.4	56	69.5	32.9	71	78	43.5	54.9	52.5	25.9
Sand content (%)	24	56.3	24	27	29.4	15	22	37.6	31.8	44.6	58.4
Silt content (%)	9	12.7	11	3	23.0	11	0	12.7	6.3	1.7	9.9
Clay content (%)	7	10.6	9	0.5	14.7	3	0	6.2	7.0	1.2	5.8
Coefficient of uniformity, C_u	685	313	1536	4	1,064	425	5.4	449	563	11	121
Coefficient of curvature, C_c	18	5	39	1	0.2	92	1.6	1	7	1	2
Atterberg Limits Test Results (ASTM D4318-17e1)											
Liquid limit	23	19	27	NP	26	26	NP	18	28	NP	NP
Plastic limit	13	14	14	NP	16	17	NP	13	14	NP	NP
AASHTO and USCS soil classification (ASTM D3282-15 & ASTM D2487-17e1)											
USCS Classification	GC	SC-SM	GC	GP	GC	GC	GW	SC-SM	GC	GW	SM
AASHTO Classification	A-2-4	A-1-b	A-2-6	A-1-a	A-4	A-2-4	A-1-a	A-1-b	A-2-6	A-1-a	A-1-b

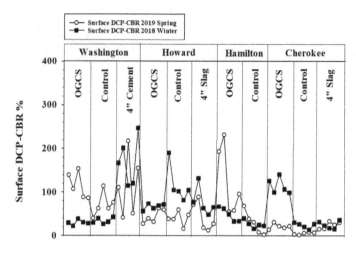

Figure 2. Results of shear strength for surface layers (five points evenly distributed in each section).

consisted of river rock, the most common surfacing material in that county, while the other counties typically used crushed limestone. Based on the data shown in Figure 2, different subgrade support conditions and inclusion of river rock can reduce the strength of the OGCS method over time, while the subgrade and surfacing materials in Washington and Hamilton counties lead to strength increases for this method.

3.2 *Falling-Weight Deflectometer (FWD) tests*

The FWD tests were conducted using a JILS model SN121 FWD with a two-piece segmented loading plate. The FWD model used nine sensors with horizontal spacings between 6 and 12 inches (152.4 to 304.8 mm) to measure deflections on the road surface and provide a measured deflection basin. Three different target dynamic pressures of 250, 275, and 300 kPa were applied on the plate. The upper limit of 300 kPa was necessary to avoid over-ranging the deflection sensors for the relatively compliant unpaved road surface. The combined Boussinesq-Odemark solution was utilized to obtain stresses, strains, and deformations at each given depth and radii for a homogeneous, linear-elastic half-space model [16]. In conformance with this combined theory, back-calculation was performed based on dynamic loads and peak deflections observed under the geophones in the two-layered system [17]–[22].

Figure 3 shows the surface elastic modulus (termed E_{FWD}) determined from FWD tests performed at five points in each section during winter 2018 and spring 2019. In Washington County, E_{FWD} for the OGCS and control sections did not exhibit any statistically significant changes from winter to spring (one point in each section was stiffer in spring). However, in the 4" Cement section the modulus underwent a considerable decrease from winter to spring. In Howard and Cherokee Counties, the maximum change of E_{FWD} occurred as decreases in the OGCS section in Howard County and the OGCS and 4" slag sections in Cherokee county. The maximum decrease in E_{FWD} among all sections was in the OGCS section in Hamilton County; this section exhibited the highest average initial E_{FWD} (~18,000 MPa) compared to all sections during winter 2018 (due to the frozen ground in accordance with the weather information that day). The modulus for this section decreased significantly during spring 2019, becoming similar to that of the other sections (~1,300 MPa). The E_{FWD} for the control section also decreased slightly from winter 2018 to summer 2019 in Hamilton.

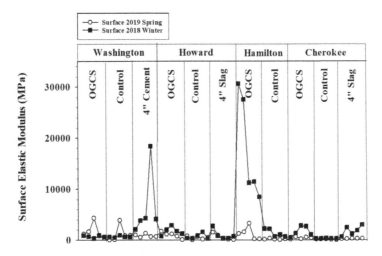

Figure 3. Elastic modulus of granular surface layers determined from FWD tests (five points evenly distributed in each section).

3.3 *Nuclear gauge & result summary*

Nuclear-gauge tests were conducted in the four counties to monitor changes in water content (ω) and dry density (γ_d) of the surface aggregates. The average results for ω and γ_d are shown in Table 2, along with the average values of E_{FWD} and DCP-CBR for all sections.

Based on the average values reported in Table 2, the cement-stabilized surface section in Washington County, the control sections in Howard and Cherokee counties, and the steel slag section in Howard County all exhibited significant decreases in CBR, indicating that these sections lost shear strength during spring thawing. The OGCS and control sections in Hamilton, Howard, and Cherokee counties, as well as all slag and cement sections, experienced significant decreases in elastic modulus during the spring thaw as well. However, the average strength and stiffness for all OGCS and cement-stabilized sections were higher than those of the corresponding control sections in all four counties after the spring thaw. The moisture

Table 2. Test result summary for Winter 2018 and Spring 2019 (average values for five tests in each section).

Section		Avg. DCP-CBR (%)		Avg. E_{FWD} (MPa)		Avg. γ_d (kN/m^3)		Avg. ω (%)	
		F[1]	S[2]	F	S	F	S	F	S
Washington	OGCS	29.7	114.9	683.9	1,672.3	21.6	21.5	6.0	5.8
	Cement	169.9	115.1	6,547.6	846.7	20.3	20.0	9.2	8.2
	Control	33.9	70.9	645.4	1,162.6	20.2	20.5	7.9	6.3
Howard	OGCS	66.6	44.4	1,728.6	984.4	19.8	19.3	7.4	6.3
	Slag	77.1	43.4	555.8	363.6	26.4	25.3	4.3	3.5
	Control	116.7	39.7	691.9	426.3	20.8	18.9	7.8	6.7
Hamilton	OGCS	48.9	126.7	17,882.7	1,316.5	20.7	22.1	5.7	3.7
	Control	26.0	109.9	1,356.4	189.05	20.2	20.0	9.5	9.1
Cherokee	OGCS	114.4	21.2	1,666.2	388.4	21.1	21.1	7.1	5.2
	Slag	24.7	23.9	1,363.7	257.1	25.04	26.3	3.6	2.4
	Control	23.6	5.2	284.9	120.9	20.5	20.7	9.9	5.4

1 Fall 2018, [2]Spring 2019

contents of all the control sections were found to be slightly higher than all other sections, with the exception of the cement-stabilized surface.

3.4 *Cost comparison*

For all the test sections, the construction and maintenance costs were tracked over the duration of the project. Figure 4 shows a summary of the construction costs.

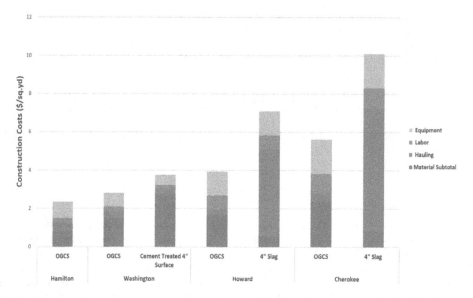

Figure 4. Construction and maintenance costs ($/sq.yd.) for stabilized sections in all four counties.

According to Figure 4, the two steel slag sections had relatively high costs compared to the other sections. The hauling cost was the most significant component for the slag sections, but this cost could be significantly lower for counties located close to a slag source. The cost for Portland cement materials comprised approximately 80% of the construction costs for the cement-stabilized section, and it should be noted that a rental cost was not incurred for the milling machine as it was owned by Washington County. Construction costs for the OGCS sections included hauling costs which depend on distance from the supplier on the eastern border of the state, and several hours of blade-mixing and rolling. The aggregate materials used in the OGCS sections came from the local quarries in each county.

After construction was finished, maintenance records including cost summaries were recorded by county engineers. Details of maintenance costs during the first year incurred by blading and spreading fresh aggregate over the test sections as needed are given in Table 3 (information from Cherokee County was not available). Maintenance costs in each section were minor compared to construction costs. In Hamilton county, the entire cumulative length of test sections was bladed instead of treating them separately. The control section in Howard county had a maintenance cost greater than four times that of the stabilized sections.

Beyond having a reasonable construction cost, to be cost-effective a stabilization method should have relatively good strength and binding properties that can be maintained for a reasonable length of time. The corresponding construction method should be easily achievable using readily-available equipment and materials. Because hauling costs will depend on distance from the stabilizer source, their contribution to the overall construction and maintenance costs should be carefully considered by secondary-road agencies.

Table 3. Construction costs for the four counties (2018-2019).

County	Detail	Section Name	Material Cost ($)	Equipment Cost ($)	Labor Cost ($)	Cost ($/sq. yd.)
Howard	1.42hrs blading	OGCS	NA	99.49	51.37	0.073
	1.9hrs blading	Control	497.60	133.01	68.68	0.340
	0.75hrs blading	4" Phoenix Slag	NA	52.48	27.10	0.077
Washington	3hrs blading	Entire 1 mile	NA	228.45	123.69	0.030
Hamilton	8hrs blading	Entire 1 mile	NA	609.20	254.88	0.060

4 CONCLUSIONS

While most of the stabilized test sections lost surface and subgrade strength and exhibited a decrease in elastic modulus due to effects of freezing and thawing, according to the test results not all counties lost strength after the winter season. The main conclusions with respect to the stabilization methods examined in this paper are as follows:

- The OGCS method exhibited good average strength that actually increased after the spring thaws in Howard and Washington counties, but decreased after spring thawing in Cherokee and Hamilton counties. The method was therefore found to be more effective in counties with clayey subgrades, and experienced significant strength loss in the county with a sandy subgrade. However, in all four counties the average strength and stiffness in the OGCS sections after the spring thaw were greater than those of the nearby control sections. The construction and maintenance costs for this method were also relatively low compared to the other two stabilization methods examined. Overall, while the OGCS method appears to be a cost-effective stabilization alternative easily achievable in Iowa, further observation is required to evaluate this method's long-term performance.
- While the steel slag sections exhibited good strength in both Howard County and Cherokee County, the cost of hauling steel slag can be significant depending on distance from available sources.
- The cement-stabilized surface course treatment exhibited a large increase in average strength that decreased after the spring thaw but remained higher than that of the control section. The stiffness was greatly improved after construction but fell below that of the control section after spring. Based on these conclusions, the 4" cement-stabilized surface is also recommended as a cost-effective alternative.

According to the test results, maintenance logs, and survey photos of the test sections, not all sections suffered from a significant strength decrease after the spring thaw. Based on cost and performance comparisons, the optimized gradation with clay slurry and 4" cement-stabilized surface methods appear to be the most cost-effective alternatives among those examined in this study.

ACKNOWLEDGEMENTS

This research was supported by the Iowa Department of Transportation through IHRB Project TR-721. This support is gratefully acknowledged. The findings and opinions expressed herein are those of the authors and do not necessarily reflect the views of the Iowa DOT. The authors wish to thank County Engineers Nick Rissman, Nicole Stinn, Jacob Thorius, and Brandon Billings as well as their crews for assistance in construction and maintenance of the test sections. The authors also thank the representatives of the slag source and the sand company providing the clay slurry for overseeing and providing input during construction of the related test sections.

REFERENCES

[1] C. Li, P. K. R. Vennapusa, J. Ashlock, and D. J. White, "Mechanistic-based comparisons for freeze-thaw performance of stabilized unpaved roads," *Cold Reg. Sci. Technol.*, vol. 141, no. June, pp. 97–108, 2017.

[2] S. Satvati, J. C. Ashlock, A. Nahvi, C. T. Jahren, B. Cetin, and H. Ceylan, "A Novel Performance-Based Economic Analysis Approach: Case Study of Iowa Low Volume Roads.," in *12th TRB International Conference on Low-Volume Roads*, 2019.

[3] C. Li, J. C. Ashlock, D. J. White, and P. K. R. Vennapusa, "Mechanistic-based comparisons of stabilised base and granular surface layers of low-volume roads," *Int. J. Pavement Eng.*, pp. 1–13, 2017.

[4] A. J. Puppala, *Estimating stiffness of subgrade and unbound materials for pavement design*, vol. 382. Transportation Research Board, 2008.

[5] E. Kausel and J. M. Roesset, "Stiffness matrices for layered soils," *Bull. Seismol. Soc. Am.*, vol. 71, no. 6, pp. 1743–1761, 1981.

[6] M. Mahedi, S. Hossain, M. Faysal, M. S. Khan, and A. Ahmed, "Prediction of Strength and Stiffness Properties of Recycled Pavement Base Materials Using Non-Destructive Impact Echo Test,"sin *International Congress and Exhibition"Sustainable Civil Infrastructures: Innovative Infrastructure Geotechnology,"* 2017, pp. 121–136.

[7] P. K. R. Vennapusa, M. Asce, D. J. White, and M. Asce, "Performance Assessment of Secondary-Roadway Infrastructure in Iowa after 2011 Missouri River Flooding," vol. 3, no. 4, pp. 1–11, 2015.

[8] D. J. White, *Low-Cost Rural Surface Alternatives : Literature Review and Recommendations.* 2013.

[9] P. K. R. Vennapusa, D. J. White, J. Siekmeier, and R. A. Embacher, "In situ mechanistic characterisations of granular pavement foundation layers," *Int. J. Pavement Eng.*, vol. 13, no. 1, pp. 52–67, 2012.

[10] T. liang Wang, Y. jun Liu, H. Yan, and L. Xu, "An experimental study on the mechanical properties of silty soils under repeated freeze-thaw cycles," *Cold Reg. Sci. Technol.*, vol. 112, pp. 51–65, 2015.

[11] A. E. Johnson, "Freeze-thaw performance of pavement foundation materials." Iowa State University, 2012.

[12] I. Qamhia, E. Tutumluer, L. C. Chow, and D. Mishra, "A Framework to Utilize Shear Strength Properties for Evaluating Rutting Potentials of Unbound Aggregate Materials," *Procedia Eng.*, vol. 143, no. Ictg, pp. 911–920, 2016.

[13] E. Tutumluer, I. I. A. Qamhia, and H. Ozer, "Field Performance Evaluations of Sustainable Aggregate By-product Applications," in *Geotechnics for Transportation Infrastructure*, Springer, 2019, pp. 3–23.

[14] J. C. Ashlock, Y. Wu, B. Cetin, H. Ceylan, and C. Li, "Construction of Chemically and Mechanically Stabilized Test Sections to Reduce Freeze–Thaw Damage of Granular Roads," in *12th International Conference on Low-Volume Roads*, 2019, p. 58.

[15] W. H. Chesner, R. J. Collins, M. H. MacKay, and J. Emery, "User guidelines for waste and by-product materials in pavement construction," Recycled Materials Resource Center, 2002.

[16] AASHTO, "AASHTO Design Guide for Pavement Structures," *Am. Assoc. State Highw. Transp. Off. Washington, D.C*, 1993.

[17] M. R. Stokoe, K.H. II, Wright, G.W., James, A.B., and Jose, M. R. "Characterization of geotechnical sites by SASW method," pp. 15–25, 1994.

[18] M. Saltan, V. E. Uz, and B. Aktas, "Artificial neural networks-based backcalculation of the structural properties of a typical flexible pavement," *Neural Comput. Appl.*, vol. 23, no. 6, pp. 1703–1710, 2013.

[19] J. Boussinesq, *Application dès potentiels a l'étude de l'équilibre et du mouvement des solides élastiques*. Gauthier-Villars, 1885.

[20] J. Grasmick, M. Voth, and C. Senseney "Capturing a Layer Response during the Curing of Stabilized Earthwork Using a Multiple Sensor Lightweight Deflectometer," *J. Mater. Civ. Eng.*, vol. 27, no. 6, pp. 1–12, 2014.

[21] N. Odemark, "Investigations as to the elastic properties of soils and design of pavements according to the theory of elasticity," 1949.

[22] C. Li, J. C. Ashlock, D. J. White, and P. K. R. Vennapusa, "Mechanistic-based comparisons of stabilised base and granular surface layers of low-volume roads," *Int. J. Pavement Eng.*, vol. 8436, no. May, pp. 1–13, 2017.

[23] C. Li, J. C. Ashlock, B. Cetin, and Jahren, C. (2018) "Feasibility of Granular Road and Shoulder Recycling", Final Report, IHRB Project TR-685, 182 pp., April 2018.

[24] Henry, S. K., Olson, P. J., Farrington, P. S., & J, Lens,. (2005). *Improved performance of unpaved roads during spring thaw* (No. ERDC/CRREL-TR-05-1). Engineer Research and Development Center, Hanover NH, Cold Regions Research and Engineering Lab.

[25] C. Li, and J. C. Ashlock (2018) *Gradation Optimization for Granular Surface Materials*, MS Excel Spreadsheet, https://intrans.iastate.edu/research/completed/feasibility-of-gravel-road-and-shoulder-recycling/.

[26] ASTM D2487 – 17 (2017). *Standard Practice for Classification of Soils for Engineering Purposes (Unified Soil Classification System)*, ASTM International, West Conshohocken, PA.

[27] ASTM D3282-15 (2015). *Standard Practice for Classification of Soils and Soil-Aggregate Mixtures for Highway Construction Purposes*, ASTM International, West Conshohocken, PA.

[28] ASTM D4318-17e1 (2017). *Standard Test Methods for Liquid, Plastic Limits, and Plasticity Index of Soils*, ASTM International, West Conshohocken, PA.

[29] ASTM D6913/D6913M-17 (2017). *Standard Test Methods for Particle-Size Distribution (Gradation) of Soils Using Sieve Analysis*, ASTM International, West Conshohocken, PA.

[30] ASTM D7928-17 (2017). *Standard Test Method for Particle-Size Distribution (Gradation) of Fine-Grained Soils Using the Sedimentation (Hydrometer) Analysis*, ASTM International, West Conshohocken, PA.

Stabilization and geogrids

Evaluating the cost effectiveness of using various types of stabilized base layers in flexible pavements

A. Francois, D. Offenbacker & Y. Mehta
Center for Research and Education in Advanced Transportation Engineering Systems, Department of Civil Engineering, Rowan University, Glassboro, New Jersey, USA

ABSTRACT: This study was initiated with the aim of determining the cost effectiveness of utilizing various types of stabilizers in the base layer of flexible pavements. Five field sections that were formerly part of a controlled study conducted by Rhode Island Department of Transportation (RIDOT) were evaluated in this study. All field sections contained a 4.5 in. hot mix asphalt (HMA) surface layer and 1 in. granular subbase. Four of the five sections were constructed using stabilized base layers (i.e., calcium chloride ($CaCl_2$), Portland cement, geogrid, and bituminous (asphalt emulsion) stabilized base layers). A control section was constructed using 100% untreated, recycled asphalt pavement (RAP) base layer. All sections were located on Rhode Island Route 165 (between utility poles 304 and 521). Falling weight deflectometer (FWD) tests were conducted on the field sections and the collected deflection data was used to back-calculate the elastic moduli for all layers on each highway segment. AASHTO-Ware Pavement ME Design simulations were conducted to determine the predicted fatigue and rutting performance of the flexible pavement sections. The results of the Pavement ME design simulations were used in conjunction with the RIDOT Pavement Structural Health Index (PSHI) to determine the total life cycle cost of each pavement section. Based on the results of the study it was determined that stabilization of base layers during full depth reclamation were not cost effective because pavement sections that contained stabilized bases had a higher total life cycle cost than the pavement section that contained an untreated base.

Keywords: Back-calculation, full depth reclamation, falling weight deflectometer testing, life cycle cost analysis

1 INTRODUCTION

The stiffness of the base layer in flexible pavement can significantly influence the tensile strains in the hot mix asphalt (HMA) layers and the compressive strains on the subgrade layer of the pavement system. Since the stiffness of the base layer depends on the material properties of that layer, the type of base layer used in a flexible pavement has a direct impact on the overall performance of flexible pavements. Two types of base layers are typically utilized in the flexible pavements: unbound aggregate bases and bound (stabilized) aggregate bases. Unbound aggregate bases consist of unmodified granular aggregates while stabilized bases consist of granular material bounded by a stabilizing agent. The main function of all base layers, regardless their type, is to provide support for the HMA layer(s) and to efficiently distribute traffic loads onto subgrade and/or subbase pavement layers.

The treatment of untreated granular aggregates with stabilizing agents provides bound bases with improved stability because stabilization increases aggregate interlock and facilitates load transfer. However, the actual stabilizing agents themselves contribute very little to the structural capacity of the bound base. There is a variety of stabilizing agents currently used to treat the base layers of flexible pavements. Some of the commonly used stabilizers in bound bases include: foamed asphalt, cutback asphalt, Portland cement concrete (PCC), geogrids, and calcium chloride ($CaCl_2$). The increased aggregate interlock provided by these stabilization agents enables load transfer in bound bases to be more efficient than the load transfer in unbound bases. There is a general consensus among researchers that the use of stabilizing agents in the base layer of flexible pavements improves the overall performance of that layer (Hungener et al. 2015, Wang et al. 2011, Abu-Farsakh et al. 2011, Tao et al. 2015, Tang et al. 2008, and Ogundipe, 2014). However, there is limited insight on how different types of stabilized base layers compare in terms of their impact on the overall performance of flexible pavements and their overall cost effectiveness.

2 GOAL

Therefore, the goal of this study is to evaluate the performance and cost-effectiveness of using stabilized bases in HMA pavement systems. For this purpose, a life cycle cost analysis was conducted on five field sections each having a respective base material—unbound recycled asphalt pavement (RAP) base, calcium chloride stabilized base, bituminous stabilized base, Portland cement stabilized base, and geogrid stabilized base. Falling weight deflectometer (FWD) testing was conducted to determine the layer moduli of the pavement sections and Pavement ME design simulations were performed to evaluate the predicted performance of the sections. The results of the Pavement ME Design simulations were subsequently utilized to assess the relative condition of the pavement sections.

3 FIELD SECTION DESCRIPTION

Five field sections located on Route 165 (between utility poles 304 and 521) in Rhode Island (RI) were evaluated in this study. The sections were part of a controlled study by Rhode Island Department of Transportation (RIDOT) which, evaluated the long-term field performance of these sections. Four of the five sections were constructed using stabilized base layers and one was constructed as a control section using an unmodified, RAP aggregates base. Four different stabilizing agents (i.e., calcium chloride, Portland cement, geogrids, and emulsified asphalt) were utilized to construct the four stabilized base layers.

Figure 1 presents the pavement structure utilized in all five sections analyzed in this study. All sections contained a 114.3 mm HMA layer that consisted of a 50.8 mm surface course and 63.5 mm intermediate course. The sections also contained a 203.2 mm, thick base layer, and 203.2 mm crushed gravel subbase layer. Additionally, the gradation of the base layers of all the pavement sections were similar. That is, 95 to 100% passing the No. 3 sieve (6.7 mm) and 2% to 15% passing the No. 200 sieve (75 µm). During construction, all field sections underwent full depth reclamation (FDR). The section which contained the $CaCl_2$ treated base, was stabilized using a $CaCl_2$ solution (35 % alkali chloride, 2% sodium chloride, and 0.1 % magnesium chloride). The PCC, stabilized base was constructed by uniformly blending 17.4 kg/m^2 of dry cement with pulverized RAP material. The geogrid stabilized base was mechanically reinforced at the interface of the base and subbase using polypropylene geogrids with triangular apertures. The bituminous stabilized base was treated with 3.62 L/m^2 of an anionic medium to rapid setting (HFMS-2) asphalt emulsion.

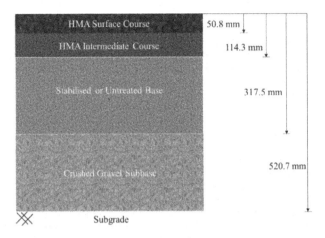

Figure 1. Pavement cross-section of five field sections.

4 TESTING AND EVALUATION PLAN

The testing plan adopted in this study to assess the cost effectiveness of the stabilized bases evaluated in this study involved: FWD testing, Pavement ME Design simulations and pavement condition assessment. FWD tests were conducted on all five field sections approximately one year after they were originally constructed. At each test location on the sections, four drop heights that corresponded to target loads of 29.4 kN, 41.2 kN, 58.8 kN, and 78.4 kN, respectively were utilized. Geophones were spaced at 0, 200 mm, 300 mm, 450 mm, 600 mm, 900 mm, and 1200 mm, respectively from the applied load. The collected pavement deflection data was then utilized to back-calculate the elastic moduli for all layers in all pavement sections. The back-calculation process was performed using the BAKFAA software which is a computer program developed and used by the Federal Aviation Administration (FAA). In this computer program, initial "seed" values for the moduli of respective pavement layers are used as inputs in BAKFAA. Deflections are then calculated by a subroutine program which runs a static-linear elastic analysis using concepts from the multilayered elastic theory. During this process, the original seed layer moduli values are altered using an iterative method in which the root mean squared error between computed and measured deflections is minimized by matching the radius of curvature of the measured and calculated deflection basins.

The Pavement ME Design simulations entailed the use of level 3 inputs. Table 1 presents a summary of all inputs utilized in conducting the Pavement ME Design simulations. Data collected from weigh in motion (WIM) stations that were in close proximity to Route 165 was used to determine a vehicle class distribution that was representative of traffic on Route 165. The back-calculated layer moduli were utilized as inputs in the Pavement ME Design simulations along with the layer thicknesses presented in Figure 1. Climate data obtained from a weather station located in Providence, RI were also used as inputs in the Pavement ME Design simulations. The predicted rutting and fatigue cracking performance of the five field sections were then employed to determine the pavement condition of the sections.

The results of the Pavement ME design simulations were used in conjunction with the RIDOT Pavement Structural Health Index (PSHI) to determine the change in condition of each of the pavement sections. The Pavement Structural Health Index is a tool used to assess and monitor the condition of flexible pavements. It is a weighted average that is determined from a comprehensive scoring system which, accounts for distresses such as: International Roughness Index (IRI), rutting, cracking (i.e. longitudinal, transverse, alligator, and block), and patch failure. The weight distribution of the pavement distresses accounted for in the

RIDOT PSHI scoring system are presented in Table 2. The scoring system for each pavement distress typically ranges from 0 to 100 and the score depends on value of the measured distress. For instance, the distress score for IRI, rutting and cracking is reliant on whether the measured roughness, cracking and rutting falls within a particular range of values. The change in PSHI over time was subsequently used to determine the total life cycle cost of each pavement section.

Table 1. Pavement ME Design Inputs.

Vehicle Class	Distribution (%)	Other Inputs	
Class 4	06.62	AADT	5800
Class 5	68.59	AADTT	240
Class 6	09.37	Traffic Growth Rate (%)	1.3
Class 7	01.64	HMA Thickness (mm)	114.3
Class 8	03.41	Base Thickness (mm)	203.2
Class 9	10.10	Subbase Thickness (mm)	203.2
Class 10	00.23	Surface Course HMA Binder Grade	PG 64-28
Class 11	00.00	Intermediate Course HMA Binder Grade	PG 64-22
Class 12	00.00	Base Modulus (kPa)	Varies*
Class 13	00.04	Subbase Modulus (kPa)	Varies*

Table 2. RIDOT Pavement Structural Health Index Scoring System Weight Distribution.

Pavement Distress	Weight Distribution of Distress Scores in RIDOT PSHI Scoring System
Alligator Cracking	16
Longitudinal Cracking	7
Transverse Cracking	7
Block Cracking	10
Patch Failure	20
Total Rutting	10
International Roughness Index (IRI)	30

5 RESULTS AND DISCUSSION

5.1 Back-calculated moduli values

The back-calculated moduli values for all pavement sections considered in this study is shown in Figure 2. Based on the results presented in this figure, the HMA layers in all sections had the highest moduli values. This trend was expected because HMA layers are usually constructed using better-controlled and better-performing materials when compared to other layers. All other layers (i.e., base, subbase, and subgrade) had relatively the same back-calculated moduli values (i.e., approximately 241 MPa) with the exception of the cement stabilized bases layer in Section 3.

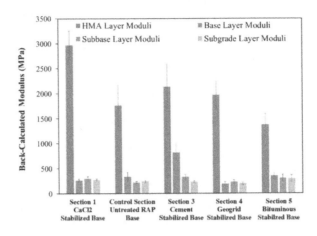

Figure 2. Back-calculated layer moduli of all five pavement sections.

5.2 *Predicted pavement performance*

The results of the Pavement ME Design simulations are shown in Figure 3. The total predicted fatigue cracking in the control section as well as sections 1 ($CaCl_2$) and 5 (bituminous stabilized base) were relatively similar (i.e., approximately 312.5 m/km). These results indicate that the control section as well as sections 1 and 5 have relatively the same susceptibility to fatigue cracking. Section 3, which contained the cement treated base layer, had lower total predicted fatigue cracking value than the control section. This observation implied that the cement treated improved the fatigue life of flexible pavements. The results in Figure 3a also show that section 4 (geogrid stabilized base) may be more prone to fatigue cracking than the control section. This is because the total predicted fatigue cracking for section 4 was approximately 56% higher than that obtained for the control section.

The results of the total predicted rutting for all sections is presented in Figure 3b. The total predicted rutting for the control section as well as sections 1 and 5 were similar. This indicated that the rutting susceptibility of these pavement sections was similar. The results in Figure 4b also show that the total predicted rutting for section 4 (geogrid stabilized base) was 23% higher than that obtained for the control section. This suggested that using geogrids to stabilize aggregate base layers may have a negative impact on flexible pavements' rutting life.

The results of the predicted IRI on all the pavement sections considered in this study is illustrated in Figure 3c. Section 4 (geogrid stabilized base) had the highest IRI after the 20 year analysis period (i.e., 2.23 m/km) followed by sections 1 ($CaCl_2$ stabilized base), 2 (untreated base), and 5 (bituminous stabilized base) which had an IRI of 2.17 m/km. Section 3 (cement stabilized base) had the lowest IRI of all sections after the 20 year analysis period (i.e., 2.15 m/km). Generally, the IRI of sections that contained the $CaCl_2$, bituminous, and cement stabilized bases had almost identical predicted IRI values as the section that contained the untreated base. However, section 4 (geogrid stabilized base) consistently had higher predicted IRI values than the control section throughout the 20 year analysis period. These results suggests that pavement ride quality deteriorated at a faster rate on section 4 when compared to all other sections.

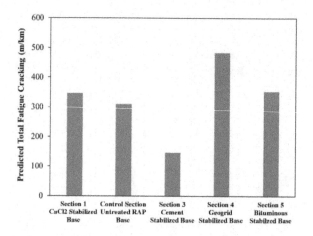

(a)

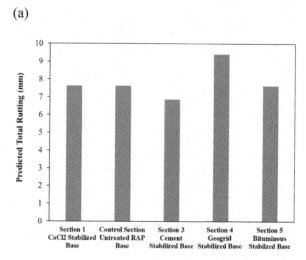

(b)

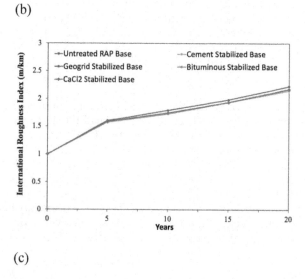

(c)

Figure 3. Pavement ME Design simulation results: (a) predicted fatigue cracking (b) predicted rutting and (c) predicted International Roughness Index.

5.3 Comparison of pavement condition

The PSHI of the five pavement sections evaluated in this study was used to qualitatively classify the pavement condition and compare the life cycle costs of the sections. The pavement condition was defined as excellent, good, fair, poor, or failed based on the PSHI score. PSHI scores typically range from 0 to 100. However, the range of the PSHI scores in this study ranged between 0 and 70 because the scope of the study did not account for block cracking and patch failure. The final PSHI scores were therefore calculated as percentages; with 70 being the total or maximum PSHI score. Pavement condition was classified as excellent when the PSHI scores of the pavement section fell within the 90.5 to 100% range. Pavement condition was classified as good, fair, poor, and failed when the PSHI score fell within 84.1 to 90.5 %, 75.6 to 84.1 %, 64.5 to 75.6 %, and 0 to 64.5%, respectively.

Figure 4 presents the change in PSHI of each pavement section during the 20 year Pavement ME analysis period. The cement stabilized base section underwent the smallest reduction in PSHI (i.e., 100 to 78.4) where pavement condition transitioned from excellent to fair. The geogrid stabilized base section experienced the largest reduction in PSHI (i.e., 100 to 72.4) which reflected a change in pavement condition from excellent to poor. The sections that contained the untreated RAP, $CaCl_2$ stabilized, and bituminous stabilized bases had a relatively similar overall reduction in PSHI throughout the analysis period. These results suggested that the cement stabilized base generally improved the overall performance of the flexible pavement section while the $CaCl_2$ and bituminous stabilization had little to no effect on the overall performance of the flexible pavement section. The results also implied that addition of the geogrid reinforcement negatively impacted on the overall performance of the flexible pavement section.

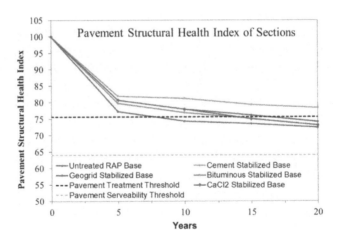

Figure 4. Change in pavement structural health index of pavement sections.

5.4 Comparison of pavement section life cycle costs

A life cycle cost analysis was performed to gain insight about the cost effectiveness of performing FDR of the flexible pavements sections using the stabilizing agents considered in this study. The cost of initial construction and rehabilitation were the only costs considered in the cost analysis. No discounting interest rate was factored in the life cycle costs calculations to maintain a conservative estimate regarding future rehabilitations. The pavement rehabilitation costs was determined using the time required for the PSHI of the pavement sections to reach the pavement treatment threshold during the analysis period. This is because preservation treatments are typically applied to the surface layers of flexible pavements when this occurs. A two-inch mill and overlay was selected as the treatment of choice when the condition of the pavement sections became "poor" because it adds more years of pavement serviceability (5 years) when compared to other treatments used in RI. The total pavement rehabilitation cost for the sections was determined by multiplying the amount of treatment applications required during the 20 year Pavement ME analysis period by the cost of conducting a mill and overlay ($142,000 per lane mile).

The total life cycle cost of each pavement section is presented in Table 3. The total life cycle cost of the pavement sections was determined by adding the initial pavement construction costs to their respective rehabilitation costs over the 20 year analysis period. Based on the results, the geogrid stabilized base section had the highest total life cycle cost of all sections. This is because the pavement section required three mill and overlay interventions in the 8th, 13th, and 18th years of the analysis period. The untreated base section had the lowest total life cycle cost because it had the lowest initial construction costs of all sections and only required one mill and overlay application in the 15th year of the analysis period. The bituminous, $CaCl_2$ and cement stabilized base sections had a relatively similar total life cycle cost which, was slightly higher than that of the control section. This is because the bituminous and $CaCl_2$ stabilized base sections had similar initial construction costs and required two mill and overlay interventions during the analysis period. On the other hand, the cement stabilized base section required no rehabilitation during the analysis period. However, the comparatively higher initial construction cost offset the total life cycle cost of that pavement section. These results suggested that geogrid, $CaCl_2$ bituminous and cement stabilization was not cost effective. This is because the geogrid, $CaCl_2$ and bituminous stabilized base section performed worse than the control section at a slightly higher life cycle cost. In addition, it is noted that the cement stabilized base section provided more years of serviceability than the control section However, the high initial cost of cement stabilization makes it less cost effective than the control section.

Table 3. Total life Cycle cost of Pavement Sections.

Pavement Section	Initial Construction Cost ($ in millions) per lane mile	Total Life Cycle Cost ($ in millions) per lane mile
Untreated RAP Base	3.69	3.84
Bituminous Stabilized Base	3.78	4.06
$CaCl_2$ Stabilized Base	3.80	4.08
Cement Stabilized Base	4.02	4.02
Geogrid Stabilized Base	3.88	4.31

5.5 *Conclusions*

This study was initiated with the aim of determining the cost effectiveness of utilizing various types of stabilizers in the base layer of flexible pavements. Based on the results of this study it was determined that:

- The section that contained the geogrid stabilized base had the highest predicted fatigue cracking and rutting of all sections while the predicted rutting and cracking on the cement stabilized base section was lowest.
- Geogrid, $CaCl_2$, or bituminous stabilization of the base layer did not significantly increase the overall stiffness of that layer. In fact, full depth reclamation with these stabilizing

agents caused the pavement sections to have less years of serviceability than the control section, at a higher life cycle cost.
- Cement stabilization of the base layer during full depth reclamation is also not a cost effective option during full depth reclamation. This is because the high initial construction cost of cement stabilization makes it less cost effective than untreated bases even though it provides more years of serviceability.

REFERENCES

Abu-Farsakh, M., Souci, G., Voyiadjis, G. and Chen, Q. 2012. *Evaluation of Factors Affecting the Performance of Geogrid Reinforced Granular Base Material Using Repeated Load Triaxial Tests*. Journal of Materials in Civil Engineering, Vol 2, No. 2, pp.72–83.

Hungener, M., Partl, M. and Morant M., 2014. *Cold Asphalt Recycling with 100% Reclaimed Asphalt Pavement and Vegetable Oil-based Rejuvenators*. Road Material and Pavement Design, Vol.15, No. 2, pp. 239–258.

Ogundipe, O. 2014. *Strength and Compaction Characteristics of Bitumen-Stabilized Granular Soil*. International Journal of Scientific & Technology Research. Vol. 3, No. 9,

Tang, X., Chehab, G. and Palomino, A. 2008. *Evaluation of Geogrids for Stabilizing Weak Pavement Subgrade*. International Journal of Pavement Engineering. Vol. 9, No 6, pp. 413–429.

Tao, M., Wang, H., Zhao, Y. and Huang, X. 2015. *Laboratory Investigation on Residual Strength of Reclaimed Asphalt Mixture for Cold Recycling*. International Journal of Pavement Research and Technology. Vol. 8, No. 1.

Wang, Y., Ma, X. and Sun, Z. 2010. *Shrinkage Performance of Cement-Treated Macadam Base Materials*. Proceedings of Traffic and Transportation Studies, pp. 1378–1386.

Measurement of modulus of subgrade reaction for geogrid stabilized roadways

Prajwol Tamrakar, Mark H. Wayne & Garrett Fountain
Tensar International Corporation, Alpharetta, GA, USA

David J. White & Pavana Vennapusa
Ingios Geotechnics, Northfield, MN, USA

ABSTRACT: Performance of rigid pavements depends primarily upon uniformity in the slab support condition and a composite response of unbound aggregate layers and subgrade. Several pavement design and analysis methods (e.g., AASHTO, PCA) consider modulus of subgrade reaction (k-value) as one of the key parameters representing pavement foundation strength. The k-value is represented by the reaction pressure required for per unit settlement and is directly influenced by the characteristics of unbound pavement layers and subgrade strength. AASHTO T222 provides guidance for measuring k-value in the field for ensuring that the compacted aggregate layers have achieved the target stiffness. However, such a method of measuring in-situ k-value is not common in practice. As an alternative, some empirical approaches are available for estimating "theoretical" k-value that include back-calculating FWD deflection data or correlating with material stiffness parameters. This paper presents k-value test results from static plate load testing on geogrid stabilized and unstabilized roadways at two sites. Two types of multi-axial geogrid were selected for constructing two separate stabilized aggregate layers over the existing subgrade. An Automated Plate Load Testing (APLT) system was used for conducting the static plate load tests to measure k-value, by following the AASHTO T222 protocol. At the first site, the aggregate base course (ABC) thicknesses of the stabilized and unstabilized sections were 125 mm and 287 mm, respectively. Similarly, the ABC thicknesses of the stabilized and unstabilized sections were 118 mm and 212 mm at the second site. Despite this reduction in thickness, the geogrid stabilized sections exhibited higher k-values in comparison to the unstabilized sections.

Keywords: Geogrid, stabilization, k-value, rigid Pavement, AASHTO T222

1 INTRODUCTION

Rigid pavements (e.g., cement concrete slab) are different from flexible pavements (e.g., asphalt concrete layer) in terms of the durability of the material, method of construction and distribution of vehicle loads to the underlying pavement layers. In the rigid pavement, the vehicle loads are distributed in a wider area and at the shallow depth through the concrete slab. On the contrary, the loads are distributed narrowly and deeply under the asphalt concrete layer in flexible pavements. Further, the stiffness or rigidity of cement concrete is significantly higher than that of asphalt concrete. Therefore, the performance of rigid pavements depends primarily upon uniformity in the slab support condition and a composite response of unbound aggregate layers and subgrade (White et al. 2004).

DOI: 10.1201/9781003222880-16

Rigid pavements are commonly represented by a slab-spring model where the interaction between the slab and foundation is characterized by the spring component (Westergaard 1926). Winkler (1867) proposed such a model consisting of a linear elastic spring. Several other researchers had contributed to improve the slab-spring model by considering the variants such as shear interactions of slab-foundation interface (Shi et al. 1994), realistic contact stresses and the nonlinear stress-dependent elastoplastic characteristics of unbound pavement materials (Farouk and Farouk 2014), cross anisotropic behavior of aggregate base course (Saha et al. 2019).

Although several sophisticated slab-spring models are available to date, most pavement design and analysis methods (e.g., AASHTO, PCA) depend on the modulus of subgrade reaction (k-value) based on the Winkler model. The k-value is considered as one of the key parameters representing pavement foundation strength and is represented by the reaction pressure required for per unit settlement which is directly influenced by the characteristics of unbound pavement layers and subgrade strength. AASHTO T222 provides guidance for measuring k-value in the field for ensuring that the compacted aggregate layers have achieved the target stiffness. However, such a method of measuring in-situ k-value is not common in practice. As an alternative, some empirical approaches are available for estimating "theoretical" k-value that include back-calculating FWD deflection data or correlating with material stiffness parameters (Darter et al. 1995).

This paper presents a direct measurement of the k-value for geogrid stabilized and unstabilized roadways. Two types of multi-axial geogrid were selected for constructing two separate stabilized aggregate layers over the existing subgrade. An Automated Plate Load Testing system was used for measuring k-value by following the AASHTO T222 protocol.

2 BASE COURSE STABILIZATION

Base course stabilization is the process of constructing a mechanically stabilized layer (MSL) with the use of geogrid (see Figure 1). The term "stabilization" is different from "reinforcement" because the former one is referring to "stiffness enhancement" as well as "stiffness retention for a longer period." In contrast, the term "reinforcement" implies "adding force" (Giroud and Han 2016). This mechanism is only effective if the forces are large which in turn implies that large vertical permanent deformations exist in the aggregate overlying the geogrid.

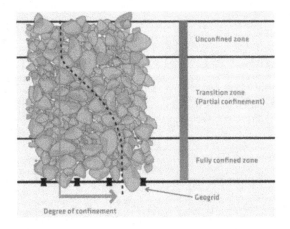

Figure 1. Confinement Zones in an MSL.

When a geogrid is incorporated into a granular material, the aggregate particles interlock with the geogrid and are confined within the geogrid apertures. As a result, the particles are restrained from moving laterally. The lateral restraint provided by the geogrid contributes to reducing induced strain due to traffic loading and thereby increases the stiffness of the granular layer (Sun et al. 2018; Wayne et al. 2013, 2019; Rakowski and Kawalec 2020). In addition to the lateral restraint, granular particles immediately adjacent to the interlocked particles are themselves restrained by the particle to particle interlock. Thus, the influence of the geogrid inclusion extends beyond the geogrid particle interface. Cook and Horvat (2014) demonstrated the existence of such variation in particle interlocking using a multi-level shear box. The authors found that the influence of the geogrid on the stiffness of a granular layer will decrease in relation to the distance from the geogrid. This can be represented as zones of confinement from fully confined to unconfined as illustrated in Figure 1. Behaviors of confinement due to different geogrid and material types were also investigated through the use of numerical modeling (Jas et al. 2015; Konietzky et al. 2004; Lees 2017; McDowell et al. 2006; Stahl et al. 2014), bender element shear wave measurement technology (Kang et al. 2020; Kim et al. 2020) and measurement of aggregate particle motion using SmartRocks (Liu et al. 2017).

3 AASHTO T222 TESTING

Static plate load tests were conducted in accordance with AASHTO T222 (2012) to determine the modulus of subgrade reaction (k-value) using a 762 mm diameter loading plate setup shown in Figure 2b. The plate stress was applied up to 103.4 kPa at 17.2 kPa increments. Plate deformations were measured and the uncorrected k-value was determined using Equation 1.

$$k'_u = \frac{69 \; kPa}{\delta_o} \tag{1}$$

Figure 2. Automated Plate Load Testing (APLT) Setup.

where, k'_u = uncorrected modulus of subgrade reaction (kPa/mm), δ_o = deformation corresponding to the 69 kPa loading stage (mm). The k'_u value was then corrected for plate bending to determine the corrected modulus of subgrade reaction (k_u) following the procedure described in the test standard and Equation 2 for $k'_u \geq 27.14$ kPa/mm and ≤ 271.44 kPa/mm (note that the unit of k_u and k'_u in Equation 2 is pci). However, the correction for saturation was not performed in this study.

$$k_u = -39.91 + 5.50 \times [k'_u]^{0.702} \qquad (2)$$

After the first load sequence, a reload sequence was applied in three loading steps. The reloading curve was used to determine the reload modulus of subgrade reaction (k'_{ur}) at 69 kPa. The k'_{ur} value was then corrected for plate bending to determine the corrected reload k-value (k_{ur}) using Equation 2.

4 AUTOMATED PLATE LOAD TESTING (APLT)

The APLT system (see Figure 2) consists of an advanced electronic-hydraulic control system for applying cyclic and static load pulses through circular steel plates and high-resolution sensors for measuring vertical ground displacements (White and Vennapusa 2017). Compared to FWD, the APLT has the advantage of applying a conditioning loading prior to testing and measuring peak, resilient and permanent deformations for each loading cycle. The APLT system had been used for developing permanent deformation models for geogrid stabilized pavement (White and Vennapusa 2017; Tamrakar et al. 2019) and measuring composite and layer-specific moduli for various pavement layers (Vennapusa et al. 2018). In this study, the APLT system with static stresses was used for measuring k-values by following the AASHTO T222 protocol. However, the other studies had utilized the APLT system for measuring the stress-dependent properties of unbound aggregate layers by using repetitive loads with various stress levels (Tamrakar et al. 2019) or loading frequencies (Tamrakar et al. 2021). The load sequences used for measuring k-values are shown in Table 1.

Table 1. Load Sequences for APLT.

Load Sequences	Stage	Load Step	Target Applied Load (kN)	Target Applied Stress (kPa)
0	Seating	0	3.14	6.89
1	Load	1	7.86	17.24
	Load	2	15.72	34.47
	Load	3	23.58	51.71
	Load	4	31.44	68.95
	Load	5	39.30	86.18
	Load	6	47.16	103.42
	Unload	7	31.44	68.95
	Unload	8	15.72	34.47
	Unload	9	7.86	17.24
2	Load	10	15.72	34.47
	Load	11	31.44	68.95
	Load	12	47.16	103.42
	Unload	13	7.86	17.24
	Unload	14	0.00	0.00

5 PROJECT DETAILS

The project sites were located in Los Angeles and Santa Clarita, CA. The first project site (Site 1) was located at an intermodal facility and the second site (Site 2) was located on a pavement section of Interstate 5. For Site 1, the geogrid stabilized section consisted of 125-mm-thick aggregate base course (ABC) and the unstabilized section consisted of 287-mm-thick ABC (see Figure 3). For Site 2, the stabilized pavement section consisted of 118-mm-thick ABC and the unstabilized section consisted of 212-mm-thick ABC. The ABC for both sites was a crushed miscellaneous base consisting of concrete, asphalt and aggregate. For Site 1, the material consisted of a maximum particle size of 37 mm with about 5% fines (particle passing sieve size 200). For Site 2, the material consisted of a maximum particle size of 25 mm with about 10% fines. Other properties of ABC are reported in Table 2. The stabilized sections were prepared by placing a layer of geogrid at the ABC/subgrade interface. The properties of multi-axial geogrids are reported in Table 3. Both sites consisted of considerably firm foundations. The subgrade CBR values, estimated from Dynamic Cone Penetrometer (DCP) test, were higher than 100 for Site 1 and 30 for Site 2 (see Table 2).

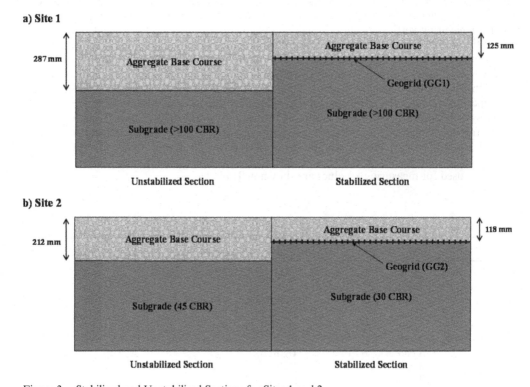

Figure 3. Stabilized and Unstabilized Sections for Sites 1 and 2.

6 PRESENTATION OF RESULTS

Figure 4 shows average plate deformation versus applied stress for the two loading cycles using the loading sequences described in Table 1 for Sites 1 and 2. The first load sequence consisted of applying load up to 103 kPa, and unloading the plate to achieve 17 kPa contact stress. And, the second load sequence consisted of reloading up to 103 kPa, and complete

Table 2. Section Details.

Project Site			Site 1	Site 2
Location			Los Angeles	Santa Clarita
ABC thickness (mm)		Unstabilized	287	212
		Stabilized	125	118
ABC Properties	Material Type		Crushed Miscellaneous Base consisting of concrete, asphalt and aggregate	
	Particle size distribution (mm)	D10	0.27	0.07
		D30	2.45	0.57
		D50	7.33	2.20
		D60	10.16	4.17
		D85	21.65	11.28
	Maximum size (mm)		37	25
	Soil classification	AASHTO	A-1-a	A-1-a
		USCS	GW	SW
	Fines content		5%	10%
Existing subgrade CBR (%)		Unstabilized	>100	45
		Stabilized	>100	30

Table 3. Summary of Geogrid Properties.

Parameters	GG1	GG2
Product	TX8	TX130s
Rib shape	Rectangular	Rectangular
Aperture shape	Triangular	Triangular
Rib pitch (mm)	33	33
Mid-rib depth (mm)	$1.6^d, 1.2^t$	–
Mid-rib width (mm)	$0.4^d, 0.7^t$	–

^{d} *Diagonal direction;* ^{t} *Transverse Direction*

unloading (i.e., achieving 0 kPa contact stress). Results indicated that the rate of increase in deformation is lower for the stabilized section than that for the unstabilized section (see Figure 4), even though the unstabilized section had a thicker ABC than the stabilized section. Further, at the end of the unloading cycle, the unstabilized section showed more deformation than that of the stabilized section. The unstabilized section exhibited 1.76 and 1.51 times more deformation than the stabilized section for Site 1 and Site 2, respectively.

For the second loading cycle, the rate of increase in the deformation of both stabilized and unstabilized sections was similar which are illustrated in Figure 5 and Figure 6. However, there is a difference in deformation between the first and second loading cycles at various applied stresses. The difference in deformation is higher at lower stress. Further, such deformation difference is higher for the unstabilized section than the stabilized section for both sites. Table 4 provides a summary of deformation occurring at various applied stresses.

Table 5 summarizes the results (k values) from the first and second loading cycles. For site 1, k_u for stabilized and unstabilized sections were 107 kPa/mm and 75 kPa/mm. Similarly, k_u for stabilized and unstabilized sections were 54 kPa/mm and 42 kPa/mm for site 2. Despite the difference in the ABC thickness, the stabilized section showed a higher k_u (a.k.a. modulus of subgrade reaction or k-value) than that for the unstabilized section. Table 6 provides a summary of the modulus of subgrade reaction values, including k-value ratios, for Site 1 and Site 2. The results indicate that the stabilized section having 56% thinner ABC than that of the unstabilized section demonstrates 43% increase in k-value for site 1. Similarly, for Site 2, the stabilized section shows 28% higher k-value despite 44% less ABC thickness.

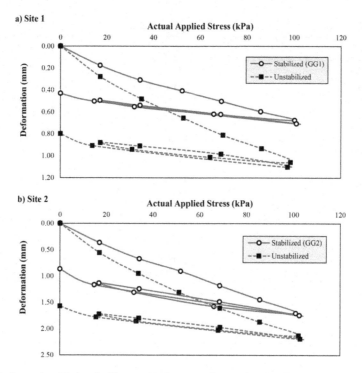

Figure 4. Deformation Under the Plate at Various Loading and Unloading Cycles.

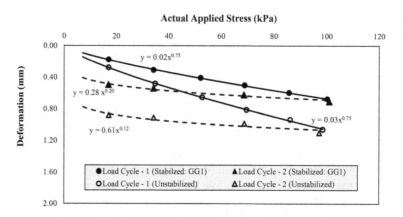

Figure 5. First and Second Loading Cycle for Site 1.

The reload modulus of subgrade reaction (k_{ur}) values, and the ratio k_{ur}/k_u are provided in Table 5. The ratio provides the level of "compactability" of the material and degree of achievable stiffness of the pavement foundation layers. The lower ratio means a better compaction level was achieved and a lower potential improvement in stiffness with additional compaction effort. Hence, this indicates the material is stiffer and more elastic for a better construction platform. Some European specifications (e.g., ATB VÄG 2005) recommend the ratio of 300 mm diameter two-cycle quasi static plate load tests to be ≤ 2.8 for concrete pavement foundation layers in the top 250 mm below the pavement and ≤ 3.5 for the underlying 250 to 500 mm. The higher k_{ur}/k_r ratios for unstabilized sections presented in Table 5 suggest that the

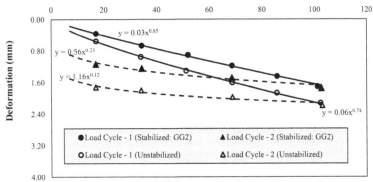

Figure 6. First and Second Loading Cycle for Site 2.

Table 4. Deformation at Various Applied Stresses.

	Applied Stress	Deformation (mm) for Stabilized			Deformation (mm) for Unstabilized			Ratio=
	kPa	LC 1	LC 2	Δ_{st}	LC 1	LC 2	Δ_{un}	Δ_{un}/Δ_{st}
Site 1	20	0.20	0.50	0.30	0.32	0.88	0.56	1.85
	40	0.33	0.57	0.24	0.54	0.95	0.42	1.72
	60	0.45	0.62	0.17	0.73	1.00	0.27	1.58
Site 2	20	0.42	1.14	0.73	0.63	1.68	1.05	1.44
	40	0.75	1.34	0.59	1.06	1.83	0.76	1.29
	60	1.06	1.48	0.41	1.44	1.92	0.48	1.16

Note: LC=Loading Cycle; Δ=Difference in deformation of LC1 and LC2

Table 5. Summary of Results from the First and Second Loading Cycles.

Loading	Parameters	Site 1		Site 2	
		Stabilized	Unstabilized	Stabilized	Unstabilized
First Loading Cycle	Deformation (mm)	0.50	0.80	1.18	1.60
	k_u (kPa/mm)	107	75	54	42
Second Loading Cycle	Deformation (mm)	0.17	0.11	0.45	0.31
	k_{ur} (kPa/mm)	245	325	116	155
	k_{ur}/k_u	2.29	4.35	2.13	3.73

Table 6. Modulus of Subgrade Reaction (k-value).

	Stabilized		Unstabilized		
	ABC Thickness (mm)	k-value (kPa/mm)	ABC Thickness (mm)	k-value (kPa/mm)	Ratio*
Site 1	125	107	287	75	1.43
Site 2	118	54	212	42	1.28

* *Ratio= k-value$_{stabilized}$/k-value$_{unstabilized}$ which indicates the improvement in the k-value with geogrid stabilization*

stiffness of the tested pavement foundation layer could have been improved with additional compaction effort prior to testing.

7 SUMMARY AND CONCLUSION

Field studies, using an Automated Plate Load Testing (APLT) system, were performed over geogrid stabilized and unstabilized sections at two sites. For Site 1, the aggregate base course (ABC) thicknesses of the stabilized and unstabilized sections were 125 mm and 287 mm, respectively. Similarly, the ABC thicknesses of the stabilized and unstabilized sections were 118 mm and 212 mm for Site 2. The stabilized section consisted of a layer of multi-axial geogrid placed at the interface of ABC/subgrade. Static loads consisting of loading and unloading cycles were applied in accordance with the AASHTO T222 protocol for determining the modulus of subgrade reaction (k-value) of stabilized and unstabilized sections.

The key findings of this study are:

- The deformation under the plate increased with the increase in the applied stress. For the first loading cycle, the rate of increase in deformation was higher for the unstabilized section.
- At the end of the first unloading cycles, the unstabilized sections exhibited 1.76 and 1.51 times more deformation than the stabilized sections for Site 1 and Site 2.
- The change in deformation between the first and second cycles with increasing stress was more for the unstabilized section than that for the stabilized section. Therefore, the stabilized section is well compacted and has more uniform stiffness distribution. For this reason, aggregate bases underlain by a multi-axial geogrid could be considered very erosion resistant for the design of rigid pavements.
- The geogrid stabilized sections exhibited a higher k-value than the unstabilized sections despite the reduction in ABC thickness.

REFERENCES

AASHTO T222. (2012). *Standard method of test for nonrepetitive static plate load test of soils and flexible pavement components for use in evaluation and design of airport and highway pavements.* American Association of State Highway and Transportation Officials, Washington DC.

ATB VAG. (2005). *Technical Specifications for Roads.* Swedish Road Administration, Borlänge, Sweden.

Cook, J., and Horvat, F. (2014). "Assessment of particle confinement within a mechanically stabilised layer." *10th International Conference on Geosynthetics. Berlin.*

Darter, M. I., Hall, K. T., and Kuo, C.-M. (1995). *Support under Portland cement concrete pavements.*

Farouk, H., and Farouk, M. (2014). "Calculation of subgrade reaction modulus considering the footing-soil system rigidity." *Vulnerability, Uncertainty, and Risk: Quantification, Mitigation, and Management,* 2498–2507.

Giroud, J. P., and Han, J. (2016). "Mechanisms governing the performance of unpaved roads incorporating geosynthetics." *Geosynthetics,* 34(1).

Jas, H., Stahl, M., te Kamp, L., Konietzky, H., and Oliver, T. (2015). "Discrete element simulation: Modelling and analysis of a geogrid stabilized sub-base while loaded with a moving wheel." *XVI European Conference on Soil Mechanics and Geotechnical Engineering.*

Kang, M., Kim, J. H., Qamhia, I. I. A., Tutumluer, E., and Wayne, M. H. (2020). "Geogrid Stabilization of Unbound Aggregates Evaluated Through Bender Element Shear Wave Measurement in Repeated Load Triaxial Testing." *Transportation Research Record,* SAGE Publications Sage CA: Los Angeles, CA, 2674(3), 113–125.

Kim, J. H., Kang, M., Byun, Y.-H., Qamhia, I. I. A., Tutumluer, E., and Wayne, M. H. (2020). "Bender Element Shear Wave Measurement Based Local Stiffness Characteristics Related to Permanent Deformation Behavior of Geogrid-Stabilized Aggregate Specimens." *Geo-Congress 2020: Geotechnical Earthquake Engineering and Special Topics,* 517–526.

Konietzky, H., te Kamp, L., Groeger, T., and Jenner, C. (2004). "Use of DEM to model the interlocking effect of geogrids under static and cyclic loading." *Numerical modeling in micromechanics via particle methods*, AA Balkema, Rotterdam, 3–12.

Lees, A. (2017). "Simulation of geogrid stabilization by finite element analysis." *Proceedings of 19th International Conference on Soil Mechanics and Geotechnical Engineering, Seoul*, 1377–1380.

Liu, S., Huang, H., Qiu, T., and Kwon, J. (2017). "Comparative evaluation of particle movement in a ballast track structure stabilized with biaxial and multiaxial geogrids." *Transportation Research Record*, SAGE Publications Sage CA: Los Angeles, CA, 2607(1), 15–23.

McDowell, G. R., Harireche, O., Konietzky, H., Brown, S. F., and Thom, N. H. (2006). "Discrete element modelling of geogrid-reinforced aggregates." *Proceedings of the Institution of Civil Engineers-Geotechnical Engineering*, Thomas Telford Ltd, 159(1), 35–48.

Rakowski, Z., and Kawalec, J. (2020). "The Technology of Mechanically Stabilized Layers for Road Structures in Cold Regions." *Transportation Soil Engineering in Cold Regions, Volume 2*, Springer, 63–70.

Saha, S., Gu, F., Luo, X., and Lytton, R. L. (2019). "Development of a modulus of subgrade reaction model to improve slab-base interface bond sensitivity." *International Journal of Pavement Engineering*, Taylor & Francis, 1–12.

Shi, X. P., Tan, S. A., and Fwa, T. F. (1994). "Rectangular thick plate with free edges on Pasternak foundation." *Journal of engineering mechanics*, American Society of Civil Engineers, 120(5), 971–988.

Stahl, M., Konietzky, H., Te Kamp, L., and Jas, H. (2014). "Discrete element simulation of geogrid-stabilised soil." *Acta Geotechnica*, Springer, 9(6), 1073–1084.

Sun, X., Han, J., Parsons, R. L., and Thakur, J. (2018). "Equivalent California Bearing Ratios of Multiaxial Geogrid-Stabilized Aggregates over Weak Subgrade." *Journal of Materials in Civil Engineering*, American Society of Civil Engineers, 30(11), 4018284.

Tamrakar, P., Wayne, M. H., and White, D. J. (2019). "Permanent and Resilient Deformation Behavior of Geogrid-Stabilized and Unstabilized Pavement Bases." *Geo-Structural Aspects of Pavements, Railways and Airfield*.

Tamrakar, P., Wayne, M. H., White, D. J., and Vennapusa, P. K. R. (2021). "In Situ Assessment of Geogrid Stabilized Flexible Pavement Using Automated Plate Load Testing." *Geosynthetics Conference*.

Vennapusa, P. K. R., White, D. J., Wayne, M. H., Kwon, J., Galindo, A., and Garcia, L. (2018). "In situ performance verification of geogrid-stabilized aggregate layer: Route-39 El Carbón–Bonito Oriental, Honduras case study." *International Journal of Pavement Engineering*, Taylor & Francis, 1–12.

Wayne, M., Fraser, I., Reall, B., and Kwon, J. (2013). "Performance verification of a geogrid mechanically stabilized layer." *The 18th International Conference on Soil Mechanics and Geotechnical Engineering, Paris*, 1381–1384.

Wayne, M. H., Fountain, G., Kwon, J., and Tamrakar, P. (2019). "Impact of Geogrids on Concrete Highway Pavement Performance." *Geosynthetics Conference*.

Westergaard, H. M. (1926). "Stresses in concrete pavements computed by theoretical analysis." *Proceedings of Highway Research Board*, 90–112.

White, D. J., Rupnow, T. D., and Ceylan, H. (2004). "Influence of subgrade/subbase non-uniformity on PCC pavement performance." *Geotechnical engineering for transportation projects*, 1058–1065.

White, D. J., and Vennapusa, P. K. R. (2017). "In situ resilient modulus for geogrid-stabilized aggregate layer: A case study using automated plate load testing." *Transportation Geotechnics*, Transportation Geotechnics, 11, 120–132.

Winkler, E. (1867). *Die Lehre von der Elasticitaet und Festigkeit: mit besonderer Rücksicht auf ihre Anwendung in der Technik für polytechnische Schulen, Bauakademien, Ingenieue, Maschinenbauer, Architecten, etc.* Dominicus.

Waste paper ash as an alternative binder to improve the bearing capacity of road subgrades

J.J. Ceprià & R. Orejana
ACCIONA Construction, Madrid, Spain

R. Miró, A. Martínez, M. Barra, D. Aponte & H. Baloochi
Universitat Politècnica de Catalunya, Barcelona, Spain

ABSTRACT: The interest on alternative hydraulic road binders to replace cement is increasing these days aiming at reducing the carbon footprint of road projects. The PaperChain project, an EU funded project under the H2020 programme, tackles the use of one of these binders, Waste Paper Fly Ash (WPFA), a waste stream coming from the energy recovery of paper rejects and sludge, as alternative binder for subgrade stabilisation and cement-modified subbase layers. WPFA has been extensively tested at laboratory scale and demonstrated at real scale in three field tests covering the three types of stabilised soils recognised in the Spanish Road Regulations, complying with all technical requirements. Nonetheless, the complete replacement of cement by WPFA is a challenge from the design point of view. This paper focuses on the construction and monitoring of one of those cases, specifically the one allocated for the pavement subgrade with the highest bearing capacity. Pairs of cement and WPFA-stabilised laboratory specimens have been tested showing a different hardening pattern and different reaction modulus evolution. Field-testing (load plate tests) and compressive resistance results confirmed this trend. These differences can result in positive effects on cracking development thanks to the slower hardening speed during the first days but the current design values should be reviewed when applied to WPFA, given the notable differences expected over the long term.

Keywords: Waste paper ash, alternative binder, stabilized soil, road subgrade, monitoring

1 INTRODUCTION

The subgrade is a stable and uniform platform for the overlying pavement structure of a road. To meet these properties of stability and uniformity, the subgrade depends on the bearing capacity and the resistance to volumetric changes of the soil used. Subgrades are built, in principle, with the native soils of the road area, but when they do not meet adequate minimums, they must be avoided or improved to ensure good performance of the pavement. Among the most common solutions is the replacement and stabilization of poor quality soils.

The Guide prepared by Jones et al (2010) for the State of California, recognizes 4 types of stabilization: mechanical, cementitious, asphalt and non-traditional with additives. Mechanical stabilization consists of improving the resistance of the structure by compaction, although it can also be done by mixing with other better quality soil. Cementitious stabilization consists of adding cement or lime to increase the bearing capacity of the poor soil (using cement) or to reduce the high plasticity of a clay (using lime). Other alternative cementitious stabilizers are

DOI: 10.1201/9781003222880-17

fly ash, cement kiln dust, lime kiln dust, and ground-granulated blast furnace slag. Stabilization with asphalt binders (emulsions, foamed asphalt and cutbacks) is not so common because it is more often used with well graded soils with a low percentage of fines, with the purpose of building flexible and water resistant platforms. And there are also the non-traditional additives, such as chlorides, polymers, resins, synthetic emulsions and sulfonated oils. Some of them act as dust palliatives and do not provide enough strength to be considered as a soil stabilizer.

In the specific case of fly ash, the object of study in this paper, the pozzolanic reaction when mixed with the soil increases the strength although this increase cannot be generalized since the composition of fly ash is variable depending on its source. There are previous studies on the use of fly ash for road stabilization, thanks to the high content of calcium oxides and silicates that provide pozzolanic properties and increased compressive strength to the soil, such as those conducted by Mulder (1996), Lahtinen (2005), Tuncan (2000), Zhou (2000), Zuber (2013), Ohenoja (2020). It has also been confirmed that thaw resistance and bearing capacity are improved when fly ash is used, Vestin (2012), Zhang (2019), Wang (2016), Arm (2014).

When talking about fly ash stabilization, some researchers like Lahtinen, (2001) classify fly ash (FA) with different abbreviations according to the fuel used. It is usually called CFA for coal fly ash, PFA for peat fly ash, WFA for wood fly ash, and MFA for miscellaneous fly ash (e.g., fly ash from the combustion of mixed fuel like fibre sludge and wood).

One of the sources of fly ash is the paper industry, which currently produces a significant amount of municipal solid waste. While the demand for printed books and newspapers has recently decreased due to digital media substitution, (Hänninen, 2014), the consumption of other types of products, such as tissues or packaging, have greatly increased (Cherian, 2019).

Much of the ash from this industry (WPFA) goes to landfill, causing a significant economic impact for producers. For this reason, the pulp and paper manufacturing sector in Europe has been developing many alternative products in order to achieve a circular economy. Among the most outstanding studies, it is worth mentioning the research program called Environmentally friendly use of non-coal ashes, also known as "the ash programme" developed by the Swedish Thermal Engineering Research Institute (Ribbing, C., 2007), in order to increase knowledge on the by-products of energy producers and their application. The results obtained allowed the elaboration of a handbook for using these ashes in unpaved roads.

The PaperChain project, an EU funded project under the H2020 programme, tackles the use of Waste Paper Fly Ash, a residue coming from the energy recovery of paper rejects and sludge, as alternative binder for subgrade stabilisation and cement-treated pavement layers.

This paper focuses on the construction and monitoring of one of the three Spanish field trials executed in Spain under the Paperchain framework (The paperChain Project, 2017), which consists of a stabilised soil for a subgrade of a paved periurban road.

2 OBJECTIVE

The study aims at describing and construction and monitoring of one of the pilots constructed to demonstrate the technical, environmental and economic feasibility of using WPFA as an alternative hydraulic road binder instead of cement commonly used in the different stabilized soil layers foreseen in the Spanish Road Pavement Catalogue (Ministerio de Fomento, 2003).

This pilot was built in Villamayor de Gállego (Zaragoza, Spain), in a section of a periurban road of 1,000 m long and 5.7 m width with no records of heavy traffic intensity but busy with heavy agricultural vehicles and trucks trafficking to avoid the centre of the village to access the local service area. The solution consisted of a layer stabilized with ash as a substitution of a conventional stabilized soil with cement type 3 (S-EST3, according to the Spanish Road Regulations, Ministerio de Fomento, 2015). In order to provide a more comfortable and at the same time, economical surface, a double bituminous treatment was performed over the improved subgrade, Figure 1.

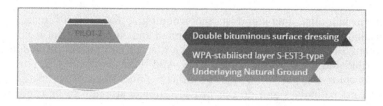

Figure 1. Schematic solution for the pilot built in Villamayor de Gállego.

3 MATERIALS AND JOB MIX FORMULA

The soil was characterised prior to its treatment and then, mixed with different rates of WPFA in order to define the job mix formula. This soil met the specifications to be stabilized according to the Spanish requirements. Maximum dry density from Modified Proctor test (EN 103501) was 2.047 g/cm3 t/m^3 and optimum moisture content was 10%. The WPFA composition is described in section 5.1. The Spanish requirements for the stabilization with cement and the results obtained with the WPFA are collected in Table 1.

At laboratory scale, compressive strength tests allowed to obtain this formula with 5% of WPFA and 8.2% of water, taking into account a different compaction procedure with respect to that used in the cement stabilization. In this case, a 30-minute delay time is necessary to be applied before compaction and the compaction degree must reach the maximum Modified Proctor density, instead of the 98% required for the conventional solution. (Baloochi et al., 2020). Complementary leaching and durability tests (wetting/dry cycles) were also carried out with the objective of ensuring the adequate long-term behaviour of the layer.

A mixture of soil and cement was also designed for the reference section, with 3% of cement type CEM IV/B (Q) 32.5N and 7% of water. These specimens were compacted up to 98% of the corresponding maximum Modified Proctor density and no delay time was applied.

Table 1. Spanish requirements for S-EST3 stabilised soil with cement and results with WPFA.

Characterization	Unit	Standard	S-EST3 (with cement)	Stabilized with WPFA
Unconfined compressive strength at 7 days	MPa	UNE-EN 13286-41	≥ 1.5	≥ 1.5
Density (Modified Proctor)	% of max density	UNE 103501	≥ 98	98
Organic matter	% of mass	UNE 103204	< 1	0.08
Soluble sulphate	% of mass	UNE 103201	< 0.7	0.04
Atterberg limits	-	UNE 103103 + UNE 103104	Liquid Limit ≤ 40 Plasticity Index ≤ 15	Non plastic
Particle size	% pass through #sieve	UNE-EN 933-1	#80mm = 100% #2mm > 20% #0.063mm <35%	#80mm = 100% #2mm = 32% #0.063mm = 9%

4 CONSTRUCTION

Due to the heterogeneity of the materials in the road section, it was necessary to remove 10 cm of the original surface layer and replace it with the borrowed pit soil, which was levelled and compacted simulating an *in situ* soil for S-EST3 stabilisation with cement (Figure 2).

Figure 2. Milling procedure of the original surface and levelling process of the borrow soil.

Then, the experimental field trial was executed following the job mix formula determined in the laboratory. The WPFA was spread out with a dosing machine and mixed with the underlying soil (in a depth of 25 cm) by the stabilizer (Figure 3), after which the compaction and levelling tasks went on. Four passes of a standard vibratory roller compactor were applied and a grader passed to level the surface. Then, a final pass with the roller compactor on static mode was applied to get smoothness. Immediately after the final compaction, a double bituminous treatment was laid out generating a protective surface layer.

Figure 3. Stabilizer and dosing machinery.

In order to evaluate the quality achieved in the section, samples were taken from the mix spread and compacted. Table 2 collects the results of the Unconfined Compressive Strength (UCS) after 7 days of specimens moulded with these samples at 98 % of the maximum dry density (modified proctor test) and the modulus calculated after performing the load plate test on the field trial.

Table 2. Quality control results of the experimental section.

Test	Requirement	Stabilized soil with WPFA
Unconfined compressive strength after 7 curing days (MPa).	≥ 1.5	1.8 (average)
Load plate test at 14 days (diameter 300 mm) as a function of the subgrade type (MPa)	Ev2 > 60 for subgrade E1 Ev2 > 120 for subgrade E2 Ev3 > 300 for subgrade E3	Ev2 > 334 (average) ⇒ subgrade E3

5 MONITORING

5.1 Environmental monitoring

The Environmental Monitoring Plan (EMP) was stated according to the results of a preliminary environmental risk analysis based on the following findings:

a) Determination of the environmental base line, taking samples from vegetation, soils and surficial waters from the road surroundings (Figure 4). Gathering of relevant information about the groundwater and irrigation water quality.
b) Chemical composition of WPFA, paying attention to the resulting concentrations of elements of concern at the defined WPFA dose rates (5 %) in the stabilised soil. The results were compared against the Generic Levels of Reference (GLR) for the soils of Aragon (Departamento de Medioambiente Aragón 2008), in order to assure that the threshold limits for the "urban uses" category were not surpassed.
c) Pathways of WPFA distribution. Air dispersion during WPFA spreading works and at a lesser extent, water transport through runoff along the roadsides during the service life.
d) Potential receivers of the WPFA along the road life cycle, mainly workers during building.

According to these premises, the EMP included sampling and analysis of construction materials (borrowed soil and WPFA), surrounding natural soils, vegetation (dust on leaves) and finally, ground and surficial waters before and after the pilot construction. Samples from the stabilised layer were also taken one month later to determine their leaching properties and the trace elements content. A piezometric network was also built in order to control the leaching properties evolution beneath the stabilised layer (3 piezometers, 2m deep) and three more boreholes were drilled up to 10 m deep aiming at controlling potential impacts in groundwater levels originated by irrigating waters (Ebro River Quaternary alluvial is too distant, 60-70 m below the road surface). Table 3 compiles the material tested, the purpose of the testing and the type of sampling conducted.

Table 3. Summary of the EMP. Sampling and testing strategy.

Material	Purpose	Sampling & testing strategy
WPFA	Comparison of the used WPFA against the historic chemical variation identified.	Total content of elements of concern. Daphnia toxicity test and leaching tests
Borrowed soil	Soil baseline. Chemical characteristics of the untreated soil.	Total element content (elements of concern) and leaching tests according to (EN-12457-4. 2 samples taken before building works.
Stabilised soil	Chemical characterisation of the stabilised layer and its variability.	Concentration of elements of concern. Leaching test according EN-12457-4. 1 sample from the stabilised layer
	Forecast over the long term of the stabilised soil layer, risk assessment.	3 boreholes drilled along the road. 1 sample from each one.
Soil (roadsides and surrounding crop areas)	Wind dispersion of WPFA particles to the nearby soils. Before building (baseline) and after	Total element content. Soil quality control along roadsides (<2 m). 10 samples in total.
Vegetation	Impact control by WPFA dust on the vegetation.2 sampling campaigns before and after the stabilisation works	Total element content. Sampling of most interesting vegetation for humans (food). 2 points per roadside.
Underground waters	Control of the potential affection to the underground waters.	Total elements of concern content in waters. Three control wells up to 10m depth.
Surficial waters (canals)	Impact control of WPFA dust on surficial waters in irrigation canals near the pilot, sampling before (baseline) and after construction.	Total element content Two sampling points (upstream and downstream) in waters from two canals crossing the pilot road.

The analysis of the trace elements concentration in soils and in dust collected from the vegetation were similar before and after construction works, showing no conclusive differences. Antimony and Vanadium increased around 15 %, Aluminium, Cobalt, Copper in 10 % and Manganese, Barium and Zinc raised a 5 %. The rest of elements remained constant in concentration or diminished. Some of the increments (Antimony, Aluminium, Copper, Barium and Zinc) could be consistent with the relative abundance in WPFA compared with natural soils, although in general, they are below the standard deviation, not being conclusive to prove WPFA dispersion occurred during the construction stage by means of the identified pathways.

Figure 4. Sampling soils after works (baseline definition) and sampling of vegetation before.

The total concentrations of these elements were below the threshold limits set for "urban uses" according to the GLR (Department of Environment of Aragon, 2008).

Leaching of the original components of the stabilised layer and the resulting blend were assessed according to EN-12457-4. The results have been compared with the threshold limits for landfilling laid down in the Royal Decree 646/2020. Finally, a representative leachate could be sampled from two piezometers on January 2020 after a heavy rainfall event (60 mm), being possible to classify the material as inert according to the aforementioned Decree (Table 4).

The WPFA-stabilised material complies with the inert waste criteria except for Antimony, however, the leachate collected in the piezometers showed very low metal mobility. All metals remained below the threshold limits for inert waste when compared with the percolation test.

Table 4. Leaching behaviour of borrowed soil, WPFA and stabilised layer compared against the reference values for inert waste according to Royal Decree 646/2020.

Anions/cations (Mg/kg dry mass)	Borrowed Soil (mean)	WPFA	Stabilised Layer (SL)	Cores from SL (mean)[1]	Leachate piezometer (mean)[2]	Reference for inert waste Agitation Percolation[3] L/S=10 l/Kg (mg/l)	
Chlorides	<50	12000	761	911±117	272	800	450
Fluorides	<5	6.2	<5	<5	0.17	10	2.5
Sulphates	<50	62.2	<50	78.5±3.1	629	1000	1500
Antimony	<0.01	0,041	0.19	0.23±0.02	0.009	0.06	0.01
Arsenic	<0.01	<0,05	<0.01	<0.01	0.009	0.5	0.06
Barium	0.335	98	1.4	3.2±2.3	0.05	20	4
Cadmium	<0.01	<0.004	<0.01	<0.01	< 0.001	0.04	0.02
Cobalt	-	-	-	-	< 0.02	-	-
Copper	<0.2	0.26	0.37	0.23±0.04	0.16	2	0.6
Chromium	<0.2	0.13	0.31	0.30±0.02	<0.005	0,5	0.1
Mercury	<0.01	<0.0005	<0.01	<0.01	< 0.0005	0.01	0.02
Molybdenum	<0,2	0,18	<0,2	<0,2	< 0.02	0,5	0.2
Nickel	<0,2	<0,1	<0,2	<0,2	<0.005	0.4	0.12
Lead	<0,2	0,40	<0,2	1.13±0.67	0.018	0.5	0.15
Selenium	<0,05	<0,039	<0,05	0.07±0.01	-	0.1	0.04
Zinc	<0,2	0,21	<0,2	<0,2	0.10	4	1.2

1) Mean of 3 core samples from the stabilised layer. 2) Mean of 2 samples from piezometers. 3) EN 14405.

5.2 Technical monitoring

5.2.1 On-site inspection

Two inspections were performed from October 2018 to February 2020. The first one (6 months after completion) showed no damages related to a WPFA malfunction. Some defects were occasionally observed associated with joints. The pavement was observed in good condition 15 months after completion during the second inspection, although some minute defects like potholes and loose aggregates not related with the stabilization itself were seen (Figure 5). Some transverse cracks were observed every 4 m in the reference section of stabilized soil with cement. In the section with WPFA no cracking and swelling were seen

Figure 5. Cement stabilized stretch (1) and WPFA stretch (2) showing a joint. Bituminous treatment integrity (3) and potholes in the joint (4).

Two years after the road construction, new plate load tests were carried out in the same points showing an average raise of 17 % of the static modulus in the WPFA stretch, up to of 336 Mpa on average. The static modulus of the cement-treated stretch was 711 Mpa, half of the modulus obtained two years before (1500 Mpa), most probably, an anomalous value.

5.2.2 Laboratory testing

The laboratory technical monitoring comprised three types of testing with specimens manufactured with the materials taken from the construction site and under the placement conditions. These specimens were preserved in a climatic chamber for curing and then tested at different ages in order to evaluate: a) the strength evolution of the stabilized soil; b) durability analysis: swelling/shrinkage and c) X-Ray diffraction analysis.

5.2.3 Strength evolution of the stabilised soil

To study the strength evolution of the mixtures of soils and WPFA or cement, specimens were prepared and tested at different ages, according to standards EN 13285-41, after curing them in a moist room with 95% humidity at 20°C.

Considering the mechanical property required by the Spanish Specification, the results show that both stabilized soils, with WPFA and cement, were able to achieve a compressive strength higher than 1.5 MPa after 7 days. In the case of the stabilization with WPFA, the effect of curing time on the compressive resistance was not significant. The soils used for this study had no plasticity but it is important to highlight that the organic matter content of the stabilized soil was considerable high, what could be a potential obstacle for the strength increase.

5.2.4 Durability of the stabilized soil

Durability can be affected when the stabilized soil is exposed to a sulfate source. This source may come from runoff through the pavement surface or from the groundwater if the underlying soil is rich in sulfate. Calcium and aluminum oxides present in the binder (cement or ash) react with water containing sulfate leading to expansion. To carry out the durability analysis of the soil using WPFA, a new experiment was carried out.

The Spanish Specifications for Roads and Bridges (PG3) indicate an analysis of the free swelling of soil prior to the stabilization, according to UNE 103601 standard; however, this type of test is specific for plastic soils with fine particles and so, not appropriate for the soil used in this pilot, which did not show any plastic behavior. Therefore, a novel test was developed for measuring the vertical swelling or shrinkage under horizontally confined conditions when subjected to dryness/humidity cycles as well.

This experiment is carried out to measure the displacement in vertical axis in a confined system using PVC molds. The job mix formula is reproduced with a representative soil sample, 5% WPFA, 8.2% water and the application of a delay time of 30 minutes. After the delay time passes, the material is poured into a cylindrical PVC mold and compacted in 3 layers with a vibrocompactor until reaching the reference density. Then, a metal lid and small balls are placed inside it to have a uniform surface. Then the samples are cured at 20°C and 95% RH for 7 days and after that, put in a bath with water at 20°C to allow water absorption, as shown in Figure 6.

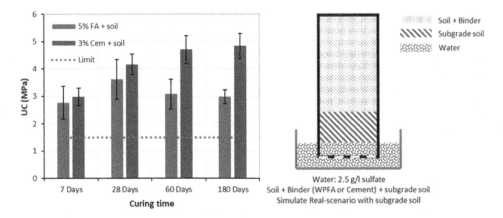

Figure 6. UCS evolution (left). Test scheme for evaluation of swelling or shrinkage (right).

The water in the bath contained a sulfate concentration of 2.5 g/l, simulating on site conditions. The height and weight increases were measured up to 200 days. Figure 7 shows the results obtained for the samples with WPFA and cement, subjected to the aforementioned sulfate concentration solution, placed over a subgrade layer. Despite the weight increase due to water absorption, no swelling was detected and a small shrinkage was measured instead.

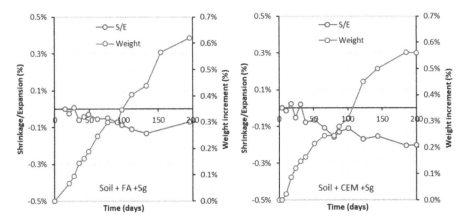

Figure 7. Weight and displacement evolution for samples with FA (left) and cement (right).

5.2.5 *X-ray diffraction analysis of stabilized soil in presence of sulfates*

This study puts in contact the components of the stabilised soil with a sulfate source and analyze by XRD at 14, 30, 90 and 180 days at 20°C the minerals resulting from this interaction which may be damaging for the overall structure of the road pavement. For this purpose, the mixtures of soil, cement and WPFA are ground to a particle size smaller than 63 µm with the aim of accelerating the chemical reactions when they are subjected to the sulfate attack. The sample conditioning uses the same amount of sulfate and same temperature of the previous test.

The slow formation of ettringite (common cause of swell), is observed in WPFA and cement samples. As can be seen in Figure 8, the XRD spectrum intensity in the ettringite angle are similar in both samples. However, perhaps samples with WPFA have higher available Ca and Al, which could foster the reaction. Another reason for the slower ettringite formation rate in the cement specimens could be due to the cement type IV used, which is high in pozzolans, preventing from sulfate attack.

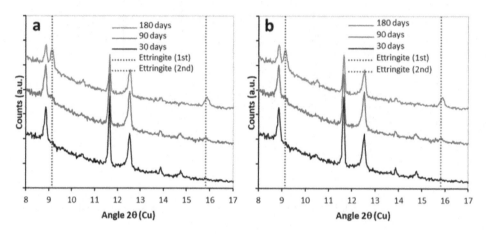

Figure 8. XRD difractograms for samples with FA (a) and cement (b).

6 CONCLUSIONS

After confirming the cementitious properties of WPFA, it was used as a substitute of cement in a stabilized soil aiming at improving the subgrade bearing capacity of a road. An experimental section was built and monitored from a technical and environmental point of view.

The results achieved in laboratory and onsite monitoring showed higher strength and modulus than those specified in the Spanish regulations, which allows to state that WPFA is an alternative hydraulic binder. However, it is important to avoid mixing with soils with a high sulphate content, since the formation of ettringite is highly favourable.

Regarding the environmental monitoring, the mixture with WPFA meets the criteria to be considered inert except for Antimony; although the leachate collected onsite indicated that all metals remained below the threshold limit for inert classification.

ACKNOWLEDGEMENTS

This project has received funding from the European Union's Horizon 2020 research and innovation program under grant agreement No 730305.

The authors would like to thank SAICA and the Laboratorio de Carreteras de Aragón de la Dirección General de Movilidad e Infraestructuras del Gobierno de Aragón for their support.

REFERENCES

Arm, M., Vestin, J., Lind, B. B., Lagerkvist, A., Nordmark, D. and Hallgren, P., 2014. *Pulp mill fly ash for stabilization of low-volume unpaved forest roads—field performance*. Canadian Journal of Civil Engineering, 41(11), 955–963.

Baloochi, H., Aponte, D., Barra, M., Martínez, A., Miró, R. M., Cepriá, J., Orejana, R. and Oleaga, A., 2020. *Alternative secondary raw materials for road construction based on pulp and paper industry reject. Paperchain project*. Routes/Roads: Roads, (383), 19–23.

Cherian, C. and Siddiqua, S., 2019. *Pulp and Paper Mill Fly Ash: A Review*. Sustainability, 11(16), 4394.

Departamento de Medioambiente Gobierno de Aragón (2008). Orden del 5 de mayo por la que se procede al establecimiento de los niveles genéricos de referencia para la protección de la salud humana de metales pesados y otros elementos traza en suelos de Aragón. (in Spanish).

Hänninen, R., Hetemäki, L., Hurmekoski, E., Mutanen, A., Näyhä, A., Forsström, J., Viitanen, J. and Koljonen, T., 2014. *European forest industry and forest bioenergy outlook up to 2050: A synthesis*. Cleen Oy, Helsinki.

Jones, D., Rahim, A., Saadeh, C. and Harvey, J., 2010. *Guidelines for the stabilization of subgrade soils in California*. University of California Pavement Research Center.

Lahtinen, P., Jyrävä, H., Maijala, A. and Mácsik, J., 2005. *Fly ashes as binders for the stabilization of gravel*. Laboratory tests and preparations for a field test.

Lahtinen, P., 2001. *Fly ash mixtures as flexible structural materials for low-volume roads*. Helsinki University of Technology.

Ministerio de Fomento, 2003. *Norma 6.1 IC. Secciones de firme. Instrucción de Carreteras*. Dirección General de Carreteras. Centro de publicaciones, Madrid (in Spanish).

Ministerio de Fomento, 2015. *Pliego de Prescripciones Técnicas Generales para obras de Carreteras y Puentes. PG-3*. Dirección General de Carreteras, Centro de publicaciones, Madrid (in Spanish).

Mulder, E., 1996. *A mixture of fly ashes as road base construction material*. Waste Management, 16(1-3), 15–20.

Ohenoja, K., Pesonen, J., Yliniemi, J. and Illikainen, M., 2020. *Utilization of Fly Ashes from Fluidized Bed Combustion: A Review*. Sustainability, 12(7), 2988.

Ribbing, C., 2007. *Environmentally friendly use of non-coal ashes in Sweden*. Waste Management, 27(10), 1428–1435.

The paperChain Project, 2017. New niche markets for the Pulp and Paper Industry waste based on circular economy approaches. https://paperchain.eu

Tuncan, A., Tuncan, M. and Koyuncu, H., 2000. *Use of petroleum-contaminated drilling wastes as sub-base material for road construction*. Waste management & research, 18(5), 489–505.

Vestin, J., Arm, M., Nordmark, D., Lagerkvist, A., Hallgren, P. and Lind, B., 2012, May. *Fly ash as a road construction material*. WASCON 2012 Conf. proceedings, ISCOWA and SGI.

Wang, S. L., Lv, Q. F., Baaj, H., Li, X. Y. and Zhao, Y. X., 2016. *Volume change behaviour and microstructure of stabilized loess under cyclic freeze–thaw conditions*. Canadian Journal of Civil Engineering, 43(10), 865–874.

Zhang, Y., Johnson, A. E. and White, D. J., 2019. *Freeze-thaw performance of cement and fly ash stabilized loess*. Transportation Geotechnics, 21, 100279.

Zhou, H., Smith, D. W. and Sego, D. C., 2000. *Characterization and use of pulp mill fly ash and lime by-products as road construction amendments*. Canadian Journal of Civil Engineering, 27(3), 581–593.

Zuber, S. Z. S., Kamarudin, H., Abdullah, M. M. A. B. and Binhussain, M., 2013. *Review on soil stabilization techniques*. Australian Journal of Basic and Applied Sciences, 7(5), 258–265.

Experiences with the use of stabilisation geogrids in demonstrating an improvement in bearing capacity of recycled materials

P. Mazurowski
Tensar Polska Sp. z o.o., Gdańsk, Poland

K. Zamara & C. Gewanlal
Tensar International Ltd., Blackburn, UK

J. Kawalec
Technical University of Silesia, Gliwice, Poland

ABSTRACT: The use of recycled granular materials for road construction should be a necessity in current times; both for economic and environmental reasons. By the use of such materials, the road construction industry helps to protect natural resources and reduce its carbon footprint, while lowering the cost of pavement construction. However, the use of recycled materials can often raise concerns, viz. design compliance, low homogeneity, perceived lower quality and an onerous production process when compared to the use of virgin materials. It may be common practice for the variability in the engineering parameters of these materials to influence designers to specify thicker pavement layers in order to ensure that the required bearing capacity and/or design life of the road is achieved.

The use of hexagonal stabilisation geogrids may provide a method of mitigating against the loss of homogeneity and improve the engineering parameters of recycled granular materials. The case-studies presented herein demonstrate that the mechanical stabilisation of recycled granular materials with a hexagonal geogrid, results in an increased stiffness of the layer and significantly aids compaction. When aggregate is placed and compacted on a stiff hexagonal geogrid, aggregate particles interlock within the geogrid apertures and are confined by its stiff ribs. A composite mechanically stabilised layer (MSL) is created, in which the geogrid and aggregate act in unison. Lateral restraint, provided by the geogrid, reduces strain within the aggregate skeleton and thereby increases the stiffness of the layer. The result is that the required performance parameters, be it a bearing capacity of a layer and/or life of the pavement, can be successfully achieved with the use of, and possible reduction in the required layer thickness of, recycled granular materials.

The concept of mechanical stabilisation of aggregate layers by the use of geogrids is detailed in this paper. Furthermore, several case studies are presented which demonstrate the use of hexagonal stabilisation geogrids to improve the bearing capacity of various recycled materials such as recycled asphalt, recycled concrete, and coal mine shales.

Keywords: Recycled aggregate, hexagonal geogrid, mechanical stabilisation, recycled asphalt, RA

DOI: 10.1201/9781003222880-18

1 INTRODUCTION

Construction and demolition waste (CDW) accounts for approximately 25% - 30% of all waste generated in the EU (European Commission, 2019). Currently, the level of recycling of CDW varies in different countries, ranging from 10 to 90%. The goal of the EU is to reach a minimum 70% level of recycling by the end of 2020.

According to Symonds, 2019 "the largest single consumer of aggregates in most Member States is the road construction industry". There are a lot of economic and environmental benefits associated with the use of recycled aggregates in the road construction industry.

Arguably the most important environmental benefit is resource conservation. As the availability of virgin aggregate material becomes limited, every opportunity to reduce its use generates financial and environmental benefits. Another benefit is a substantial reduction of greenhouse gas emissions, obtained due to reduced transportation and energy consumption. Recycling can eliminate the need for disposal of hazardous debris, like old asphalt with tar; these can be utilized in mixes with cement and bitumen emulsion instead. As recycled material does not have to be disposed to landfill, recycling helps to preserve landfill capacity. Recycling also helps with conservation of the landscape – both by avoiding the creation of new stockpiles of demolition rubble and by negating unsightly changes to the landscape cased by quarrying of virgin aggregates.

Significant economic benefits can be realised through the reduction of material, plant and labour project costs. Snyder et al. 2018 gives examples of several projects showing substantial savings achieved due to the use of recycled concrete; one such project achieved as much as 5$ million in savings. Cost reduction results not only from the lower price of recycled aggregate (which is not always the case), but also from reduced haul distances, construction program and disposal costs. Haul distance can be reduced because in most cases the recycling facility is located much closer to the construction site than any virgin aggregate quarry. In the case of certain materials, such as concrete and asphalt, it is also possible to recycle these on site, eliminating the need for haulage completely. The reduction in haul distance, material volumes and construction time results in reduced damage to existing access roads. A secondary benefit is the reduced negative impact that construction activities often have on local communities VIZ. travel time increase, noise, dust, and plant exhaust fume emissions to name a few.

The use of hexagonal stabilisation geogrids can add further benefits to projects by improving the effective engineering properties of recycled aggregates and thereby enabling the use of reduced layer thicknesses while achieving comparable performance of pavements.

2 AGGREGATE STABILISATION WITH HEXAGONAL GEOGRID

2.1 *Concept of stabilisation with geogrid*

Geogrids in civil engineering structures have been used for decades. The intended use of geogrids is to mechanically stabilise granular layers in order to minimise deformation during trafficking, to improve the load-bearing capacity and to increase the design life of roads. The combination of the geogrid and aggregate creates a mechanically stabilised composite layer with significantly improved properties and performance capabilities in response to dynamic and static loading when compared with the aggregate layers alone.

Research on the effect of geogrids on pavements aggregate began over 30 years ago. Early research at Oxford University dated in the late 70s and 80s investigated mechanisms of geogrid's improvement of aggregate layers (e.g. Milligan and Love 1984). This research in the field of geogrid mechanical stabilisation concluded that aggregate layers combined with geogrids exhibited a higher bearing capacity than those without. Further the mechanism associated with that behaviour was defined as interlock which resisted tension at the level of the geogrid. Also, the tensioned membrane effect, that these days is attributed to high strength products, was considered to have a limited effect on performance and this too only at high deformation levels within the fill base. The difference between the interlock/confinement mechanism and the tensioned membrane mechanism is presented in Figure 1.

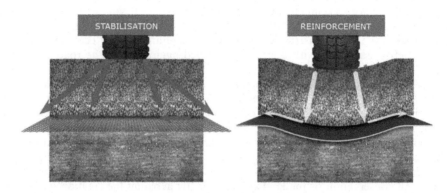

Figure 1. Mechanisms of geosynthetic support in pavement, Left: mechanical stabilisation by geogrid; Right: reinforcement at high deformation level.

Further laboratory research by Haas et al. 1988 concluded that for optimal long-term pavement performance, geogrids should operate within a zone of moderate elastic strain (i.e. 0.05 to 0.2%) and for the overall design life this plastic strain should not exceed 1 to 2% (this being linked to pavement rutting criteria). The above conditions ensure geogrid insitu performance within its elastic limits which contributes to the effective confinement of aggregates particles.

The early and insightful research by Haas et al. created a foundation for further research and development into the field of geogrid mechanical stabilisation in pavements. The studies progressed into full scale experiments that confirmed the conclusions of laboratory studies.

Cook et al. 2016 provided a summary of published full scale trafficking trials where the mechanisms of geogrid stabilisation were investigated. A series of Transport Research Laboratory (TRL) trafficking trials were analysed and the mechanisms of geogrid stabilisation versus geogrid reinforcement were discussed. A total of eight trials undertaken in the UK TRL laboratories were summarized in the study with different geosynthetic products tested for their stabilisation performance.

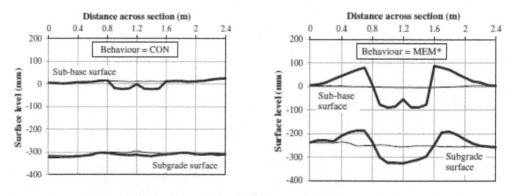

Figure 2. Transverse surface profiles, Left: full confinement ("CON"), negligible subgrade deformation; Right: full tensioned membrane ("MEM"), with surface rut and subgrade deformation of much the same size and geometry (Cook et al. 2016).

Cook et al. 2016 analysed in detail the cross-sections of aggregate layers within the pavement structures. Deformation at surface were compared with deformations at the geosynthetic level (Figure 2). The conclusions of Cook et al. 2016 echoed the early studies of Milligan and Love 1984 and Haas et al. 1988, by attributing the benefit of geogrid inclusion to particles confinement and effective interlock. If the true stabilising mechanism is achieved, this further manifest itself in improved load distribution due to lateral restraint and reduced aggregate deterioration. Geogrids allow the aggregate to lock itself in the geogrid apertures, which results in the formation of a stiff layer at the base of the aggregate. Monitored geogrid stabilisation and reinforcement (tensioned membrane) effect is presented in Figure 2.

The strong international, scientific evidence supporting the argument of the geogrid mechanical stabilisation function (*i.e.* interlock and confinement) being different to the reinforcing function (*i.e.* tensioned membrane), resulted in acceptance of the geogrid mechanical stabilisation function definition. ISO 10318-1:2015 Amendment 1-2018 defines the geogrid mechanical stabilisation function as 'Improvement of the mechanical behaviour of an unbounded granular material by including one or more geosynthetic layers such that deformation under applied loads is reduced by minimizing movement of the unbound granular material'.

2.2 *Concept of pavement optimisation*

Aggregate layers stabilised with hexagonal geogrids have increased stiffness compared to non-mechanically stabilised layers. This influences the performance of the whole pavement and can be utilised in two ways: (1) it is possible to increase the life of a pavement compared to a non-mechanically stabilised pavement of similar thickness, and (2) it is possible to reduce the thickness of individual pavement layer(s) and achieve similar performance to thicker, non-mechanically stabilised pavements. This layer thickness reduction need not necessarily be limited to the mechanically stabilised layer; the other granular layers and even the asphalt thickness may be reduced to offset the improved performance provided by the mechanically stabilised layer. This way of using geogrids in pavements was introduced several years ago by the producer of hexagonal geogrids under the term "Pavement Optimisation".

A series of Accelerated Pavement Tests (APTs) were performed at the U.S. Army Engineering Research and Development Centre to quantify the benefits of hexagonal geogrids. Tests were conducted between 2010 and 2016, in three phases. In total, 8 sections were tested.

The granular base of test sections consisted of either 15 or 20cm of crushed limestone. Four of the sections had a base layer stabilised with hexagonal geogrids. Six sections had surfacing of 5, 7.5 or 10cm of dense-graded hot mix asphalt (HMA), and two sections had a double bituminous surface treatment (DBST). The subgrade consisted of locally available clay prepared to achieve a bearing capacity of either 3 or 6% CBR.

A Heavy Vehicle Simulator (HVS) device was used to traffic test sections. Sections on 3% CBR subgrade were subjected to 100 000 axle loads, while sections on 6% CBR up to 800 000 axle loads. Throughout the tests, rut depth measurements were collected on all test sections.

Full tests results are presented in reports: Jersey et al. 2012, Norwood et al. 2014, Robinson et al. 2017. All sections stabilised with hexagonal geogrids substantially outperformed their non-mechanically stabilised control equivalents including the sections with thicker asphalt and/or base layers. Based on these results, modifications to pavement design methods were developed by the manufacturer of the hexagonal geogrids. Basic assumptions of the modification to the mechanistic-empirical pavement design method were presented by Mazurowski et al. 2019.

3 RECYCLED CONCRETE AGGREGATE

Recycled Concrete Aggregate (RCA) is one of the most used recycled aggregates in pavement construction. It is commonly used as a substitute for virgin aggregates in capping, sub-base or base layers. In the US, about 65% of RCA is used as aggregate base material in road construction (Snyder et al. 2018).

RCA can be produced to meet all the requirements of typical aggregate base layer specifications. This, however, affects its price – the cost of good quality RCA can be comparable to the cost of virgin aggregate (Snyder et al. 2018). Lower quality RCA, which can contain crushed bricks and other contaminations (wood, plastic etc.) is usually much cheaper, but often does not meet the requirements and its use is limited.

One possible way of improving parameters of RCA is by stabilisation with hexagonal geogrids. This is especially useful in the case of low-quality RCA.

RCA usually has lower resistance to abrasion mass loss compared to virgin aggregate (American Concrete Pavement Association, 2020). Abrasion of aggregate in pavement is a result of movement of particles under cycling loading. This can result in the deterioration of engineering parameters over time, especially in low quality RCA containing crushed bricks. Liu et al. 2017 confirmed reduction of particle movement under cycling loading in tests performed with a device called Smart Rock. It was a 3D-printed artificial stone, in which accelerometers were installed, which allowed for real time movement tracking of the stone. The device was installed within test sections (both stabilised with geogrids and control) of 25cm of rail ballast loaded with 1 Hz frequency cyclic loading. Stabilisation of aggregate with geogrid resulted in a substantial reduction of both displacement and particle rotation under cyclic loads, and thus reduced aggregate abrasion.

Hexagonal geogrids were used in numerous projects to stabilise RCA, both high and low quality. Two case studies are described below.

Św. Ducha is a street in the historical centre of Gdańsk, Poland; in an area known for poor insitu soil conditions. It was constructed on a subgrade of non-controlled embankment fill consisting of different soils with the addition of organic soils and debris from buildings damaged during WWII.

The original design of the reconstructed pavement included a layer of crushed aggregate stabilised with a hexagonal geogrid and a layer of cement stabilised sub-base material. The addition of a second layer of hexagonal geogrid allowed for optimisation of the pavement, which resulted in a thickness reduction of 5cm of the base layer, removal of cement stabilised sub-base, and substitution of virgin crushed aggregate caping layer with RCA (Figure 3). This resulted in substantial savings for the contractor, who had a big quantity of good quality RCA available from another nearby construction site.

Hexagonal geogrids were used to stabilise unpaved access roads to Kretinga Wind Farm in Lithuania Figure 5. Roads were constructed in areas of low bearing capacity soils, and were required to handle traffic of trucks carrying aggregate, concrete, cranes and turbine parts.

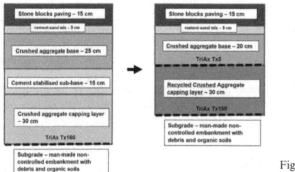

Figure 3. Original and optimised structure. Tx5, Tx150 and Tx160 describe different geogrids types.

Figure 4. Installation of geogrid.

Locally available RCA was used to construct the pavements. The RCA was of a very low quality, with high content of crushed bricks. The use of hexagonal geogrids not only allowed for the installation and compaction of the RCA aggregate on weak soils, but also for the reduction of the thickness of the pavements.

Figure 5. RCA layer stabilised with hexagonal geogrid at Kretinga Wind Farm.

4 COAL MINE SHALES

Deep coal mining results in the production of large volumes of waste material stored in large heaped stockpiles. As coal is typically interlayered with other type of soft rock, once brought to the surface special mechanical processes are implemented to separate the coal from the low-calorie material which cannot be used for energy production. Currently environmental aspects are forcing most coal producers to reduce/eliminate such type of fuel from energy production, but historical mining activities have already resulted in billions of tons of excavated material being stockpiled.

In most cases, coal mining waste consists of clay coal shales. Natural coal shales are not frost resistant and as such cannot be used within the frost depth of any engineering structure. However due to their good mechanical properties and ease of compaction they are often used to form lower parts of road embankments as depicted in Figure 6.

Figure 6. Stabilisation of expressway embankment base constructed from unburnt clay shales – Road 902 project in Gliwice.

Figure 7. Typical burn coal shale stored in an open heap during excavation.

Storage in high, loose heaped stockpiles (uncompacted, especially during the early days of mining when no compaction equipment was available) often resulted in self-ignition. If temperatures in the stockpile reached over 700-800 ^{O}C, the clay shales became a fully burnt red coloured material of different physical and mechanical properties as depicted in Figure 7. The material becomes frost resistant and as such is acceptable for use within the frost dept of the road structure (subbase and base). Secondly, due to its low cost and the large volumes available, it becomes a preferred material by Contractors wanting to keep their project budgets within limits. Coal mine shales stabilised with hexagonal geogrids were used in numerous road projects in Silesia, Poland over the last 12 years. Some examples of these are depicted in Figures 8 and 9.

Figure 8. Example of geomattress with two layers of burnt coal mining shales (red colour) under motorway pavement - A1 motorway project in Zabrze.

Figure 9. Example of mining waste layer stabilised by geogrid for road construction – road 86 project in Katowice.

5 RECYCLED ASPHALT

Two rehabilitation projects in South Africa have shown that the stabilisation effect produced by hexagonal geogrids can also be realised with the use of bound treated base materials. Recycled asphalt and granular base materials were used in these projects to save time, money and the environment. The Harvey and Hanger Roads project was constructed in Bloemfontein in 2012 and used a bitumen stabilised material (BSM) base with 25% milled recycled asphalt in the mix design. The McGregor Street project was constructed in Bloemfontein in 2016 and used a BSM foam base with 30% milled recycled asphalt and 60% milled granular base/subbase materials. Both mix designs are detailed in Table 1.

Table 1. BSM mix designs for Harvey and Hangar Roads as well as McGregor Street.

Harvey and Hangar Roads	McGregor Street
• 75% Natural crushed stone; • 25% Recycled asphalt surfacing; • 2.5% Bitumen (SS60 emulsion); • 1.2% Cement	• 30% Recycled asphalt surfacing • 20% Milled crushed stone base • 40% Milled gravel subbase • 10% Imported crushed stone • 1% Cement • 2.2% Bitumen (50/70 pen)

As both projects are busy arterial roads carrying in excess of 10,000 vehicles per day, the traffic disruption needed to be minimized (Raman, 2017). The ground investigations revealed the occurrence of large hard packed stone in some sections, like a Macadam pavement. The railway station that Harvey Road serves was completed in 1890. The type of Macadam pavement in sections of Harvey and Hanger Roads most probably dated back to the early 1900's (Pretorius et al. 2013). The most significant site constraint was that all the sites had shallow services at a depth of roughly 300mm beneath the existing surface. The analyses for these roads produced a required repair depth well in excess of the depth of the existing services.

Being located in the midst of a well-established Central Business District with existing drainage, kerbs and shop entrances, the existing surface level could not be raised. The stabilisation factor produced by the use of a hexagonal geogrid with a bound bitumen modified base was therefore utilised to provide the required performance at a depth which did not disturb the existing services. The pavement layerworks designs, showing the theoretical traffic carrying capacity in terms of Million Equivalent Standard Axle Loads (MESAL's), are depicted in Figure 10.

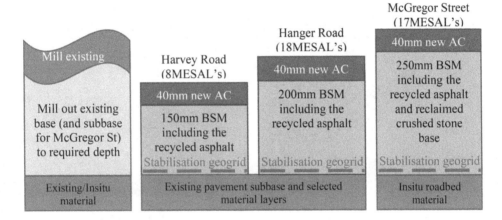

Figure 10. Layerworks designs.

Harvey and Hangar Roads: Prior to rehabilitation the existing pavements showed severe signs of distress and, at the time of analysis, had no residual life. After 8 years in service after reconstruction, both roads show only limited signs of deterioration.

Some of the milled granular base material was reused as base materials in the pedestrian sidewalks and the remainder was stockpiled by the Client for use on future projects; there was thus very little spoil generated in this project. The bitumen emulsion stabilised material was mixed and stockpiled off-site due to the site constraints of the busy built-up area. A pug mill was specially designed for this project with the ability to be used thereafter for similar applications or for pre-coating of aggregates in sealing applications.

The construction sequence for Harvey and Hangar Road is depicted in Figures 11 to 14.

Figure 11. Milling existing asphalt. Figure 12. Installation of stabilisation geogrid.

McGregor Street: Prior to rehabilitation, McGregor Street exhibited signs of fatigue and crocodile cracking as depicted in Figure 15. At the time of analysis, the road had a residual life of 0.2 to 0.7 MESAL's which equated to 1 to 2 years.

Figure 13. Installation of BSM with paver. Figure 14. Rolling of base.

McGregor Street after reconstruction shows no sign of deterioration after some 4 years in service as depicted in Figure 16. The construction sequence for McGregor Street was similar to that of Harvey and Hangar Roads depicted in Figures 11 to 14.

Figure 15. McGregor Street prior to rehabilitation. Figure 16. McGregor Street in 2020.

The McGregor Street project maximized the use of recycled milled materials. These were sorted and tested in a laboratory to optimize the performance of the pavement. Ultimately only 10% of imported crushed stone was required as mentioned in Table 1.

The BSM1 foam was mixed in-plant using a high-quality mobile mixer and allowed for up to 5 days of stockpiling before placing on the hexagonal geogrid using a paver.

6 CONCLUSIONS

The use of recycled aggregate brings numerous benefits, both economic and environmental, to the road construction industry. Stabilisation of aggregate with hexagonal geogrids is a way to maximize these benefits, by reducing the volume of good quality aggregate required or increasing the performance of low-quality aggregates. Extensive research carried out over the last 30 years. and numerous successful case studies have confirmed the beneficial effects of stabilisation on both virgin and recycled aggregates in pavement construction. Engineers may be reluctant to use recycled asphalt because they doubt the uniformity of the material. Mechanically stabilisation of recycled asphalt by hexagonal geogrids not only enhances its engineering properties (this is known as the modulus increase effect) but also improves its uniformity. Layer thickness reduction by a factor of 1.5 to 2 times (known as the stabilisation factor) can be achieved without any loss of pavement performance by incorporating hexagonal geogrids into the layerworks design. Therefore, less recycled and/or imported fill as well as less cut to spoil of insitu material is required, resulting in cost savings.

REFERENCES

American Concrete Pavement Association website Wikipave, 2020, Recycling Concrete Pavements website http://wikipave.org/index.php?title=Recycling_Concrete_Pavements&oldid=8728

Asphalt Academy, 2009.*TG2, Technical Guideline, Bitumen Stabilised Materials*. Committee of Land Transport Officials, Pretoria, South Africa: Department of Transport, South Africa, 1988, February, Draft UTG3 Structural Design of Urban Roads. Urban Transport Guidelines

Cook, J., Dobie, M., Blackman, D., 2016. *The Development of APT Methodology in the Application and Derivation of Geosynthetic Benefits in Roadway Design* J.P. Aguiar-Moya et al. (eds.), The Roles of Accelerated Pavement Testing in Pavement Sustainability, DOI 10.1007/978-3-319-42797-3_17© Springer International Publishing Switzerland 2016

European Commission, 2019, Environment website: https://ec.europa.eu/environment/waste/construction_demolition.htmBrussels,Belgium.

Haas R, Walls J and Carroll RG 1988 *Geogrid Reinforcement of Granular Bases in Flexible Pavements*; Transportation Research Record 1188

ISO 2015. *Geosynthetics – Part 1: Terms and definitions* 10318–1:2015 Amendment 1–2018. ISO/TC 221 Geosynthetics.

Jersey, S.R., Tingle, J.S., Norwood, G.J., Kwon, J. and Wayne, M., 2012. *Full-Scale Evaluation of Geogrid-Reinforced Thin Flexible Pavements*. Transportation Research Record, Journal of the Transportation Research Board, No. 2310, TRB of the National Academies, Washington, USA

Liu, S., Huang, H. and Qui, T. Kwon J., 2017. *Comparative Evaluation of Particle Movement in a Ballast Track Structure Stabilised with Biaxial and Multiaxial Geogrids*. Transportation Research Record: Journal of the Transportation Research Board, No. 2607, 2017

Mazurowski, P., Buckley, J., Kawalec, J., 2019, *Modification of mechanistic-empirical pavement design method to incorporate the influence of hexagonal stabilisation geogrids*. AAPA International Flexible Pavements Conference, 18th, 2019, Sydney, Australia

Milligan, G., Love, J.P., 1984. *Model testing of geogrids under an aggregate layer on soft ground*. Polymer Grid Reinforcement: Proceedings of a Conference Sponsored by the Science and Engineering Research Council and Netlon Ltd, London 22–23 March 1984

Norwood, G.J. and Tingle, J.S., 2014. *Performance of Geogrid-Stabilised Flexible Pavements*. Final Report. EDRC/GSL TR-14-28. U.S. Army ERDC, USA

Pretorius, V., Deetlefs, J., Leeuw, O., Roe, T., 2013. *Innovation In Overcoming Road Rehabilitation Challenges Faced In An Urban Environment: An In-Depth Case Study.*; Proceedings of the 32nd Southern African Transport Conference, Pretoria, South Africa.

Raman, C., 2017. *Utilising Bitumen Stabilised Material And A Triaxial Geogrid In The Rehabilitation Of A Busy Urban Arterial*. Proceedings of the 81st Institute of Municipal Engineers of South Africa Conference, Gauteng, South Africa.

Robinson, W.J., Tingle, J.S., and Norwood, G.J., 2017. *Full-Scale Accelerated Testing of Multi-Axial Geogrid Stabilised Flexible Pavements*. Draft Final Report. EDRC/GSL TR-17-X. U.S. Army ERDC, USA

Snyder, M. B., Cavalline, T. I., Fick, G., Taylor, P. and Gross, J., 2018. *Recycling Concrete Pavement Materials: A Practitioner's Reference Guide*. National Concrete Pavement Technology Center, Iowa State University, Ames, US.

Symonds, ARGUS, COWI, PRC Bouwcentrum, 1999. *Construction and Demolition Waste Management Practices, and Their Economic Impacts*. Report to DGXI European Commission.

Airport design

Introducing dynamic asphalt modulus to the design of flexible aircraft pavement structures

H. Weisser & G. White
School of Science and Engineering, University of the Sunshine Coast, Sippy Downs, Queensland, Australia

ABSTRACT: Most modern aircraft pavement thickness design methods are mechanistic-empirical, with the mechanistic element based on layered-elastic analysis of trial pavement structures, which are refined until they are optimised for the subgrade conditions and project aircraft traffic loadings. An elastic modulus and Poisson's ratio are assigned to each layer in the pavement. In the past, presumptive or standard values of modulus have been used for the various pavement layers. This includes the granular crushed rock or natural gravel, as well as the asphalt surface and sub-surface asphalt layers. However, in modern times, there is great interest in using a mixture-specific dynamic asphalt modulus. As the traffic speed increases and the temperatures reduces, the asphalt modulus increases significantly. Because the temperature of asphalt changes with the depths of the layer within the pavement and with day-night and summer-winter fluctuations, dynamic modulus is complex and is not readily incorporated in layered-elastic software. This research measured the modulus of a typical dense graded airport asphalt mixture at various temperatures and load speeds and incorporated the resulting dynamic modulus into aircraft pavement thickness design using APSDS via asphalt sub-layering. It was concluded that the difference in modulus associated with summer and winter, or day and night, temperature profiles changed the predicted life of the pavement by orders of magnitude, suggesting that all significant pavement damage occurs on the hottest few days of the pavement's life. Although dynamic asphalt modulus was relatively efficiently incorporated into APSDS, its use in routine layered elastic design is unlikely to be justifiable and is it expected to introduce more challenges than it is likely to solve.

Keywords: Airport, Asphalt, Dynamic modulus, Thickness design

1 INTRODUCTION

Modern flexible aircraft pavements are generally designed using the mechanistic-empirical, layered elastic method (AAA 2017). A static load is applied to the pavement and the stresses and strains are calculated at critical points in the pavement using layered elastic theory. The maximum strains at the critical points are then related to the theoretical number of allowable repetitions of that load until failure is predicted, using empirically derived transform functions, also known as performance criteria or failure criteria (Huang 1993). The pavement is modelled as layers of homogenous and infinitely wide elastic materials which are characterised by an elastic modulus and a Poisson's ratio (Huang 1993). The modulus values have a great impact on pavement thickness and for routine design, modulus values are usually sourced from literature or jurisdiction-based design guidance. The modulus assigned to the asphalt is generally constant or static. However, it is well established and broadly known that the modulus of asphalt depends on the temperature and load frequency imposed on it (Mounier et al. 2015).

The exact modulus of asphalt at a certain temperature and loading speed is known as the dynamic modulus and can be determined by laboratory testing. This includes testing over a range of load frequencies and temperatures, then creating a master curve of modulus (Mounier et al. 2015). It is expected that using dynamic modulus for a pavement design provides a more accurate pavement design as different climates that effect the prevailing modulus values are considered. Furthermore, it is clear that asphalt modulus is material specific, even for the same asphalt type. For example, polymer modified bitumen (PMB) is associated with higher or lower modulus values than otherwise comparable asphalt, depending on the polymer used (White & Embleton 2015). Furthermore, different aggregate sources provide different contributions to mixture stiffness, depending on their shape, interparticle friction and affinity for bituminous binder (Pan et al. 2005). The use of a mixture-specific dynamic modulus allows these differences in asphalt mixtures to be accounted for in pavement design.

The aim of this research was to compare the standard design approach to a design based on dynamic asphalt modulus values. The modulus values were determined based on realistic pavement temperature profiles and the dynamic modulus characterisation of a typical aircraft pavement asphalt mixture. Different aircraft and a typical range of subgrade conditions were considered, allowing general trends to be explored across various practical design scenarios.

2 BACKGROUND

2.1 Flexible aircraft pavements

Similar to road pavement, flexible aircraft pavements consist of four primary courses, the asphalt surface, a base, subbase and the supporting subgrade. The asphalt surface layer generally consists of a hot mix asphalt produced with crushed and graded hard rock aggregate and natural sand (FAA 2016). There are often two asphalt layers, the wearing, or surface layer, and the supporting upper base layer, which transmits the load to the underlying granular layers. The wearing layer is usually made up of hard aggregate which can resist the high shearing forces associated with aircraft take-off and braking. The wearing layer also provides an impermeable surface, which prevents water from penetrating into the base layers. Base layers can be the same as the surface layer but commonly have a larger aggregate size and lower binder content, to increase the stiffness of the base.

Generally, airport pavements in Australia have a relatively thin asphalt surface, usually 50-100 mm, over a layer a high quality and well compacted crushed rock base course (AAA 2017). In contrast, aircraft pavements in the USA usually have thicker asphalt surfaces and a bound upper base course. The upper base course is often asphalt, resulting in a total of approximately 250 mm of asphalt. Almost opposite to Australia, some European countries prefer only a thin lower subbase to act as a working platform, with the majority of the pavement built using high modulus asphalt layers (White et al. 2020). These full depth asphalt pavements are generally thinner, but more expensive to construct, than other pavements of equivalent structural strength. Despite these differences in the composition of aircraft pavements, the thickness is universally determined to protect the subgrade from the loadings associated with the projected aircraft traffic.

2.2 Aircraft pavement thickness design

As stated above, the layered elastic method is a mechanistic-empirical approach to design and is the most commonly used method for modern flexible pavement design. It uses a cumulative damage factor (CDF), which is the ratio of the applied load repetitions compared to the allowable load repetitions that are predicted to cause the pavement to fail. Failure is generally defined in terms of asphalt fatigue cracking and/or permanent vertical subgrade deformation, commonly known as pavement rutting. A CDF below one means that the pavement has not yet reached its structural limit. Once the CDF reaches 1.0, the pavement is predicted to have failed (Mincad 2010).

Several layered elastic based pavement design softwares have been developed by different jurisdictions. The Federal Aviation Administration (FAA) has developed FAARFIELD (FAA 2017) which is commonly used in the USA and many other countries around the world. In France the software Alize is used (Balay et al. 2009), while in Australia, Airport Pavement Structural Design Software (APSDS) (Wardle & Rodway 1998) is common. APSDS is similar to its parent software for road pavement design, known as Circly (Wardle 1977). Despite differences in user interfaces, flexibility, conservatism and some internal calculation processes, all these softwares are based on the same principles:

- The pavement layers and materials are selected.
- The layer thickness, material modulus and Poisson's ratio are assigned to each layer.
- A subgrade is selected for each pavement profile, characterised by a CBR and converted to an elastic modulus.
- Aircraft traffic spectra are defined by nominating one of more aircraft types, weights, tyre pressures and the number of load repetitions of each.
- The damage indicators, usually stress or strain, are calculated at the critical points.
- The maximum value of the critical damage indicator is converted to the number of allowable load repetitions using the applicable transform function.
- Using the CDF concept, the sum of the damage over all loads is calculated.
- The pavement thickness is iteratively increased or decreased until the CDF is equal to 1.0.

APSDS was chosen for this research as it is widely used in Australia and provides more flexibility for material inputs, such as modulus and layer thickness, compared to FAARFIELD. This makes APSDS particularly useful for pavement thickness determination of non-standard materials and non-routine pavement structures (White et al. 2020).

2.3 Pavement layer modulus

As discussed above, APSDS allows for the modification of all the required pavement design input parameters. Most variables can be chosen from the APSDS internal library (Mincad 2010). This include the subgrade, which usually has a modulus equivalent to ten times the CBR, and the base and subbase, which use automated sub layering known as the Barker and Brabston method (Barker & Brabston 1975).

The asphalt modulus has traditionally been based on published static values for base (3,300 MPa) and surface (1,500 MPa) layers, which reflects the different temperature and its effect on asphalt modulus. The surface layer is exposed to the sunshine and solar radiation, increasing its temperature beyond that of the air. In contrast, sub-surface layers are insulated and their temperature changes are much lower, with asphalt more than 300 mm below the pavement surface generally considered to experience a constant temperature state (Gray et al. 2015). The traditional static or constant asphalt modulus values are expected to represent asphalt produced with unmodified binders, during a typical summer day in most temperate climates. However, road pavement design has introduced material, vehicle speed and temperature specific modulus values, generally known as the dynamic modulus. This approach has been adopted for this research.

2.4 Dynamic modulus of asphalt

The dynamic modulus can be determined via different laboratory tests, such as the flexural and compressive methods detailed in Austroads AGPT-T274-16 and the American Association of State Highway and Transportation Officials (AASHTO) T 342-11 respectively. These tests measure the modulus at several temperature and frequency combinations. Using these results, master curves can be constructed and fitted to best match the test results, as detailed in the Queensland Department of Transport and Main Roads (TMR) Technical Note 167. The master curves depend on the reduced frequency, a temperature shift factor (α_t) and several master curve fitting parameters (C_1, C_2, α, β, γ and δ) (Equation 1). The temperature shift factor can be calculated using several methods. The Williams–Landel–Ferry Equation (WLF Equation) was chosen for this research, as it was found to be the most accurate option for asphalt materials (Equation 3).

$$E_{pred} = 10^{\delta + \frac{\alpha}{1 + e^{\beta + \gamma * \log f_{red}}}} \qquad (1)$$

Where: E_{pred} = Predicted modulus (MPa).
α, β, γ and δ = master curve fitting parameters.
f_{red} = reduced frequency as per (Equation 2).

$$f_{red} = \alpha_t * f \qquad (2)$$

Where: α_t = temperature shift factor as per (Equation 3).
f = frequency (Hz).

$$\log \alpha_t = \frac{-C_1 (T - T_{ref})}{C_2 + (T - T_{ref})} \qquad (3)$$

Where: α_t = temperature shift factor.
C1 and C2 = empirical constants.
T = temperature at testing (°C).
T_{ref} = temperature of interest (°C).

Although Equation 3 is convenient, it applies only to a single temperature value. Because the temperature of the pavement varies with depth, a new version of Equation 3 must be used for each depth of interest, based on the estimated temperature at that depth. Accordingly, the usefulness of the predicted modulus is dependent on the accuracy of the temperature profile used to determine the pavements temperature at different depths.

2.5 *Temperature profiles*

The temperature of the asphalt is critically important to implementing dynamic modulus implementation and can either be measured or predicted. The temperature of asphalt depends on several factors including the depth of the measurement, the daily temperature and the solar radiation (Gray et al. 2015). This makes temperature predictions complex and several temperature models are available to predict the temperature of asphalt at different depths and in different locations.

Gedafa et al. (2014) developed and validated a model that predicts the temperature in asphalt pavements. They found a linear relationship between the temperature in the middle of each pavement layer (T_{pavt}) at a certain depth (D) and the pavements surface temperature (T_{sur}), as well as the average air temperature of the previous day (T_{sur}) and the time of day (t_d) at which the temperature has been measured. The model (Equation 4) had a coefficient of determination (R^2) of 0.94 when correlated to the measured temperatures, indicating a high level of agreement with the real world.

$$T_{pav} = -5.374 - 0.752\, T_{sur} + 0.022 T_{sur}^2 + 2.016\, T_{avg} - 0.032\, T_{sur} * T_{avg} + 1.549\, t_d - 0.022\, D \qquad (4)$$

Ariawan et al. (2015) also developed a linear regression temperature model for tropical climates experiencing dry and wet seasons, as opposed to winter and summer. Due to the change in seasons, the pavement temperature depends on the air temperature (T) and humidity (H). Separate equations were developed for different pavement depths, for example the model for 70 mm below the surface is at Equation 5.

$$T_{70mm} = 1.965 + 0.755\ T_{air} + 0.331\ T_{surface} \tag{5}$$

For this research, the temperature model developed by Islam et al. (2015) was used. The model is based on the solar radiation and daily temperature data and includes two formulae, one for predicting the minimum temperature (night time) and one for the maximum temperature (day time). For the minimum temperature, the only independent variables are the depth at which the temperature is predicted and the daily minimum temperature (Equation 6). The maximum temperature is also dependent on the intensity of solar radiation (Equation 8). The models were validated by comparing the predicted temperatures to measured pavement temperatures during various prevailing weather conditions (Islam et al. 2015).

$$y_{min} = 1.84 +\ x_{min} + 20x \tag{6}$$

Where: y_{min} = predicted daily minimum temperature at any depth (°C).
x_{min} = daily minimum surface temperature (°C) as per (Equation 7).
x = depth (m).

$$x_{min} = 1.33\ \alpha_{min} + 3.21 \tag{7}$$

Where: x_{min} = daily minimum surface temperature (°C).
α_{min} = daily minimum air temperature (°C).

$$y_{max} = 2.5 + 2.91 x_{max} - 25.6\ x - 0.004 S_{max} \tag{8}$$

Where: y_{max} = predicted daily maximum temperature at any depth (°C).
x_{max} = daily maximum surface temperature (°C) as per (Equation 9).
x = depth (m).
S_{max} = daily maximum solar radiation (W/m^2).

$$x_{max} = 1.33\ \alpha_{max} + 3.21 \tag{9}$$

Where: x_{max} = daily maximum surface temperature (°C).
aα_{max} = daily maximum air temperature (°C).

The predicted temperatures during summer, autumn, winter and spring days and nights in the Sunshine Coast region (Queensland, Australia) are shown in Figure 1. The daily temperature and solar radiation values used for these predictions were sourced from the Bureau of Meteorology. The reversal of the temperature profiles from day to night is clear. This reflects the exposure of the surface to solar radiation and the prevailing air temperature as well as the insulation provider deeper in the pavement. Consequently, the difference between the minimum and maximum temperatures is much greater near the pavement surface than the difference predicted deeper in the pavement.

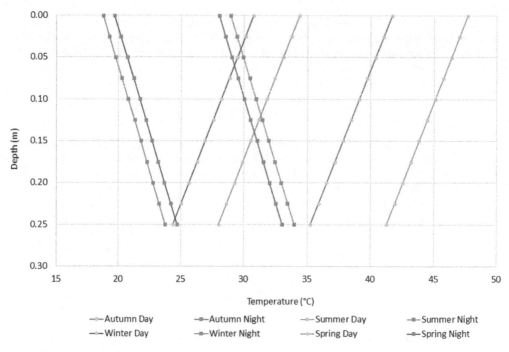

Figure 1. Comparative temperatures predicted by Islam et al. (2015).

3 METHODS AND RESULTS

3.1 *Methods*

The effect of using a dynamic modulus in aircraft pavement design was determined by designing a range of pavements based on the standard static asphalt modulus values and then calculating the change in CDF when the pavement thickness was retained, but the asphalt modulus was changed to a dynamic value. To investigate trends across a range of realistic design scenarios, different pavement structures, subgrade conditions and aircraft were considered. Three pavement structures were analysed, intended to be representative of pavements used in the USA (standard pavement with asphalt base course), in Australia (thin asphalt on a granular base) and in France (full depth asphalt), as shown in Figure 2. Each pavement was analysed for four subgrade conditions, intended to be representative of the full range of subgrade conditions encountered in most practical design scenarios (Table 1). All pavements were designed to separately accommodate 100,00 total passes of the B737-800, and A330-300 and B777-300ER aircraft at their maximum mass and standard tyre pressure, intended to be representative of typical two, four and six wheeled aircraft.

Table 1. Subgrade conditions.

CBR (%)	Modulus (MPa)	Representative of
3	30	Reactive clay or silt
6	60	Stiff clay of silt
10	100	Gravel
15	150	Well compacted sand

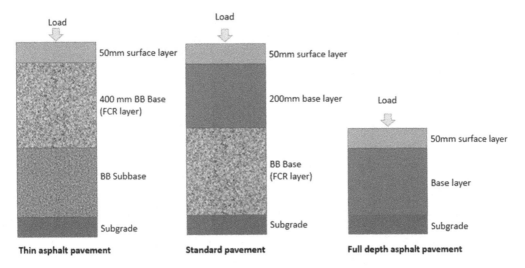

Figure 2. Pavement structures. Note: BB is Barker & Brabston sub-layered FCR or subbase, while FCR is fine crushed rock, usually 20 mm sized.

All asphalt layers were assigned a static modulus of 1,500 MPa (surface) or 3,300 MPa (base). The dynamic modulus values were determined from a master curve generated by laboratory testing of a typical airport used for the resurfacing of Rockhampton Airport (Queensland, Australia) in 2018 (Table 2). The asphalt was a nominal 14 mm sized, densely graded mixture containing a proprietary polymer modified binder and was generally typical of airport asphalt used in Australia (White 2018).

Table 2. Asphalt mixture properties.

Parameter	Test Method	Value
Binder Content (% by mass)	AS 2891.3.1	5.3
Air Voids (%)	AS 2891.8	2.9
Marshall Stability (kN)	AS 2891.5	15.6
Marshall Flow (mm)	AS 2891.5	2.7
Percentage Passing (by mass) Australian Standard Sieves (according to AS1141.11.1)		
Sieve size (mm)	-	Percentage passing (%)
19.0		100
13.2		97
9.5		82
6.7		70
4.75		58
2.36		45
1.18		33
0.600		24
0.300		13
0.150		8.8
0.075		6.1

The asphalt modulus was measured at all combinations of temperature and load frequency levels detailed in Table 3. The asphalt modulus was measured by repeated flexure of rectangular samples (390 mm wide by 63.5 mm high) under four point bending, known in Australia as the complex modulus test, as detailed in test method AG:PT/T274.

Table 3. Asphalt modulus temperature and load frequency levels.

Temperature (°C)	Load frequency (Hz)
5	0.1
15	0.2
25	0.5
30	1
40	2
	5
	10
	20

3.2 *Results*

The laboratory test results (Figure 3) were used to develop a master curve of asphalt modulus, allowing the determination of modulus values at any temperature and load speed/frequency combination. The master curve was generated with reduced frequency as the independent variable (Figure 4) but an equivalent master curve with reduced temperature as the independent variable is equally valid, because the two parameters (temperature and frequency) are combined into the single reduced frequency (or reduced temperature) (Austroads 2013).

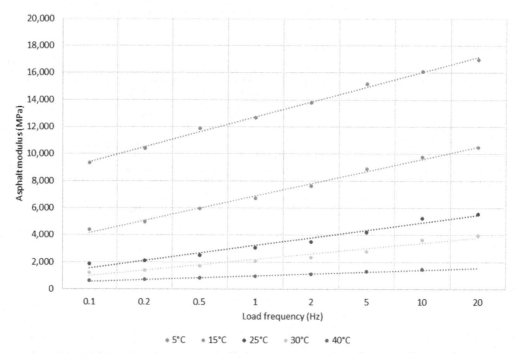

Figure 3. Modulus results at various test temperatures and frequencies.

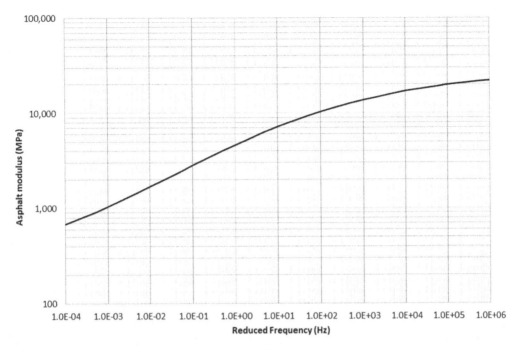

Figure 4. Master curve developed from asphalt modulus results.

As discussed previously, the traditional static asphalt modulus values were used to calculate the required thicknesses of the pavement layer above the subgrade for all pavement types, subgrades and aircraft combinations (Table 4). These are subsequently referred to as the reference thicknesses.

Table 4. Total pavement thickness of pavements (mm).

Pavement type	Subgrade	B737-800	A330-300	B777-300ER
Thin asphalt pavement	CBR 3	1,369	1,907	2,580
	CBR 6	951	1,098	1,550
	CBR 10	674	776	848
	CBR 15	506	593	603
Standard pavement	CBR 3	1,084	1,410	1,869
	CBR 6	762	904	1,190
	CBR 10	528	628	707
	CBR 15	340	445	454
Full depth asphalt	CBR 3	493	654	856
	CBR 6	399	477	595
	CBR 10	333	380	422
	CBR 15	274	313	322

4 DISCUSSION

The full depth asphalt required the thinnest thickness, whereas the thin asphalt pavement had the greatest thickness for any subgrade and aircraft combination. All pavement types required a greater thickness when for a lower subgrade CBR value and the larger aircraft required thicker pavements. These trends are well established and are not considered further.

4.1 Effect of asphalt sub-layers

Because the introduction of variable asphalt modulus values with depth requires many sub-layers to be generated, it was first necessary to determine the influence of sub-layering the asphalt within the APSDS pavement model. Nine pavement-subgrade-aircraft combinations were modelled with 25 mm sub-layers, with each sub-layer assigned the static modulus value (1,500 MPa or 3,300 MPa) used to determine the reference thicknesses (Table 4). The CDF with sub-layering was calculated for each reference thickness and compared to the CDF of the same combinations modelled as single layers. The selected pavement-subgrade-aircraft combinations and CDFs are summarised in Table 5.

Table 5. Comparison of single layer and sub-layered CDF values for reference thicknesses.

Pavement type and subgrade	CDF using single layers			CDF using sub-layers		
	B737	A330	B777	B737	A330	B777
Standard pavement CBR 3	1.000	1.000	1.000	1.000	1.000	1.000
Thin asphalt pavement CBR 6	1.000	1.000	1.000	1.000	1.000	1.000
Full depth asphalt pavement CBR 10	0.998	0.996	0.999	0.998	0.997	0.999

All CDF values associated with sub-layering were within 0.5% of the reference thickness CDF values. The slight deviations from exact agreement reflects the precision to which APSDS adjusts a pavement thickness to achieve a CDF of 1.0, and are of no practical consequence. It was concluded that the addition of sub-layers and interfaces did not introduce any change in the pavement modelling. Consequently, the effect of dynamic modulus on the pavement structures could be determined by comparing the pavement structures with sub-layered asphalt courses directly to the reference pavement thicknesses in Table 4.

4.2 Effect of aircraft speed on asphalt modulus

The modulus of the asphalt increased with test frequency (Figure 3) and consequently the master curve shows that dynamic modulus increased with aircraft speed. For example, at 25°C the modulus ranges from 3,250 MPa at 1 km/hr, up to 9,410 MPa at 300 km/hr (Figure 5). This indicates that pavements of equal thickness will have less structural strength in parking areas, where aircraft move slowly, than on taxiways, where 40 km/hr is a common speed, or for runways, where aircraft land at around 250 km/hr. This difference in modulus has significant implications for pavement thickness and/or predicted life when using dynamic modulus in pavement design.

4.3 Effect of temperature on asphalt modulus

Temperature also has a significant effect on asphalt modulus and this is reflected in the test results (Figure 3). Figure 6 shows the asphalt modulus at various temperatures, compared to the modulus at 25°C, which is a common reference temperature for asphalt modulus values. On average, the modulus at 5°C was 256% greater than at 25°C, while at 40°C, the average modulus was just 30% of the 25°C modulus values. That is, that modulus approximately doubles with every 4°C decrease in temperature. Since the temperature at the surface of the pavement is expected to differ from the temperature 250 mm below the surface by more than 5°C, the effect on pavement thickness and/or predicted life is expected to be significant.

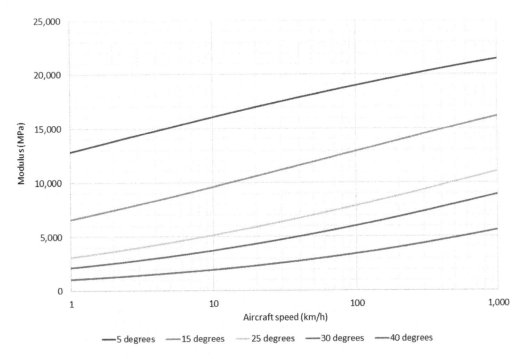

Figure 5. Effect of aircraft speed on asphalt modulus.

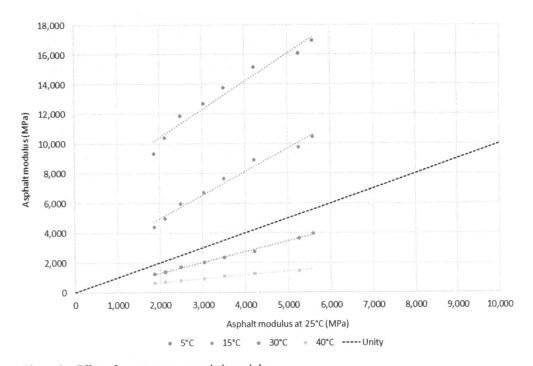

Figure 6. Effect of temperature on asphalt modulus.

4.4 Effect of dynamic modulus on pavement

The CDF value associated with dynamic asphalt modulus was calculated for the reference pavement thickness, for every combination of pavement type, aircraft and sub-grade CBR. The dynamic modulus values were determined from the modulus master curve (Figure 4) based on the 25 mm thick sub-layers, for each of the four temperature with depths models shown in Figure 3. The CDFs are shown in Figure 7. The difference in the seasons is clear. On average the summer CDF values are approximately 918,000 times higher than the corresponding winter CDF values. The range of CDF ratios was 1.1 to 13,000,000. The broad range of summer to winter CDF ratios also highlights the difference in trends for the three pavement types.

CDF value summary statistics are in Table 6. The thin asphalt pavement was associated with only moderate differences in CDF, indicating that the incorporation of dynamic modulus had only modest influence on the predicted pavement lives. This reflects the relatively low portion of asphalt (50 mm) in the overall pavement composition (506 mm to 2,580 mm). In contrast, the full depth asphalt pavement, which is comprised entirely of asphalt, was very sensitive to the inclusion of dynamic modulus, with an average CDF of 0.91, but a range of CDF values of 0.00014 to 65,500. This indicates that based on day time summer temperature profiles, the pavement would fail after just 0.0015% of the design traffic, which is 11 days of a 20 year design life. However, under night time temperature conditions in winter, the 20 year design life would be extended to over 140,000 years. The standard (FAA) type of pavement, with 250 mm of asphalt, was part-way in between, with CDF values of 0.00023 to 5,440.

Overall, the thicker the asphalt layer, the greater the temperature and modulus variation and the greater the effect on the predicted pavement life. Following the same trend, the temperature at which the CDF using the dynamic modulus method was equal to that for the static modulus values, was constant for each load and subgrade strength combination of the thin asphalt pavement, but varied for pavements with thicker asphalt layers. The fixed modulus modelling method overestimated the thin asphalt pavement's strength up to an outside maximum temperature of 30°C. For temperatures above 30°C, the traditional static modulus approach overestimated the strength of thin asphalt pavements, compared to the dynamic modulus approach. This could result in pavement failures on extremely hot summer days. For the standard (FAA) pavement, the temperature at which dynamic modulus CDF values were equal to 1.0, ranged modestly from 25.5°C to 26.1°C, depending on the aircraft and subgrade combination. For the full depth asphalt pavement, the temperature of parity ranged even more, from 25.8°C to 32.4°C. This indicated that introducing dynamic modulus to pavement design becomes more complex as the asphalt depth increases, due to the dependence of the asphalt modulus on the asphalt temperature.

Further complicating the issue, this analysis has only considered the asphalt modulus as a function of asphalt temperature, with a constant aircraft speed of 10 km/hr adopted. However, to introduce dynamic modulus to routine pavement thickness design, the effect of loading speed must also be considered, with asphalt modulus values generally higher for runways than for taxiways and parking aprons. However, aircraft also line-up on runway as they await take-off clearance and in many regional airports, aircraft perform u-turns at the runways ends before taxiing back up the runway to the exit taxiway. Consequently, even the loading speed on a typical runway varies greatly, from stationary to landing speed. Similar complexity exists in highway pavements, which might be posted as a 100 km/hr speed zone, but may regularly have stationary truck traffic during peak hour congestion.

5 CONCLUSION

It was concluded that the thicker asphalt layers within a pavement caused greater temperature and modulus sensitivity through the pavement profile. This occurred because a greater portion of the pavement profile was temperature dependent. The fixed modulus method is therefore more appropriate to model thin asphalt pavements than full depth asphalt pavements, as it

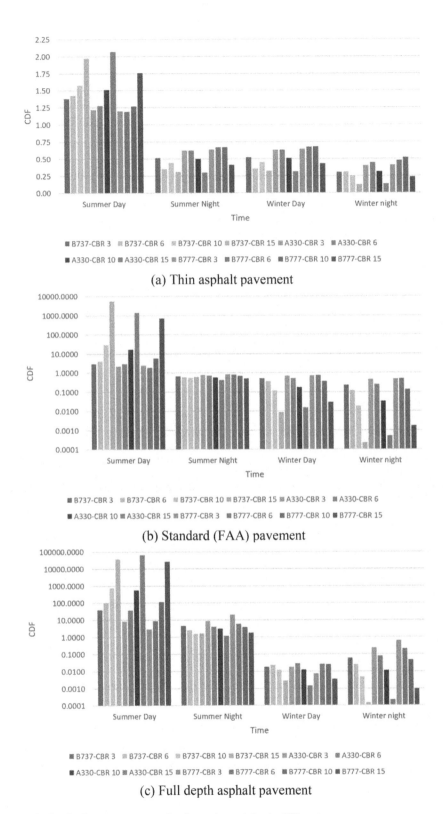

Figure 7. CDF of reference pavement for dynamic modulus in different seasons.

Table 6. Summary statistics of reference pavement CDF values.

Statistic	Thin asphalt	Standard (FAA)	Full depth asphalt
Minimum	0.12	0.00023	0.00014
Medium	0.52	0.55	0.91
Maximum	2.07	5,440	65,500

does not account for large modulus variations. For pavements with a thicker asphalt layer, the dynamic modulus method was more precise and may be more representative of field conditions. It was also concluded that the difference in modulus associated with summer and winter, or day and night, temperature profiles changes the predicted life of the pavement by orders of magnitude, suggesting that all significant pavement damage occurs on the hottest few days of the pavement's life. This research demonstrated that dynamic modulus can relatively efficiently be incorporated into aircraft pavement design. However, many challenges remain, including the analysis of the traffic volumes that occur during various representative temperature conditions and how to cumulatively account for the low damage that occurs in colder weather, as well as the excessive damage predicted to occur during hot weather. In light of the fact that very few aircraft pavement failures are attributed to inadequate pavement thickness, the introduction of dynamic modulus to routine aircraft pavement thickness design is unlikely to be justifiable and is expected to introduce more challenges than it is expected to solve.

REFERENCES

AAA 2017, *Airfield Pavement Essentials*, Airport Practice Note 12, Australian Airports Association, Canberra, Australia, April.

Ariawan, IMA, Subagio, BS & Setiadji, BH 2015, 'Development of asphalt pavement temperature model for tropical climate conditions in West Bali region', *Procedia Engineering*, Vol 125, pp.474–480.

Austroads 2013, Improved Design Procedures for Asphalt Pavements: Pavement Temperature and Load Frequency Estimation, AP-T248-13, 27 September.

Austroads 2016, Characterisation of Flexural Stiffness and Fatigue Performance of Bituminous Mixes, accessed 26 April 2020 <https://austroads.com.au/publications/pavement/agpt-t274-16>.

Balay, JM, Caron, C & Lerat, P 2009, 'Alize-Lcpc Airfield pavement, a new software for the rational design of airport pavement', 2nd *European Airport Pavement Workshop*, Athens, Greece, 13-14 May.

Barker, WR & Brabston, WN 1975, *Development of a Structural Design Procedure for Flexible Aircraft Pavements*, Report No S77-17, US Army Corps of Engineers, Waterways Experiment Station, Vicksburg, USA.

FAA 2016, Advisory Circular 150/5320-6F, accessed 10 May 2019 http://www.faa.gov/documentlibrary/media/advisory_circular/150-5320-6f.pdf.

FAA 2017, *FAARFIELD*, Version 1.42, Federal Aviation Administration, Washington, District of Columbia, USA, 18 September, accessed 17 December 2017 <http://www.airporttech.tc.faa.gov/Download/Airport-Pavement-Software-Programs/Airport-Software-Detail/ArtMID/3708/ArticleID/4/FAARFIELD-142>.

Gedafa, DS, Hossain, M & Romanoschi, SA 2014, 'Perpetual pavement temperature prediction model', *Road Materials and Pavement Design*, Vol. 15, 1, pp. 55–65, DOI:10.1080/14680629.2013.852610.

Gray, C, Yeaman, J & Tighe, S 2015, 'Using innovative in-situ measuring tools to better understand asphalt performance', *AAPA International Flexible Pavements Conference*, Gold Coast, Queensland, Australia, 13-16 September.

Huang, YH 1993, *Pavement Analysis and Design*, Prentice-Hall, New Jersey, USA.

Islam, MR, Ahsan, S and Tarefder, RA 2015, 'Modeling temperature profile of hot-mix asphalt in flexible pavement', International Journal of Pavement Research and Technology, vol. 8, no. 1, pp. 47–52, doi: 10.6135/ijprt.org.tw/2015.8(1).47.

Mincad 2010, *APSDS5: Airport Pavement Structural Design System: User Manual*, Revision 5.0.055, 9 November, accessed 30 November 2017, <https://www.mincad.com.au/downloads/access/?download>.

Mounier, D, Broutin, M and Bost, R 2015, 'Mechanistic-Empirical Procedure for Flexible Airfield Pavement Design: The New French Technical Guidance', *Airfield and Highway Pavements*, pp. 720–729. Doi: 10.1061/9780784479216.064.

Pan, T, Tutumluer, E & Carpenter, SH 2005, 'Effect of coarse aggregate morphology on the resilient modulus of hot-mix asphalt', *Transportation Research Record*, no. 1929, pp. 1–9.

TMR 2017, *A New Approach to Asphalt Pavement Design*, accessed 26 April 2020, <https://www.tmr.qld.gov.au/-/media/busind/techstdpubs/Technical-notes/Pavements-materials-geotechnical/TN167.pdf>.

Wardle, LJ, 1977, *Program CIRCLY User's Manual*, Commonwealth Scientific and Industrial Research Organisation, Australia, Division of Applied, Geomechanics, Geomechanics Computer Program No 2.

Wardle, L & Rodway, B 1998, 'Recent developments in flexible aircraft pavement design using the layered elastic method', *3rd International Conference on Road and Airfield Pavement Technology*, Beijing, April.

White, G 2018, 'State of the Art: Asphalt for Airport Pavement Surfacing', *International Journal of Pavement Research and Technology*, vol. 11, no. 1, pp. 77–98.

White, G & Embleton, K 2015, 'Next generation binder for airport asphalt', *16th AAPA International Flexible Pavement Conference*, Gold Coast, Queensland, Australia, 13-16 September.

White, G, Kelly, G, Fairweather, H & Jamshidi, A 2020, 'Theoretical socio-enviro-financial cost analysis of equivalent flexible aircraft pavement structures', *99th Annual Meeting of the Transportation Research Board*, Washington, District of Columbia, USA, 12-16 January.

Practical implications for the implementation of the new international airport pavement strength rating system

G. White
School of Science and Engineering, University of the Sunshine Coast, Sippy Downs, Queensland, Australia

ABSTRACT: Since 1981, most airports have used an internationally standardized method of airport pavement strength rating, which is based on simple pavement analysis models and categorical subgrades. With the broad implementation of mechanistic-empirical layered elastic pavement design methods, a mis-match between pavement thickness determination and pavement strength rating has evolved. To address this, a new system is being implemented that uses the same layered elastic model for strength rating as is used for pavement thickness determination in the United States and new subgrade categories, taken from French road design. Case studies on four Australian airports indicate how the new system may be implemented in countries that do not mandate prescriptive methods of pavement strength rating, rather using engineering judgment exercised by specialist engineers based on historical using aircraft, historical pavement performance and reverse engineering of existing pavement structures. The transition to the new system will create some challenges, mostly for those airports that do not know the basis of their current strength rating and can not locate historical existing pavement composition and subgrade bearing strength data. The cost of transition is estimated to be significant and is not justified because the new system provides no practical benefit to most airports in Australia.

Keywords: Runway, strength, ICAO, PCN, PCR

1 INTRODUCTION

Since their first introduction in the early 1900s, aircraft have become larger and heavier. Particularly since WWII, aircraft wheel loads and tyre pressures have increased significantly (Roginski 2007; Fabre et al. 2009) and this requires ever-stronger pavements to be provided by airports. In 1958, the USA Federal Aviation Administration (FAA) implemented a policy to restrict the development of new aircraft that stressed pavements more than a DC8-50 aircraft operating at 159 t. In 2009 the FAA pavement protection policy was withdrawn, effectively allowing more demanding aircraft to be developed without restriction (Rodway 2009). Consequently, the provision of airport pavements capable of accommodating any more demanding future aircraft became the responsibility of airport owners. The trend of increasing aircraft tyre pressures and individual wheel loads has continued since that time and is not expected to abate in the future.

To assist airport owners and operators to control and manage the impact of different aircraft on their pavements, the International Civil Aviation Organisation (ICAO) introduced an internationally recognised system for airport pavement strength rating in 1981 (CROW 2003). The system is known as the Aircraft Classification Number (ACN)-Pavement Classification Number (PCN) system and is subscribed to by the approximately 200 member States of ICAO

(ICAO 2020). As detailed further below, every aircraft has a mathematically exact ACN, based on the weight of the aircraft, its tyre pressure and the bearing capacity of the subgrade over which it operates. The pavement subgrade bearing capacity is important because it determines the approximate depth of the covering pavement and therefore the degree of interaction between the 2, 4 or 6 wheels on any given aircraft landing gear leg or truck. The ACN of a particular aircraft is compared to the PCN assigned to a particular pavement. If the ACN is equal to or less than the PCN, the aircraft is permitted to operate without restriction. When the ACN is greater than the PCN, the airport operator must seek the airport owner's permission to operate, under what is known as a Pavement Concession.

Unlike the ACN of an aircraft, the PCN of a pavement is not mathematically exact and is subject to the discretion of the airport owner. Some ICAO member States prescribe the method of PCN determination for the airports under their jurisdiction. For example, the FAA of the USA has a formal process for PCN determination (FAA 2014b). This reflects the centralised role of the FAA in providing significant funding towards airport pavement construction and upkeep, via the Airport Improvement Program (FAA 2020). In contrast, Australia provides only minimal guidance regarding methods for PCN determination (CASA 2011). This reflects the Australian Government's divestment from civilian airport ownership in the 1990s and allows private airport owners to set the PCN of the runway at any value they wish, based on the aircraft they wish to attract in order to increase the associated level of income. Importantly, in Australia, a private airport owner also pays for airport pavement construction and maintenance, meaning the private airport owner must consider the increased pavement damage associated with a higher PCN value, against the increase in revenue that it allows.

In the USA and some other States, where a prescriptive and formal process of PCN determination is followed, challenges have occurred due to a difference in the mathematical models used for pavement thickness design and those used for PCN determination. As explained below, modern airport pavement thickness design is based on layered elastic and/or finite element calculation of strain as the indicator of damage within a realistic pavement structure (Wardle & Rodway 2010). In contrast, the software prescribed for aircraft ACN calculation, and therefore PCN assignment, uses Boussinesq-based deflection as the damage indicator within a standard pavement composition (FAA 2014a). Non-standard compositions, which are common, must be converted to an equivalent standard thickness, using material equivalent factors (FAA 2014a). This difference can lead to a pavement being designed for a specific aircraft and then having a PCN assigned that requires that same aircraft to operate under a Pavement Concession. That is, the pavement is assigned a PCN that is lower than the ACN of the aircraft that it was designed for, which is illogical.

To avoid this challenge in the future, ICAO has developed a replacement for the ACN-PCN system. The ACN is replaced by an Aircraft Classification Rating (ACR) and the PCN is replaced by a Pavement Classification Rating (PCR). The ACR-PCR system is intended to operate on the same principles as the ACN-PCN system with one primary difference. ACR values are calculated using the critical strain in the same layered elastic model that is used in the FAA's pavement thickness design software FAARFIELD (FAA 2017). However, to replace the ACN-PCN system with the ACR-PCR system requires all aircraft manufacturers to determine and publish ACR values. It also requires every airport in every ICAO member State to change their PCN to an equivalent PCR.

This paper presents examples of the transition from ACN-PCN to ACR-PCR for different airports that are typical of the 400 significant and paved airports in Australia (AAA 2017). First the two pavement strength rating systems are compared and trends in the relationship between ACN and ACR values are explored. Four examples are then presented as case studies of the transition. Finally, transitional and implementation issues and challenges are compared to the benefit that the new pavement strength system will provide. Although this paper is presented in the context of flexible Australian airport pavements, the findings also apply to other ICAO member States that generally allow airports to determine the basis of their pavement strength rating.

2 BACKGROUND

2.1 Australian Airport pavements

Like road and highway pavements, airport pavements are designed to protect the natural or imported subgrade from the traffic loads they are expected to support over their design life. Furthermore, airport pavements are generally either rigid or flexible in nature, although some composite pavements are also used (Deilami & White 2020). Traditionally, flexible airport pavements were designed to comprise thick layers of well compacted and high quality fine crushed rock (FCR) base over uncrushed gravel sub-base with a thin bituminous surfacing. The pavement thickness was determined using the methods developed by the US Army Corps of Engineers (Corps) between the 1940s and 1970s (Ahlvin 1991). Since that time, many counties have developed significantly different pavement styles, with much thicker asphalt, bound or stabilised sub-base layers and some use full depth asphalt structures (White et al. 2020). However, Australia largely retains the Corps approach to airport pavement design and all Australian runways are comprised of a flexible pavement structure (AAA 2017). One aspect of airport pavement practice that is generally peculiar to Australia is the significant use of a sprayed bituminous surface directly over granular base course (White 2019). This reflects the large number of Australian airports that are associated with smaller commercial aircraft in remote areas and the high cost of providing asphalt surfaces in these remote locations (AAA 2017). Importantly, all Australian airports subscribe to the ACN-PCN system of pavement strength rating and all rated airports publish their PCN in the public domain.

2.2 ACN-PCN system

As stated above, the ICAO airport pavement strength rating system is known as ACN-PCN. Aircraft loads are expressed by an ACN which allows no discretion. For a specific aircraft at a given operating mass and tyre pressure, there is only one ACN per pavement subgrade category. The ACN is defined as twice the wheel load (in tonnes) which on a single wheel, inflated to 1.25 MPa, causes vertical pavement deflection (calculated at the top of the subgrade) equal to that caused by the actual multi-wheel aircraft gear, at its actual gear load and its actual tyre pressure. The interaction between multiple wheels of a specific landing gear changes with pavement depth. This means that two aircraft with different landing gear configurations, but the same ACN for a particular subgrade category, will cause relatively different damage to pavements with different thickness. Pavement thickness is also significantly affected by subgrade stiffness, usually expressed as the California Bearing Ratio (CBR). The application of ACN-PCN therefore changes with subgrade CBR. Rather than a continually varying ACN across all possible subgrade CBR values, subgrades are categorised and a representative CBR adopted (ICAO 2013):

- Category A. High strength. Represented as CBR 15.
- Category B. Medium strength. Represented by CBR 10.
- Category C. Low strength. Represented by CBR 6.
- Category D. Ultra-low strength. Represented by CBR 3.

The PCN is presented as a multi-element expression, such as the example in Equation 1. The main element is the number against which the ACN is compared. It is intended to protect the pavement structure from overloading, primarily based on permanent subgrade deformation, also known as pavement rutting. The tyre pressure limit is secondary and is compared to the operating tyre pressure of the aircraft and is intended to protect the surface from high near-surface stress.

$$PCN\ 58/F/A/X/T \tag{1}$$

Where: 58 is the numerical element against which the ACN is compared.
F is to indicate a Flexible pavement, rather than R for Rigid.

B is the category of subgrade detailed above.
X is the tyre pressure limit category detailed above.
T is to indicate a Technical assessment rather than U for a Usage based assessment.

The tyre pressure limits are categorical in nature and are inherently empirical. Aircraft with tyre pressures less than the assigned category limit are permitted to operate without specific approval from the airport owner. Aircraft with higher tyre pressures require a Pavement Concession. Some countries, such as Australia, have adopted airport-specific tyre pressure limits rather than tyre pressures categories and category limits (ASA 2020).

In 2008 increases in the tyre pressure limits were proposed by aircraft manufacturers (Rodway 2009). After some full-scale testing, the proposed increase in the categorical tyre pressure limits of the ACN-PCN system was approved by ICAO in 2013 (Roginski 2013). Table 1 provides the original and revised tyre pressure limits. This change allowed an increase in aircraft tyre pressure from 1.50 MPa to 1.75 MPa to operate on airports with a Category X tyre pressure limit.

Table 1. ACN-PCN Tyre Pressure Category Limits.

Category	Original Tyre Pressure Limits	Revised Tyre Pressure Limits
W	Unlimited	Unlimited
X	1.50 MPa	1.75 MPa
Y	1.10 MPa	1.25 MPa
Z	0.50 MPa	0.50 MPa

As stated above, some jurisdictions publish standard procedures for the formal calculation of the PCN. In the USA, FAA (2014) requires an airport to determine a runway PCN by generally:

- Converting the actual pavement structure to an equivalent thickness of a standard structure. This is often complicated because actual pavement structures in the USA are generally different to the standard pavement composition in COMFAA, which is 75 mm of P-401 (asphalt) on 150 mm of P-209 (crushed rock) on the required thickness of P-154 (uncrushed gravel).
- Determining the critical aircraft, which is the aircraft that is predicted to consume the largest portion of the pavement's design life, based on the actual number of passes included in the traffic mix for pavement thickness design. This is often the aircraft with the highest ACN, but can be an aircraft with a lower ACN when the traffic mix includes a large number of the less damaging aircraft, compared to the aircraft with the higher ACN value.
- Determining the weight of the critical aircraft that would result in it causing the same total pavement damage in the predicted number of passes that all the aircraft caused when they were converted to an equivalent number of passes of the critical aircraft.
- Determine the ACN of the critical aircraft, at its increased/decreased weight, that produces the same damage as all of the aircraft in the aircraft traffic spectrum at the weights and load repetitions that were used in pavement design.
- Assigning that ACN value as the PCN of the pavement, which can be higher than all the actual ACN values of all the aircraft that the pavement was designed for.

When an aircraft with a relatively high ACN is predicted to operate at relatively low frequency, this process can result in an PCN that is unreasonably high and the FAA requires that a minimum 1,000 coverages be adopted over a 20 year design life (FAA 2014b). Although this formal method reduces the ability for an airport to arbitrarily over-rate or under-rate the strength of a runway pavement, it still requires some discretion when an irregular aircraft has the highest ACN. Furthermore, in some instances, an aircraft that is less damaging during design can be found to be more damaging during PCN determination. This reflects the difference in relative damage when the damage is determined using maximum deflection in a Boussinesq model of equivalent standard thickness, compared to using strain in a layered elastic based realistic pavement model. For example, 1,000 annual departures of the aircraft in Table 2 all require the same 428.6 mm thickness of flexible deep asphalt pavement on CBR 6 subgrade. However, the associated flexible ACN values for subgrade C are only approximately equal, meaning that a pavement designed for the B767-300 and assigned a PCN of 50, would not allow all the other aircraft to operate without a Pavement Concession, even though all require the exact same pavement thickness and are therefore considered to be equally damaging. Although the practical significance of a ACN value of 52, compared to a value of 50, is likely to be low, the rigid system adopted in the USA, and the inexperienced and sometime conservative nature of airport owners and managers, often results in Pavement Concessions and PCN assignment being viewed in a black and white manner. This makes small mismatches between airport pavement thickness design and airport pavement strength rating/management, a much more significant issue in practice than it should be.

Table 2. Example ACN (C) values for equal FAARFIELD pavement thickness.

Aircraft	Mass	ACN (C)
B737-800	79.2 t	50.3
A321-200	82.0 t	51.6
B767-300	156.1 t	49.5
A350-900	193.9	50.8
B777-300ER	242.3	49.7
A350-1000	236.5	51.8

Note: Aircraft masses adjusted to require the same thickness as the B737-800 at standard mass. Pavement modelled as the required thickness of P401 (asphalt) over 150 mm of P154 (sub-base) over a CBR 6 subgrade.

3 ACR-PCR SYSTEM

To overcome the discrepancy between layered elastic based pavement thickness determination and Boussinesq based PCN calculation, ICAO has developed a replacement system (Fabre 2018). As stated above, the new ACR-PCR system was designed to be more rationale by using a critical strain calculated by layered elastic methods as the indicator of damage. The new system was developed by ICAO's Airport Pavement Expert Group over the period 2014-2018 and approved by ICAO in 2019 (Fabre 2019). The new system became applicable in July 2020 and will be effective in November 2024. That means that between July 2020 and November 2024, all ICAO member States will need to transition from ACN-PCN to ACR-PCR. Similarly, all aircraft manufacturers will need to publish ACR information for their aircraft and must publish both ACR and ACN data during the transition period, during which time some ICAO member States will be using ACR-PCR, but others will still be using ACN-PCN.

The ACR-PCR system was developed to be based on the same principles and mechanics as the ACN-PCN system. Despite the apparent similarities in the two systems, ACR-PCR system includes significant changes (Fabre 2019):

- Strain used as the relative damage indicator, rather than deflection.
- All wheels considered explicitly, rather than being converted to an equivalent single wheel.
- Actual pavement materials and composition considered explicitly, rather than being converted to a standard composition that is rarely used in many countries.

- Load repetitions, tyre pressures and pavement structures selected to be more comparable to typical modern airport pavement structures.
- Rigid and flexible pavement subgrade categories use the same elastic modulus (or CBR) ranges and removal of the largely discontinued k-value (modulus of subgrade reaction) for rigid pavement subgrades.

The term ACR-PCR was adopted to avoid confusion with ACN or PCN values. Furthermore, the ACR is defined as twice the equivalent wheel load in hundreds of kilograms, rather than in tonnes. This means that ACR values generally range from 50 to 1000, compared to ACN values which generally range from 5 to 100. This change was also designed to avoid confusion between the two systems.

3.1 Comparing the two systems

As stated above, the ACR-PCR system was developed to be based on layered elastic methods and to be more representative of modern typical pavement structures and aircraft. However, ACR-PCR was developed to operate in a similar manner to ACN-PCN. That is the aircraft ACR is compared to the pavement PCR. If the PCR exceeds the ACR, then the aircraft can operate without restriction. However, when the ACR exceeds the PCR, a Pavement Concession is required. Also similar to the ACN-PCN system, the tyre pressure limit check is also required, and this is effectively unchanged by the transition to ACR-PCR.

The main differences between ACN-PCN and ACR-PCR relate to the basis on which the equivalent wheel load is determined, and include (Fabre 2019):

- Standard tyre pressure.
- Standard pavement structures.
- Subgrade categories.
- Calculated indicator of relative damage.

The standard wheel load, to which other landing gear loads are converted, now has a 1.50 MPa tyre pressure to better reflect large modern aircraft, compared to the 1.25 MPa tyre pressure used to calculate ACN values. The flexible standard pavement structure has greater asphalt thickness and now depends on the number of wheels in the landing gear being considered. Table 3 shows the two flexible pavement structures. The rigid pavement structure is not affected by the number of wheels in the landing gear, as shown in Table 4.

Table 3. ACR-PCR standard flexible pavement structures.

Layer	ACN-PCN thickness	ACR-PCR thickness for 1-2 wheels	ACR-PCR thickness for 3 or more wheels
Asphalt surface (P401/403)	75 mm	76 mm	127 mm
Crushed rock (P209)	150 mm	As required	As required
Uncrushed gravel (P154)	As required	Not used	Not used
Subgrade	Infinite	Infinite	Infinite

Table 4. ACR-PCR standard rigid pavement structures.

Layer	ACN-PCN thickness	ACR-PCR thickness
Concrete base (P501)	As required	As required
Crushed rock (P209)	Combined with subgrade	200 mm
Subgrade	Infinite	Infinite

The standard subgrade categories have been adjusted to be the same for rigid and flexible pavements and now correspond to subgrades categories used in France for road and highway pavement design. Although the alignment of rigid and flexible pavement subgrade rating is supported, the selection of French roads and highways as the basis is arbitrary and will result in many aerodromes needing to change from one subgrade category to another, which will complicate the transition from ACN-PCN to ACR-PCR. The current and new subgrade categories are summarised in Table 5. Figure 1 then compares the subgrade CBR categories and highlights the representative values used for each category in the respective systems. Figure 1 also highlights the CBR values that will necessitate airports changing from one subgrade category to another.

Table 5. ACN-PCN and ACR-PCR flexible subgrade categories.

Subgrade Category	ACN-PCN system		ACR-PCR system	
	Nominal CBR	CBR Range	Nominal CBR	CBR Range
A	15	13 and above	20	15 and above
B	10	8-12	12	10-14
C	6	4-8	8	6-9
D	3	4 and below	5	5 and below

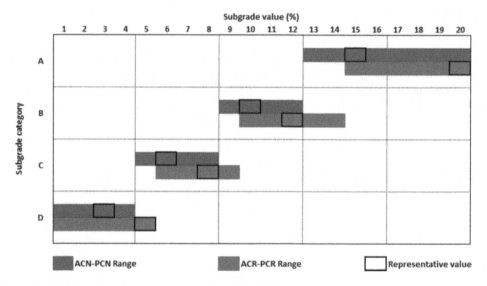

Figure 1. Flexible subgrade category ranges for ACN-PCN and for ACR-PCR.

The ACR-PCR system actually uses the elastic modulus of the subgrade (expressed in MPa) to reflect the input into modern pavement thickness design software, but Table 5 shows equivalent CBR values using a simply linear conversion of 10 times. The use of elastic modulus avoids the need to estimate k-values for rigid pavements, which simplifies the ACR-PCR system for rigid pavements. The category D increase from CBR 3 to CBR 5 reduces the representativeness of the system for many Australian aerodromes that have old and poor natural subgrades with very low CBR values (White et al. 2021).

The ACR-PCR indicator of relative damage caused by different aircraft is vertical strain at the top of the subgrade, instead of maximum deflection at the top of the subgrade. Furthermore, the layered elastic models in FAARFIELD (FAA 2020) are used to calculate the magnitudes of strain, rather than the simpler (Boussinesq) models used in COMFAA. This change reflects the more sophisticated computer power that is now readily accessible and greatly reduces the anomalies between pavement thickness design and strength rating in the USA. However, it means that the relationship between ACN and ACR is not fixed.

3.2 Comparing ACN and ACR values

Figure 2 Shows the ACR and ACN values for 17 common commercial and General Aviation (GA) aircraft on each of the four subgrade categories, ranging from the Super King Air 350 (6.9 tonnes) to the A380 (575 tonnes). On average, the ACR values were 9.5 times the ACN values for the same aircraft, with the ratios between ACN and ACR ranging from 7.7 to 12.0. It is these minor deviations away from ACR being 10 times

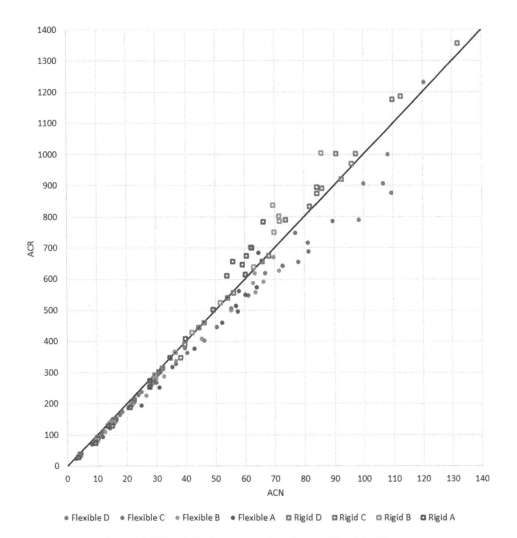

Figure 2. Comparison of ACN to ACR for various aircraft on rigid and flexible pavements.

the ACN that will reduce the discrepancies between FAARFIELD designed aerodromes pavement thickness and COMFAA based pavement strength rating and PCN assignment.

It is important to understand that the conversion is not fixed and therefore some aircraft have an ACR value that is less than 10 times the ACN, while other aircraft will have an ACR value that is more than 10 times the ACN. This effectively means that the transition from ACN-PCN to ACR-PCR implies a difference in the relative damage caused by common aircraft. This has the potential to change the aircraft that is deemed to be the critical aircraft at any particular airport. Furthermore, Figure 2 generally indicates that:

- ACR-PCR predicts that on average, the same aircraft will do more damage to a rigid pavement than to a flexible pavement with the same PCN. That effectively requires rigid pavements to be thicker, relative to flexible pavements, than currently predicted by the ACN-PCN system.
- ACR-PCR predicts that smaller aircraft (one and two wheels) will be even less damaging, compared to larger (four and six wheels) aircraft, implied by ACN-PCN. The practical effect of this will vary depending on the mix of aircraft that operate at any particular airport. However, large aircraft always govern the strength rating of any airport pavement, so this will only affect airports with a strong main runway and a weak secondary runway, because one runway will likely have an PCR that is more than 10 times the PCN, while the other runway will likely have a PCR that is less than 10 times the PCN.
- The subgrade bearing strength (CBR indicated by the category) had no impact on the relationship between the ACN value and the ACR value.

4 CASE STUDIES ON TRANSITION

Other researchers and practitioners have presented examples of the application of ACR-PCR (Fabre & Vaurs 2017; Fabre 2018; Fabre 2019). However, these examples have generally been for large airports with complex aircraft traffic spectra. Most examples have also continued to be based on the prescriptive method of strength rating following in the USA (FAA 2014b; FAA 2020) and have focussed on the mechanics of that process. To balance these previous examples, this research presents the practical PCR determination for four different airports in Australia, where the prescriptive process for strength rating is not mandated. The airports were selected to demonstrate the range of complexities that are expected to arise as the system is imposed on all Australian airports. Note that Australia uses tyre pressure limits, rather than the categories shown in Table 1. However, the PCN and PCR expressions in the following examples are based on the applicable tyre pressure categories for simplicity and because the transition from ACN-PCN to ACR-PCN has no practical impact on the tyre pressure.

4.1 Example 1 – Mudgee airport

The PCN for Mudgee airport is shown in Equation 2. The 'U' is important because it indicates that the strength rating is based on aircraft usage and not on a technical evaluation. This means that the basis of the strength rating is less likely to be clearly documented.

$$Mudgee\ current\ PCN\ 12/F/C/Y/U \qquad (2)$$

Review of their historical records indicated that Mudgee airport had been subject to two previous pavement investigations. Analysis of the records generally indicated that:

- The characteristic subgrade bearing strength was CBR 6%.

- Current regular aircraft are the JetSteam 41, which has an ACN on subgrade category C of 4.9.
- Historical aircraft use included JetStream 32 (ACN 6.5), Saab 340B (ACN 8.2) and Fokker F27 (ACN 11.6).

The current regular aircraft operate with an ACN well below the PCN. It was therefore surmised that the current usage-based PCN value was based on the historical use of the F27, which is no longer in service. To determine an appropriate PCR, three options were considered:

- Set the PCR to the ACR of the JetStream 41.
- Set the PCR to the ACR of the Fokker F27.
- Perform a technical assessment of an appropriate PCN and convert that PCN to a PCN value, based on the adopted critical aircraft.

Using the ACR of the JetStream 41 was not recommended because it would likely underrate the pavement based on its historical rating and usage. Simply setting the PCR to the ACR of the F27 was considered to be unreasonable because the F27 no longer operates into Mudgee airport and is no longer a common aircraft in Australia. Therefore, the PCN was first determined using the information in the historical records, supplemented by a visual condition assessment. The assessment found that the Falcon 7X aircraft was suitable at an operating mass of 16.3 tonnes, which results in an ACN of 10. The equivalent ACR value is 92, resulting in the strength rating in Equation 3 (PCN) or equivalent rating in Equation 4 (PCR). Because subgrade CBR 6 is a category C subgrade under both strength rating systems, there was no requirement to change subgrade categories.

$$Mudgee\ recommended\ PCN\ 10/F/C/Y/T \qquad (3)$$

$$Mudgee\ recommended\ PCR\ 92/F/C/X/T \qquad (4)$$

Mudgee airport was fortunate that historical records provided enough information to allow the PCN to be reviewed on a technical basis and the equivalent PCR value to be determined. However, the assessment took two days to be performed by a specialist engineer that was required to travel from inter-State to conduct the condition assessment, costing around AUD 5,000.

4.2 Example 2 – Birdsville airport

The current PCN for Birdsville airport is in Equation 5. Again, the basis was not known to the airport staff and the 'U' indicates that it was historically based on using aircraft. The only significant and regular aircraft is the Saab 340 B, which has an ACN (on subgrade B) of 6.5, which is significantly below the current PCN.

$$Birdsville\ PCN\ 10/F/B/Y/U \qquad (5)$$

Birdsville airport did not know its characteristic subgrade strength or the basis of its current strength rating and did not have a good understanding of its historical aircraft traffic that might be associated with the current PCN value. To understand these issues and to enable a transition to PCR, investigations were required, including:

- Condition assessment by a specialist engineer.
- Geotechnical testing.
- Analytical evaluation of the current pavement strength.

It was determined that the characteristic subgrade CBR value was 9. Although that is a category B subgrade under the ACN-PCN system, it is a category C under the ACR-PCR system, as shown in Table 5. The pavement was found to be adequate for projected operations of a CL 604 at 18.2 tonnes. The ACN of the CL 604 is 9.8 (on subgrade B) indicating that the current usage based PCN (Equation 5) is appropriate.

When converting this PCN to an equivalent PCR, the different subgrade category also needed to be accounted for. At 18.2 tonnes, the CL 604 has an ACR of 88 (subgrade B) and 98 (subgrade C). Because of the change in the subgrade category changes, the recommended PCR was based on subgrade category C, as shown in Equation 6.

$$Birdsville\ PCR\ 98/F/C/Y/T \tag{6}$$

In retrospect, Birdsville airport could have simply picked any aircraft with an ACN of 10, on subgrade C, and determined the ACR of that aircraft. However, that would rely on retaining the category C subgrade because the characteristic subgrade CBR was not known. Such an approach would have resulted in a PCN value of anywhere between 90 and 110, depending on the aircraft selected. However, this range of potential PCR values would not be of practical significance because the ACR of the Saab 340 B is just 43 (for subgrade category B).

The investigation of the existing pavement structures, and the determination of an accurate PCR, took six months and cost around AUD 80,000. Although the capacity of the existing pavement was determined and future upgrade options were better understood, the practical return on the investment was negligible.

4.3 Example 3 – Proserpine airport

Proserpine airport was upgrade by partial reconstruction and structural asphalt overlay in 2017. Prior to the upgrade the PCN (Equation 7) was lower than the ACN of the using A330-300 aircraft, which is 58 on flexible pavement with a category A subgrade. Following the upgrade, the PCN was increased to reflect the A330-300, as shown in Equation 8, although the B737-800 remained the dominant aircraft based on frequency, but was not the critical aircraft for pavement thickness determination. Interestingly, during the upgrade design, the subgrade was found to have a characteristic CBR value of 18, resulting in the change from PCN subgrade category B to category A.

$$Proserpine\ pre - upgrade\ PCN\ 46/F/B/X/T \tag{7}$$

$$Proserpine\ post - upgrade\ PCN\ 58/F/A/X/T \tag{8}$$

Because the pavement was recently upgraded, which included significant geotechnical investigations, the existing pavement structures were well known. The pavement upgrade design and construction records were also available, meaning that the basis of the current strength rating (Equation 8) was clearly documented. Consequently, the conversion from, PCN to PCR only required the calculation of the A330-300 ACR. Although the prescriptive process could be followed, the intent of the strength rating remained to allow the A330-300 to operate in an unrestricted manner, and the change from ACN-PCN to ACR-PCR did not change this. The resulting PCR for Proserpine airport is shown in Equation 9.

$$Proserpine\ PCR\ 754/F/A/X/T \tag{9}$$

Despite this being the simplest of all the examples presented in this paper, the airport was unable to determine the PCR without assistance from a specialist engineer. This required the specialist engineer to review the documentation associated with the 2017 pavement upgrade and calculate the PCR of the A330-300 and recommended the published PCR in a short report. In total, one day of a specialist engineer was required, costing around AUD 2,000.

4.4 Example 4 – Brisbane airport

Brisbane airport is a major international airport with two main runways of comparable strength. The original runway was constructed in the 1980s and the new runway was completed in 2020. The airport is also unusual in that it is constructed on a deep sand fill, typically 2.0-2.5 m thick, over low CBR marine clay subgrade. Although the strength ratings reflect the clay as the subgrade (category D), the pavement structures were designed to protect the sand fill, which has a typical CBR value greater than 20 (category A). Prior to the new runway's construction, the old runway accommodated all current domestic and long haul aircraft, up to and including the B777-300ER, despite having a strength rating (Equation 10) lower than the ACN of the B777-300ER, which is 120 for a flexible pavement on a category D subgrade. The new runway has a higher PCN (Equation 11) and the basis of the new runway strength rating is not known, although the new runway was designed to carry the same aircraft as currently uses the old runway.

$$Brisbane\ old\ PCN\ 108/F/D/X/T \qquad (10)$$

$$Brisbane\ new\ PCN\ 124/F/D/X/T \qquad (11)$$

It is expected that the two Brisbane airport runway PCN values resulted from a technical evaluation using the current FAA method of PCN determination using COMFAA. Despite the old runway PCN being rated significantly lower than required for the B777-300-ER, the old runway has supported B747-400, A380-800, A350-900 and B777-330ER operations multiple times per day in recent years. When the two runway structures were evaluated under the new FAA PCR determination method (FAA 2020) the results were similarly different, as shown in Equation 12 and Equation 13.

$$Brisbane\ old\ PCR\ 1112/F/D/X/T \qquad (12)$$

$$Brisbane\ new\ PCR\ 1272/F/D/X/T \qquad (13)$$

The rating of the new runway to a level that exceeds all practically expected aircraft is questioned. Regardless of whether the strength rating is PCN 124 (compared to an ACN of 120) or PCR 1272 (compared to an ACR of 1233), over-rating the pavement simply because that is the outcome of the prescriptive FAA strength rating process is unjustified. Furthermore, maintaining a lower rating for the old runway, whether PCN 108 or PCR 1112, is also not recommended, because the runway has been supporting aircraft up to the B777-300ER (ACN 120 or ACR 1233) multiple times a day for some years. Consequently, it is recommended that both runways be re-rated to reflect the ongoing regular use of the B777-300ER, as shown in Equation 14.

$$Brisbane\ recommended\ PCR\ 1233/F/D/X/T \qquad (14)$$

5 TRANSITIONAL IMPLICATIONS

The ACN-PCN system has been in place for almost 30 years, since 1981, and during that time has only been subject to minor variations, as described above. Consequently, ACN-PCN is well entrenched in the strength rating and aeronautical information publications (AIP) of ICAO member States around the world. Consequently, the change to ACR-PCR will be significant for airports across the globe and this will undoubtably create some transitional challenges.

ICAO are permitting member States to transition from ACN-PCN to ACR-PCR anytime from July 2020 until November 2024. This will require aircraft manufacturers to publish both

ACN and ACR values for their aircraft until at least the end of 2024. It will also require the FAA and other organisations that produce ACN calculators, such as COMFAA (FAA 2014a) and PCN strength rating guidance (FAA 2014b) to update their guidance and eventually withdraw the ACN-PCN based documentation. On a national scale, each ICAO member State will either need to transition all airports on a single day, or allow a similar period of optional strength rating. Because it is impractical for AIP producers to run parallel systems and for an aircraft to take-off from Brisbane on a PCN and land in Sydney on a PCR, it is expected that all Australian airports will transition on a single day, which will create additional challenges.

5.1 Re-publication

There are approximately 700 registered airports in Australia. Of these, around 480 have a strength rating, of which around 350 have a PCN (White et al. 2021). The others are generally for general aviation and have a 5,700 kg aircraft mass limit. Each airport publishes its PCN online and in print, in a document known as the ERSA (ASA 2020). The ERSA is updated every three months and hard copies are distributed to more than 1,000 registered users. If all 350 airports with a current PCN are required to transition to a PCR on the same date, then 350 (of the 480) ERSA entries will need to be updated in the same revision of the ERSA. Such a significant revision is expected to put substantial strain on the publishers of the ERSA.

5.2 Subgrade categorization

Many of the changes associated with the ACR-PCR system will not affect the users greatly. For example, an airport does not need to understand the differences between the simple Boussinesq model used by COMFAA and the more complex layered elastic model used by FAARFIELD. However, airports do need to understand their subgrade bearing strength in order to appropriately publish their strength rating based on the correct subgrade category. Many airports know their subgrade category, from the current PCN, but not the actual characteristic subgrade CBR value.

The subgrade category of any given airport may change in the transition from ACN-PCN to ACR-PCR. Not all airports will have a subgrade bearing strength that will change their subgrade category when transitioning to ACR-PCR. In fact, only airports with a characteristic subgrade bearing capacity of:

- CBR 5. Will move from category C to category D.
- CBR 9. Will move from category B to category C.
- CBR 13-14. Will move from category A to category B.

However, if all an airport knows is that they are currently in subgrade category C (CBR 5-8), they will no longer know if they remain a category C (CBR 6-9) or move to a subgrade category D (CBR 5 or less). To determine whether they need to change subgrade categories, they first need to know their characteristic CBR value and this is an expensive process for many regional airports. The irony is that the new subgrade CBR values are of no tangible benefit and do not contribute to the desire to reduce the mismatch between strength rating and the basis of pavement design. Instead, ICAO could have removed the subgrade categories all together and used actual subgrade CBR values. The original categories were only provided to simplify the system, which was originally based on ACN charts and tables, meaning that many different subgrade categories would create an unwieldy system. However, in modern times, all ACN/ACR calculations are performed using tablet applications and computer software, meaning that any number of subgrade categories could be included in the system without causing any issue. It would have improved the accuracy of the system by simply using each CBR value from 3 to 15, with a high and low bucket at each end, represented by CBR 2 (for values 2 or less) and CBR 20 (for values 16 and greater).

5.3 Existing pavement strength ratings

The transition to ACR-PCR will highlight that the basis of many regional airport current PCN values is not known. Many Australian airports were assigned a usage based PCN at the time the airport was transferred from the Commonwealth to local government agencies. In some cases, the historical strength rating is not reflected by the current airport operations and in many cases an expensive geotechnical investigation and engineering assessment will be required. Furthermore, these assessments may result in strength ratings being downgraded, despite historically good performance under significant aircraft loading. Therefore, these airports will feel obliged to use the services of a specialist engineer to make a usage based assessment or judgment and that will add to the expense. In many cases, the ACN-PCN system has served these airports well and they are generally not challenged by the mis-match between the strength rating system and pavement thickness design. However, they will still incur the cost of transitioning from ACN-PCN to ACR-PCR.

5.4 Cost to industry

The cost of transitioning to the ACR-PCR system is significant. The cost was estimated for Australia's 350 airports with a current PCN, based on a range of approximations:

- 15% of airports undertake geotechnical investigations and use a specialist engineer to determine a PCR value based on a balance of the prescriptive FAA process, reverse engineering of their existing pavement structures and observed pavement performance under longer term using aircraft.
- 20% of airports use a specialist engineer to determine a PCR value based on historical records and reverse engineering of their existing pavement structure.
- 60% of airports use a specialist engineering to determine a PCR value that is equivalent to their current PCN value, on the grounds that the basis of the current pavement strength rating is well known.
- 5% of airports convert their current PCN to an equivalent PCR based on their using aircraft or the known basis of the current pavement strength rating and they perform this task internally.

Based on the examples presented in this study and the time required to perform the various analyses, the cost to each airport is expected to be:

- Geotechnical investigation and reverse engineering by a specialist engineer. AUD 50 k.
- Reverse engineering by a specialist engineer based on historical records. AUD 10 k.
- Calculation of an equivalent PCR and review of using aircraft by a specialist engineer. AUD 2 k.
- Calculation of an equivalent PCR value internally. No cost.

When the cost per airport is applied to the portion of the 350 airports that will undertake this assessment, the total cost to the industry is estimated to be AUD 3.75 million. This may be justified if there was a tangible benefit to the airports, in the form of allowing bigger aircraft to operate or the better protect pavements from overloads or to enable better quantification of the implications associated with allowing Pavement Concessions. However, none of those factors will be improved because the actual strength of the pavements will not increase and the current system is rating the pavement strength appropriately. In Australia, and other countries where the prescriptive processes detailed in the FAA guidance for determining airport pavement strength ratings are not mandated or commonly applied, the cost to industry will be incurred without any tangible benefit whatsoever.

6 CONCLUSION

A robust and reliable airport pavement strength rating system is essential for the management of airport pavements around the world. The current system has served airports well since 1981 but will be replaced by a new, more rational system, for all ICAO member States by 2024. Case studies on four Australian airports indicate how the new system will be implemented in countries that do not mandate prescriptive methods of PCN determination, rather deferring to engineering judgment exercised by specialist engineers based on historical using aircraft, historical pavement performance and reverse engineering of existing pavement structures. The transition to the new system will create some challenges, mostly for those airports that do not know the basis of their current strength rating and can not locate historical existing pavement composition and subgrade bearing strength data. The cost of transition is estimated to be significant and is not justified because the new system provides no practical benefit to the majority of Australian airports.

REFERENCES

AAA, 2017, *Airfield Pavement Essential*, Airport Practice Note 12, Australian Airports Association, Canberra, Australian Capital Territory, Australia, April, accessed 17 September 2020, https://airports.asn.au/airport-practice-notes/.

Ahlvin, R. G. 1991, *Origin of Developments for Structural Design of Pavement*, Technical Report GL91-26, US Army Corps of Engineers, Waterways Experiment Station, Vicksburg, Mississippi, USA.

ASA 2020, *Aeronautical Information Package, En-Route Supplement Australia*, Airservices Australia, 5 November 2020, accessed 18 September 2020, https://www.airservicesaustralia.com/aip/aip.asp?pg=40&vdate=05NOV2020&ver=2.

CASA 2011, *Strength Rating of Aerodrome Pavements*, Advisory Circular AC139-25(0), August, accessed 31 July 2015, https://www.casa.gov.au/sites/g/files/net351/f/_assets/main/rules/1998casr/139/139c25.pdf.

CROW 2003, *The PCN runway strength rating and load control system*, CROW report 04-09, CROW.

Deilami, S. & White, G. 2020, 'Review of reflective cracking in composite pavements;, *International Journal of Pavement Research and Technology*, no. 13, pp. 524–535.

FAA 2014a, *COMFAA 3.0*, Federal Aviation Administration, 14 August, accessed 18 September 2020, https://www.airporttech.tc.faa.gov/Products/Airport-Pavement-Software-Programs/Airport-Software-Detail/ArtMID/3708/ArticleID/10/COMFAA-30

FAA 2014b, *Standardized method of reporting Aircraft Pavement Strength – PCN*, Advisory Circular 150/5335-5C, Federal Aviation Administration, Washington, District of Columbia, USA, 14 August.

FAA 2017, *FAARFIELD*, Version 1.42, Federal Aviation Administration, Washington, District of Columbia, USA, 18 September, accessed 17 December 2017, http://www.airporttech.tc.faa.gov/Download/Airport-Pavement-Software-Programs/Airport-Software-Detail/ArtMID/3708/ArticleID/4/FAARFIELD-142.

FAA 2020, *Airport Improvement Program*, Federal Aviation Administration, accessed 18 September 2020, https://www.faa.gov/airports/aip/.

Fabre, C. 2018, 'The Aircraft Classification Rating – Pavement Classification Rating ACR-PCR', *ALACPA 2018*, Quito, Ecuador, 28 May to 1 June.

Fabre, C. 2019, 'The ACR-PCR Method', *International Coordinating Council of Aerospace Industries Association*, 15 September.

Fabre, C., Balay, J., Lerat, P. & Mazars, A. 2009, 'Full-scale aircraft tire pressure test', *Eighth International Conference on the Bearing Capacity of Roads, Railways and Airfields*, Urbana-Champaign, Illinois, USA, 29 June - 2 July, pp. 1405–1413.

Fabre, C. & Vaurs, G. 2017, 'Development of rational ACN/PCN system', *10th International Conference on the Bearing Capacity of Roads, Railways and Airfields*, Athens, Greece, 28–30 June.

ICAO 2013, *Aerodrome Design and Operations*. Annex 14, Volume 1, to the Conventions on International Civil Aviation, International Civil Aviation Organization, Montreal, Canada, February.

ICAO 2020, *Member States, International Civil Aviation Organisation*, 1 October 2019, accessed 18 September 2020, https://www.icao.int/about-icao/Pages/member-states.aspx.

Roginski, M. J. 2007, 'Effects of aircraft tire pressures on flexible pavements', *Advanced Characterisation of Pavement and Soil Engineering Materials*, Athens, Greece, 20–22 June, Taylor and Francis, pp. 1473–1481.

Roginski, M. J. 2013, 'ICAO update – status of high tyre pressure revision to Annex 14', *FAA Working Group Meeting*, Atlantic City, New Jersey, USA 15–17 April, Federal Aviation Administration.

Wardle, L. J. & Rodway, B. 2010, 'Advanced design of flexible aircraft pavements', *24th ARRB Conference*, Melbourne, Victoria, Australia, 12–15 October, ARRB Transport Research.

White, G. 2019, 'A standardised sprayed sealing specification for Australian airports', *12th Conference on Asphalt Pavements for Southern Africa*, Sun City, South Africa, 13–16 October, pp. 806–821.

White, G., Kelly, G., Fairweather, H. & Jamshidi, A. 2020, 'Theoretical socio-enviro-financial cost analysis of equivalent flexible aircraft pavement structures', *99th Annual Meeting of the Transportation Research Board*, Washington, District of Columbia, USA, 12–16 January.

White, G., Farelly, J. & Jamieson, S. 2021, 'Estimating the Value and Cost of Australian Aircraft Pavements Assets', *International Airfield and Highway Pavements Conference*, 8–10 June, a virtual event.

Disclaimer:

The views expressed in this paper are solely those of the author and are not necessarily supported or endorsed by the airports that own, operate and manage the runways that have been used as the case study examples. In some cases, existing pavement and aircraft data has been amended to improve the educational value of the examples presented in the case studies. Airports should not rely on the calculations and values presented in this paper to determine the PCN or PCR of their runway.

Analyzing the bearing capacity of materials used in arresting systems as a suitable risk mitigation strategy for runway excursions in landlocked aerodromes

M. Ketabdari, E. Toraldo & M. Crispino
Transportation Infrastructures and Geosciences section, Department of Civil and Environmental Engineering, Politecnico di Milano University, Milan, Italy

ABSTRACT: Runway-related incidents/accidents based on aircraft operations can be categorized as incursions and excursions. Among all excursion events, landing and take-off overruns are responsible for the major portions. In fact, safety efforts are persistently needed by discovering these types of events, assessing the consequent risks and developing befitting mitigation strategies. There are numerous mitigation strategies to reduce the severity of consequences of runway overrun events. One of the most common strategies is to expand the geometry of the Runway End Safety Area (RESA), which is effective but beside the fact that it is noticeably expensive, many airports are landlocked, and it is difficult to expand their lands. On the other hand, Engineered Materials Arresting Systems (EMAS), as precast blocks or cast in situ slabs, are emerging as a possible interesting solution. Although this is not a new-discovered solution, many aspects are still unknown and are needed to be studied. Therefore, this paper reports the achievements of a study in which a numerical code was developed to simulate the behavior of EMAS materials and evaluate their bearing capacity and efficiency in reducing the aircraft braking distance in case of runway overrun. Thanks to this numerical code, it is possible to achieve not only the exact aircraft stopping location, but also the critical drag and uplift forces acting on the aircraft main gear and the deceleration rate (tire-material interface) after striking the arresting blocks. Additional outcome is a risk contour map overlapped on the EMAS layout that assigns the probability of aircraft stoppage location to the related coordinates on the arrestor bed. In order to discover an appropriate EMAS material with enough bearing capacity, low-density concrete with different natural, mineral and synthetic additives were considered in the evaluations. Boeing B747-8 aircraft was selected as the case study in all the computations.

Keywords: Material bearing capacity, arresting systems, mitigation strategy, aerodrome risk analyses, runway end safety area

1 INTRODUCTION

Aircraft ground operation events assigned to the runway are classified according to their location of occurrence. They can be divided into incursion and excursion events. Runway overruns (in particular, those that occur during landing) are responsible for the major portion of runway excursion events (Ketabdari et al., 2020). Numerous boundary conditions can convert a normal landing operation into an overrun event (e.g. poor aircraft braking potential conditions due to its low maintenance level, unfavorable weather conditions and pilot errors).

DOI: 10.1201/9781003222880-21

In order to mitigate the severity of consequences of runway overrun incidents/accidents and to protect involved passengers, specific areas after runway end borders are designed which are called Runway End Safety Areas (RESAs) (EASA, 2014). RESA must be capable of supporting the aircraft in occasional overruns, in dry pavement condition, without causing aircraft structural damage or injury to its occupants (Crispino et al., 2018). International Civil Aviation Organization (ICAO) recommended to increase the length of standard RESA from 90 m to 240 m, starting from the end of the runway strip (which itself is 60 m from the end of the runway) (ICAO, 2010). This new dimension is recommended for designing new runways and the existing ones. Although this strategy may mitigate the severity of incidents/accidents, not all the airports have enough land to accommodate these standard recommendations (Ketabdari et al., 2018).

Engineered Materials Arresting System (EMAS), which is approved by Federal Aviation Administration (FAA) (FAA, 2004, FAA, 2005), is an alternative mitigation strategy for the consequences of landing overrun events for those airports with landlocked circumstances. Therefore, there is no necessity to resize the length of a runway or declare its length to be less than the actual length if there is an adverse operational influence on the airport.

EMAS is defined as a bed of pre-cast blocks, consisting of FAA-approved materials, that is placed at the area after runway end threshold to decelerate an overrunning aircraft in an emergency. No external energy source is required for this system since it is a passive mitigation action. These blocks will predictably crush under the weight of an aircraft to provide one gentle aircraft deceleration. Drag forces develop at the tire-arrestor material interface to decelerate the aircraft. The successful deceleration reduces stopping distance considerably. An EMAS is positioned within the RESA set back area that is located after the runway end border and this setback length protects the arrestor bed from the aircraft intrusion during an under shoot, a short overrun, or material degradation due to jet blast.

In 1985, soft ground arrestor materials (such as clay, sand and water ponds) were considered in designing an arresting bed in order to evaluate their capability in stopping the aircraft in the shortest distance. Investigation was carried out for these materials at their best performance conditions to determine their functionality. The arresting bed simulations indicated that in clay and sand, the stopping distance would be 198m (650ft.) and 183m (600ft.) respectively. For water ponds, aircraft deceleration would not be neither constant nor gentle, therefore the damages to the aircraft would be more severe. In any case, due to the limitations due to the characteristics of these materials, they were discarded (Cook, 1987).

In 1986, the ARRESTOR computer code was used by FAA to predict aircraft stopping distances after entering RESA (Heymsfield, 2009). This code is an extension of the FITER 1 which was developed in 1985 by Cook (Cook, 1985) to predict fighter plane movements on soft ground. The Arrestor code uses EMAS geometry and material properties, and limited aircraft characteristics. This code could analyze the behavior of arrestor for only three types of aircraft - Boeing 707, 727 and 747. In order to evaluate other types of aircraft, a new computer code, SGAS (Soft Ground Arresting System) was developed in 1987 (Cook, 1987). Though it did not have user friendly pop-up windows, it required the input of only those aircraft parameters which are involved during the tire-material interaction. The outputs of these programs are the stopping distance, deceleration pattern, gear loads, tire penetration and the EMAS deformation. The sensitivity analysis with the aircraft type showed that the stopping distance increases for the heavier aircraft. The stopping distance increases as the material strength is increased and for a higher bed depth configuration, the aircraft decelerates faster because of the increase in the drag force.

In 2013, four different types of aircraft (B737-900ER, B767-400ER, B757-300, B747-400ER) were investigated. The SGAS that was used for the computations of these aircraft categories, was not included in the FAA arrestor code. Furthermore, sensitivity analysis was carried out considering different EMAS configurations and compressive strength of the material (Heymsfield, 2013). Typically, an EMAS is designed for the most critical (i.e. heaviest) aircraft operating at the airport. In 2016, lightweight regional jets have been investigated (Heymsfield, 2016). Aircraft behavior after entering an installed EMAS at the RESA for CRJ200ER, B727-100, and B737-900ER aircraft have been studied.

The scope of this study is to evaluate the functionality of EMAS in mitigating the consequences of landing overruns with the principal aim of safety for the passengers on board. The basic concepts for modelling the aircraft tire-arrestor material interface within the EMAS, are developed based on the previous studies in this field (Heymsfield, 2009).

2 METHODOLOGY: MAIN GEAR FOOTPRINT WITHIN AN EMAS

Aircraft behavior within the EMAS area can be simulated respect to the aircraft braking distance. In other words, in case of landing overrun, the distance from the runway end threshold at which the aircraft decelerates to a full stop within the EMAS installation (Ketabdari et al., 2019). To compute arresting distance, it is necessary to the correct EMAS entry velocity should be selected based on system design requirements laid down by the FAA. Since the entity of the desired output is based on Probabilistic Risk Analysis (PRA), aircraft touchdown velocity and EMAS entry velocity are both selected as normal Probability Density Functions (PDFs).

In order to simulate the behavior of aircraft within EMAS installation, footprint of aircraft main gear and its interaction with EMAS material after entry should be analyzed. Tire-material interface model is required to couple the forces generated as the result of smashing arresting materials under aircraft landing gear. Consequently, deceleration rate, drag and vertical forces and ultimately stopping distance can be obtained. The drag and the vertical forces are functions of material mass density, crushing stress, EMAS entry velocity, and tire footprint area. In this regard, it is necessary to consider that tire deformation has dynamic characteristic (Pasindu et al., 2011) and it changes according to the velocity of the aircraft that leads to variation in tire footprint, as presented in Figure 1.

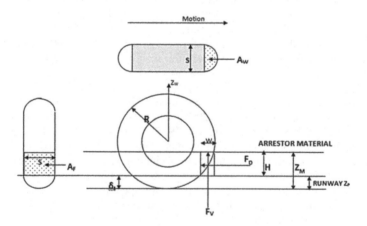

Figure 1. Wheel interface model to define drag and vertical forces (Cook, 1987).

As an aircraft enters an arresting bed, material crushes under the aircraft's self-weight and creates an interface between the tire and arresting material. The forces that are generated at the interface are vertical force, which supports the aircraft weight, and drag force, which aids in aircraft deceleration. Vertical and the drag forces are equal to the pressures on the horizontal (A_W) and vertical tire footprints (A_F), as shown in Figure 1. For this matter, a conceptual model (Cook, 1987) is implemented to involve these forces into principal equilibrium of deceleration rate and consequent arresting distance simulations, as explained in the following.

$$F_D = \underbrace{0.5pAV^2C_D}_{\text{Aerodynamic Drag Force}} + \{\underbrace{(P_c + 0.5P_cV^2)}_{\substack{\text{Crushing Strength + Dynamic Stress}\\ \text{as a function of entry Velocity of Aircraft}}} \times \underbrace{S \times (Z_M - \delta_t)}_{\text{Projected Area of the Tire } (A_F)}\} \times C_d \quad (1)$$

$$F_V = \underbrace{0.5pAV^2C_L}_{\text{Aerodynamic Lift Force}} + \{\underbrace{(P_c + 0.5P_cV^2)}_{\substack{\text{Crushing Strength + Dynamic Stress}\\ \text{as a function of entry Velocity of Aircraft}}} \times \underbrace{S \times W \times 0.66}_{\text{Projected Vertical Area } (A_W)}\} \times C_L + \underbrace{K \times \delta_t}_{\text{Vertical Spring Force}} \quad (2)$$

$$W = R^2 - (R - Z_m + \delta_t)^2 - R^2 - (R - \delta_t)^2 \quad (3)$$

Where, A is the aircraft wing area [m²]; V is aircraft cruise speed [m/s]; R is landing main gear radius [m]; S is tire width [m]; δ_t is the tire deformation; ρ is the air density [kg/m³]; C_D is the coefficient of drag; C_L is the coefficient of lift for the concerned aircraft; C_d is coefficient of drag at zero lift, which for light aircraft can be from 0.02 to 0.04 and for subsonic aircraft can be from 0.013 to 0.020; Z_m is EMAS thickness [m]; K is half the tire stiffness; P_c is the material crushing stress [Pa]; F_D is the generated drag force; F_V is the generated vertical force; W is the main gear footprint after entering the EMAS area (Cook, 1987).

By adopting above equations, an incremental distance and a new reduced speed due to deceleration is computed at every step which becomes the initial conditions for the computations in the next step. The distances are added cumulatively to get the final arresting distance when the aircraft comes to a full stop. Consequently, the computation continues till the aircraft velocity reaches a zero at final stage.

3 DEVELOPMENT OF A MATLAB APPLICATION

The model described in the previous paragraphs has been implemented in MATrix LABoratory (MATLAB®) in order to compute the stopping distance of the aircraft within different EMAS materials. The developed application named ABIAS (Aircraft Behavior in Arresting System) can be used for any airport with any type of land restrictions.

The aircraft characteristics can be numerically inputted, and runway geometry can be reported in coordinates in the referring mesh calculation, pointing the threshold at which the runway ends and EMAS installation starts. The application then simulates the aircraft behavior inside the EMAS by means of the following inputs: Maximum Take-Off Weight (MTOW), average wingspan, main gear load, mean touchdown velocity, headwind/tailwind velocities, runway length, EMAS dimensions, material with desired crushing stress (Pc) to be used in EMAS, etc. The outputs are aircraft full-stop PDF with/without EMAS installation in case of landing overrun, aircraft braking distance contour map overlapped on EMAS layout.

General Aviation with MTOW less than 5.7 t will normally stop within the runway without entering EMAS installation, therefore, mainly commercial aircraft (with MTOW≥5.7 t) are considered in order to develop ABIAS application. More assumptions that are adopted in development of this application are as following:

- reverse thrust is not being considered as the aircraft exits the runway;
- aircraft braking system is in well-maintenance condition;
- there is no aircraft braking or use of reverse thrust once an aircraft enters the EMAS area.

Finally, it should be noted that ABIAS application, developed by the authors, is compatible with multiple wheels and dual-gear aircraft.

4 CASE STUDY: BOEING 747-8 AIRCRAFT

Boeing 747-8 is selected to be studied by ABIAS, since B747-8 is amongst top five heaviest commercial aircraft with MTOW equal to 450 t and wingspan of 68.4 m. Therefore, analyzing this case study will lead to considering longest braking distance in case of landing overrun accident. The simulation is performed with a typical runway with an average length of 2200 meters, EMAS length equal to 150 m with 0.3 m thickness. The selected material of the EMAS is Low-Density Concrete (LDC) containing natural additives with crushing (threshold) stress Pc, 930 kPa. Pc is the ultimate stress at which the highest strains are produced and after which greater stresses are required to increase the marginal strains.

The probability distributions of aircraft full stop have been calculated for with and without EMAS installation in order to verify and demonstrate the functionality of arresting materials in decreasing the aircraft braking distance in case of landing overrun, as presented in Figure 2.

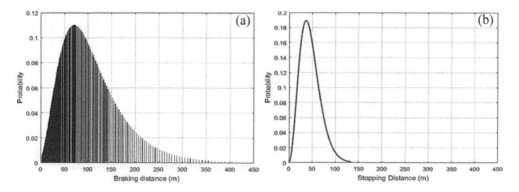

Figure 2. Primary output; Aircraft arresting distance PDF of B747-8 in case of landing overrun, (a) for paved RESA, (b) for RESA enhanced with EMAS.

Figure 2 is demonstrating the behaviour of aircraft in case of landing overrun after passing the runway end threshold with and without presence of EMAS. According to graph (a) of Figure 2, the tail of arresting distance PDF (extreme rare scenarios) indicates that, B747-8 aircraft can reach to the full stop up to 450 m after runway end, with the median of 75 m within the paved RESA. On the other hand, as presented in graph (b), by enhancing the RESA by EMAS installation, the peak of the calculated braking distance PDF is around 45 m after EMAS starting origin and the tail of this PDF demonstrates that the aircraft will reach to the full stop before 135 m from the EMAS origin.

Cumulative arresting distance PDF for B747-8 aircraft have been calculated in second step for similar conditions, as presented in Figure 3. By comparing graph (a) and graph (b) of Figure 3, it can be observed that in presence of EMAS, the arresting distance probability accumulates faster up to 75 m from the EMAS entry and then changes slightly till the full arresting distance. These PDFs are implemented in developing the final probability contour intervals overlapped on under study EMAS layout.

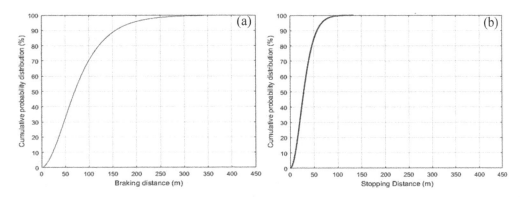

Figure 3. Primary output; Cumulative aircraft arresting distance PDF of B747-8 in case of landing overrun, (a) for paved RESA, (b) for RESA enhanced with EMAS.

Probabilistic contour intervals of aircraft arresting distances, as the secondary output, is basically a representation of the spatial probability distribution of aircraft arresting distances within the EMAS layout or within the paved RESA area, that is plotted at the end of under-study runway (Figures 4 and 5).

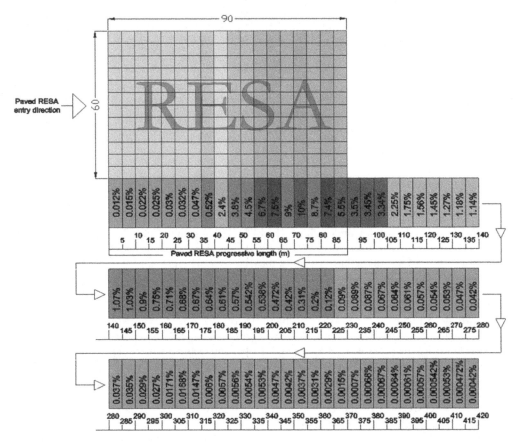

Figure 4. Secondary output; Probabilistic contour intervals for B747-8 arresting distances in case of Paved RESA.

Figures 4 and 5 determine critical locations on paved RESA and on EMAS geometry (in case of RESA enhanced with EMAS) with higher probabilities of aircraft reaching the full stop. In the calculation of paved RESA without EMAS installation, the length and the width of the RESA are considered 90 m and 60 m respectively. The reason of selecting a RESA with mentioned dimension is to evaluate how effective it would be in decreasing the probability of aircraft accident in case of landing overrun. There are numerous airports with their runways surrounded with catastrophic obstacles such as residential area, mountains, sea, etc.

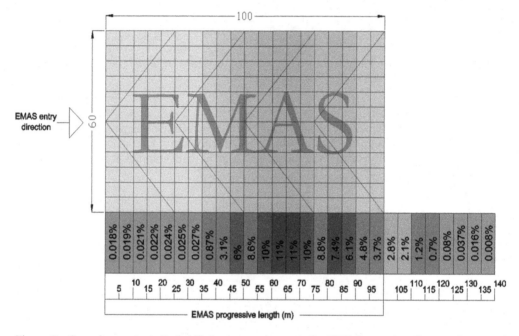

Figure 5. Secondary output; Probabilistic contour intervals for B747-8 arresting distances in case of RESA enhanced with EMAS.

An EMAS is positioned within the RESA. It is positioned within the Runway Safety Area (RSA), sets back from the runway threshold and this setback length protects the bed from aircraft intrusion during an under shoot, EMAS entry during a low velocity overrun and material degradation due to jet blast. Upon the available area at the RESA and the selected material of the EMAS this setback distance may be vary. Therefore, to plot the results, stopping distance zero is related to the starting point of aircraft entering the EMAS materials. Since deceleration rate is directly proportional to the velocity of the aircraft, where velocity decreases the deceleration rate is also decreases.

According to Figure 5, the average probabilities of arresting distances (in term of percentage) are higher in the middle of the EMAS installation. The probability that the B747-8 overruns the total EMAS area is approximately 7% beyond 100 m, which means in 93% of the landing overrun events, EMAS can arrest an overrunning B747-8 within its area without any catastrophic consequences.

The width of the EMAS is necessary since it should cover the longest Outer Main Gear Wheel Span (OMGWS) of the critical aircraft. Moreover, solver parameters (e.g. discretization of time, probability bar interval and the properties of touchdown normal PDF) are required to be considered in the numerical computation.

5 FUNCTIONALITY ASSESSMENT FOR DIFFERENT EMAS MATERIALS

Aircraft behaviors within LDC EMAS containing natural, mineral and synthetic additives with different levels crushing stress levels, are investigated by ABIAS application in order to determine a material with better functionality to be used in pre-cast blocks.

These materials consist of three LDCs with maximum crushing stress levels (P_C) of 172500, 345000 and 930000 Pa. Low crushing stress level causes low durability for the concrete which can be an issue in maintenance of EMAS infrastructure. Selection of LDCs with various crushing stress levels does not confirm that they are economical materials to be adopted for EMAS since this statement needs corresponding cost-benefit analysis. Therefore, these LDCs are adopted in order to study the behavior of EMAS materials with lowest, medium and highest crushing strengths.

Figure 6 presents typical low-density crushable concrete behavior. The concrete exhibits a strain increase as stress increases. Beyond yielding, the material exhibits a large strain increase at a nominally constant stress (within 0.2 to 0.5 strain range). Beyond 0.5 strain, additional stress is required to develop larger strains. As the material approaches a fully crushed state (0.8 strain), a substantial stress increase is required to develop marginal strain increases.

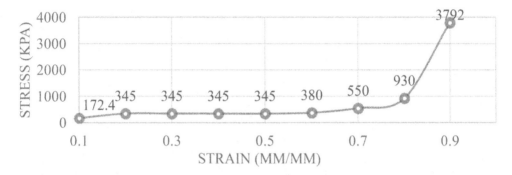

Figure 6. Low-density concrete stress/strain relationship (Heymsfield and Halsey, 2008).

The same Boeing 747-8 aircraft for the sensitivity analysis of EMAS materials is selected. The behavior of this aircraft after entering the EMAS is investigated for different arrestor bed materials with their specific crushing strengths. As presented in Figure 6, within LDCs, the strain increases initially respect to low variation in the stress values but after the material yield stress, there is a significant increase in stress but limited in strain values.

The drag force induced by the material has a correlation with the material's crushing strength and its mass density. The greater crushing strength material manages to exert, the greater drag force against the landing main gear will be generated, consequently aircraft reach to a full stop at a shorter distance, as presented in Table 1.

Table 1. Low-density concretes to be used in arresting bed with their properties.

No.	Arrestor materials	Mass density [kg/m^3]	Crushing stress [N/m^2]	Arresting distance [m]	Maximum drag force [N]
1	Low-density concrete (type A)	2300	175000	144.6	87800
2	Low-density concrete (type B)	2300	345000	139.7	92300
3	Low-density concrete (type C)	2300	930000	135.5	116750

It can be observed from Figure 7 that among the LDC mixes, there is a decrease in the aircraft arresting distance by increasing the crushing stress of the material that is used in EMAS. The arresting distance for the LDC with maximum crushing stress levels (P_C) of 172500 is the longest among the others since lower drag force is generated at the tire-material interface. Therefore, the aircraft travels a longer distance before coming to a full stop.

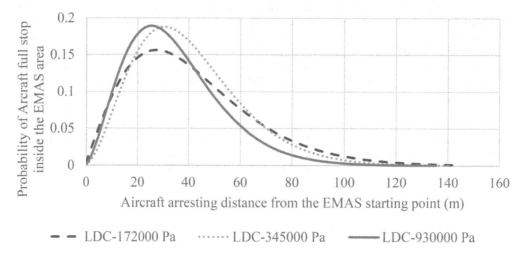

Figure 7. B747-8 arresting distance PDF in materials with different crushing stress levels.

It can be interpreted from the results that LDC with higher crushing stress level can arrest an overrunning aircraft in shorter distance since it generates higher deceleration rate respect to others.

6 CONCLUSION

An application named ABIAS (Aircraft Behavior in Arresting System) was developed by the authors in the platform of MATLAB® in order to simulate the behavior of aircraft within engineered material arresting system. Thanks to this application it is possible to compute the stopping distance of an aircraft within different EMAS materials. ABIAS is principally applicable for those airports with land restrictions.

Moreover, aircraft behavior is analyzed for materials with different crushing stress levels to be used in EMAS. Outputs of the calculations are plotted as arresting distances contour intervals, alongside the arrestor bed configuration (100 × 60m) for low-density concrete mixes with crushing stress levels of 172500 (type A), 345000 (type B) and 930000Pa (type C).

Aircraft arresting distance within the EMAS area decreases when the crushing strength of the material increases. Arresting bed formed by LDC type A causes the longest aircraft stopping distance because of its lower drag force that can generate respect to others. Drag force has a direct correlation with the material's crushing strength and mass density. Materials with greater crushing strengths cause greater drag forces against the landing gear, therefore, aircraft reaches to a full stop at a shorter distance.

To conclude, an EMAS is proposed to be considered as the most rational and feasible alternative solution for land-locked airport in order to respect the FAA and ICAO RESA recommendations and to mitigate the consequences of landing overrun events.

REFERENCES

Cook, R. F., 1985. *Aircraft operation on soil prediction techniques*. Technical Report ESL-84-04, Vol. 1 and 2, U.S. Air Force Engineering and Services Centre, Tyndall Air Force Base, FL.

Cook, R. F., 1987. *Soft-Ground Aircraft Arresting Systems*. UNIVERSAL ENERGY SYSTEMS INC DAYTON OH.

Crispino, M., Toraldo, E. and Giustozzi, F., 2018. *Improving Runway Strip Performance to Fulfill International Requirements Through Eco-Efficient Soil Treatments: Case Study of a Major Italian Airport*. Environmental Engineering & Management Journal (EEMJ), 17 (6).

European Aviation Safety Agency, 2014. Final report. *Study on models and methodology for safety assessment of Runway End Safety Areas*.

FAA, 2004. Order 5200.9. *Financial Feasibility and equivalency of Runway Safety Area Improvements and Engineered Material Arresting Systems*.

FAA, 2005. Advisory Circular 150/5220-22A. *Engineered Materials Arresting Systems (EMAS) for Aircraft Overruns*.

Heymsfield, E. and Halsey, T. L., 2008. *Sensitivity analysis of engineered material arrestor systems to aircraft and arrestor material characteristics*. Transportation Research Record, 2052(1), pp 110–117.

Heymsfield, E., 2009. *Performance Prediction of Strong Company's Soft Ground Arrestor System using a Numerical Analysis*. Arkansas, USA, Mack Blackwell Transportation Center.

Heymsfield, E., 2013. *Predicting aircraft stopping distances within an EMAS*. Journal of Transportation Engineering, 139(12), pp 1184–1193.

Heymsfield, E., 2016. *Jet Stopping Distance and Behaviour in a Regional Airport EMAS*. Journal of Performance of Constructed Facilities 30.5: 04016010.

International Civil Aviation Administration, 2010. ICAO Annex 6 - Part 1. *Operation of Aircraft - International Commercial Air Transport – Aeroplanes*.

Ketabdari, M., Giustozzi, F. and Crispino, M., 2018. *Sensitivity analysis of influencing factors in probabilistic risk assessment for airports*. Safety science, 107, pp 173–187. http://dx.doi.org/10.1016/j.ssci.2017.07.005

Ketabdari, M., Crispino, M. and Giustozzi, F., 2019. *Probability contour map of landing overrun based on aircraft braking distance computation*. Pavement and Asset Management, CRC Press, pp. 731–740. http://dx.doi.org/10.1201/9780429264702

Ketabdari, M., Toraldo, E. and Crispino, M., 2020. *Numerical Risk Analyses of the Impact of Meteorological Conditions on Probability of Airport Runway Excursion Accidents*. In International Conference on Computational Science and Its Applications (pp. 177–190). Springer, Cham. https://doi.org/10.1007/978-3-030-58799-4_13

Pasindu, H. R., Fwa, T. F. and Ghim P. O., 2011. *Computation of aircraft braking distances*. Transportation research record 2214.1, 126–135.

Highway design

Impact of introducing longer and heavier vehicles on the bearing capacity of pavement subgrades

M.S. Rahman
Pavement Technology, Swedish National Road and Transport Research Institute (VTI), Linköping, Sweden

S. Erlingsson
Faculty of Civil and Environmental Engineering, University of Iceland, Reykjavik, Iceland

ABSTRACT: Increasing the maximum permissible gross weight of vehicles from 64 tons to 74 tons on Swedish highways may significantly increase deformation in subgrades requiring extensive and expensive measures for maintenance deep down in the structures. Even though the individual axle load remains the same, longer and heavier vehicles will contain higher proportions of tandem and tridem axles and the total load will be superposed on to the subgrade resulting in higher deformations. In this study, the relative impact of single, tandem and tridem axles on subgrade deformation was analyzed based on calculations in a pavement design software called ERAPave. Three structures with different layer thicknesses and varying material properties during the spring-thaw and summer period were analyzed. In general, it was observed that the relative impact is dependent on the thicknesses and material properties of the pavement structure and the subgrade. Weaker subgrade is more affected by the tridem axle followed by tandem and single axle. The tridem axle is more damaging to the sub-grade for thicker and stiffer structures. In certain cases, the single axle showed more impact on the top of the subgrade. The tridem axle showed more vertical deflection in all cases that may induce increased fatigue cracking of the asphalt concrete (AC) layer. Duration of loading was higher for the tridem axle which has more damage potential. Thus, for the whole vehicle, containing more tridem axles, the effects may be superposed resulting in greater damage. On the other hand, the higher load carrying capacity of the tandem and tridem axle configurations will reduce the number of trips required to carry the same amount of goods. This may compensate for the increased damage to the subgrade and the impact on the life cycle of the pavement.

Keywords: Longer and heavier vehicle, subgrade, axle configuration, deformation, damage

1 INTRODUCTION

Recent trends show that longer and heavier vehicles (LHVs) benefit the economy and the environment by reducing fuel consumption and the costs of road freight transport (Ludec, 2009, Gleave et al., 2013). Historically, LHVs have been permitted to operate in Sweden. Since 1996 and as of the beginning of 2015, the maximum gross vehicle weight (GVW) of 60 tons and a length of up to 25.25 m have been permitted. In 2015, the GVW was increased to 64 tons. In 2018, the Swedish Transport Agency decided to further increase the maximum GVW to 74 tons on parts of the road network. It will probably be allowed on the entire road network if studies and experience indicate positive impact of the LHVs on traffic safety, economy and the infrastructure (Saliko and Erlingsson, 2021). Furthermore, a special permission has been granted to an iron ore producer in Northern Sweden to transport the ore using 25 m long LHVs having a gross weight of 90 tons on a road linking the mining site to the nearby train station (Erlingsson and Carlsson, 2014). Within the European Union (EU), however, the maximum permissible gross vehicle weight is 40 tons except that intermodal transports using 40-foot containers are allowed a maximum weight of 44 tons (Directive 96/53/EC) (Žnidarič, 2015).

One of the consequences of introducing 74 tons trucks in Sweden is that multiple axles (tandem and triple axles) are becoming more common. In this case, the maximum allowable load for an axle group is 24 tons. This means that the individual axle loads are reduced compared to that permitted for the drive axle (11.5 tons). As a result, stresses and strains in the upper layers of pavements are reduced. However, as the load spreads with depth, the stresses from the closely spaced axles start to superimpose and deep down in the subgrade, the total stress can exceed that permitted for the drive axle. This effect is schematically shown in Figure 1. This can be seen in field measurements as well (Saliko and Erlingsson, 2021). The consequences of this are not well-known. The conventional pavement analysis method of converting the traffic with all the different combinations of axles to the number of passes of an equivalent standard axle load (single axle with dual wheels) (ESALs) does not take this effect into consideration.

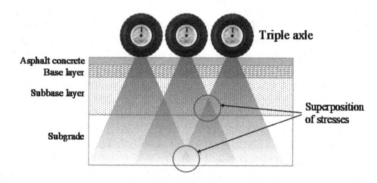

Figure 1. Schematic of stress distributions and superposition of stresses from the sub-axles of a triple-axle configuration with depth inside a typical pavement structure.

In the light of increasing the maximum GVW to 74 tons and a probable development towards even higher loads, deformations in subgrades and subbase layers have been identified as a matter of concern. Damages due to deformations in these layers can lead to very large costs in the future for maintenance and point-by-point reinforcement measures. The high costs are associated with the damages found deep down in the pavement structure requiring extensive measures. The Swedish road network largely consists of pavements that were built for completely different loads than what is experienced today. The knowledge about the detailed distribution of stresses deep in the pavement structure due to several point loads is limited, especially regarding time-dependent and non-linear responses from the asphalt concrete (AC) and unbound road materials. Indeed, several studies have been conducted to evaluate the impact of the LHVs on transport infrastructures (Leduc, 2009, Glaeser, 2006, Akerman and Jonsson, 2007, Aurell and Wadman, 2007, Knight et al., 2008, Chatti et al., 2009, Weissmann et al., 2013). In this article, the risk of damage down in the subgrade caused by the tandem and triple axles of GVW of 74 tons was assessed. This was done by calculating the stresses and strains induced in the subgrade due to tandem and triple axles and comparing the same due to a permissible drive axle (single axle of 11.5 tons with dual tires). The study was based on the theoretical analyses of the impact of these different axle configurations on the responses of subgrade of three typical hypothetical pavement constructions during the spring-thaw and summer period.

2 METHODOLOGY

In this project, three typical pavement sections, shown in Figure 2, were studied that vary from a thin structure with 5 cm AC layer, an intermediate construction with 10 cm AC layer, to a relatively thick structure with 15 cm AC layer. Two seasons were considered critical: (a) the spring-thaw period and (b) the hot summer period. During the spring-thawing period, the AC layers are quite stiff due to the low air temperature, but the unbound layers are relatively weak since the high water-content reduces the stiffness. Stresses from the heavy axles are, therefore, well dispersed through the AC layer, but the load-spreading is relatively small in the unbound layers. This means that the risk of developing high stresses in the subgrade is high. During the summer period, the high temperature reduces the stiffness of the AC layer. On the other hand, the water-content of the unbound layers is lower for which the stiffness is higher compared to the spring-thaw period. Thus, these two seasons represent two completely opposite scenarios.

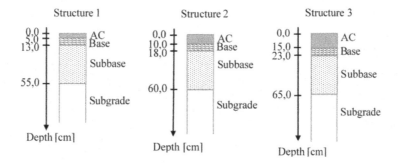

Figure 2. The three structures studied.

The calculations were conducted using the pavement analysis software ERAPave, developed at the Swedish National Road and Transport Research Institute (VTI). The program is based on multi-layer elastic theory (MLET) and considers the load to be axi-symmetrical, i.e., the contact area between the tire and the surface is circular. The impact of dual wheels and multiple axles are summed according to the theory of superposition (Ahmed, 2014).

Table 1. The three studied load cases.

Load case	Axle load [kN]	Number of sub-axles	Tire pressure [kPa]	Distance between the wheels [cm]	Distance between the axles [m]
Drive axle	115	1	800	35	-
Tandem axle	190	2	800	35	1.4
Triple axle	240	3	800	35	1.4

For the study, three load cases were considered: (a) a drive axle (single axle), (b) a tandem axle and (c) a triple axle - all with dual wheels. Table 1 provides an overview of the load cases. The speed of the vehicles was assumed to be 80 km/h.

The material parameters assumed for the summer and spring seasons are presented in Table 2 and Table 3, respectively. For each season and all structures, 2 and 3 different values of the stiffness modulus, respectively for the subbase layer and the subgrade were considered. The higher subbase modulus values represent a crushed rock material of good quality that has a low frost susceptibility and the lower modulus values correspond to an uncrushed natural aggregate of slightly lower quality. The 3 cases of subgrade materials represent a sandy soil, silty soil and a soil with high organic content. Thus, 12 structural conditions were analyzed in total. With the three load cases, the total number of scenarios is then 36.

Table 2. Selected stiffness moduli for the summer period [MPa].

	AC layer thickness [cm]		
	5	10	15
AC layer	4 000	3 000	2 500
Unbound base layer	450	450	450
Unbound subbase layer	450/240	450/240	450/240
Subgrade	50/10/5	50/10/5	50/10/5

Table 3. Selected stiffness moduli for the spring-thawing period [MPa].

	AC layer thickness [cm]		
	5	10	15
AC layer	12 000	10 000	9 000
Unbound base layer	300	300	300
Unbound subbase layer	450/100	450/100	450/100
Subgrade	50/10/5	50/10/5	50/10/5

3 RESULTS

Using ERAPAve, the calculations were conducted for the three axle load configurations (Table 1) for the three structures (Figure 2) with the combinations of the material properties for spring and summer, as presented in Table 2 and Table 3. This article presents some of the typical results necessary for the discussions. All the results presented here are along the vertical longitudinal section through the middle of the dual wheels.

3.1 Variation of stresses and strains along the depth

In Figure 3, a typical distribution of stresses and strains along the depth of the structure (for the different layers) as induced by the different axle configurations are shown. This figure shows that the single axle produces higher stresses and strains in the upper layers which is gradually superseded by the tandem and tridem axles with depth. This study focuses on the subgrade only where it is seen that the maximum values of stresses and strains occur on the top of the subgrade which diminishes with depth.

From analyses of the results for all the scenarios, it is seen that, in some cases, the single axle may produce higher stresses and strains at the upper part of the subgrade depending on the structure and layer properties. However, in all cases, the single axle is superseded by the tandem and tridem axles at lower depths. This situation is shown in Figure 4. In other cases, the tandem and tridem axle produce higher stresses and strains all through the subgrade layer. This situation is presented in Figure 5.

Thus, in terms of absolute maximum strain, which is induced at the top of the subgrade, the single axle may dominate in some situations. However, considering the overall strain along the depth of the subgrade, the tridem axle dominates in most of the situations which will be reflected in the total vertical deflection of the subgrade.

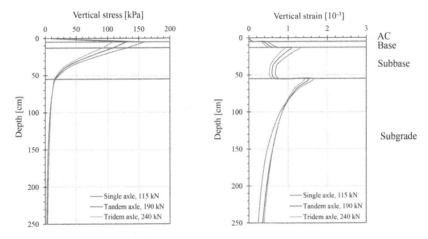

Figure 3. Distribution of vertical stresses and strains along the depth of structure 1 (springtime with subbase stiffness of 100 MPa and subgrade stiffness of 5 MPa).

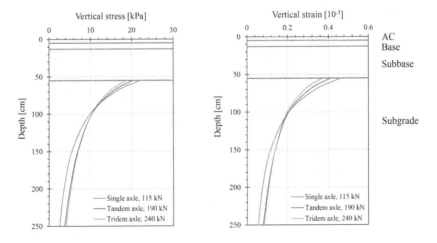

Figure 4. Distribution of vertical stresses and strains along the depth of structure 1 (springtime with subbase stiffness of 450 MPa and subgrade stiffness of 50 MPa).

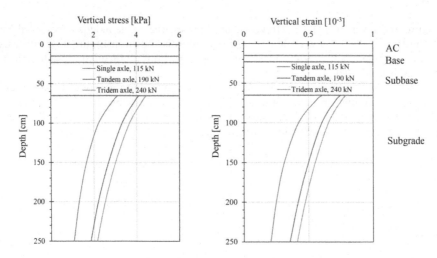

Figure 5. Distribution of vertical stresses and strains along the depth of structure 3 (summertime with subbase stiffness of 450 MPa and subgrade stiffness of 5 MPa).

3.2 Vertical deflection at the top of the subgrade

The total vertical deflection at the top of the subgrade was calculated for all cases. Here, the two typical cases as presented in section 3.1 are shown (Figure 6). It is noted that for all cases, tridem axle induces the maximum vertical deflection followed by the tandem and single axles. Thus, in this respect, the tridem axle may result in more damage to the upper AC layers due to increased fatigue cracking.

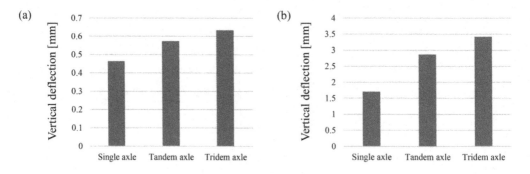

Figure 6. Total vertical deflection at the top of the subgrade: (a) structure 1, springtime with subbase stiffness of 450 MPa and subgrade stiffness of 50 MPa. (b) structure 3, summertime with subbase stiffness of 450 MPa and subgrade stiffness of 5 MPa.

3.3 Duration of loading

Vertical stresses and strains induced at two depths (one in the upper part and the other in the lower part of subgrade) with respect to time for a typical case are shown in Figure 7 and Figure 8, respectively. As can be expected, duration of loading is maximum for the tridem axle followed by the tandem and single axles. At the upper part of the subgrade (at a depth of 55.1 cm), the peaks due to the passing of each axle can be seen. This effect diminishes with depth as the load is distributed and superposed. As a result, no peaks can be seen at a depth of 250 cm. The duration of loading is dependent on the speed of the vehicle. The longer duration of loading for the tandem axle may result in higher damage depending on the moisture and drainage condition and porosity of the subgrade. If the porosity of the subgrade is high, moisture may get expelled resulting in some settlement or permanent deformation. The longer duration may also induce more damage due to viscoelastic nature of the AC layer.

3.4 *Damage ratio*

The relative damage of the subgrade caused by the tandem and tridem axles compared to the single axle can be calculated based on the Asphalt Institute Method (Huang 2004). According to the permanent deformation criterion, the allowable number of load repetitions to control permanent deformation can be expressed as

$$N = a(\varepsilon_v)^b \quad (1)$$

where, N is the allowable number of load cycles, ε_v is the vertical compressive strain at the top of the subgrade a and b (b is usually taken as -4) are material constants. A relative damage ratio (D) for a single passage can be defined as the ratio between two allowable number of load repetitions. Thus, in this study, the damage ratios for the tandem and tridem axles in relation to the single axle were computed as:

$$D = \frac{N_s}{N_t} = \left(\frac{\varepsilon_{vt}}{\varepsilon_{vs}}\right)^4 \quad (2)$$

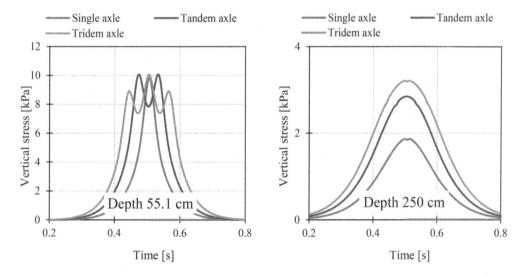

Figure 7. Vertical stress with respect to time at two depths (structure 1, springtime, subbase modulus 100 MPa and subgrade modulus 5 MPa).

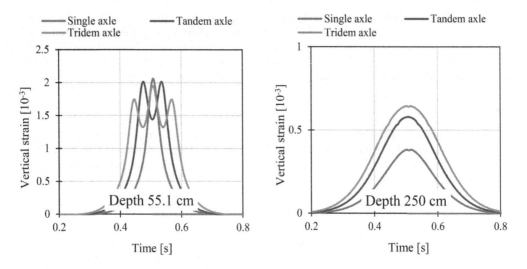

Figure 8. Vertical strains with respect to time at two depths (structure 1, springtime, subbase modulus 100 MPa and subgrade modulus 5 MPa).

where, N_s and N_t are the allowable number of load repetitions for the single and tandem or tridem axles, respectively. ε_{vs} and ε_{vt} are the vertical strains at the top of the subgrades for the single and tandem or tridem axles, respectively. The calculated damage ratios for all the scenarios studied here are presented in Table 4. Here, the values of D less than one (shaded grey) means that the single axle is more damaging and vice versa.

It is observed that, out of the 36 cases, single axle is more damaging in 13 cases (in one case, $D = 1$, shaded light grey), compared to the tandem axle and in 17 cases compared to the tridem axle. It can be expected that the relative damage will depend on the stiffnesses and thicknesses of the upper layers which influences the amount of load spreading and superposition. The data shows that the single axle is more damaging when the AC layer is thinner, and the base and subbase layers are weaker. The reason is that in those cases, the load spreading is relatively lower and the amount of superposition of stresses on the top of the subgrade for the multiple axles is less. Thus, the heavier weight of the single axle causes more strain on the top of the subgrade. It is again more damaging when the subgrade stiffness is higher. This implies that for similar pavement structures, the tridem axle will cause more damage to the subgrade if the subgrade is weaker. According to Table 4, D ranges from 0.27 to 3.49. The most critical situation with $D = 3.49$ (marked with box) appears to be for structure 3 loaded with the tridem axle during the springtime when the subbase and subgrade moduli are 450 MPa and 5 MPa, respectively. On the other hand, the most favorable situation with $D = 0.27$ (marked with box) occurs in structure 1 for tridem axle during the springtime with the subbase and subgrade moduli of 100 MPa and 50 MPa, respectively.

In terms of load carrying capacity, the tridem axle configuration can transport much higher load than the single axle configuration. This means that the number of trips required to transport the same amount of goods will be much fewer for the tridem axle compared to the single axle. In this respect, a much higher damage ratio (which is the ratio of the allowable number of passes) than 1 for the tridem axle can have the same impact on the life cycle of the pavement compared to the single axle.

It should be noted that the damage ratios (with respect to the single axle) were calculated at the top of the subgrade. At greater depths, D will always be the highest (and greater than 1) for the tridem axle followed by the tandem axle, as can be seen from the plots in Figure 4 and 5. Thus it may turn out that even though the tridem and tandem axles appear to be less

Table 4. Damage ratio.

		Layer stiffness [MPa]			Damage ratio (with respect to single axle)		Peak vertical strain (ε_{vt}) [10^{-3}]
		Base	Subbase	Subgrade	Tandem axle	Tridem axle	Single axle
Structure 1 (5 cm AC layer)	Spring	300	100	5	0.91	0.80	2.06
				10	0.72	0.53	1.64
				50	0.50	0.27	0.77
			450	5	1.32	1.48	0.87
				10	1.02	0.98	0.74
				50	0.61	0.40	0.46
	Summer	450	240	5	1.12	1.13	1.34
				10	0.86	0.74	1.11
				50	0.55	0.32	0.62
			450	5	2.12	2.52	0.71
				10	1.03	0.98	0.79
				50	0.61	0.39	0.48
Structure 2 (10 cm AC layer)	Spring	300	100	5	1.71	1.73	1.31
				10	1.31	1.12	1.02
				50	0.78	0.46	0.47
			450	5	2.22	2.68	0.63
				10	1.72	1.79	0.53
				50	0.98	0.69	0.33
	Summer	450	240	5	1.83	1.97	1.00
				10	1.40	1.28	0.83
				50	0.83	0.52	0.46
			450	5	2.12	2.52	0.71
				10	1.63	1.65	0.60
				50	0.94	0.64	0.36
Structure 3 (15 cm AC layer)	Spring	300	100	5	2.33	2.68	0.92
				10	1.75	1.75	0.72
				50	0.98	0.66	0.33
			450	5	2.70	3.49	0.49
				10	2.07	2.35	0.42
				50	1.12	0.86	0.25
	Summer	450	240	5	2.08	2.40	0.83
				10	1.59	1.56	0.69
				50	0.90	0.60	0.38
			450	5	2.36	2.95	0.60
				10	1.81	1.93	0.51
				50	1.00	0.72	0.31

harmful than the singe axle for certain cases in Table 4, they can have the opposite consequences for the pavement depending on the overall structural conditions.

4 CONCLUSIONS

In this article, the impact of LHVs on the subgrades of pavement structures was theoretically analyzed. Since LHVs will consist of more tandem and tridem axles than single axles, this study focused on the relative impacts of the single, tandem and tridem axles. Three typical pavement structures were investigated here with varied material properties during the spring and the summer. The analyses were based on layered elastic theory carried out with the aid of

ERAPave which provided an overall estimate of the relative impacts of the different wheel configurations.

In general, it was observed that the relative impact is dependent on the thicknesses and material properties of the pavement structure and the subgrade. Weaker subgrade is more affected by the tridem axle followed by tandem and single axle. On the other hand, for thicker and stiffer pavement structure, the tridem axle is more damaging to the subgrade. In certain cases, the single axle showed more impact on the top of the subgrade. The tridem axle showed more vertical deflection in all cases that may induce increased fatigue cracking of the AC layer. The speed of the vehicle also has some impact on the damage. Duration of loading was higher for the tridem axle which has more damage potential. Thus, for the whole vehicle, when there are more tridem axles, the effects may be superposed resulting in greater damages. On the other hand, since tandem and tridem axle configurations have higher load carrying capacity, the number of trips required to carry the same amount of goods will be reduced compared to the single axle. This might compensate for the increased damage to the subgrade and the impact on the life cycle of the pavement.

Since this study showed that the overall impact of an LHV on a pavement structure depends on the axle configuration, structural and climatic conditions and the number of trips, it was hard to draw any generalized conclusion. The assessment should be conducted for specific cases including life cycle cost analysis. Furthermore, similar studies should be carried out for other types of structures and climatic conditions with the whole vehicle configurations. The actual damage to the subgrades and the effect on the life span of a pavement structure should as well be investigated using full scale tests or field observations. Evaluation of pavement structural responses and performance are valuable to improve the mechanistic or mechanistic-empirical pavement design and maintenance procedures under heavy traffic loading.

ACKNOWLEDGEMENT

This work was sponsored by the Swedish Transport Administration (Trafikverket).

REFERENCES

Ahmed, A.W., 2014. *Mechanistic-Empirical Modelling of Flexible Pavement Performance*. Doctoral thesis, KTH Royal Institute of Technology, Stockholm, Sweden.
Akerman, I. and Jonsson, R., 2007. *European Modular System for Road Freight Transport – Experiences and Possibilities*. TFK Report.
Aurell, J. and Wadman, T., 2007. *Vehicle Combinations Based on the Modular Concept*. Committee 54: Vehicles and Transports, Report 1/2007.
Chatti, K., Manik, A., Salama, H., Mohtar, C.E. and Lee, H.S., 2009. *Effect of Michigan Multi-Axle Trucks on Pavement Distress*. Final Report, Volume II, Project RC-1504, Department of Civil and Environmental Engineering, Michigan State University, East Lansing, MI 48823, USA.
Erlingsson, S. and Carlsson, H., 2014. *The Svappavaara Road Test Sections – Instrumentation*. VTI notat 12A-2014, Swedish National Road and Transport Institute (VTI), Linköping, Sweden.
Glaeser, P.K. et al., 2006. *Effects of New Vehicle Concepts on the Infrastructure of the Federal Trunk Road Network: Final Report*. Federal Highway Research Institute, BAST.
Gleave, S.D. et al., 2013. *A Review of Mega-Trucks: Major Issues and Case Studies*. Directorate General for Internal Policies, European Parliament.
Huang, Y.H., 2004. *Pavement analysis and design*. 2nd ed. Englewood Cliffs, NJ: Prentice Hall.
Inge, V., Håkan, B., John, M. and Mattias, H., 2008. *The effects of long and heavy trucks on the transport system: report on a government assignment*. VTI Report 605. Swedish National Road and Transport Research Institute (VTI), Linköping, Sweden.
Knight, I., Newton, W., McKinnon, A., et al., 2008. *Longer and/or Heavier Good Vehicles-A Study of the Likely Effects if Permitted in the UK: Final Report*. TRL Report PPR285.
Koskinen, O.H. and Sauna-aho, J., 2002. *Nordic vs. Central European vehicle configuration; fuel economy, emissions, vehicle operating costs and road wear*. 7th International Symposium on Heavy Vehicle Weights and Dimensions. Delft, the Netherlands, June 16-20, pp. 241–254.

Leduc, G., 2009. *Longer and Heavier Vehicles: An Overview of Technical Aspects*. JRC Scientific and Technical Report.

Saliko, D. and Erlingsson, S., 2021. *Damage investigation of thin flexible pavements to Longer Heavier Vehicle loading through instrumented road sections and numerical calculations*. Submitted to Road Materials and Pavement Design.

Weissmann, A.J., Weissmann, J., Papagiannakis, A. and Kuni-setty, J.L., 2013. *Potential Impacts of Longer and Heavier Vehicles on Texas Pavements*. Journal of Transportation Engineering, Vol. 139, No. 1.

Žnidarič, A., 2015. *Heavy-Duty Vehicle Weight Restrictions in the EU: Enforcement and Compliance Technologies*. Scientific advisory group report, 23rd ACEA, ZAG – Slovenian National Building and Civil Engineering Institute Ljubljana, Slovenia.

Post-Modern Pavement: Theory, concept, performance and challenges

A. Jamshidi & G. White
School of Science, Technology and Engineering, University of the Sunshine Coast, Queensland, Australia

ABSTRACT: The Post-Modern Pavement (PMP) is a new concept in pavement design, construction, utility and recycling. It is conceptualised to address current and future challenges for various pavement types, such as asphalt, concrete and concrete block. The PMP is a novel method for categorising and evaluating pavements that serves as an interface between structural/environmental design, social requirements and human physiology in different service conditions. In this paper, the proposal of this category is premised upon a state-of-the-art study of pavement engineering history. The theories and concepts of PMP are presented, and the principles of PMP design and the challenges ahead are discussed. Results show that PMP is a multi-role pavement system that connects elements of transportation infrastructure assets into the ecosystem and public health.

Keywords: Structural performance, Sustainability, Multi-role infrastructure, Climate change, Urban design

1 INTRODUCTION

The evaluation of pavement performance is changing due to evolving demand of governments and the associated expectations of modern society. Pavements are no longer hard surfaces made of concrete and bituminous materials for transporting goods and passengers. As part of transportation infrastructure assets, pavements are the backbone of socioeconomic development. Hence, governments have invested huge capital in the construction, maintenance and rehabilitation of pavements. However, construction standards are being continuously upgraded to meet new requirements. For pavement engineers, the pavement system should adequately resist structural stress/strain induced by traffic loads and temperature gradients. Furthermore, pavements are inspected regularly to record structural and functional failures for planning maintenance and overhaul. In this regard, many laboratory tests, predictive models and field investigation protocols are carried out to analyse the bituminous/concrete mix in different transportation assets, including roads, ports and airports. Although the structural/functional performance of pavements is a key characteristic in such analyses, more variables should be considered due to the evolving attitudes in infrastructure development.

For many years, environmental engineers have criticised pavement engineers for paying insufficient attention to the ecosystem, thereby resulting in many natural disasters. As a result, eco-friendly materials and technologies have been developed to consider environmental concerns. New technologies and strict environmental regulations cannot fix all the consequences, but their development would be a great step taken towards sustainable practice in pavement design and construction.

DOI: 10.1201/9781003222880-23

The development of pavement networks to support the rapid expansion of urban sprawl has resulted in a cost- and time-effective mode of transportation and, therefore, increased convenience and safety. In terms of social effects, feedback from pavement users plays a key role in the successful design and utility of pavements. Consequently, a new pavement categorisation method, termed the *post-modern pavement* (PMP) is proposed in this paper. The theories and concepts behind the PMP are also discussed briefly. The reason for introducing this category of pavement is explained via a review of pavement technology and science in different ages of urban design and human civilisation.

2 HISTORY OF PAVEMENT

2.1 *Birth and growth of pavement*

Binding materials such as bitumen and cement have been used since the dawn of civilisation. One of the first references to the use of bitumen was in Genesis 6:14, where the Lord told Noah, 'So make yourself an ark of cypress wood; make rooms inside it and coat it with pitch inside and out' (Roberts et al., 2002). Pitch is a sort of natural bitumen that floats to the surface of water bodies from fissures in the Earth's crust that leak crude oil. The crude oil exposed to sunlight loses its light oils and resins, resulting in a hard black or brown material by-product called bitumen. Bitumen was extensively used in Mesopotamia, Persia and Elam as mortar in the construction of buildings and coating roadways. It was likewise used to waterproof containers (baskets, earthenware jars, storage pits), wooden posts, flooring, reserves of lustral waters, bathrooms and palm roofs (Moorey, 1996). Mats, sarcophagi, coffins and jars used for funeral practices were often sealed using bituminous materials (Connan, 1999). In ancient Egypt, bituminous materials were used for mummification (Nissenbaum and Buckley, 2013, Clark et al., 2016). Therefore, people from the ancient age were already aware of the binding and sealing characteristics of bitumen as a construction and building material.

However, as bitumen was not an easily available material elsewhere, local materials such as cobblestone and brick were used as paving materials instead (ICPI, 2003). Relics of pavements can be found in the ancient cities of the Roman and Persian empires (Knapton, 1996; Hassani and Jamshidi, 2005; Zoccali et al., 2017). In this era, basic knowledge about natural raw materials for pavement construction was obtained from experience. The criteria for pavement design were based on the availability of materials and the application of roads. The main motivation for road construction was to develop safe and efficient routes for business, rapid and expedient communication across empirical territories and military movement. For example, the 2,669 km Royal Road was constructed to enable couriers to transmit administrative messages across the Persian Empire's territory in nine days (Axworthy, 2008). Labelling the Royal Road as the first highway in civilisation would not be an exaggeration. Another example is the Salt Road in the Roman Empire (Faktorovich, 2015). The road was constructed to transport salt, which was a strategic product because it was used as a food preservative to save people from famines.

The technology of pavement construction was mainly based on human power and animals, such as horses, mules and camels. There is no available detailed cost analysis of pavement construction from the dawn of civilisation to the 18th century. Note that the durability and structural resistance of pavement were the main criteria for pavement engineering.

2.2 *Enlightenment era of the pavement*

From the 18th to the 19th century, the empirical methods were developed that were based on the personal experiences gained by artists, architects, masons and craftsmen. In this timeframe, bigger carts rolled out, which required more durable road surfaces for efficient operation (Lilley, 1991).

The development of maritime transportation technology for military and civilian intercontinental transportation also necessitated the advancement of pavement technology. That is, the construction of ocean cruise ships was cutting-edge technology and played a key role in the exploitation and exploration of other continents (Uddin et al. 2013). To support the gigantic ships, new infrastructure assets were developed in the United Kingdom, Spain, Portugal, the Netherlands and France. One of the assets was the wharfs, which were often paved with wood and timber. Thus, the port pavement emerged to support evolving marine transportation.

At that time, paved roads were often made of cobblestone. Owing to the basic knowledge obtained and the initial understanding on the role of pavement as vital infrastructure to support different transportation modes, this age can be regarded as the enlightenment era of the pavement. Similar to the 18th century, the durability and structural resistance of pavement were the main concerns in pavement construction.

2.3 Renaissance of pavement engineering

In the 19th century, improvements in material technology accelerated the trend of pavement engineering. In addition, increased as a result of the rapid growth of manufacturing industries as a consequence of the Industrial Revolution. To prepare the roads for cars, more durable and resistant pavements were essential. Therefore, pavement engineers tried to identify the key design parameters and select the relevant design criteria.

The engineers likewise attempted to find the interrelations between design criteria and parameters through field and laboratory tests. In this regard, some initial laboratory tests were developed to simulate loads on the paving materials. Furthermore, the rapid development of railroad as another transportation mode helped in the development of testing facilities. For example, the Micro-Deval test (ASTM D6928) used to analyse the abrasion of ballast was primarily developed to analyse the abrasion characteristics of aggregate materials.

The theories developed by Winkler, Terzhagi, Westergard and Boussinesq had key roles in developing the initial analytical procedures for pavement design (Yoder and Witczack, 1975; Huang, 2005). It suggested that this age be referred to as the renaissance of pavement engineering because of the relatively rapid development of basic analytical knowledge on the characterisation of pavement materials and their engineering properties.

2.4 Modern age of pavement engineering

2.4.1 Phase I (From 1900s to 1970s)

The rapid development of oil fields in the United States and the Middle East resulted in the cost-effective production of petrol for vehicles. In the early 20th century, the dominant transportation modes were railway and marine, both which were still powered by coal. The aviation industry was restricted to some commercial balloons and light aircraft developed by military sectors. After the second post-war period, demand for intercontinental air transportation increased.

Furthermore, the rate of car ownership skyrocketed in the United States and Europe in the 1950s (Croney & Croney, 1991). As the axle weight of vehicles increased, it resulted in the higher stress/strain on pavement structures. Technological development achieved during the jet age also changed the gross weight of the main gear of aircraft and the wheel configuration. Owing to these changes, it was necessary to build more durable and stronger pavements for roads and airports. To tackle this problem, transportation agencies strove to develop guidelines for the structural design of pavements and the production of more durable materials. Polymer-modified binder, for example, was one of the products created to reduce distress on pavement surfaces. More detailed guidelines were developed to determine the thickness of various layers of bituminous and concrete pavements, eventually leading to the development of empirical-mechanistic design methods. In this regard, many administrative institutes and agencies were established in different countries to standardise materials and construction methods.

Although material technology and structural design methods were improved, empirical methods, such as the California bearing ratio, continued to play a core role in engineering judgement. Given the variety of axle loads and types in vehicle and truck design, American Association of State Highway and Transportation Officials (AASHTO) undertook a long-term field test to estimate the damaging effects of various axles. The main objective of the test was to determine the relationship between the number of repetitions of a set of specific axle loads of various magnitudes and their arrangement. The output of the test was a set of proximity factors that convert all axle loads to a single-axle load of 8.20 tons (standard axle load). The factors were very useful for estimating the traffic load throughout the life span of a road pavement. The results of the AASHTO test serve as the backbone of pavement design in many countries.

Similarly, in airport pavements, the US Army Corps of Engineers (the Corps) conducted full-scale testing of trial aircraft pavements near Stockton, California. The result was an empirical pavement thickness design method to allow pavements to be developed to cater for the larger aircraft that were developed as a result of the WWII and the following Cold War period (White, 2017). This work later formed the basis of modern empirical-mechanistic design methods of aircraft pavements and full-scale testing continues under the direction of the Federal Aviation Administration (FAA) near Atlantic City, New Jersey, as aircraft continue to evolve.

In light of the breakthrough in computing method and data storage technology, some analytical methods and computer packages were developed to calculate the stress/strain and displacement of pavement layers. Therefore, pavement engineers were able to predict pavement response in terms of the stress/strain values and distribution patters. Additionally, the effect of axle type could be characterised through analysing the interaction between different wheels. ILLYSLAB, KENPAVE, BISAR and JSLAB are examples of the computer packages developed for pavement analysis.

In this age, new standards of testing materials were developed for selecting appropriate aggregate materials. Various tests were also used to characterise asphalt ageing, cement curing and rheology. Furthermore, performance-related tests were developed to characterise the structural response of asphalt and cement concrete mixtures. For example, new parameters such as resilient modulus and dynamic modulus were used to characterise the viscoelastic behaviour of asphalt mixture.

The development of mechanistic methods of analysis and design based on elastic theories and finite element resulted in more accurate numerical models to predict structural behaviour of pavements. In particular, structural design methodologies in Phase 1 became mature enough to address structural requirements under various utility conditions. Besides, practice standards and quality control protocols provided reliable outputs in the construction of flexible and rigid pavements.

2.4.2 Phase II (From 1970s to 2020)

The abundance of raw materials such as aggregates, along with the low price of crude oil, the main source of industrial fuel and bitumen resulted in the rapid development of transportation infrastructure assets. Although most concerns in the structural design and construction of pavements were addressed by the development of analytical approaches, a new challenge emerged in the 1970s. The oil shock that arose because of political challenges increased crude oil prices dramatically. Therefore, the initial activities to produce less energy- or material-intensive pavements began. For example, reclaimed asphalt pavement (RAP) technology was developed to produce cost-effective structural and surface layers. In other words, the development and implementation of RAP technology was largely a consequence of the energy crisis.

Environmental concerns were also raised in the public discourse because of global warming. To tackle the problem, stricter environmental regulations were prescribed to the construction industry by environmental policy makers and authorities. This means that structural design was no longer the single greatest challenge in pavement technology, being overtaken by global environmental concerns. The environmentalists encouraged pavement engineers and technologists to construct eco-friendly pavements that not only show high structural performance but also have lower environmental burdens in terms of greenhouse gas emission. To improve sustainability, a wide variety of industrial by-products and waste materials were used in pavement construction. For example, fly ash, blast furnace slag, cement kiln dust, demolished concrete and rubble were used in the concrete production.

Development of sustainable pavement technology accelerated during the first energy crisis in the 21st century. Unlike what happened in the 1970s, the second energy crisis was not caused by political challenges but by the lack of oil supply because the peak point of global oil production was expected to occur in the near future, or may have already occurred (Armstrong and Blundell, 2007). To address the energy crisis, warm mix asphalt (WMA) emerged as a sustainable asphalt technology that results in lower emissions and energy consumption. WMA production has increased because of promising results in the laboratory and the field. However, it should be noted that the trend of research and development on WMA depends significantly on the price of crude oil (Jamshidi and White, 2020).

There are various WMA additives and technology that reduce the mixing and compaction temperatures of asphalt through different mechanisms. For example, Sasobit®-WMA reduces total particulate emissions, SO_2, NOx, CO, CO_2, volatile organic compounds and benzene soluble matter levels by 74%, 83%, 21%, 63%, 42%, 51% and 80%, respectively, in comparison with hot mix asphalt (Hurley et al., 2009). Moreover, the whole life emission of pavements, from the cradle to the grave, can be evaluated via rating systems such as life cycle analysis (LCA) and leadership in energy and environmental design (LEED). The results of LCA and LEED can be used to rank different pavement types and construction materials based on norms targeted by environmental and pavement agencies.

In Phase II of the modern age, a new category of pavement came on the stage that showed satisfying structural and functional performance as well as lower carbon footprint compared to traditional pavements. Therefore, the environmentally friendly pavement system emerged. This pavement category bridges the chasm between pavement and environment engineering. Environmental impact assessment became an inevitable part of paving projects worldwide, but it is impossible to embrace all environmental requirements. This consensus between pavement technology and the environment paves the way towards more sustainable practice.

In Phase II, the price of bitumen had again increased because oil refineries prefer to produce high value-added products over low-quality materials such as bitumen. Thus, the production of alternative binders is a practical strategy to reduce energy and raw material consumption in the pavement industry.

In the modern age, great improvements were achieved in the analysis of the engineering properties of paving materials. For example, Superpave and balanced mix design were developed to address new requirements of asphalt concrete based on field requirements. In this regard, new testing protocols and equipment were produced. For instance, Brookfield viscometer was produced to measure high-temperature properties of asphalt binders, while dynamic shear rheometer was designed on the premise of rheometers used in the plastic industry to evaluate binder properties from 0°C to 100°C, where the binder displays viscoelastic behaviour.

Given that pavements are distressed over their life span, it is necessary to develop a pavement management system (PMS) for their maintenance and rehabilitation. PMS is a subsection of the asset management system, which is a useful support system for decision making by administrators, policy makers and engineers to consider the possible economic, social and environmental impacts of their investment decisions. Predictive models of distress were developed in the modern age to establish the optimum time frame of maintenance. Some computer packages were developed, such as the highway development and management model (HDM-4) and MicroPAVER. Note that PMSs and pertinent technologies vary in different countries because of administrative changes. For instance, monitoring airport construction has not been the responsibility of the Department of Housing and Construction in Australia since 1982 (White, 2016). Instead, the councils and the government of states are in charge of the operation, construction and design of airport pavements.

Pavements are impervious surfaces that decrease the permeability of catchments. Thus, the design of the drainage system in the vicinity of the pavement system is crucial. Construction of a permeable pavement is a practical measure integrating the pavement system and drainage. A permeable pavement system is a multi-role infrastructure asset that not only withstands the stress/strain induced by traffic loads but also drains the water that infiltrates. Therefore, it decreases the water head of the drainage system, which results in huge capital savings in urban development. The microbiological emission of runoff infiltrates through the permeable pavement can be decreased via specific geotextiles and fabrics (Imran et al. 2013; Scholz, 2013; Jiang et al. 2015). In other words, the pavement system can also act as hydraulic infrastructure for water treatment and storm water management.

However, permeable pavements have some drawbacks. For example, the porosity of the porous asphalt (PA) is relatively high, resulting in its lower structural capacity in comparison with the dense mix. Moreover, the binder creep phenomenon may lead to clogging (Hamzah et al., 2012). PA is usually used in parking lots and low traffic loads. Clogging is one of the factors that decrease the service life of permeable concrete and concrete block pavements. It depends on the hydrologic properties of catchment, hydraulic characteristics of aggregate materials, rainfall intensity, slope of the surface and amount of debris and organic waste materials on the pavement surface.

As a result, modern pavements are the interface between structural and environmental design (Figure 1). The weights of structural challenges and environmental concerns are almost identical. To think that current structural design and material characterisation methods do not need further updates and improvements is misleading and unrealistic. Pavement engineers and researchers still need to improve mechanistic methods and develop new materials with high structural consistency. More reliable predictive structural models of paving materials must be developed as well. For example, the Long-Term Pavement Performance (LTPP) programme can yield useful data for the evaluation of pavement performance.

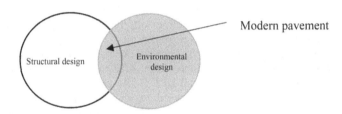

Figure 1. Definition of modern pavement as an interface between structural and environmental requirements.

2.5 *Post-modern pavement*

2.5.1 *Theory*
Although the structural performance of any pavement is a key factor in the life span of a pavement system, its functional performance, such as skid resistance, noise emission and evenness of the surface, is imperative for pavement users. Users do not care that pavement layers are determined through elastic or finite element approaches. Therefore, the concept of user-friendly pavements was developed. This kind of pavement describes a broad range of technologies, materials, monitoring system and construction methods that result in higher-quality road surfaces.

To take the user-friendliness of pavements into account, in addition to the structural requirements and environmental necessities, the PMP was proposed. In other words, The PMP is a high strength and environmental-human friendly pavement.

2.5.2 *Concept*
Although there is some useful information on the utility condition of pavement surfaces, they are not able to characterise the effects of the environment on the users. Such characterisation

is a new demand because urban sprawl has led to the urban heat island phenomenon, which results in higher human thermal load. Heat illness has been increasing in inner cities (Japan's Ministry of the Environment, 2017). To meet the demand, a new concept called *human-friendly pavement* has been proposed. The human thermal load (T_L) is characterised according to human physiological responses (Figure 2) and calculated through Equation (2).

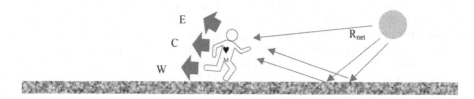

Figure 2. Schematic illustration of TL.

$$T_L = M - W + R_{net} - C - E \qquad \text{Aoki et al.}(2019) \qquad (2)$$

T_L: the human thermal load (W/m^2)

M: metabolic rate (W/m^2) which is calculated from body surface area, oxygen intake and carbon dioxide production

W: work load (W/m^2) which is the amount external work the human body carries out

R_{net}: net radiation (W/m^2) which is amount of solar and infrared radiation absorbed and gives by human body

C: sensible thermal loss (W/m^2) which is due to difference between skin and ambient temperature as well as thermal exchange via respiration;

E: latent heat loss (W/m^2) which is due to sweating, expiration and insensible perspiration

The user-friendly pavement is not a new concept, originally dating back to the AASHTO tests in the late 1950s. In addition, Carey and Irick (1960) conceived the pavement serviceability rating (PSR) as a key criterion of pavement to serve pavement users. Later, PSR was replaced with the pavement serviceability index (PSI), which ranges from 0 to 5. The highest score denotes a new constructed pavement, which decreases over time, while the lowest score denotes a fully distressed surface. PSI and international roughness index (IRI) are two quality indicators of the pavement surface. The main objective of user- or human-friendly pavements is to make a comfortable and safe environment for pavement users (Figure 3). This comfort can be in terms of a healthy and safe environment for pavement users. Another example is age-friendly pavement, which was proposed by the World Health Organization (WHO, 2007) to prepare cities for global population ageing.

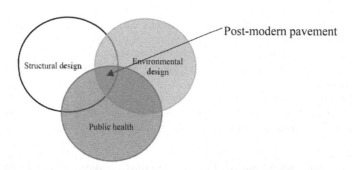

Figure 3. Definition of PMP as an interface between structural, environmental and social requirements.

2.5.3 *Performance and challenges*

To improve the utility condition of users, different technologies and materials have been developed for asphalt, concrete and concrete block pavement systems. For example, a water-retaining block pavement surface was proposed to decrease T_L. Figure 4 shows that the T_L of water-retaining blocks was lower than that of asphalt. Therefore, concrete block pavement is human-friendly pavement. In contrast, the construction of concrete block pavement is costly, and the level of noise emission of such pavement is higher than those of asphalt and concrete pavements (Karasawa et al., 2000; Ishai, 2003). The PMP concept allows these advantages and disadvantages to the objectively compared to those associated with other pavement or surface types.

To increase the reflection of the pavement surface, different types of high reflectance coating are used. The reflection of the coating results in less thermal energy being absorbed by the pavement surface. An estimated 50-60% of roofs and pavements in urbanised areas can be treated using coating (Li and Xie, 2020). In an experimental study, fluorinated acrylate coating increased the albedo of samples by 79%, and the sample temperature was reduced by 13°C when the normal asphalt mixture approached 60°C (Cao et al., 2017). Although coating is a cost-effective technology that can be adopted for various alternative pavements, its durability is a matter of concern. In addition, some coating materials decrease surface friction, which affects the level of safety. Lower friction would be underscored when the surface of the coated pavement becomes wet.

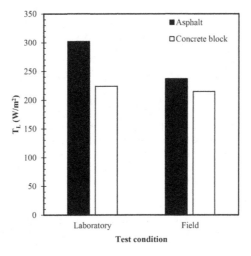

Figure 4. T_L of different alternative pavements, plotted based on the data reported by Aoki et al. (2019).

3 CONCLUSION

The structural performance of pavements was a serious engineering issue for many years. The experiences and technical knowledge gained during the renaissance, enlightenment and modern ages of pavement history resulted in breakthroughs in the material characterisation and structural design of different alternative pavements. However, environmental disasters and energy challenges opened a new frontline. In the modern age, pavement engineers must consider both structural and environmental requirements. In the future, the user-friendliness of pavement must also be considered and the PMP has been conceptualised to enable this. The PMP recognises pavements as a multi-role infrastructure or an integrated system that not only has high structural capacity against traffic loads but also has high sustainability due to its use of recycled and porous materials for water drainage and treatment. In addition, the PMP can improve public health via the enhancement of surface reflectivity and the like. Therefore, it is necessary to design a PMP based on the new concept called *human–infrastructure interface*, which is similar to the concept of human–machine interface based on ergonomic principles. The PMP is a new terminology in pavement history that opens new horizons in research and engineering. As a result, it is necessary to propose standards and practice codes to consider structural, environmental and social requirements for PMP, using LCA and LEED concepts calibrated against LTPP models.

REFERENCES

American Society for Testing and Materials (ASTM) D6928, 2014., *Standard test method for resistance of coarse aggregate to degradation by abrasion in the Micro-Deval apparatus*, Conshohocken, Pennsylvania, USA, 2014.

Aoki, M., Nakamura, T., & Shimazaki, Y., 2019. *Experimentally investigation of the effect water retentive sidewalks block pavement on improving thermal environment*, the 12th International Conference on Concrete Block Paving, Seoul, South Korea,150–160.

Armstrong, F.A., and Blundell, K.M., 2007. *Energy… Beyond Oil*. Oxford University Press, UK.

Axworthy, M., 2008. *Iran: Empire of the mind: A history from Zoroaster to the present day*, Penguin UK.

Cao, X., Tang, B., Luo, T., & Guo, P., 2017. *Preparation of fluorinated acrylate coating with high albedo and its cooling effect on asphalt mixture*, Road Materials and Pavement Design, 18(2), 464–476.

Carey, W.N., & and Irick. P.E., 1960. *The pavement serviceability performance concept*, Highway Research Board. Record 250, Washington, USA.

Clark, K. A., Ikram, S., & Evershed, R. P., 2016. *The significance of petroleum bitumen in ancient Egyptian mummies*, Philosophical Transactions of the Royal Society A: Mathematical, Physical and Engineering Sciences, 374(2079), 20160229.

Connan, J., 1999. *Use and trade of bitumen in antiquity and prehistory: molecular archaeology reveals secrets of past civilizations*, Philosophical Transactions of the Royal Society of London. Series B: Biological Sciences, 354(1379), 33–50.

Croney, D., & Croney, P., 1991. *The design and performance of road pavements*, McGrow Hill, New York, USA.

Faktorovich, A., 2015. *SPQR: A History of Ancient Rome*, Pennsylvania Literary Journal, 7(3), 61.

Hamzah, M. O., Hasan, M. R. M., & van de Ven, M. 2012., *Permeability loss in porous asphalt due to binder creep*, Construction and Building Materials, 30, 10–15.

Horvli, I. and Garba, R. 2002., *Permanent deformation properties of asphalt concrete mixture*, Proceedings of the Sixth International Conference on the Bearing Capacity of Roads, Railways, and Airfields, Lisbon, Portugal.

Huang, Y.H., 2004. *Pavement Analysis and Design*, Pearson, New York, USA.

Hurley, GC., Prowell, BD., Kvasnak, AN., 2009. *Michigan field trial of warm mix asphalt technologies: construction summary*, National Center for Asphalt Technology, Auburn, AL, USA.

ICPI., 2003. *Structural design of interlocking concrete pavements for roads and parking lots*, Tech.Spec. 4, Interlocking Concrete Pavement Institute, Washington, DC, U.S.A.

Imran, H. M., Akib, S., & Karim, M. R., 2013. *Permeable pavement and stormwater management systems: a review*. Environmental technology, 34(18), 2649–2656.

Ishai, I. (2003. *Comparative economic-engineering evaluation of concrete block pavements*, Road Materials and Pavement Design. 4 (3), 251–268.

Jamshidi, A., & White, G., 2020. *The Challenges of warm mix asphalt as a mature technology*. In Proceedings of the 9th International Conference on Maintenance and Rehabilitation of Pavements—Mairepav9. Springer, Cham.

Jamshidi, A., and Hassani, A., 2006. *The principles of performance, design and construction of concrete block pavements*, Aradmehr Publication, Tehran, Iran.

JIEPA.,2017. *Japan interlocking block pavement design and construction manual*, 2017. Japan Interlocking Pavement Engineering Association, Tokyo, Japan, (In Japanese).

Jiang, W., Sha, A., Xiao, J., Li, Y., & Huang, Y. 2015. *Experimental study on filtration effect and mechanism of pavement runoff in permeable asphalt pavement*, Construction and building materials, 100, 102–110.

Karasawa, A., Kagata, M., & Ezumi, N., 2000. *Performance evaluation of the low noise drainage interlocking block pavement system applied to the driveway*, the 6th International Conference on Concrete Block Paving, Tokyo, Japan, pp. 552–561.

Knapton, J., 1996. *The Romans and their roads-The original small element pavement technologists*, in: 4th International Conference on Concrete Block Paving, Delft, Netherland, pp. 17–52.

Li, H., & Xie, N., 2020. *Reflective coatings for high albedo pavement*, In Eco-Efficient Pavement Construction Materials. Woodhead Publishing, New York, USA.

Lilley, A. A., 1991. *A handbook of segmental paving*, E & FN Spon. London, UK.

Ministry of the Environment of Japan. 2017. *Heat illness prevention information site*, Health manual on heat illness environment, http.wbgt.env.go.jp/heatstroke_manual.php.

Moorey, R., 1994. *Ancient Mesopotamian materials and industries*, Oxford: Clarendon Press.

Nissenbaum, A., & Buckley, S. 2013. *Dead sea asphalt in ancient Egyptian mummies—why?*. Archaeometry, 55(3), 563–568.

Roberts, F. L., Mohammad, L. N., and Wang, L. B., 2002. *History of hot mix asphalt mixture design in the United States*, Journal of Materials in Civil Engineering, 14(4), 279–293.

Scholz, M. 2013. *Water quality improvement performance of geotextiles within permeable pavement systems: A critical review*, Water, 5(2), 462–479.

Uddin, W., Hudson, W. R., & Haas, R. C., 2013. *Public infrastructure asset management*, McGraw Hill Professional, New York, USA.

White, G., 2017. *Airfield pavement essentials*. Australian Airport Association, Airport practice note, Canberra, Australia, 1–96. PDF file. September 18, 2020: http://airports.asn.au/wp-content/uploads/2018/04/Airport-Practice-Note-12_05_lowres.pdf

White, G., 2016. *Challenges for Australian flexible airport pavements*, Aust. Geomech, 51(3), 39–46.

World Health Organization., 2007. *Global age-friendly cities: A guide*, World Health Organization. Geneva, Switzerland.

Yoder, E.J. and Witczak., M.W., 1975. *Principles of Pavement Design*, John Wiley & Sons, New York, USA.

Zoccali, P., Loprencipe, G., & Galoni, A., 2017. *Sampietrini stone pavements: distress analysis using pavement condition index method*, Applied Sciences, 7(7), 669.

Concrete pavements

Crack behavior of continuously Reinforced Concrete Pavements (CRCP) in Germany

M. Moharekpour*
Research Assistant, Institute of Highway Engineering, Aachen, Germany

S. Hoeller
Research Engineer, German Federal Highway Research Institute (BASt), Bergisch Gladbach, Germany

M. Oeser
Professor and Head of Institute, Institute of Highway Engineering, Aachen, Germany

ABSTRACT: The traffic volume on German motorways increased steadily and this trend is expected to continue in the future. To guarantee mobility in the future and reduce the national economic consequential costs, road construction with a maximum service life and a minimum of necessary maintenance are needed. Rigid road pavements with continuous reinforced concrete pavement (CRCP) are extremely durable in terms of use and maintenance. The behavior of CRCP is influenced by a number of specific characteristics such as the thickness and the quality of the concrete, the design of the longitudinal and transversal reinforcement, the base layer and the environmental conditions at the time of construction and during service life. These aspects influence the crack pattern crack distance and crack widths. Identified positive crack pattern are one important indicator of positive long-term behavior. In Belgium, the continuously reinforced concrete pavement has been a standard construction for many years. In Germany and Poland CRCP is in the stage of field testing. From 1997 to today, a total of 8 sections with many variations have been built. A detailed comparative study of these sections has so far been lacking. As part of a research project, the RWTH university of Aachen and the German Federal Highway Research Institute (BASt) are investigating these sections with CRCP with and without an asphalt surface course in Germany, and Poland and compare it to the Belgium standard constructions. Load capacity measurements, georadar measurements, visual and automated condition assessments and crack recordings are carried out. The aim is to evaluate the different designs in the sections in terms of their behavior in service, to quantify achievable service lifes, necessary maintenance and availability. From this, a preferred variant of the construction is designed, presented in a concept and implemented on a motorway in Germany as part of a trial site. The Project is financed by the German Federal Ministry of Transport and digital Infrastructure (BMVI).

Keywords: continuously reinforced concrete pavements, crack spacing, crack behavior, FWD

1 INTRODUCTION

The two most common materials used for pavement construction are asphalt and concrete. Besides the advantages of asphalt roads like low noise, high skid resistance of a surface layer

*Corresponding author

DOI: 10.1201/9781003222880-24

(M.Asi, 2005), easy maintenance and replacement, asphalt pavement has apparent drawbacks as well, including frequent maintenance, colossal demand for repairing, and short service life (Sungun Kim, 2015). Due to its high stiffness, concrete is able to withstand the stresses from traffic and service over a very long period without permanent deformation. Approximately 9641 km (nearly 73%) out of the 13141 km of Germany's motorways are flexible pavements (asphalt wearing surface). Besides, 3500 km (nearly 26 %) are built in the rigid type (Portland cement concrete pavement with or without a bituminous wearing surface (Infrastruktur, 2020). According to the traffic forecasts from the German Federal Ministry of Transport and Digital Infrastructure, by 2030, road freight transport in Germany will increase by 17% compared to 2010 (GmbH, 2014). Due to the increase in traffic congestion and the consequent reduction in economic costs, there is a demand for road construction with maximum durability, minimal maintenance, and as few traffic restrictions as possible (Tatiana Petrova, 2018). Continuously reinforced concrete pavement (CRCP) is a representative type of concrete pavement with no transversal joints constructed, and no preventative measures for transverse expansion or contraction joints (Christopher Robinson, 1996) and the role of transverse joints in jointed concrete pavement (JCP) is replaced by suppressing the widening of cracks through longitudinal steel in a concrete slab (Young-Chan Suh, 2020). Therefore, longitudinal bars in CRCP are constructed to limit the crack width and improve the load capacity of transverse cracks (YANG Cheng-cheng, 2020). CRCP offers adequate structural strength throughout the service life because of the ability of concrete to gain strength with time (Sidney Mindess, 1981). CRCP's structural strength is also predicated on its ability to keep cracks tight and thus greatly minimize the development of distress at the cracks (Richter, 1998). The performance of the CRCP is quite satisfactory, with a lower number of distresses than in JCP for pavements with a comparable age .(Pangil Choi S. R., 2014) At the same time, the ride quality of CRCP is consistently better than JCP ((Pangil Choi D.-H. K.-H., 2016). CRCP performs better than JCP or Asphalt pavement and has longer service life and low maintenance cost even under heavy traffic loadings and challenging environmental conditions (Han Jin Oh, 2016). The initial investment cost of CRCP is higher than that of JCP and asphalt pavement, and the repairs of distresses in CRCP are more complicated, expensive, and time-consuming than those in JCP and asphalt pavement (Diependaele, 2018). conducted a life cycle cost (LCC) of CRCP (performance life of 50 years) and asphalt (performance life of 36 years) along a 14.2-km-long carriage way in each direction. Each carriageway consists of a minimum of 4 traffic lanes + 1 emergency lane. The initial cost of CRCP was 150% higher than that of asphalt. In contrast, the estimated maintenance cost and rehabilitation cost of CRCP were estimated to be as low as 9% and 78% those of asphalt, respectively. (Hu, 2015) analysed the LCCs of CRCP and JCP along a 1.6-km-long section of a 2-lane highway in 2012 with a 50-year analysis period.. The total cost of CRCP was 142.3% higher than that of JCP. In contrast, the maintenance cost and rehabilitation cost of CRCP was calculated to be 38.5% and 57.8% less than that of JCP. Major distress found to impact the performance of CRCP were wide transverse and longitudinal cracks, punch-outs, Y-cracking, and spalling (Jeffrey Stempihar, 2020). The American Association of State Highway and Transportation Officials (AASHTO) Design Guide recommends a minimum crack spacing of 107 cm to minimize punch-out or spalling (Blaschke, 1993). The desired crack spacing in Belgium is defined between 0.8 m and 1.3 m based on the technical reports (references).

2 RESEARCH SIGNIFICANCE

Concrete pavements on freeways are constructed in Germany as standardized non-reinforced dowelled slabs on different base and intermediate layers. There are initial indications that this standard construction method will not achieve the planned service life of 30 years. This raises the question as to which construction methods will be able to meet the requirements of modern infrastructure in the future and ensure the necessary availability of the freeway network. The construction method with a continuously reinforced concrete pavement can be an option. In Germany, this construction method has been used for over 20 years. The developments and applications abroad have been followed with great interest. In Germany, 5 test and trial sections have been installed on federal roads and freeways from 1997 to the present day. Until the introduction as a new standard construction method, these routes will be investigated in detail and compared with Belgian standards. The question is to be answered whether this construction method can guarantee the availability of the highway network under the demands of the future.

3 METHODOLOGY

In this study, three sections with different design parameters on Federal Road B 56 near Dueren, Access Road near Geseke, and Highway A 94 near Munich are investigated, and the performance of CRCP for selected sections are evaluated. Figure 1 shows the location of the considered trial sites in Germany.

Figure 1. Location of trial sites in Germany.

BASt set up the first trial site in CRCP in Germany on the Federal Road B56 near Dueren in a length of 800 m in 1997. The construction consisted of a frost protection layer, an asphalt base layer, and a reinforced concrete surface. The thicknesses of the asphalt base course is 15 cm and CRCP is 22 cm. The bottom layer had a compressive strength of 45 N/mm² and the top layer with aggregates 0/7 mm had a compressive strength of 55 N/mm², and the reinforcement was laid in the middle. In the longitudinal direction, the proportion of reinforcement was 0.61% of cross-section. Four reinforced anchor lugs were built at both ends, which ensured that the concrete slab was saved without shifting. Due to the limited space available, the 7.5 m wide carriageway was installed in strips and anchored together. Figure 2 illustrates the longitudinal section of the three trial sites. The CRCP with a thickness of 22 cm on Geseke and B56 and 24.5 cm on A94 rested on an asphalt base with a thickness of 10 cm, 15 cm, and 10 cm, respectively.

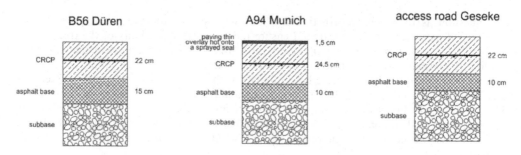

Figure 2. Federal Road B56 near Dueren // Highway A94 near Munich // Access Road near Geseke.

In 2009, Heidelberg Cement AG completed an access road between the Milke cement plant and the Elsa quarry south of Geseke in a length of 1.030 m in CRCP. With the new road construction in CRCP, Heidelberg Cement has optimized the transport of raw materials and clinker between the two production sites and, at the same time, relieved the public road network by driving heavy goods vehicles on a separate route. The road is loaded with around 450 heavy trucks of Heidelberg Cement every day. The width of the road, including the shoulder, is 7.50 m. It was built in two construction lanes of 3.5 m and 4.0 m with a slipform paver and connected with tye bars. The concrete (C35/45) was installed on a 10 cm thick asphalt base layer (AC 22 TS). At both ends, four reinforced endanchors were installed to absorb the shear stresses or longitudinal forces from thermal deformations of the construction. The section's reinforcement ratio is 0.75%, whereby longitudinal reinforcement with a diameter of 20 mm was laid on a transverse reinforcement with a diameter of 16 mm at intervals of 17.5 cm. The separately installed lanes were anchored with adhesive tye bars (Φ20 mm, length of 650 mm) at a distance of 67 cm to prevent the lanes from moving apart in the longitudinal compression joint area. The road surface was then textured with grinding to improve evenness, skid resistance, and drainage. The general information and the average number of vehicles per day, and the number of heavy trucks in these three sections are illustrated in Table 1.

Table 1. General information about considered tracks.

section	unit	Highway A 94 near Munich	Federal Raod B 56 near Dueren	Access road near Geseke
length	m	4000	800	1030
construction date	year	2011	Oct. 1997	2009
ADT	veh/d (in year)	10887 (2018)	3200 (1997)	-
ADT-Truck	veh/d (in year)	2276 (2018)	100 (1997)	450 (2009)

In August 2011, the trial site on Highway A 94 east of Munich was paved in standard road concrete C30/37 in a length of 4 km. It was constructed in one layer with 0/22 aggregates, and longitudinal joints were arranged lane wise. In the end areas, reinforced end anchors were realized. The concrete surface was then covered with a thin overlay of hot mixed asphalt on a sprayed seal some weeks after concrete paving. Two criteria were checked on the surface before paving asphalt: resistance against abrasion > 1.0 N/mm² and residual moisture of concrete < 3.0 %. The C30/37 grade concrete was placed on a 10 cm thick asphalt base layer (AC 22 TS) using a slipform paver. The evaluation of the measured temperatures shows that on hot summer days the asphalt-covered concrete surface becomes 4°C to 6°C warmer than directly trafficked concrete pavement.

Table 2. Requirements for the reinforcement.

steel grade/section		Highway A 94 near Munich	Federal Raod B 56 near Dueren	Access road near Geseke
longitudinal reinforcement	reinforcement ratio	0.75 %	0.61 %	0.75 %
	diameter	20 mm	16 mm	20 mm
	axle spacing	17.5 cm	15 cm	17.5 cm
	Concrete cover	central	9 cm	9 cm
transverse reinforcement	reinforcement ratio	0.12 %	0.09 %	0.15 %
	diameter	16 mm	12 mm	16 mm
	axle spacing	70 cm	60 cm	60 cm
	angle to the longitudinal reinforcement	60°	65°	60°

In CRCP, longitudinal steel bars installed inside the concrete pavement are primarily intended to restrain a significant volume change in the concrete due to temperature variations and traffic load. The longitudinal-reinforcement ratio of the Highway A94 near Munich and the Access Road near Geseke are 0.75% and Federal Road B 56 near Dueren 0.61%, respectively.

4 RESULTS AND SHORT DISCUSSION

In Belgium, the best long-term behavior was found for sections with free cracks at intervals of 0.8 m to 1.3 m. In the USA, a crack spacing of 1.07 m is aimed for. Based on these two experiences, the following classification was made for further investigations. In order to better evaluate crack spacing in all sections, the following classification was made. Crack spacing smaller than 0.3 m, which is shown in red, and these are the critical crack spacing as there is a risk of punch-out. Crack spacing between 0.3 m and 0.7 m are shown in yellow. There is no direct danger of punch-out, but danger could arise if a further division takes place here due to an additional crack. The third class is shown in light blue, which belongs to the crack spacing between 0.7 m and 1.4 m. The fourth class, shown in blue, includes all crack spacing between 1.4 m and 2.5 m. In dark blue are the crack distances larger than 2.50 m. These are less critical but more significant than the desired crack spacing. The third and fourth classes also still belong to the desired crack spacing. Figure 3 illustrates the classification of crack spacing, which were carried out in 2019 and 2020 for all three considered sections in the number of cracks per 100 meters.

On Federal Road B56 near Dueren during the crack survey on 02/11/2008, 93.6 cracks per 100 m were determined in both directions. This results in an average crack spacing of 1.08 m. Approx. 43 cracks have been added from 2001 to 2008. The last crack survey was carried out on November 2019, and 125 cracks per 100 m and an average crack spacing of 0.80 m were recorded. Approx. 20% of all recorded cracks in the driving direction (N) were Y-cracks. 80 cracks per 100 m (63%) belong to the first and second classes, and only 43 cracks per 100 m (35%) are assigned to the desired classes. On A94 near Munich, 79 days after the concrete construction and before asphalt overlaying on 07/05/2011, a manual crack survey recording the location and opening widths was carried out. During this time, 38 cracks per 100 m and an average crack spacing of 2.6 m have formed. On 06/2014, 72 cracks per 100 m and an average crack spacing of 1.40 m were recorded. The current Monitoring was

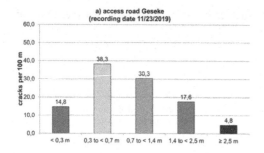
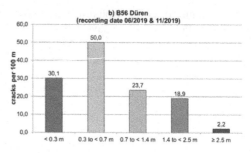
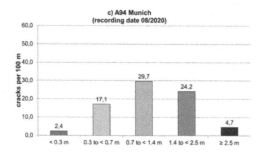

Figure 3. Classification of crack spacing in number of cracks per 100 m, a) Access Road near Geseke, b) Ferdal Road B56 near Dueren, c) Highway A94 near Munich.

carried out on 08/2020, and 78 cracks per 100 m and an average crack spacing of 1.3 m were recorded. 2.4 cracks per 100 m (3%) are the critical cracks, which have a crack risk of punch-out. 54 cracks per 100 m (69%) are assigned to the desired classes. Figure 4 illustrates the classification of crack spacing in percent. The current survey on the B56 was carried out in two different days due to a construction site in one part of the route.

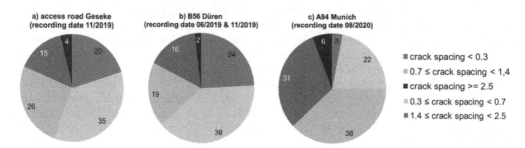

Figure 4. Classification of crack spacing in percent: a) Access Road near Geseke, b) Ferdal Road B56 near Dueren, c) Highway A94 near Munich.

4.7 cracks per 100 m (6%) belong to the fifth class with a crack spacing ≥ 2.5 m. All tracks visual condition in the last survey was good, and no sections exhibited punch-outs and other damages. Evaluating the structural condition of pavement sections is a complex task.

One of the more feasible means for structural condition evaluations is the use of falling weight deflectometer (FWD) (Pangil Choi D.-H. K.-H., 2016) with three drops per point and a minimum impulse of 50 kN. The load capacity of the sections was measured using FWD to evaluate the mechanical response of the field pavement, estimate the load capacity, to provide better pavement maintenance (Huang, 2020) and future performance of the road surface. The results in the load center were shown in Figure 5.

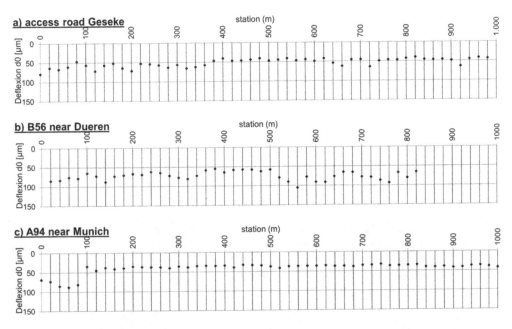

Figure 5. Load capacity resulting from FWD-values: a) Geseke, b) B56 Dueren, c) A94 Munich.

On the Y-axis, the average deflection was shown, and the zero value was at the top. The deflections run downwards. With a drop weight of 50 kN, the deformations were determined at regular intervals of 20 m. All values were between 30 μm and 100 μm. The evaluation of the FWD measurements shows an average FWD-value of 48 μm (mean temp: 24.4°C), 53 μm (mean temp: 26.5°C), and 78 μm (mean temp: 17.6°C) for the A94, Geseke, and Dueren, respectively. Thus all deflections were low, which means good load capacity (Pangil Choi D.-H. K.-H., 2015). Meanwhile, the best load-bearing section is A94 near Munich due to a 24.5 cm CRCP and reinforcement ratio. The concrete slab's interface bonding strength and concrete compressive strength were determined on drill cores and are represented in Figure 6.

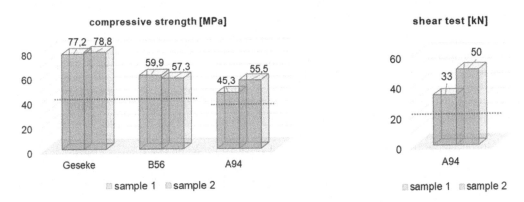

Figure 6. The results of tests on drilling cores: left: compressive strength on drill cores, right: Leutner shear test.

269

The results of the compressive strength test on drill cores show better results on access road Geseke compared to other sections. The section on Highway A94 near Munich was overlaid with a thin hot mix asphalt on a sprayed seal, and the shear test, which developed in Germany in the late 1970s, was carried out to measure the bond strength of the interface between asphalt and concrete layer. During the shear test at the standard loading rate of 50 mm/min, a constant rate of shear displacement was applied across the interface, and the resulting shear force was monitored (Asphalt, 1999). From the measured data in Figure 6, τ_{max} is the peak shear stress of the interface, and δ_{max} is the displacement at peak shear stress. The shear stress of the interface between concrete and asphalt shows a good result. According to Additional technical contract conditions and guidelines for the construction of asphalt pavements Asphaltbauweisen, 2013, the peak shear stress of the interface between the surface layer and binder layer should be higher than 15 kN (Asphalt Z., 2007). In order to investigate and evaluate the development of crack characteristics in concrete pavements, the crack width changes along the drill cores were recorded with laser scanners from LMI Technologies with a resolution of 150 μm (x-y) and 19 μm (z). The device scans the crack path on the drill core surface and visualizes the measuring object in 3D. The results were imported and evaluated in MATLAB.

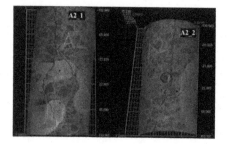 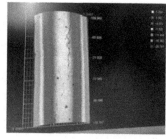 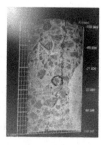

Figure 7. Evaluation of crack paths on drill cores; left: laser scanner, right: recorded images.

A crack length of 3 cm is measured for drill core No. 21. The mean crack width is 0.737 mm and is clearly above the maximum width of 0.5 mm. The crack width at the pavement surface is 0.958 mm, which is below the limit of 1.0 mm. Drill core No. 22 shows two wider cracks along the drill core depth, forming a continuous crack. The first crack has a length of 5 cm and extends in depth into the bottom concrete. The average crack width is 0.527 mm and thus slightly exceeds the maximum crack width. In the lower part of the drill core, the second crack has a length of 2.5 cm and a significantly higher mean crack width of 0.951 mm. Overall, the crack length extends along the drill core length with different crack widths. The crack becomes clearly narrower in the transition area from crack no. 2 to no. 1. While the crack width at drill core 21 increases up to the maximum crack width and then decreases again, crack No.2 at drill core 22 has a funnel-shaped opening. The following table lists the relevant measured values for crack path on two cores of the B56.

The highly loaded access road near Geseke shows many short cracks. Due to the minimal crack lengths, most of the cracks are surface cracks. Remarkable are the slight spalling at the crack widths, which are clearly visible on most drill core surfaces. There are different crack widths in drill core depth on ten drill cores. Table 4 shows the mean crack widths resulting from the measured crack widths. For the investigation of the crack widths, drill cores with a crack length of at least 3 cm are considered. At the drill cores G7, G10-1 and E2-1 the crack widths decrease from the crack mouth to the crack root. While crack no. 2 at drill core G13 runs with a funnel-shaped crack opening, crack no. 1 has developed narrower over the maximum crack width to the end of the crack. A similar crack path as for crack no. 1 can be seen on drill core G11.

Table 3. Evaluated crack development on the drill cores of the B56 near Düren.

core no.:		21	22-1	22-2
Crack length [cm]		3,0	2,5	5,0
Crack path [mm]	start	0,958	0,619	0,654
	max. crack width	1,143	1,459	0,654
	end	0,398	1,459	0,502
Position of the max. crack width [cm]		0,670	17,84	0,170
Average crack width [mm]		0,737	0,951	0,527

Table 4. Evaluated crack development on the drill cores of the Geseke.

core no.:		G7	G10-1	G10-2	G11	G13-1	G13-2	G15	E2-1	E2-2
Crack length [cm]		6	4	3	3	3,5	3	2,5	3	2
Crack path [mm]	start	1,676	1,861	2,059	1,054	0,766	1,347	1,081	1,022	0,645
	max. crack width	1,676	1,861	2,059	1,325	0,926	1,347	1,144	1,022	0,645
	end	0,591	0,336	0,565	0,099	0,393	0,904	1,144	0,320	0,276
Position of the max. crack width [cm]		0,22	1,33	0,72	2,01	2,88	0,49	12,68	0,21	0,32
Average crack width [mm]		1,140	0,777	0,749	0,651	0,631	0,686	1,013	0,616	0,449

5 DISCUSSION

All sections considered in this study were executed with longitudinal and transverse steel reinforcement. This construction method has shown positive behavior. The tests did not indicate that the yield point of the material was exceeded or that corrosion of the reinforcement occurred. Nevertheless, it should be considered which alternatives to steel reinforcement are available. One option is the use of profiled glass fiber or carbon rods. In 2019, a short test section was set up in Poland with a concrete ceiling with continuous glass fiber bar reinforcement. The investigation and evaluation of both standard and innovative construction methods must consider the entire life cycle. This also includes dismantling and recycling. The processes for recycling concrete pavements that exist in Germany can in principle also be applied to continuously reinforced concrete road surfaces. However, this can be made more economical. One option here is the Resonant machine. Here, the concrete slab is crushed with a high-frequency pressure stamp. Steel and concrete can then be separated with little effort.

6 CONCLUSIONS

In order to guarantee the necessary transport of goods on motorways and especially with heavy loads, the previous standard construction methods in asphalt and concrete may no longer be sufficient. A construction with a continuously reinforced concrete pavement with an optional thin asphalt surface layer could be an alternative. In Germany, 8 trial and test sections have been built using this construction method from 1997 to the present. In the context of this study, three sections with a continuously reinforced concrete paving slab on B56 near Dueren, access road near Geseke, and A94 near Munich, which were built in 1997, 2009, and 2011 are analyzed. The visual condition in the three sections is acceptable. There are no concrete punch-outs or slab displacements. Free transverse cracks have appeared, which, on the basis of the Belgian experience, are predominantly positive. Load capacity measurements with the FWD showed small deflections in all sections. The shear strength and adhesive tensile strength between asphalt and concrete were tested on drill cores. All results met the required values. Comparing the considered section show that section A94 with 1.5 cm paving thin overlay hot onto a sprayed seal on 24.5 cm CRCP with a 0.75% reinforcement ratio gives the best performance. In this section, most cracks belong to the desired crack spacing between 0.7 m and 2.4 m. The load capacity is much better than other sections, the bond strength of the interface between asphalt and concrete layer is higher than the suggested value, and the compressive strength of concrete is acceptable. The research project aims to find out whether the construction with a continuously reinforced concrete pavement and optionally with a thin surface layer of asphalt or alternative building material can be a new standard construction method for highly loaded motorways with loads of the future. The results of investigations to date are positive, and potential is evident.

ACKNOWLEDGEMENTS

This study was sponsored under the research project "FE 08.0248/2016/CGB "Asphalt surface course on continuously reinforced concrete pavement; scientific monitoring of the test sections in composite construction during the operating phase", these are being investigated by the RWTH Aachen University together with the German Federal Highway Research Institute (BASt). The authors would like to thank the German Federal Ministry of Transport and Digital Infrastructure (BMVI) for supporting this research.

REFERENCES

Asphalt, A. z. (1999). *Teil 4: Prüfung des Schichtenverbundes nach Leutner*. Cologne: FGSV.
Asphaltbauweisen, A. (2013). *Zusätzliche Technische Vertragsbedingungen und Richtlinien für den Bau von Verkehrsflächenbefestigungen aus Asphalt*. Cologne: FGSV.
Blaschke, B. C. (1993). *Aashto Guide for Design of Pavement Structures*. Washington D.C.: The Amercian Association of state highway and transportation officials.
Christopher Robinson, M. A. (1996). Distress as function of age in continuously reinforced concrete pavements: models developed for Texas pavement management information system. *Transportation Research Record*, pp. 145–151.
Diependaele, M. (2018). *A guide on the basic principles of Life-Cycle Cost Analysis (LCCA) of pavements*. Brussels: European Concrete Paving Association.
GmbH, I. C. (2014). *Verkehrsverflechtungsprognose 2030*. Berlin: Federal Ministry of Transport and Digital Infrastructure.
Han Jin Oh, Y. K.-M. (2016, July 01). Experimental evaluation of longitudinal behavior of continuously reinforced concrete pavement depending on base type. *Construction and Building Materials*, pp. 374–382.
Hu, J. e. (2015). Optimizing concrete pavement type selection based on life cycle cost analysis. *International Concrete Sustainability Conference*. Miami: NRMCA.
Huang, Y. Z. (2020, April 16). A method for evaluating CRCP performance based on edgeloaded FWD test . *Materials and Structures*, pp. 53–46.
Infrastruktur, D. B. (2020, Januar). *BMVI*.Retrieved from https://www.bmvi.de/SharedDocs/DE/Artikel/G/infrastruktur-statistik.html
Jeffrey Stempihar, N. W. (2020, July 24). Assessment of California's Continuously Reinforced Concrete Pavement Practice and Performance. *Transportation Research Record*, pp. 832–842.
M. Li, W. K. (2014, May 30). "Investigation on material properties and surface characteristics related to tyre-road noise for thin layer surfacings. *Construction and Building Materials*, pp. 62–71.
M.Asi, I. (2005, september 25). Evaluating skid resistance of different asphalt concrete mixes. *Building and Environment*, pp. 325–329.
Pangil Choi, D.-H. K.-H. (2015, November 6). Evaluation of structural responses of continously reinforced concrete pavement (CRCP) using falling weight deflectometer. *Canadian Journal of Civil Engineering*, pp. 28–39.
Pangil Choi, D.-H. K.-H. (2016, 02 02). Evaluation of structural responses of continuously reinforced concrete pavement (CRCP) using falling weight deflectometer. *NRC Research Press*, pp. 28–39.
Pangil Choi, S. R. (2014). *Project Level Performance Database for Rigid Pavements in Texas, II*. Texas: Texas Department of Transportation.
Richter, C. (1998). *What makes portland cement concrete pavements rough?* Washington D.C.: FHWA.

Sidney Mindess, J. F. (1981). *Concrete*. Englewood Cliffs, N.J.: Prentice-Hall.

Sungun Kim, S. H. (2015, August 05). Estimation of service-life reduction of asphalt pavement due to short-term ageing measured by GPC from asphalt mixture. *Road Materials and Pavement Design*, pp. 153–167.

Tatiana Petrova, E. C. (2018, December 31). Methods of road surface durability improvement. *Transportation Research Procedia*, pp. 586–590.

YANG Cheng-cheng, W. X.-f.-h.-j. (2020, March 01). Study on the Load Transfer of Transverse Cracks of Continuously Reinforced Concrete Pavements. *Journal of Highway and Transportation Research and Development*, pp. 10–17.

Young-Chan Suh, D.-H. J.-W. (2020, January 12). Comparison of CRCP and JCP based on a 30-year performance history. *International Journal of Pavement Engineering*, pp. 1–8.

Modelling

A model for the permanent deformation behavior of the unbound layers of pavements

M.S. Rahman
Pavement Technology, Swedish National Road and Transport Research Institute (VTI), Linköping, Sweden

S. Erlingsson
Pavement Technology, Swedish National Road and Transport Research Institute (VTI), Linköping, Sweden
Faculty of Civil and Environmental Engineering, University of Iceland, Reykjavik, Iceland

A. Ahmed & Y. Dinegdae
Pavement Technology, Swedish National Road and Transport Research Institute (VTI), Linköping, Sweden

ABSTRACT: This article presents a model for the permanent deformation (PD) behavior of unbound granular materials (UGMs) used in the base and subbase layers of pavement structures. The model was developed based on multistage (MS) repeated load triaxial (RLT) testing. This is essentially a modified version of a previously developed model to better suit to field conditions in a simple and effective manner. The model was calibrated for eight commonly used UGMs using MSRLT tests with a range of moisture contents. For validation, the calibrated models were used to predict the PD behavior of three of the UGMs in MSRLT tests with stress levels and moisture contents different from those used during the calibrations. This model showed better quality of fit when compared with another widely used PD model. The model was further tested successfully for field conditions by capturing the PD behavior of an instrumented pavement test section in a controlled environment using a heavy vehicle simulator (HVS) based accelerated pavement testing (APT). Inputs for calibrating the model were based on the readings from the instrumentations. The parameters of the model were adjusted to match the measured data with the predictions. Based on these results for various design conditions, some ranges of values of the material parameters of the model were suggested.

Keywords: Unbound granular materials, permanent deformation, model, triaxial test, moisture

1 INTRODUCTION

Gradual accumulation of permanent deformation (PD) in unbound granular materials (UGMs) used in the base and sub-base layers of flexible pavements may lead to rutting and eventual failure of the structure (Hornych and El Abd, 2004). Mechanistic-empirical (ME) design of pavements requires predicting the deformation behavior of the UGMs used in different layers, for the expected traffic load and environmental conditions, using constitutive models (Ramos et al., 2020). The aim is to control rutting and to implement it in pavement management systems and in life cycle cost analysis (Di Graziano et al., 2020). Ideally, the constitutive models should be able to reliably predict the deformation behavior of the materials considering the major influencing factors on the behavior. However, this is relatively difficult to achieve because of the complex and nonlinear behavior of the materials (Gidel et al., 2001, Hornych and El Abd, 2004).

Several of the PD models for UGMs predict the accumulation of PD with the number of load cycles by combining the influence of stress levels (Lekarp, 1999, Gidel et al., 2001, Korkiala-Tanttu, 2005). In a study by Rahman and Erlingsson (2014), the models proposed by Gidel et al. (2001) and Korkiala-Tanttu (2005) showed satisfactory performance in Multi-Stage (MS) Repeated-Load Triaxial (RLT) tests. However, these models relate the amount of permanent strain to the shear strength properties of the material. The shear strength parameters are determined using static failure triaxial (SFT) tests. At least three SFT tests are required for each UGM to reliably obtain these parameters. Then the RLT test is used to evaluate the rest of the parameters of the models. This is quite time consuming and expensive. Again, these models do not explicitly include the important factors, such as moisture content (w), particle size distribution (PSD) and the degree of compaction that affect the PD behavior. This becomes even more tedious process if the influences of these factors need to be investigated. Moreover, several researchers have criticized the idea of predicting the behavior of UGMs in cyclic loading based on SFT tests, arguing that the behavior of UGMs is very complex and the structural response of the materials may not be the same in these two kinds of tests (Lekarp, 1999). To overcome these issues, Rahman and Erlingsson (2015) proposed a simpler model that relates the PD to the applied stresses and the number of load cycles. This model can be calibrated using a single RLT test and showed good performance. However, for field conditions and for application in a pavement design software, it is more convenient to use a model that relates the PD to the resilient strain instead of stress levels directly. In this regard, the Mechanistic-Empirical Pavement Design Guide (MEPDG) model, proposed by Tseng and Lytton (1989) has certain advantages. However, this model showed some limitations in performance when used for MS RLT tests of UGMs (Rahman and Erlingsson, 2014) and filed application (Fladvad & Erlingsson, 2021). Hence, in this study, the Rahman and Erlingsson (2015) model was modified to include resilient strain as the input parameter instead of stress levels.

2 PERMANENT DEFORMATION PROPERTIES OF UGMS

UGMs are inhomogeneous and anisotropic in nature. Mechanical resistance of UGMs derive mainly from particle interlocking and friction between the particles (Lekarp, 1999). UGMs in pavements are subjected to cyclic stresses of varying magnitudes from the moving traffic load. These stress pulses contain vertical, horizontal and shear components. For UGMs, the vertical and horizontal stresses can only be compressive. Because of the moving wheel load, a rotation of the principal stress axes also occurs (Lekarp, 1999). The total deformation due to compressive cyclic stresses in a UGM consists of two parts: (a) elastic or recoverable or resilient deformation (RD) and (b) irreversible or plastic or permanent deformation (PD). Although small compared to the RD, the PD accumulates in the material for each load cycle and may become significantly large to cause failure of the pavement (Lekarp, 1999, Ramos et al., 2020).

The amount of PD is dependent on the magnitude of the stresses (Lekarp, 1999, Gidel et al., 2001). It is found to be directly related to deviator stress and inversely related to confining pressure. Several researchers have linked the amount of PD to some form of stress ratio consisting of both deviator stress and confining pressure. It has been also reported that reorientation of the principal stresses in pavement structures results in increased permanent deformations (Lekarp, 1999). Other factors governing the amount of permanent deformation in UGMs are stress history, moisture content, degree of compaction, PSD, aggregate type, etc. Permanent strain resulting from a certain stress level may be significantly reduced if the UGM had been subjected to another stress cycle previously (Brown and Hyde, 1975).

Permanent deformation is generally reported to increase with increasing moisture where materials with higher fines content are more affected (Rahman and Erlingsson, 2013, Lekarp, 1999).

According to the shakedown range (SDR) theory, the evolution of PD with load applications can be classified into three types which are dependent on the stress level (Werkmeister et al., 2001). SDR A occurs for relatively low stress levels where the accumulation of PD ceases after a certain number of load cycles. For higher stress levels, SDR B occurs where the accumulation of PD continues at a steady rate. For even higher stress levels, SDR C behavior is observed where PD accumulates rapidly, leading to failure. Suggested criteria to distinguish between the different SDR behaviors can be found in Werkmeister (2003) and CEN (2004a).

Several models have been proposed for PD of UGMs (Ramos et al., 2020). One of the most referred to PD models is the one implemented in the Mechanistic Empirical Pavement Design Guide (MEPDG) (ARA, 2004), originally proposed by Tseng and Lytton (1989). This model implicitly considers the influence of the stress level, assuming a direct relationship between the resilient deformation and applied stresses and combines the influence of the number of load applications. This model is expressed as:

$$\hat{\varepsilon}_p(N) = \varepsilon_r \varepsilon_0 e^{-(\frac{\rho}{N})^\beta} \quad (1)$$

where, ε_0, ρ and β are material parameters and ε_r is the resilient strain at N^{th} load cycle. The parameter β is estimated using the gravimetric moisture content, w (%) as:

$$\log \beta = -0.61119 - 0.017638w \quad (2)$$

3 EXPERIMENTAL PROCEDURE

The model proposed in this article was developed based on MS RLT tests in the laboratory, conducted following the European standard EN-13286-7 (CEN, 2004a). The MS RLT test, in contrast to the single stage (SS) RLT test, applies several stress paths on a single specimen. This approach inherently demonstrates the effect of stress history and enables for a more comprehensive and realistic study of the material behavior. The sizes of the cylindrical specimens were 150 mm in diameter and 300 mm in height. The specimens were prepared using the vibro-compaction method. The axial deformations were measured using three linear variable displacement transducers (LVDTs), placed 120° apart and anchored to the middle third of the specimen. Average readings from the LVDTs were used in the analyses. Haversine loading pulses with a frequency of 10 Hz and no rest period were applied. The tests were carried out under free drainage conditions. The confining pressure was applied using compressed air. The moisture contents reported in this study refer to the target or initial gravimetric moisture contents. The tests were replicated for better reliability and to account for the experimental dispersions usually encountered in RLT tests on UGMs. Here, the average measurements are reported.

For the MS loading approach, the European standard presents two sets of stress levels, termed as 'high stress level' (HSL) and 'low stress level' (LSL). Each set is divided into five sequences. Each of these sequences contains several stress paths with a constant confining pressure and different deviator stresses (total 28 stress paths for the HSL and 30 stress paths for the LSL). Each stress path is applied for 10,000 cycles. For the tests carried out here, the sequences were applied consecutively.

The MS RLT tests were conducted on eight different UGMs with different PSDs and a range of moisture contents. Some of the materials were tested applying both the HSL and LSL. One set of stress levels was used for fitting the model and the other set was used to validate the model by comparing the prediction obtained using the fitted model with the measured

PD. Also, the influence of moisture on the model parameters was evaluated and used in the predictions.

One of the materials tested here was crushed rock aggregate obtained from Skärlunda in Sweden. Three different PSDs of this material were investigated, derived using the Fuller's equation with n = 0.62, 0.45 and 0.35. Three other materials tested, referred to as Hallinden, VKB and SPV, were crushed rock aggregates obtained from different road construction sites in Sweden. The other two materials tested, referred to as SG1 and Siem 25, were natural aggregates dug out of gravel pits in Denmark where the fractions of crushed material were produced by crushing the oversized particles from the gravel pits. The PSDs of these materials are shown in Figure 1. The maximum particle size used for the tests was 31.5 mm. The optimum moisture contents (w_{opt}) and the maximum dry densities were determined using the modified Proctor method according to the European standard EN 13286-2 (CEN 2004b). Properties of these materials and the specific test conditions are summarized in Table 1.

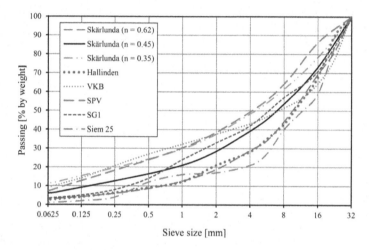

Figure 1. PSDs of the UGMs used for this study.

4 THE PROPOSED MODEL

In a previous study, the following regression based empirical model for the PD behavior of UGMs was proposed by the authors (Rahman and Erlingsson, 2015):

$$\hat{\varepsilon}_p(N) = AN^{BS_f} S_f \quad (3)$$

where $\hat{\varepsilon}_p(N)$ is the accumulated permanent strain after N number of load cycles and A and B are regression parameters related to the material. The term S_f describes the effect of stress condition on the development of PD which is expressed as:

$$S_f = \frac{\left(\frac{q}{p_a}\right)}{\left(\frac{p}{p_a}\right)^\alpha} \quad (4)$$

where, q is the deviator stress, p is the hydrostatic stress (one-third of the sum of the principal stresses) and α is a parameter determined using regression analysis. The term p_a = 100 kPa (reference stress taken equal to the atmospheric pressure) was used to make the expression non-dimensional.

For a series of MS RLT tests with different UGMs with different moisture contents and degrees of compaction, the model showed reliable and satisfactory performance (Rahman and Erlingsson, 2015). However, for application of the model for field conditions using any layered elastic theory-based pavement analysis software, it becomes problematic if any of the calculated stresses turns out to be negative. For these kinds of applications, it is obvious that a resilient strain (ε_r) based model would be more convenient and appropriate. Thus Equation (5) was modified by simply replacing S_f with ε_r as follows:

$$\hat{\varepsilon}_p(N) = aN^{b\varepsilon_r}\varepsilon_r \qquad (5)$$

To use this model for MS RLT test conditions where several stress paths are consecutively applied to a single specimen, the time hardening approach (Erlingsson and Rahman, 2013) was adopted. According to this method, Equation (5) can be reconstructed for MS RLT tests as:

$$\hat{\varepsilon}_{p_i}(N) = a(N - N_{i-1} + N_i^{eq})^{b(\varepsilon_r)_i}(\varepsilon_r)_i \qquad (6)$$

where the suffix i refers to the ith stress path, $\hat{\varepsilon}_{p_{i-1}}$ is the accumulated permanent strain at the end of $(i-1)^{th}$ stress path. N_i^{eq} can be calculated as:

$$N_i^{eq} = \left[\frac{\hat{\varepsilon}_{p_{i-1}}}{a(\varepsilon_r)_i}\right]^{b^{-1}(\varepsilon_r)_i^{-1}} \qquad (7)$$

5 MODEL FITTING

The model was fitted to the MS RLT test data using the least square curve fitting method. The initial values of the parameters were estimated based on boundary conditions and engineering judgement. For the parameter a, the initial value was set equal to the accumulated permanent strain times 1000 at the end of the first stress path and was restricted to be positive. The value of the parameter b was restricted between 50 and 500. The quality of fit achieved with the model was evaluated by the coefficient of determination R^2. Also, the correct matching of the SDR between modelled and measured data for each stress path was evaluated and expressed as percentage. For most of the cases, a value of b close to 250 provided reasonably good fit. Thus, to avoid complications due to multiple possible combinations of parametric values, b was fixed to 250. Using this approach, reasonable qualities of fits were achieved by only optimizing the parameter a. Figure 2 shows an example of the measured versus modelled accumulated permanent strain as a function N for one material with different moisture contents. The values of the parameters and the estimates of R^2 and SDR matching are presented in Table 1. Figure 2 and Table 1 show that the model fitted reasonably well with satisfactory values of R^2 and SDR matching. The missing data for some of the tests are because of termination of the tests due to excessive deformations.

The values of a for the different specimens were plotted against w in Figure 3. It shows that a can be expressed as a linear function of w (within the certain range used in this study) as follows:

$$a = c_1(w) + c_2 \qquad (8)$$

where, c_1 and c_2 are the regression parameters specific to each specimen. The values of these parameters for the different materials and the corresponding R^2 values are given in Table 5.

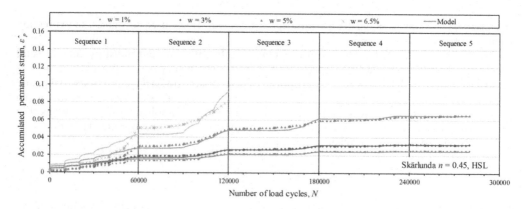

Figure 2. Measured and modelled accumulation of permanent strain for a series of w (Skärlunda $n = 0.45$, HSL).

Table 1. Material properties, test conditions and the calibrated model parameters.

| Material | Specific gravity [-] | w_{opt} [%] | Max. dry density [ton/m^3] | Test condition | | Stress level (used for model fitting) | Parameter a ($b = 250$) | R^2 | SDR matching [%] |
				w [%]	Dry density [ton/m^3]				
Skärlunda ($n = 0.62$)	2.64	5.5	2.11	1	2.05	HSL	3.78	0.88	61
				3		HSL	5.20	0.94	64
				5		HSL	7.79	0.91	86
				7		HSL	8.89	0.84	89
Skärlunda ($n = 0.45$)	2.64	6	2.26	1	2.19	HSL	15.96	0.92	93
				3		HSL	19.00	0.97	89
				5		HSL	26.00	0.97	75
				6.5		HSL	34.00	0.84	93
Skärlunda ($n = 0.35$)	2.64	6.5	2.22	1	2.15	HSL	6.70	0.98	93
				2		HSL	10.69	0.93	79
				3.5		HSL	17.72	-1.15	86
				5		HSL	23.15	0.98	100
				6		HSL	29.00	0.90	100
Hallinden	2.63	5.5	2.075	1	2.01	LSL	2.55	0.99	89
				3.5		LSL	2.62	0.93	89
				5.5		LSL	2.76	0.99	93
				6.5		LSL	3.10	0.98	71
VKB	2.54	6	2.21	2	2.19	LSL	4.19	0.73	68
				4.5		LSL	4.66	0.95	82
				6		LSL	9.73	0.90	82
SPV	2.68	6.9	2.35	2	2.23	LSL	4.11	0.99	82
				4		LSL	7.14	0.95	68
				7		LSL	12.58	0.97	79
SG1	2.49	7.5	2.13	3.5	2.02	HSL	1.60	0.88	82
				5.5		HSL	3.78	0.97	75
				7.5		HSL	6.70	0.97	89
				8.5		HSL	7.79	0.98	75
				9.2		HSL	8.10	0.87	100
Siem 25	2.61	5	2.16	1	2.1	LSL	1.43	0.99	93
				3.5		LSL	1.90	0.94	89
				5		LSL	2.93	0.97	93
				7		LSL	3.87	0.92	71

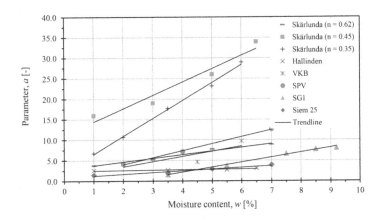

Figure 3. Parameter *a* as a function of *w* for the different materials.

Table 2. Parameters of Equation (8).

Material		Skärlunda ($n = 0.62$)	Skärlunda ($n = 0.45$)	Skärlunda ($n = 0.35$)	Hallinden	VKB	SPV	SG1	Siem 25
Parameters	c_1	0.896	3.26	4.38	0.087	1.26	1.71	1.20	0.42
	c_2	2.83	11.11	2.12	2.39	0.93	0.56	-2.62	0.81
R^2		0.98	0.94	0.99	0.76	0.69	0.99	0.99	0.94

6 PREDICTING PD USING THE FITTED MODELS

For validation, the fitted models for the different UGMs in section 5 were used to predict the accumulation of PD in different combination of stress levels other than those used for the fittings. Thus, if HSL was used for the fitting, LSL was used for the validation and vice versa. The predicted PD were then compared to the actual measurements from the MS RLT tests. The quality of predictions was evaluated using the R^2 values and SDR matching. For comparison, the MEPDG model was used in the same manner. Additionally, the parameter *a* was calculated for the specific moisture contents using Equation 8 and Table 2 and used in the model during validation which is represented here as 'proposed model (moisture based)'. An example of the test data and the predictions by the models are shown in Figure 4. The R^2 values and SDR matching for the models are presented in Table 3. These indicate that all models worked reasonably well for predicting the PD of these UGMs in MS RLT tests. However, the proposed model showed better agreement with the measurements, especially during the last two stress paths (sequence 5) for all cases. The moisture-based version of the model also worked well. Thus, it indicates that using this approach, the model may be used reliably to predict the PD behavior of UGMs with variations in stress level and moisture content.

7 APPLICATION OF THE MODEL FOR FIELD CONDITIONS

Besides for the RLT test environment, the model was employed to capture the PD behavior of UGMs in a real pavement structure. For this, data from an accelerated pavement testing (APT) of an instrumented pavement test section were used. The APT was conducted in a controlled environment using a heavy vehicle simulator (HVS). The schematic of the test section and instrumentations is shown in Figure 5 (a). During the test, the groundwater table (GWT) was raised after 48,6750 load cycles (converted to equivalent standard axle loads (ESALs)), which is shown in this figure. Details of the structure and the test can be found in Saevarsdottir and

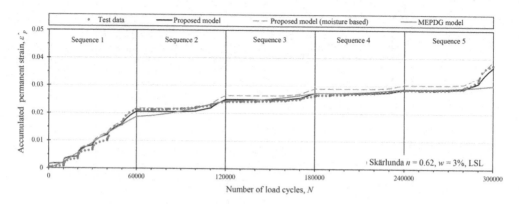

Figure 4. Measured versus predicted accumulation of permanent strain using the fitted models (Skärlunda $n = 0.62$, LSL).

Table 3. Statistical evaluation of the qualities of predictions by the models.

			R^2			SDR matching [%]		
Material	w [%]	Stress level (used for validation)	Proposed model	Proposed model (moisture based)	MEPDG model	Proposed model	Proposed model (moisture based)	MEPDG model
Skärlunda $n = 0.35$	1	LSL	0.98	0.97	0.89	96	96	64
Skärlunda $n = 0.62$	3	LSL	0.98	0.96	0.97	82	82	61
SG1	5.5	LSL	0.98	0.96	0.96	100	100	79

Erlingsson (2015). The required resilient strain (ε_r) values as inputs for calibrating the model and the actual accumulated PD of the different layers were obtained from the readings of the instrumentations. The parameters of the model were adjusted to match the measured data with the predictions. In this case, the values of the parameter a close to those for the RLT tests worked well. Like the RLT tests, the values of the parameter b as 250 worked here too except for the base layer where a value of 60 provided the best fit. The values of a were different for the moist and wet conditions (i.e. before and after raising the GWT, respectively) for the base and subgrade. The measured and modelled accumulated PD of the different layers of the test structure are shown in Figure 5(b). The values of the model parameters are presented in Table 4. Generally, good agreements between the measured and modelled responses were observed. The ranges of the model parameters were stable as well indicating reliability of the model.

8 CONCLUSIONS

In this study, a model to predict the accumulation of PD in UGMs with the number of load applications for variable stress conditions was derived. The objective was to better suit a pre-existing model for field applications and to implement it in a pavement analysis software in future. The study was based on MS RLT tests since it allows for a comprehensive study of the material behavior with minimal effort. The idea was to develop a simple and reliable model for MS loading conditions that can be calibrated with reduced effort compared to some of the existing models.

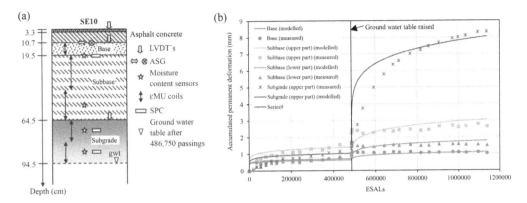

Figure 5. (a) Schematic of the test structure (Saevarsdottir and Erlingsson, 2015). (b) Measured vs. modelled PD in different layers.

Table 4. Model parameters for the different layers of the pavement section.

	Model parameters			
	Moist		Wet	
Layer	a	b	a	b
Base	2.2	60	2.5	60
Subbase (upper part)	2.5	250	2.5	250
Subbase (lower part)	4	250	4	250
Subgrade (upper part)	2	250	9.5	250

The model proposed here can be calibrated for a specific material using a single RLT test without conducting SFT tests. The model was validated using MS RLT test data applying the time hardening approach. Despite its simplicity, the proposed model showed reliable results. Some deviances in predictions compared to measured values for some cases may be considered acceptable if one allows for the experimental scatter usually encountered in MS RLT tests with UGMs. In this study, for simplicity, the parameters b was restricted to 250 leaving only parameter a to be regressed for different UGM specimens. Yet, the model provided good quality of fit. With this approach, it was possible to investigate the impact of moisture on parameter a. The model was also successfully used for an in-situ pavement condition in HVS testing. With additional MS RLT testing of several other UGMs (not presented here), the observed ranges of the parameters a and b are presented in Table 5.

Table 5. Suggested range of model parameters.

Parameter	Well-graded (w = 1% -7%)	Fine graded (w = 1% -10%)	Coarse and open graded (w = 1% -7%)
a	1-15	1-30	1-10
b	100-250	200-350	150-250

*Degree of compaction: 95-97% (modified Proctor method)

Since this was a limited study, the model developed here should be validated with more tests with a wider range of materials and different field conditions. With additional studies, it will be possible to incorporate the influence of moisture, PSD and degree of compaction into the model and establish reasonable ranges of the parameters. Further work is currently underway to validate and implement this model for real pavement conditions and in a pavement analysis software.

ACKNOWLEDGEMENTS

This work was sponsored by the Swedish Transport Administration (Trafikverket). The Danish aggregates were provided by the Danish Road Directorate.

REFERENCES

ARA, Inc., 2004. *Guide for the Mechanistic Empirical Design of New and Rehabilitated Pavement Structures*. Final report, NCHRP1-37A, Transportation Research Board, National Research Council, Washington D.C., USA.

Brown, S. F. and Hyde, A. F. L., 1975. *Significance of cyclic confining stress in repeated-load triaxial testing of granular material*. Transportation Research Record: Journal of the Transportation Research Board, Transportation Research Board of the National Academies, Washington, D.C., No. 537, 49-58.

CEN-European Committee for Standardization, 2004a. *Cyclic Load Triaxial Test for Unbound Mixtures*. Brussels: European Standard, EN 13286-7.

CEN-European Committee for Standardization, 2004b. *Test Methods for the Determination of the Laboratory Reference Density and Water Content – Proctor Compaction*. Brussels: European Standard, EN 13286-2.

Di Graziano, A., Marchetta, V. and Cafiso, S., 2020. *Structural health monitoring of asphalt pavements using smart sensor networks: A comprehensive review*. Journal of Traffic and Transportation Engineering (English Edition), 7(5), pp. 639–651.

Erlingsson, S. and Rahman, M. S., 2013. *Evaluation of Permanent Deformation Characteristics of Unbound Granular Materials by Means of Multistage Repeated-Load Triaxial Tests*. Transportation Research Record: Journal of the Transportation Research Board, Transportation Research Board of the National Academies, Washington, D.C., No. 2369, 11–19. DOI: 10.3141/2369-02.

Fladvad, M. and Erlingsson, S., 2021. *Modelling the response of large-size subbase materials tested under varying moisture conditions in a heavy vehicle simulator*, Road Materials and Pavement Design, DOI: 10.1080/14680629.2021.1883462.

Gidel, G., et al., 2001. *A New Approach for Investigating the Permanent Deformation Behaviour of Unbound Granular Material Using the Repeated Load Triaxial Apparatus*. Bulletin de Liaison des Laboratoires des Ponts et Chaussées, No. 233, July-August, 2001, 5–21.

Hornych, P. and El Abd, A., 2004. *Selection and Evaluation of Models for Prediction of Permanent Deformations of Unbound Granular Materials in Road Pavements*. Competitive and Sustainable Growth (GROWTH) Programme, SAM-05-DE10.

Korkiala-Tanttu, L., 2005. *A New Material Model for Permanent Deformations in Pavements*. In: Proceedings of the 7th International Conference on Bearing Capacity of Roads and Airfields, BCRRA '05, Trondheim, Norway.

Lekarp, F., 1999. *Resilient and Permanent Deformation Behavior of Unbound Aggregates under Repeated Loading*. Doctoral thesis. Royal Institute of Technology (KTH), Stockholm, Sweden.

Rahman, M. S. and Erlingsson, S., 2013. *Moisture Sensitivity of the Deformation Properties of Unbound Granular Materials*, Proceedings of the 9th International Conference on Bearing Capacity of Roads and Airfields, BCRRA '13, Trondheim, Norway, pp. 777–786.

Rahman, M.S. and Erlingsson, S., 2014. *Predicting permanent deformation behaviour of unbound granular materials*, International Journal of Pavement Engineering, 16:7, 587–601, DOI: 10.1080/10298436.2014.943209.

Rahman, M.S. and Erlingsson, S., 2015. *A model for predicting permanent deformation of unbound granular materials*, Road Materials and Pavement Design, 16:3, 653–673, DOI: 10.1080/14680629.2015.1026382.

Ramos, A., Gomes Correia, A., Indraratna, B., Ngo, T., Calçada, R. and Costa, P. A., 2020. *Mechanistic-empirical permanent deformation models: Laboratory testing, modelling and ranking*, Transportation Geotechnics, 23, art. no. 100326.

Saevarsdottir, T. and Erlingsson, S., 2015 *Modelling of responses and rutting profile of a flexible pavement structure in a heavy vehicle simulator test*, Road Materials and Pavement Design, 16:1, 1–18, DOI: 10.1080/14680629.2014.939698.

Tseng, K.H. and Lytton, R.L., 1989. *Prediction of permanent deformation in flexible pavement materials.* Implication of Aggregates in Design, Construction, and Performance of Flexible Pavements, ASTM STP 1016, H. G. Schrauders, and C. R. Marek, eds. American Society for Testing and Materials, Philadelphia, pp. 154–172.

Werkmeister, S., 2003. *Permanent Deformation Behavior of Unbound Granular Materials.* Thesis (PhD). University of Technology, Dresden, Germany.

Werkmeister, S., Dawson, A. R. and Wellner, F., 2001. *Permanent Deformation Behavior of Granular Materials and the Shakedown Concept.* Transport Research Record, Journal of the Transportation Research Board (TRB), No. 1757, 75–81.

Development of simplified models to assess pavement structural condition on network level

M. Pettinari & S. Baltzer
Pavement section at the Danish Road Directorate, Hedehusene, Denmark

M. Kalantari & D. Jansen
Section Design and Structure of Pavements at BASt, Bergisch Gladbach, Germany

ABSTRACT: Monitoring pavement structural condition on network level has become possible since Traffic Speed Deflectometer (TSD) was introduced. TSD data have been deeply investigated over the past years and many studies have shown relatively strong correlation with Falling Weight Deflectometer (FWD) results. Furthermore, the agreed vison, formulated by road authorities and research institutions, is represented by utilizing the TSD data directly for back-calculation of the modulus of the pavement layers. To accomplish this vision, layer thicknesses and temperature of the asphalt layer must be available. While the second can be calculated by the help of BELLS (BALTZER, ERTMAN-LARSEN, LUKANEN and STUBSTAD) or similar models, layer thicknesses are not always available on network level. Based on this aspect, the Danish Road Directorate (DRD) and the German Federal Highway Research Institute (BASt) have been working on a project with the aim to develop simplified models that could rank the remaining lifetime of a pavement structure using Surface Curvature Index (SCI) data. This paper presents the models developed by using two approaches based i) on historical FWD data processed with a standard back-calculation software and ii) on simulated FWD loads, applied on different standard pavement sections analyzed with a multilayer linear elastic (MLE) based computer program. Comparing the results obtained by the two approaches, a power type function which correlates SCI and remaining life (in ESALs) was found where its coefficients are different from country to country. Effects of relevant variables such as damage development rate of the asphaltic layer and the effect of different countries fatigue transfer functions have also been considered.

Keywords: structural evaluation, TSD, network, FWD, pavement durability

1 INTRODUCTION

Billions of euros are spent by road agencies each year on managing the transportation infrastructure assets to comply regulations and user expectations. A major component of those assets is represented by pavements and their structural rehabilitation is becoming critical due to costs and complexity. To optimize maintenance strategies and decision on this manner, pavement management system (PMS) plays a fundamental role. An optimal indicator of the pavement structural condition is given by deflection measurements (Rada et al., 2016).

Measuring deflections is standard practice in pavement engineering and nowadays it can be performed at traffic speed. The Traffic Speed Deflectometer (TSD) is the most used version of moving deflection testing devices but their data have not been implemented yet to assess the pavement remaining lifetime. To calculate the lifetime of a pavement structure and its respective layers moduli, it is standard practice of road state agencies to still use back-calculation on Falling Weight Deflectometer (FWD) data (Lytton, 1988). FWD is a non-destructive testing (NDT) and non-intrusive device but, being stationary, becomes inefficient for network level application. Measuring with FWD is time demanding, due to the stop-and-go operation, and traffic disruptive. For this reason, many road authorities have been focusing on monitoring their network using structural data collected at traffic speed. So far, interpretation of these data for pavement management is based, in most of the cases, on the Surface Curvature Index 300 (SCI300) which is directly derived from the measured deflection basin as difference between specific peaks (d0 and d300) at certain distances from the applied load (Levenberg et al., 2018; COST 336 - FWD at Network Level, 1998).

Recently, the TSD has also been implemented with three additional lasers behind the loading wheel with the intent to capture the asymmetry of the deflection basin produced by a moving load on a non-linear multilayer system. Furthermore, technological development has made the data collected by this vehicle repeatable and consistent opening to a new era for road authorities. In fact, the combination of the achievements in deflection measurements and the increased available computational power has made possible to use TSD data for back-calculation of the layer moduli and consequently the remaining ESALs (Nasimifar et al., 2017).

The implementation of this approach on project level and in the PMS is still not in place due to challenges faced: (i) accessing correct layers thickness and (ii) using asymmetric deflection basins generated by the TSD to assess remaining lifetime.

Layers' thicknesses reported in database system of road authorities are often not available at the required precision or not reliable. Consequently, thicknesses used for back-calculation of the layer moduli are conventionally taken by drilling cores or using the borescope. The ground penetrating radar (GPR) has become an alternative even if some processing limitations must be mitigated. With regards to the second point, layers moduli have been back calculated using TSD data but still some uncertainties are in place about the precision of calculated remaining equivalent single axle loads (ESALs) from stresses and strains obtained using a viscoelastic mechanistic model.

Based on these challenges, the Danish Road Directorate (DRD) and the German Federal Highway Research Institute (BASt) cooperated in a project with the objective of developing simplified models to classify the remaining lifetime of a pavement using Surface Curvature Index data. Since the developed models refers to FWD (both historical and simulated) analyzed with a multilayer linear elastic (MLE) theory and being aware that FWD and TSD loading configurations are different, these models are meant to be used only to ranking lifetime within different classes and additionally implemented in PMS to coordinate maintenance strategies.

2 OBJECTIVE AND EXPERIMENTAL PROJECT DESCRIPTION

This paper presents the primary results of a joint project between German Federal Highway Research Institute (will be referred as BASt) and Danish Road Directorate (will be referred as DRD), with the aim of developing a simple model to be able to assess the structural condition of the pavements on network level.

Among different methods of analyzing the data from FWD and TSD devices, deformation-based indexes are the most applicable parameters as they are easily calculated without needing extra parameters like the pavement layer's thicknesses. SCI 300 was selected for this research and correlated with residual life of the respective pavement. To determine the remaining life of the pavement, the mechanistic empirical (ME) method was applied in two different approaches:

I. Analysis of standard pavement sections: In this approach different standard pavement sections from Germany and Denmark were selected with material parameters as representative designs (named as model pavements). They were analyzed under the simulated FWD load to determine the surface deformations and the critical responses at desired points for calculation of the SCI and the remaining life of the sections.

II. Analysis of the existing FWD data: in this approach, the existing FWD measurements data (known as historical FWD data) were used to determine the SCI indexes and the back-calculation of layers' stiffnesses. The back calculated material characteristics were used in a ME method to determine the remaining life.

The results were used to develop models relating the SCI to the remaining life.

2.1 Simulated FWD method

A three-layer linear elastic model consisting of asphalt layer, granular layer and subgrade was selected to model the pavement sections subjected to FWD load of 50 kN (with circular contact area of 15 cm radius). When needed, different asphaltic layers were combined in a single layer using Odemark method. BISAR® software was applied for the analysis of the model pavement sections. Fatigue cracking was considered as the failure criteria and therefore horizontal tensile strain at the bottom of the asphalt layer was taken as the critical response.

To select the German pavements, the national pavement design guideline known as RStO 12 (FGSV 499, 2012) was used. The guideline is a catalogue-based pavement design which propose different thicknesses based on the pavements traffic class and the type of its material in 7 different traffic classes (known as BK 100 to BK 0.3). For higher traffic amounts, the analytical design approach RDO Asphalt 09 should be used (FGSV 489, 2009).

To determine the stiffness of the asphalt layer for the model, the results of a research study performed in Germany were used (Stöckner et al., 2020). As a part of that study, a lot of asphalt data samples from wearing, binder and base layers were collected to determine typical stiffness master curves. The results, measured at 20°C and 10 Hz, were classified in different levels and 5,000 MPa was selected as representative mean of the asphalt modulus for German pavement sections. Table 1 shows the properties of the German pavements model sections used in this study.

Table 1. Properties of the German pavements model sections.

Range	Thickness (cm)		Stiffness/Modulus (MPa)		
	Asphalt layer	Granular layer	Asphalt layer	Granular layer	Subgrade
Min.	14	30	5,000	200	90
Max.	34	30	5,000	200	90

To select the Danish side model pavements, the Danish design guidelines known as design catalogue and MMOPP guidelines (MMOPP Design Program for Road Pavements, 2017), were considered as the base.

In the design catalogue and in MMOPP, pavements are typically 4-layer structures (asphalt layer, gravel base layer, sand formation layer and subgrade), it was needed to transfer them into 3-layer structures by combining the two unbound layers. The modulus of the combined unbound layer was calculated as a weighted average. Table 2 shows the properties of the Danish pavements model sections. The stiffness of asphalt concrete was calculated from FWD measurements at the reference temperature of 20 °C.

Table 2. Properties of the Danish pavements model sections.

Range	Thickness (cm)		Stiffness/Modulus (MPa)		
	Asphalt layer	Granular layer	Asphalt layer	Granular layer	Subgrade
Min.	13	47	2,230	180	40
Max.	31	91	4,280	220	40

As mentioned before, fatigue was considered as the failure criteria and horizontal tensile strain at the bottom of the asphalt layer as the critical response. Fatigue damage was modeled by decreasing the stiffness of the asphalt material in 10% steps. To relate the response to the life, DRD has their own fatigue transfer function (see equation 1) (MMOPP Design Program for Road Pavements, 2017).

$$\varepsilon_h = -0.000250 \times (N10/10^6)^{-0.191} \qquad (1)$$

Where N is allowable number of axles [ESAL of 10 ton] and
ε_h is the largest permitted horizontal tensile strain under the asphalt layer.

In Germany, for each RDO Asphalt pavement design, asphalt samples are tested in the laboratory under indirect tensile fatigue test (at 20° C and 10 Hz) to determine the fatigue function of the relative asphalt material (see equation 2). This fatigue function is used with a shift and safety factor to determine the pavement's life in case of fatigue failure criteria (FGSV 489, 2009). In this research the proposed fatigue functions from the results of the before-mentioned research study (Stöckner et al. 2020), was used as the representative fatigue functions of the asphalt base materials in Germany. Table 3 represents the parameters of these fatigue functions. A factor of 1500 was used to shift the laboratory life to field life based on existing German analytical design guideline (FGSV 489, 2009).

$$N = a\varepsilon^b \qquad (2)$$

where N is the number of load cycles till the initiation of the fatigue crack in the sample [-],
ε: Initial horizontal elastic strain [‰]
a & b: Fatigue parameters of the material which, are determined by the regression on the results of the fatigue tests

Table 3. Parameters of different classes of HMA base material's fatigue functions in Germany.

Different categories	Fatigue function parameters	
	a	b
Upper class	6.49415208	-3.62690871
Middle class	4.92836355	-3.21161007

2.2 *Historical FWD method*

Historical FWD data were used to define a correlation between remaining lifetime of pavement structure and relative properties of the deflection basin. Since the overall objective of this study is to define a model that can be used with TSD data, deflection basin characteristics, such as SCI300 and SCI600, were used as independent variables of the regression analysis.

Table 4. Summary of the pavement structure characteristics used in the FWD analysis.

	Traffic	Thickness (cm)		Deflection characteristics (μm)		
	ESALs/year	Asphalt layer	Granular layer	SCI300	SCI600	D0
Min	2.3E+04	6	11	2.1	5.7	32.0
Max	3.4E+06	37	82.5	538.5	738.1	910.4

The sample of FWD data include approximately 700 measurements on pavement structure having a wide variety of bearing capacity levels. The table (Table 4) below includes some basic information about the deflection basins characteristics of the studied sample of data.

FWD measurements were carried out at different temperatures so all the back-calculated moduli of the asphalt layers were corrected to a reference temperature of 20°C and used to obtain the relative deflection basin. The following equation (equation 3) has been used to correct the asphalt modulus to the reference temperature of 20°C.

$$E_{20°C} = E(T) \times [1 - 2log10(20°C/T)] \qquad (3)$$

where E(T) is the modulus of the asphalt layer at the temperature T.

For each measurement, two different empirical approaches were used and compared: i) remaining lifetime is assessed using the empirical equations implemented in the Danish pavement design guide (MMOPP); ii) the lifetime is given by the back-calculation software ELMOD®. In both cases, all pavements were represented as three layers system and modeled using mechanistic model based on multilayer elastic theory (MLET). Three different failure points were evaluated based on: a) tensile strain at the bottom of the asphalt layer, b) stress on top of the granular layer and c) stress on top of the subgrade. Equation 1 is used to assess the durability of the asphalt layer while equation 4 is used to evaluate the potential failures based on the vertical stress for the failure points b) and c).

$$\sigma = 0.086 MPa \times (N/10^6)^{-0,25} \times (E/160 MPa)^{1.16} \qquad (4)$$

N is allowable number of ESALs of 10 ton
σ is the vertical stress on top of the referring layer while E is the relative modulus.

The remaining lifetime of the pavement was given by the lowest number of ESALs obtained by using the equation 1, and 4 to the three critical points. ELMOD® estimates the remining lifetime using an incremental recursive method (IRME) where the material properties for the pavement are updated in terms of damage as the pavement life simulation progresses. This procedure works in increments of time and uses the output from one increment, recursively, as the input to the next increment. The IRME design method incorporates various mathematical models to describe and predict material and pavement performances.

3 ANALYSIS OF THE RESULTS

As described in the previous paragraph, all the models, used to assess the remaining lifetime of a pavement structure, have as input a physical stress or strain calculated trough a mechanistic model of the pavement structure subjected to a specific loading condition. To provide a better understanding of pavement structures included in this investigation, distribution of the critical tensile strains, used to calculate the remaining lifetime, are shown in

Figure 1. For the sample of pavements of the historical FWD dataset, also vertical stresses on top of the granular base and subgrade are presented (Figure 2).

Considering the data processed using the simulated FWD method, almost 30% of the BASt structures included in this analysis have a peak of tensile strain at the bottom of the asphalt between 50 and 100 µstrain. Still considering BASt structure, all intervals from 0-50 to 200-250 are represented by at least 10% of the analyzed sections. The simulated DRD structures, since were selected trying to include a wide amount of traffic levels, show a wider distribution of tensile strains when compared to the BASt structures. In fact, the studied structures designed based on DRD guidelines are significantly represented also in the intervals 250-300 µstrain and 300-350 µstrain.

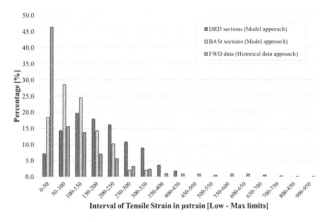

Figure 1. Summary of the horizontal tensile strains at the bottom of the asphalt layer investigated with both approaches.

Distribution of compressive stresses on top of granular layer and subgrade are represented in Figure 2. Calculated compressive stress applied on top of the granular layer reached a maximum value of approximately 0.6 MPa while in the subgrade the maximum was 0.12 MPa.

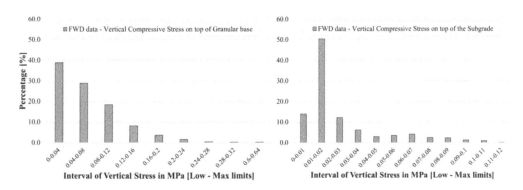

Figure 2. Stress distribution in investigated structures. Left. Stress on top of the granular layer, right. Stress on top of the Subgrade.

Based on the presented strains and stresses given by the mechanistic model of the different pavement structures subjected to 50 kN load applied on FWD plate, it was possible to develop several regression models where a surface deflection measure, such as SCI300, is used as input to define the remaining lifetime (Figure 3). It is important to select representative pavement and material properties in order to cover as many pavement types as possible.

Each dataset was used to define a reliable power regression model. While fitting the historical data back calculated using MMOPP and Dynatest Elmod®, two additional models have been added. In both cases, a 3rd degree polynomial function was defined. To be conservative, both functions were shifted by a constant value towards the lower edge of the data. All the regression models have been summarized in Table 5. For all the models, y represents the remaining number of 10-ton ESALs while x is the SCI300 at the reference temperature of 20°C.

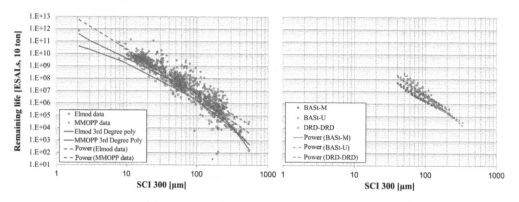

Figure 3. Left. Data and regression models using historical FWD data; right. Data and regression models using simulated FWD data.

Table 5. Summary of the regression models used to estimate remaining life.

Name	Model	Constants			
		a	b	c	d
Pow. Elmod	Power	1.45E+14	-3.8483	/	/
Pow. MMOPP	Power	7.72E+13	-3.504	/	/
Pow. DRD	Power	7.49E+16	-5.0056	/	/
Pow. BASt U	Power	1.86E+14	-3.7648	/	/
Pow. BASt M	Power	9.60E+12	-3.3395	/	/
Poly Elmod	3rd Poly	-0.1676	-0.0221	-1.7864	11.2*
Poly MMOPP	3rd Poly	-0.7519	2.4956	-5.1178	13**
Description of the regression models					
Model type	Power	3rd Poly			
Function	$y=a*x^b$	$y=a*\log10(x)^3 + b*\log10(x)^2 + c*\log10(x) + d$			

*originally 11.577, **originally 13.399

Even with the R-squared higher than 0.8 in all the studied cases, it is possible to foresee big challenges if residual life of a pavement should be assessed only based on SCI300.

4 COMPARISONS OF THE DIFFERENT METHODS/APPROACHES

The overall objective of this research is to define a model that could be used to assess residual lifetime of a pavement structure using surface measurements of the deflection basins.

Comparing all the models (Figure 4) reveals how the difference between failure functions of two countries can influence the results. When considering the simulated FWD approach, it was important to use proper pavement structures and material properties because the criteria used in that case only refers to the strain at the bottom of the asphalt layer. Even if historical FWD approach is based on a failure criterion where all layers are evaluated, the overall comparison between the two approaches show consistent results. When considering SCI300 values between 40 and 200 μm, all the power regression models are consistent, independently from the approach used. The only exception is given by the DRD-DRD model when referring to SCI300 lower than 40 μm where residual lifetime prediction seems much more optimistic than those given by the other models. The two polynomial models obtained using historical data, as specified in the previous paragraph, have been designed to be conservative and for this reason can be considered an exception.

Considering some specific SCI300 values within the above-mentioned range, the estimated remaining lifetime obtained from different regressions models are summarized in Table 6.

Table 6. Remaining pavement lifetime based on the different regression models.

SCI300 [μm]	Remaining Lifetime as number of 10-ton ESALs						
	Pow. Elmod	Pow. MMOPP	Pow. DRD	Pow. BASt U	Pow. BASt M	Poly Elmod	Poly MMOPP
40	9.91E+07	1.88E+08	7.16E+08	1.73E+08	4.29E+07	3.91E+07	1.30E+08
60	2.08E+07	4.54E+07	9.41E+07	3.76E+07	1.11E+07	1.03E+07	3.66E+07
80	6.88E+06	1.66E+07	2.23E+07	1.27E+07	4.24E+06	3.67E+06	1.31E+07

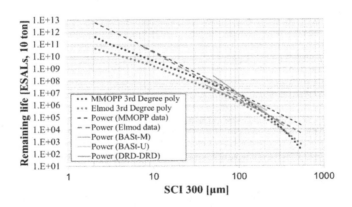

Figure 4. Comparing the results of the two approaches (historical FWD and simulated FWD).

As expected, maximum difference between predictions of the models increases with the increase in SCI. When SCI300 is equal to 80, the lowest prediction is 16 times smaller than that given by the Power DRD.

When looking at the historical data it is also possible to see that, remaining pavement lifetime might vary up to 100 times compared to the value estimated using the regression model.

The overall outcome of the comparison between residual life from models and back-calculation highlights that it is not possible to precisely predict the remaining ESALs of a pavement structure just referring to SCI300. However, considering the different regressions it was possible to define ranges of expected durability for specific SCI300 values. To define these ranges, the following conditions have been followed:

- Power DRD model was not considered;
- Lower limit defined by the mean of the most conservative regressions
- Upper limit defined by the mean of the most optimistic regressions
- Mean values were rounded to the nearest 10^6

Based on the mentioned points, the resulted residual life ranges for some SCI300 values are listed in Table 7.

Table 7. Defined thresholds for the relation of SCI 300 and remaining life.

SCI 300 (μm)	Remaining life (ESAL of 10 ton)
40	40 - 180 million
60	10 - 40 million
80	4 - 15 million

To narrow down the ranges of residual life for each SCI300, it is possible to apply a refining alternative approach based on different properties of the deflection basin. An example is given in Figure 5, where pavements having SCI300 between 0 and 30 μm are shown. Based on this sample of pavements, the deflection property that better correlates with the residual lifetime is the difference between D600 and D900. When considering pavements with SCI300 between 30 and 60 μm, the deflection property that better predict pavement durability is the difference between D300 and D600.

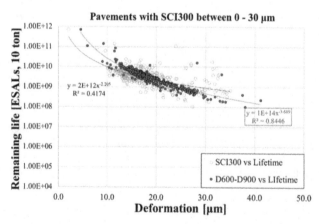

Figure 5. Alternative method for pavements having SCI300 at 20°C between 0 and 30 μm.

5 CONCLUSIONS

Introduction of the TSD made a big change in pavement's structural condition monitoring activities. Back-calculation of the layers modulus from the measured data is still the common agreed method between most of the road authorities and institutes. Back-calculation requires the correct and precise thickness of the layers which, are not always existing or appropriate to measure on network level, considering the time and human resources. Therefore, developing simplified methods to estimate remaining life from the measured surface deformation parameters are welcomed by the road authorities to be able to integrate them in their pavement management systems. Considering this need, the Danish Road Directorate (DRD) and the German Federal Highway Research Institute (BASt) defined a joint research project. The primary results of this project were presented in this paper. Two different approaches were used to develop the models; by analyzing the simulated pavement sections under the FWD load and by back-calculation of the existing FWD data measured on different road sections. Different pavement sections and material properties were taken into account when selecting representative models of pavement structures. The distribution of the calculated tensile strains at the bottom of the asphalt layer in both approaches, showed a very good coverage of the different conditions in the real field. The results of both approaches were used to construct different models to relate the remaining life of the pavement with its SCI300. Excluding the most optimistic model ("Pow. DRD" model), there is good agreement between the simulated based and measured based models. Especially by comparing the "Pow. BASt U" as the optimistic simulated based with "Pow. MMOPP" as the optimistic measured based models. The same is by comparing the "Pow. BASt M" and the "Poly Elmod". Beside this agreement, comparing the predictions of the different models to the remaining lifetime assessed by FWD shows in many cases a significant difference. This highlights that it is not possible to precisely predict the remaining life of a pavement structure just referring to the SCI300 and there is a need for i) further improvement of the models by taking other surface parameters into account, ii) integrating a reliability approach into the assessment of the remaining life rages. The research will continue to address this issue but anyhow, the research so far has shown that it is possible to use the results to define thresholds for network level decision making activities.

ACKNOWLEDGMENTS

The authors gratefully acknowledge the Danish Road Directorate and Federal Highway Research Institute.

REFERENCES

FGSV 489. (2009). *Richtlinien für die rechnerische Dimensionierung des Oberbaus von Verkehrsflächen mit Asphaltdeckschicht - RDO Asphalt.*

FGSV 499. (2012). *Richtlinien für die Standardisierung des Oberbaus von Verkehrsflächen, RStO 12.*

Levenberg, E., Pettinari, M., Baltzer, S., & Lekven Christensen, B. M. (2018). A Comparison of Traffic Speed Deflectometer and Falling Weight Deflectometer Data. *Transportation Research Record: Journal of the Transportation Research Board, 2672*(40), 22–31. https://doi.org/10.1177/0361198118768524

Lytton, R. (1988). Backcalculation of pavement layer properties. *The First Symposim on Nondestructive Testing of Pavements and Backcalculation of Moduli*, 55.

Nasimifar, B. M., Author, C., & Pike, G. (2017). Backcalculation of Flexible Pavement Layer Moduli from Traffic Speed Deflectometer Data. *Transportation Research Record: Journal of the Transportation Research Board, 2641*(1), 66–74. https://doi.org/10.3141/2641-09

Rada, G. R., Nazarian, S., Visintine, B. A., Siddharthan, R., & Thyagarajan, S. (2016). *Pavement Structural Evaluation at the Network Level: Final Report Number FHWA-HRT-15-074. September*, 286p. https://doi.org/FHWA-HRT-15-074

Stöckner, Markus; Sagnol, Loba; Brzuska, Amina; Wellner, Frohmut; Blasl, Anita; Sommer, Viktoria; Krause, Günter; Komma, C. (2020). *Abschätzung des Restwerts im PMS am Ende des Bewertungszeitraumes.*

COST 336 - FWD at Network Level, (1998).

MMOPP design program for road pavements, (2017).

Predicting asphalt pavement temperatures as an input for a mechanistic pavement design in Central-European climate

S. Cho & C. Tóth
Department of Highway and Railway Engineering, Budapest University of Technology and Economics, Hungary

P. László
Fulton Hogan Infrastructure Services, Australia

ABSTRACT: The bearing capacity of asphalt pavements is highly influenced by the modulus of the asphalt layers, which is a function of the temperature observed in depth. The temperature dependency of each layer is also influenced by the composition of the asphalt mix.

The impact of the climatic conditions is usually considered in national pavement design guidelines by the application of the equivalent temperature. This approach however may lead to incorrect pavement designs due to its simplistic nature and cannot consider the local climatic variations.

The modulus of each asphalt layer can be modelled only if the temperature in depth is known or correctly predicted. The research work presented in this paper provides a methodology for predicting in depth pavement temperatures by using ambient and surface temperatures. These input values can be obtained at a higher frequency and at lower capital investment compared to in depth pavement temperatures.

The methodology has been validated based on long-term data collected at various depths in a full depth asphalt pavement in Budapest region. The methodology provides valuable input into mechanistic pavement design, where the performance of innovative materials can be considered. With this approach real performance prediction is enabled, which would not be possible with the simplistic method of the equivalent temperature......

Keywords: Asphalt temperature estimation, central European pavement, climate input, in-depth temperature, temperature regression

1 INTRODUCTION

The bearing capacity of asphalt pavements is highly influenced by the modulus of the asphalt layers, which is a function of the temperature observed in depth. The temperature dependency of each layer is also influenced by the composition of the asphalt mix; due to the complexity of this issue this characteristic will not be examined in this paper. In mechanistic-empirical design the climatic condition is major input [1]. Therefore, the importance of dealing with temperature for pavement designers takes huge part in the design process.

The impact of the climatic conditions is usually considered in national pavement design guidelines by the application of the equivalent temperature. This approach however may lead to incorrect pavement designs due to its simplistic nature and cannot consider the local climatic variations.

DOI: 10.1201/9781003222880-27

From the application of the real weather data into estimation of asphalt pavement done for the first time by Barber [2], many studies regard to this subject is still ongoing [3] [1] [4] [5] [6]. Another study by Dempsey estimated asphalt in depth temperatures by monographs [5]. Matić et al. developed a model which describes the maximum and minimum temperature at specified depth with regression method [6].

The modulus of each asphalt layer can be modelled only if the temperature in depth is known or correctly predicted due to its viscoelastic characteristic of asphalt mixes [7]. The research work presented in this paper provides a methodology for predicting in depth pavement temperatures by regression method using on-site measured surface temperatures. These input values can be obtained at a higher frequency and at lower capital investment compared to in depth pavement temperatures.

The methodology has been validated based on long-term data collected at various depths in a full depth asphalt pavement in the Budapest region. The methodology provides valuable input into mechanistic pavement design, where the performance of innovative materials can be considered. With this approach real performance prediction is enabled, which would not be possible with the simplistic method of the equivalent temperature.

In this paper, the German asphalt temperature prediction model is firstly chosen to estimate an in-depth temperature profile [4]—secondly, a suggestion of a new asphalt temperature prediction model for the Hungarian climatic condition.

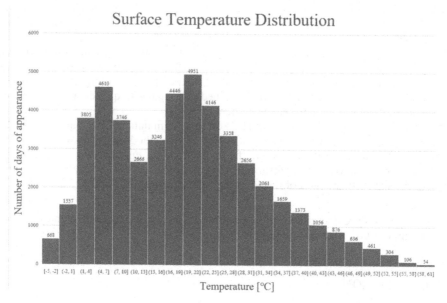

Figure 1. Surface temperature appearance.

Figure 1 shows the Hungarian asphalt's surface temperature distribution. Unlike German's asphalt surface temperatures, Hungary does not have any values under -10 °C. For a year, a Hungarian pavement structure's temperature at various depths were recorded for every 10 minutes. By utilizing these data sets, the temperatures were divided into 22 groups, which frequently appears for Hungarian pavement structure by 3°C of the interval from minimum -5 °C to a maximum 61°C.

2 TEMPERATURE ESTIMATION MODEL

2.1 *The German model*

Asphalt mixture displays a viscoelastic characteristic, which means its strength is affected by both the temperature and loading frequency [8].

To find the suitable Hungarian asphalt in-depth temperature estimation method, we reference the German design method. This German design method categorizes the asphalt surface temperature into 13 groups, by 5°C range [4]. The German temperature prediction model for the asphalt layer is predicted according to Eq. (1):

$$y = a \cdot ln(0.01 \cdot x + 1) + T \quad (1)$$

where, y is asphalt temperature (°C) at depth x (mm), T is surface temperature (°C), and a is a parameter as a function of T.

2.2 *New model*

The German temperature estimation model shows some deviation compared to the Hungarian measured data—due to the latitude and daylight hours difference—these matter on the asphalt temperature variation inside the depth. From the static data, asphalt temperature along depth does not change instantly as the air or sudden surface temperature drop. Therefore, the need to correct the function coefficient arises as per the weather condition of Hungary, Figure 3.

Table 1. Pavement system description.

	Depth from the surface[cm]	Type
Asphalt	4	AC-12/F
	6	AC-20/F
	9	AC-35/F
Base	20	CTB
Subgrade	30	Unbound

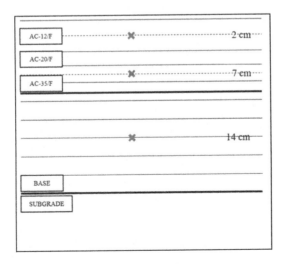

Figure 2. Sensor location.

The new model with Hungarian weather condition is prepared by using the site measured data in central region of Hungary. The pavement system is composited with Table 1 condition, and the temperature sensor is located at the point of 2, 7, 14 cm from the surface. The new model is prepared by using the mixture type described in Table 1.

One weather station in the Budapest region is selected for this study (Figure 2). Following the nonlinear temperature distribution along with asphalt depth, square root and logarithm envelope functions are considered. The 5 cm point is the critical point where the speed of divergence of those two functions creates a gap. The logarithm function, like the German model, shows much higher disagreement after this point with measured data. Thus, the square root function is chosen to represent the temperature estimation for Hungarian weather condition with more simplicity in the formation, equation (2):

$$y = a \cdot \sqrt{x} + T_0 \qquad (2)$$

where, y is asphalt temperature (°C) at depth x (cm), T is surface temperature (°C), and a is a parameter as a function of T which is shown in Table 2.

Table 2. Parameter a as a function of surface temperature.

Temperature [°C]	≤0	≤5	≤10	≤15	≤20	≤25	≤30	≤35	≤40	≤45	<45
Parameter a [-]	1.669	2.446	3.044	1.449	1.154	1.553	1.086	-0.9228	-1.3158	-2.0504	-3.5124
R^2	0.9969	0.9938	0.9773	0.9819	0.9949	0.8528	0.6896	0.7139	0.9162	0.8445	0.8948

3 VERIFICATION OF THE NEW MODEL

Due to the speed of disagreement below 5 cm point increases drastically, new model has to be developed. The 5 cm point is the critical point where the speed of divergence. The logarithm function, like the German model, shows much higher disagreement after this point with measured data. On the other hand, the root function shows good correlation with the on-site measured value (Figure 3).

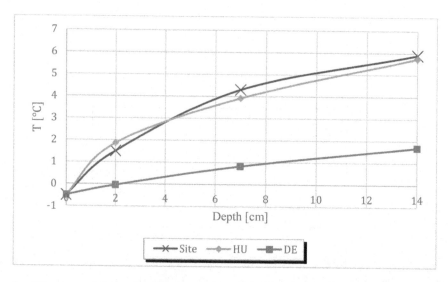

Figure 3. Site measured data and estimation models when the surface temperature is -0.5°C.

3.1 Calculation result of the Hungarian model

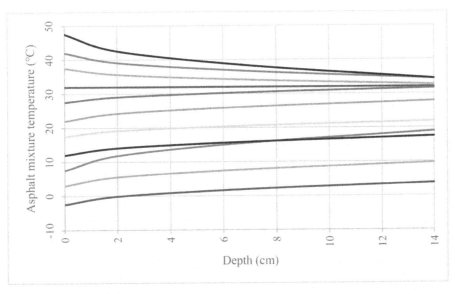

Figure 4. Temperature in-depth profile estimation by the new method with Hungarian weather data.

The Figure 4 shows the asphalt temperature distribution along the depth by the new prediction model (equation 2).

3.2 Error tolerance of the Hungarian model

The new model is prepared with a statical analysis and optimization result of Hungarian weather data. A yearly measured data is used for the method of Least Squares-a norm (L2, Euclidean Norm)-regression analysis is done. Result of the regression analysis is shown in Table 2, and the Table 3 shows the L2 loss comparison which shows how much the models are scattered from the site measured values (eq (3),(4)).

Table 3. L2 loss comparison between new model used here and German model.

	Depth [cm]				Depth [cm]		
	2	7	14		2	7	14
New model	1.460	1.616	2.245	German model	0.444	1.473	1.811
	1.026	0.925	1.683		0.680	3.800	5.000
	2.605	3.554	4.790		1.682	4.500	6.600
	1.249	1.334	1.422		0.745	2.500	4.000
	0.132	0.647	0.682		1.427	3.700	5.000
	0.496	0.109	0.089		3.483	0.578	1.921
	0.648	1.194	1.876		1.204	2.602	3.708
	0.187	3.275	4.965		0.930	0.010	0.428
	2.461	2.481	2.423		2.150	3.510	4.941
	0.900	3.075	3.828		1.886	2.928	2.308
	0.467	3.307	3.658		4.407	12.283	16.186
Sum	60.809				75.951		

The norm of these two prediction models is compared, see Table 3. Overall, the new model's norm is smaller than the German model. Thus, the statement can be delivered that the new model suits on Central European climatic condition. The model coefficient's R-squared values are given in Table 3.

$$\|x\|_2 := \left(\sum_{i=1}^{n} |x_i|^2\right)^{1/2} \tag{3}$$

$$L = \sum_{i=1}^{n} (y_i - f(x_i))^2 \tag{4}$$

where n is the element number, y_i is a real value and $f(x_i)$ is an estimated value and L means L2 loss which is Least squares error.

4 SUMMARY AND FUTURE PLAN

The bearing capacity of the asphalt is highly dependent on temperature due to its viscous nature. The stiffness is the function of temperature. Therefore, this temperature can be stated as the key input of Mechanistic-Empirical design. Much reliable temperature estimation brings a much realistic service life estimation.

This study is about a prediction of Hungarian asphalt layer's temperature. A prediction model is prepared for the asphalt layer's temperature estimation without direct investigation. Many of study were referenced, and among those studies the German mathematical method is chosen for this study due to the close geographical location. From the temperature properties of asphalt, this study puts the latitude for the first place and further developed the German model for the preparation of Hungarian model.

Throughout the few regressions of one year data set of Hungarian asphalt temperatures, a model for the Hungarian pavement is prepared. The R-squared values are generally above 0.84, except for two temperature ranges where they were found as 0.69 and 0.71; these values are considered reliable for temperature prediction models. The norm of the new model is given in Table 3. This study utilizes one sites' measurements set therefore a limitation comes for the generalization, however, with a following further modification with more sites and much longer time measurements will be expected to cover up such limitation.

REFERENCES

AASHTO, 2008. Mechanistic-empirical pavement design guide. Washington D.C.: American Association of State Highway and Transportation Officials.
AASHTO, 2011. LTPP computed parameter; Dynamic modulus, McLean: U.S. Department of Transportation.
Barber, E. S., 1957. Calculation of Maximum Pavement Temperature from Weather Reports, Highway Research Board Bulletin.
Dempsy, B. J., 1987. Characterizing Temperature Effects for Pavement Analysis and Design, Transportation Research Record.
Huang, Y., 2004. Pavement design and analysis. New York: Prentice Hall.
NCHRP, 2011. LTPP Computed Parameter: Dynamic modulus, US Department of Transportation Federal Highway Administration.
RDO, 2009. Guidelines for mathematical dimensioning of foundations of traffic surfaces with a course asphalt surface, Berlin: Research society for roads and traffic.

Stress random distributions on railway subgrade surface under train loads considering irregular shapes of ballast

J. Xiao
Shanghai Key Laboratory of Rail Infrastructure Durability and System Safety & Key Laboratory of Road and Traffic Engineering of the Ministry of Education, Tongji University, Shanghai, China

L. Xue
Key Laboratory of Road and Traffic Engineering of the Ministry of Education, Tongji University, Shanghai, China

P. Jing
College of Metropolitan Transportation, Beijing University of Technology, Beijing, China

D. Zhang
Shanghai Road and Bridge (Group) Co. Ltd, Shanghai, China

ABSTRACT: The morphological features of railway ballast, especially the edges and corners, have a great impact on the additional stress on the subgrade surface induced by train loads. In this paper, a novel shape reconstruction method was adopted to generate an arbitrary number of ballast particles. Then, a three-dimensional Discrete element-Finite element method (DEM-FEM) coupling model of the ballasted track-subgrade, in which the ballasted track was simulated by DEM and the subgrade soil was simulated by FEM, was established to analyze the contact stress at the ballast-subgrade interface. In this work, 150 groups of ballasted track-subgrade numerical models composed of irregular particles with randomly reconstructed shapes were applied to investigate the stress distributions on the subgrade surface under various train axle loads. Besides, the average stress of a 10 cm diameter circular area on the subgrade surface directly under the wheel load was calculated. The results indicated that both the peak and average additional stresses induced by train loads followed normal distributions for various numerical models (i.e., various groups of random ballast particle morphology). Compared with field testing data of the additional stress on the subgrade surface under train loads acquired by earth pressure cell, the larger average stresses were obtained in numerical models, because of direct contact between ballast particles and the subgrade along with stress homogenization by the pressure cell.

Keywords: Ballasted track, additional stress on subgrade stress, ballast shape, DEM-FEM coupling model, stress random distribution

1 INTRODUCTION

Fine-grained soil was widely used as fill on the subgrade in Chinese conventional railway, which forms the substructure of direct contact between the ballast and soil subgrade. With the growing demand of heavy-haul freight transportation, the loading cycles and subgrade dynamic stress increase. It can result in several kinds of subgrade deterioration during railway operation, posing threats to traffic safety (Xiao et al. 2020). Thus, the research on the characterization of subgrade dynamic response under heavy load condition is of theoretical significance for the reinforcement and maintenance of existing railways.

For the dynamic response inside the subgrade, the calculation method based on Boussinesq solution is widely adopted to obtain the attenuation curves of dynamic stress in the vertical direction (Chen, 2014). Hence, once the dynamic stress distribution on the subgrade surface is determined, the dynamic response of subgrade can be acquired. At present, most research works have revealed that the peak dynamic stress on the subgrade surface shows a saddle-shaped distribution in transverse direction. Relevant research work (Mei et al. 2019) has analyzed the random distribution of the subgrade dynamic stress from the aspects of stiffness change and track irregularity. However, in their studies, the ballast track was taken as the continuum, and the dispersion of stress distribution at ballast-soil interface was ignored.

With the appearance of advanced measurements, such as pressure-sensitive paper (Mchenry et al. 2015) granular material pressure cell (Liu et al. 2017), and tactile sensor (Aikawa, 2009), the contact stress inside the ballast layer was measured. These testing results indicated the angular characteristics of ballast leading to stress concentrations at the sleeper-ballast-soil interfaces. In addition to experimental methods, some scholars established the ballasted track-subgrade model through the Discrete element-Finite element method (DEM-FEM) coupling method to investigate the effect of contact dispersion and particle geometry on mechanical behavior of ballasted track. Three dimensional shapes such as spherical cluster (McDowell and Li 2016), super ellipsoid (Zhao and Zhou 2017) or polyhedron (Huang and Tutumluer 2013) were used to model ballast particles. Nevertheless, most existing reconstruction methods lack statistical analysis on the overall morphological characteristics of real ballast particles. Thus, there are gaps between the reconstructed and real ballast in shape description.

In this paper, a shape reconstruction method was adopted based on the probability density distribution of morphological indices. Through the preparatory work of laser scanning and analysis of ballast particle morphology, 150 groups of ballast samples randomly regenerated, which conform with the morphological characteristics of natural ballast. After that, a three-dimensional DEM-FEM coupling model of ballasted track-subgrade was established. The average stress of a 10 cm diameter circular area and peak stress were obtained and forecasted with statistical method. The results indicated that both kinds of stresses obey normal distributions. In comparison with field testing data acquired by earth pressure cell, the larger average stresses were obtained in numerical models.

2 STATISTICAL ANALYSIS AND RECONSTRUCTION OF BALLAST PARTICLES MORPHOLOGY

In this paper, point clouds of surface contours of ballast particle samples were obtained by 3D laser scanning and analyzed to give a quantitative description of the morphological characterization of real ballast particles. Related literature (Noura et al. 2017) has pointed out that when the particle number exceeded 400, the statistical result of the ballast particles' morphological characterization would tend to be stable. Therefore, a 50 kg ballast sample containing a total of 584 ballast particles sieved according to the current Chinese code (TB/T 2140-2008) for the first-class gravel ballast, and the grading curve of the ballast sample is presented in Figure 1.

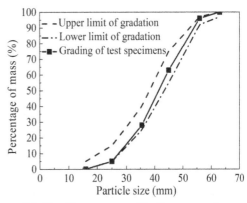

(a) Some selected ballast particles sample (b) Grading curve of ballast sample

Figure 1. Selection and grading curve of ballast sample.

In general, global and local shape features should be both considered when characterizing particle morphology (Huang and Tutumluer, 2013; Pen, 2013). Global morphological features mainly refer to the basic geometric dimensions and shape, and generally described with long axis (Φ_1), middle axis (Φ_2), short axis (Φ_3) and sphericity index (Sp^3) (Yan et al. 2016). Φ_1 indicates the distance between the two farthest surface points of the particle. Located on the maximum projection plane perpendicular to Φ_1, S, the longest dimension is called Φ_2. And Φ_3 is the maximum dimension perpendicular to Φ_1 and Φ_2. The schematic diagram of these three axes is shown in Figure 2.

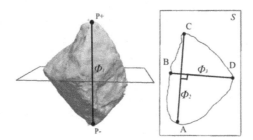

Figure 2. Schematic diagrams of Φ_1, Φ_2 and Φ_3.

Sphericity index (Sp) is defined based on the above three indices, which describes the degree of approximation between an irregular particle and a sphere, and can be solved by following Eq. (1):

$$Sp^3 = \frac{\Phi_2 \times \Phi_3}{\Phi_1^2} \qquad (1)$$

Local morphological feature mainly focuses on the edges and angularities on the surface of particles, and in this paper, it was quantified by the curvature index (*CI*). It is found that at least 20% of the points are located on the edge or vertex, and the corresponding local curvature is greater than that on the face. Therefore, the sum of the curvature from the first 20% is taken as the curvature index, as can be written as Eq. (2):

$$CI = \sum_{i=1}^{20\%N} \rho_i = \sum_{i=1}^{160} \rho_i \qquad (2)$$

where ρ_i is the discrete local curvature and is arranged in descending order; N is the number of points in each point cloud, and was normalized to 800 for each particle in this paper.

Then, the above five indices of 584 particles were statistically analyzed and their probability density distributions were calculated. However, because of the large number of original point clouds and high nonlinearity of the morphology indices, we applied the Proper Orthogonal Decomposition (POD) to find dominant eigenvectors for describing the main morphological features of ballast particles, and then took these eigenvectors and corresponding shape expansion coefficients ($C_0, \ldots, C_{34}$) to reconstruct equivalent particles. After that, the mapping relationship between the shape expansion coefficients and indices of morphological feature was obtained by Radial Basis Function (RBF) Neural Network, so that the specific coefficients could be calculated according to the desired probability distribution of the morphological feature indices. In this way, it's possible to model any number of ballast particles that conform to the desired probability density distribution. The flow chart of ballast regeneration was shown in Figure 3. For this part, more details can be found in the previous literature (Xiao et al. 2020).

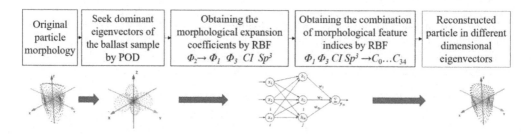

Figure 3. Flow chart of ballast shape regeneration.

3 BALLASTED TRACK-SUBGRADE MODEL

In this work, the ballasted track was simulated using DEM, whereas the sleeper and the subgrade soil were modeled by FEM. For the sake of calculation efficiency and accuracy, the surface coupling method was applied to calculate the interaction between two domains. If there are no contact between ballasts and the sleeper, the analysis of DEM and FEM is conducted separately. Once the contact occurs, the FEM interface mesh is copied into the DEM domain as facet elements and the displacement boundary conditions are introduced and the contact forces acting on the facet elements are transferred into FEM domain (Song et al. 2019).

The model size was determined in terms of the current railway track and subgrade design specifications in China. Only half of each structure was modelled along the cross-sectional. Type □ sleeper was used in simulation, with dimensions of 1.25 m×0.3 m×0.23 m (length× width× height). The height, upper and lower surface widths of track bed were 0.35 m, 1.5 m and 2.2 m respectively. The longitudinal length of track bed and subgrade was 0.6 m. For the widths of the subgrade, the upper and lower surface were 3.05 m and 4.65 m separately. In the vertical range, the thickness of the surface layer was 0.6 m, and the bottom layer was 0.4 m. In order to reduce the reflection of stress wave under cyclic loading, viscoelastic damping boundaries with the grid elements (normal stiffness was 1.2×10^8 N/m and damping coefficient was 0.6) of 5 cm thickness were set around the model. The specific dimensions are indicated in Figure 4.

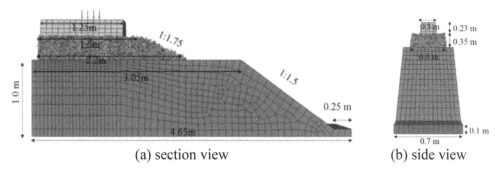

Figure 4. A three-dimensional DEM-FEM coupling model of sleeper-ballasted track-subgrade.

In this work, 150 groups of ballasted track-subgrade numerical models were generated. The number of ballast particles N_i in each sample ranged from 7373 to 7590. Based on the probability density distribution of the morphological feature indices gained by 3D laser scanning, corresponding index values were obtained in the range of each morphological index with the uniform distribution random number. Adopting the particle shape reconstruction method established above, polyhedral particles were modelled in the discrete element program YADE.

The modeling of the sleeper and subgrade were conducted by open-source code OOFEM (Patzák 2012) as FEM parts. The concrete sleeper was simulated with the linear elastic constitutive model, whereas the Drucker-Prager constitutive model was used for the subgrade. Some parameters used in the DEM-FEM modeling are presented in Table 1.

Table 1. Structural parameters in the DEM-FEM coupling model of the sleeper-ballasted track-subgrade.

DEM	Density ρ/(kg·m^{-3})	Normal volume stiffness k_n/(N·m^{-3})	Tangential volume stiffness k_s/(N·m^{-3})	Friction coefficient	Damping coefficient	Time stepΔt/(step·s^{-1})	
Ballast	2600	1.82×10^{10}	1.6×10^{10}	0.65	0.03	1×10^{-6}	
FEM	Density ρ/(kg·m^{-3})	Modulus E/Pa	Poisson's ratio μ	Friction angle /°	Damping coefficient	Cohesion / kPa c	Time stepΔt/ (step·s^{-1})
Sleeper	2600	3.0×10^{10}	0.2	-	0.03	-	1×10^{-6}
Surface layer of subgrade	1950	1.2×10^{8}	0.33	30	0.3	30	1×10^{-6}
Bottom layer of subgrade	1900	6.0×10^{7}	0.35	20	0.3	20	1×10^{-6}

The train load applied on the rail pad, with a length of 0.305 m, as shown in Figure 4 (a). About 5 sleepers bore the single axle load at a ratio of 0.1:0.2:0.4:0.2:0.1. The load magnitude was calculated as 0.4 times the wheel weight. Therefore, the amplitudes under the axle load of 25 t, 27 t and 30 t were 50 kN, 54 kN and 60 kN respectively. For the sake of computational cost, the loading frequency was fixed at 1 Hz and the duration was one cycle of half-sinusoidal wave.

4 PROBABILITY ANALYSIS OF DYNAMIC STRESS ON THE SUBGRADE SURFACE

4.1 Random characteristics of dynamic stress on the subgrade surface

The case of the model under 30 t axle load was selected as an example for the following analysis. The distribution of sleeper stress and force chains of ballast particles at the peak of the train load were shown in Figure 5. As can be seen, the distribution of contact stress between the sleeper and ballast was discrete, and stress concentration at the end of the sleeper was induced. The results are consistent with field measurements (Mchenry et al. 2015).

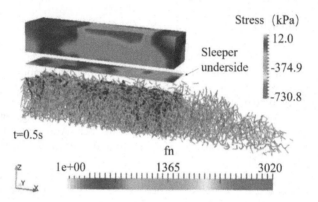

Figure 5. The distribution of sleeper stress and force chains inside ballasted track under 30 t train load.

As illustrated in Figure 5, main force chains were randomly generated inside the ballasted track below the end of the sleeper, which bore the main load and transferred it to the subgrade surface. Within these regions, the interactions between the irregular ballast particles and the subgrade surface were dominated by face–face contact, while in the other areas were edge–face or vertex–face contact. It can be seen that the random formation of the main force chains and the discrete contact at the interface contributes to the random characteristics of dynamic stress on the subgrade surface.

Figure 6 shows the stress nephogram on the subgrade surface under the 30 t axle load, which reveals that, under the loading area, the dynamic stress on the subgrade surface was significantly higher than that at other positions, and the peak dynamic stress was 2165.2 kPa. Besides, in the field experiment, the earth pressure cell with diameter no less than 10 cm is often used to measure the subgrade stress, so the average stress of a 10 cm diameter area on the subgrade surface directly under the wheel load was calculated (hereinafter referred to as "stress at D10 area"). Under the 30 t axle load, the stress at D10 area was 135.8 kPa, far less than the peak dynamic stress on the subgrade surface. Besides, as depicted from 150 groups of test results, under all axle loads, the peak dynamic stresses were found in the area under the sleeper.

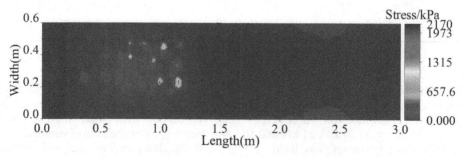

Figure 6. Stress nephograms on the subgrade surface under 30 t train load.

4.2 Normality test on peak dynamic stress and stress at D10 area

For 150 groups of test results, the peak dynamic stress and stress at D10 area on the subgrade surface under different axle loads are depicted in Figure 7.

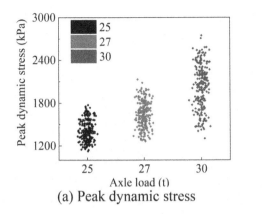
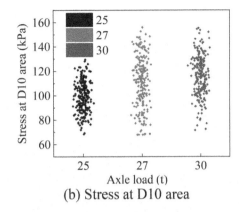

Figure 7. Peak dynamic stress and stress at D10 area on the subgrade surface under different axle load.

It is generally considered that the wheel-load is normally distributed (Chen et al. 2016). In this study, the Kolmogorov test is performed to verify the normal distribution of peak dynamic stress and stress at D10 area on the subgrade surface. The specific steps are as follows (Chen et al. 2016):

Assuming the distribution function of population X is $F(x)$, $F(x)$ is a continuous function but unknown, $(X1, X2, \ldots Xn)$ is a sample from the population, and $F_n(x)$ is a better estimate of $F(x)$. Since $F(x)$ is a nondecreasing function, the following test statistic is constructed:

$$D_n = \max_{0<i<n} |F_n(x_i) - F_0(x_i)| \quad (3)$$

$$F_n(x_i) = i/n \quad (4)$$

$$F_0(x) = \Phi\left(\frac{x - \bar{x}}{s^*}\right) \quad (5)$$

where $(x_1, x_2, \ldots x_n)$ are observation values of order statistics $(X1, X2, \ldots Xn)$, μ is the mean of the observation values, $\mu = \bar{x}$, and σ is the standard deviation, $\sigma = s^*$.

Corresponding to significance level α, there exists a critical value $D_{n,\alpha}$. When D_n is less than $D_{n,\alpha}$, $F_n(x)=F_0(x)$ can be regarded as true, that is, at the significance level α, the population obeys a normal distribution.

The normality analyses of peak dynamic stress and stress at D10 area on the subgrade surface under various axle loads are performed according to the above method. Table 2 lists the D_n values at various conditions.

Table 2. Statistical table of the normality test of the peak dynamic stress and stress at D10 area on the subgrade surface.

Axle load (t)	D_n		$D_{n,\alpha}$	
	Peak dynamic stress	Stress at D10 area	$\alpha=0.05$	$\alpha=0.01$
25	0.048	0.033	0.058	0.069
27	0.028	0.035	0.058	0.069
30	0.035	0.042	0.058	0.069

It can be seen that D_n is less than $D_{n,\alpha}$ at significance levels α of 0.05 and 0.01. Therefore, the peak dynamic stress and stress at D10 area can be considered to obey the normal distribution. Taking the statistical results of 30 t axle load as an example, the histogram of the peak dynamic stress and stress at D10 area is shown in Figure 8.

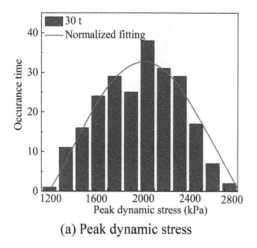
(a) Peak dynamic stress

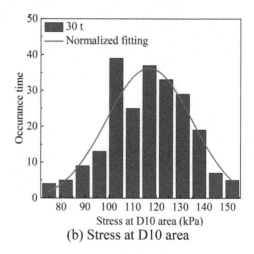
(b) Stress at D10 area

Figure 8. Histogram of peak stress and stress at D10 area under 30 t axle load.

4.3 *Maximum forecast of peak dynamic stress and stress at D10 area*

According to the 3σ principle of normal distributions, the interval (μ-3σ, μ+3σ) is regarded as the possible range for true values of the population. That is to say, μ+3σ is the maximum value of peak dynamic stress and stress at D10 area if they obey the normal distribution. Table 3 summarizes the inferred maximum value of the above two stresses under different axle load.

Table 3. Maximum peak dynamic stress and stress at D10 area on the subgrade surface.

Axle load (t)	Dynamic stress	Mean (kPa)	Standard deviation (kPa)	Maximum (kPa)
25	Peak dynamic stress	1418.35	150.61	1870.2
	Stress at D10 area	99.17	12.70	137.28
27	Peak dynamic stress	1648.94	190.17	2219.45
	Stress at D10 area	112.14	15.41	158.37
30	Peak dynamic stress	2021.79	313.64	2962.71
	Stress at D10 area	116.17	16.56	165.85

To verify the reliability of the probability analysis result, we compared the maximum stress at D10 area, the field test result and the calculation result obtained from the empirical formula (Zhang 2005), as shown in Table 4.

Table 4. Comparison of probability analysis result, test results and empirical calculations.

Analysis method	Thickness of ballasted track (cm)	Axle weight (t)	Speed (km/h)	Dynamic stress (kPa)
Field test result in Shuo-Huang railway	≥50	25	<100	93~110
		30	<100	100~123
Maximum stress at D10 area	35	25	120	137
		30	120	166
Empirical calculations result	35	25	120	104
		30	120	125

It can be seen that the maximum value of stress at D10 area inferred from the coupling model results is higher than the test value and empirical calculation value. The main reasons are as follows: (1) In the coupling model calculation, the stress concentration caused by the irregular shapes of ballast particles is considered, while the stress distribution is relatively uniform due to the stiffness of the earth pressure cell; (2) The thickness of ballasted track in the numerical model is smaller than that in the field test, which leads to the different degree of stress diffusion; (3) Particles in the numerical model simulate the clean ballasts with more obvious edges and angularities characteristics, whereas the ballast on the operating railway is fouled and worn, leading to the reduction of the contact stress.

5 CONCLUSION

In this paper, a novel shape reconstruction method was adopted to randomly regenerate 150 groups of ballast samples conforming to the statistical regularity of overall morphological characteristics of natural ballast. Based on it, a ballasted track-subgrade coupling model was established to investigate the stress distribution on the subgrade surface under three axle loads. The main conclusions are as follows:

The shape reconstruction method based on the probability density function of morphological indices can ensure that the modeled ballast has the same probability distribution as the real ballast. In addition, increasing the number of particles does not lead to the difference in the probability distribution.

The random formation of the main force chain inside the ballasted track and the discrete contact at the interface caused by the irregular shape of the particles are the major reasons for the random distribution of the dynamic stress on the subgrade surface. The peak dynamic stress was much higher than the stress at D10 area. No matter what the axle load was, the peak dynamic stress occurred under the sleepers.

Both the peak stress and stress at D10 area follow normal distributions. Due to the influence of factors such as ballast fouling, filling thickness and stress homogenization by the pressure cell, the larger average stresses were obtained in numerical models compared with field testing data.

ACKNOWLEDGEMENTS

This research is supported by the NSFC (National Natural Science Foundation of China) Program Grant NO. 51678447, which is greatly appreciated.

REFERENCES

Aikawa, A., 2009. *Techniques to Measure Effects of Passing Trains on Dynamic Pressure Applied to Sleeper Bottoms and Dynamic Behavior of Ballast Stones*. Quarterly Report of Rtri, 50(2): 102–109.

Chen, R., 2014. *Recent research on the track-subgrade of high-speed railways*. Journal of Zhejiang University, 15(12): 1034–1038.

Chen, R., Jiang, P., Duan, X., et al, 2016. *Probability Distribution of Dynamic Stress of High-speed Subgrade under Slab Track Irregularity*. Journal of the China Railway Society, 38 (9).

Huang, H., Tutumluer, E., 2013. *Image-Aided Element Shape Generation Method in Discrete-Element Modeling for Railroad Ballast*. Journal of Materials in Civil Engineering, 26(3): 527–535.

Liu, Q., Lei, X., Rose. J., et al, 2017. *Pressure Measurements at the Tie-Ballast Interface in Railroad Tracks Using Granular Material Pressure Cells*: V001T01A003.

Mcdowell, G. and Li, H., 2016. *Discrete element modelling of scaled railway ballast under triaxial conditions*. Granular Matter, 18(3).

Mchenry, M., Brown, M., Lopresti, J., et al, 2015. *Use of Matrix-Based Tactile Surface Sensors to Assess Fine-Scale Ballast–Tie Interface Pressure Distribution in railroad Track*. Journal of the Transportation Research Board, 2476(2476): 23–31.

Mei H, Leng W, Nie R, et al, 2019. *Random Distribution Characteristics of Peak Dynamic Stress on the Subgrade Surface of Heavy-Haul Railways Considering Track Irregularities*. Soil Dynamics and Earthquake Engineering, 116: 205–214.

Noura, O., Charles, V., Guillaume, P., et al, 2017. *3D Particle Shape modelling and optimization through proper orthogonal decomposition application to railway ballast*. Granular Matter, 19 (4): 86-1–86-14.

Patz´ak, B., Rypl, D. and Kruis, J, et al, 2012. *MuPIF – A Distributed Multi-physics Integration Tool*. Advances in Engineering Software., 60–61: 89–97.

Song, W., Huang, B., Shu, X., et al, 2019. *Interaction between Railroad Ballast and Sleeper: A DEM-FEM Approach*. International Journal of Geomechanics, 19(5): 04019030.1–04019030.10.

Xiao, J., Wang, Y., Zhang, D., et al, 2020. *Testing of Contact Stress at Ballast Bed-Soil Subgrade Interface under Cyclic Loading Using the Thin-Film Pressure Sensor*. Journal of Testing and Evaluation, 48 (3): 20190171.

Xiao, J., Zhang, X., Zhang, D., et al, 2020. *Morphological reconstruction method of irregular shaped ballast particles and application in numerical simulation of ballasted track*. Transportation Geotechnics, 100374.

Yan, P., Zhang, J., Fang, Q., et al, 2016. *3D numerical modelling of solid particles with randomness in shape considering convexity and concavity*. Powder Technology, 301: 131–140.

Zhao, S., Zhou, X, 2017. *Effects of particle asphericity on the macro- and micro-mechanical behaviors of granular assemblies*. Granular Matter, 19(2): 38.

Zhang, Q., 2005. *Dynamic Stress Analysis on Speed-increase Subgrade of Existing Railway*. China Railway Science, 026(005): 1–5.

Modelling and predicting asphalt deflection values with artificial neural networks

M. Rahimi Nahoujy
Ruhr-Universität Bochum, Bochum, Germany
The Federal Highway Research Institute, Bergisch Gladbach, Germany

M. Radenberg
Ruhr-Universität Bochum, Bochum, Germany

ABSTRACT: The falling weight deflectometer (FWD) is an internationally used device for measuring deflections and evaluating the quality of pavement. But FWD can only take measurements of a limited number of kilometers of road per day. On a large scale, this method is expensive in terms of time and resources. Moreover, in some cases, data sets obtained from FWD are not sufficient, or they may be limited or missing, which makes the planning of maintenance and rehabilitation measures more error-prone. Aiming to overcome these limitations and to improve the method, in this work a new artificial neural network (ANN) approach was developed in order to calculate the deflection at any arbitrary point on the entire route to complement and substitute experimental measurements. A feed-forward ANN model was developed in the MATLAB computing environment based on backpropagation by a multilayer perceptron (MLP) network for asphalt pavement. A MLP neural network model was chosen, because it can deal with those cases, in which FWD deflection data is not sufficiently available. These networks are particularly suited to modeling complex data, as provided by FWD, due to the capability of ANN to learn complex nonlinear behavior. With at least 150 data sets, a model can be trained through ANN, that has a mean square error of less than 1 percent. Fewer measuring points are needed because the missing data can be calculated from the formulation. Thus, this method offers a great potential for the optimization of the traditional measurements in two ways: First, it optimizes measurement costs, and second, it significantly enhances the accuracy of road maintenance planning. Besides, it helps solving the problem of limited or missing data sets.

Keywords: Artificial neural network, multilayer perceptron, falling weight deflectometer, non-destructive testing, deflection

1 INTRODUCTION

Road maintenance costs are a considerable factor in managing public budgets, which means that finding efficient methods for the maintenance of these roads is of utmost importance. For economically optimized, technically adapted and demand-oriented planning, regular evaluations at appointed times are fundamental in every method of pavement maintenance.

Non-Destructive Tests (NDTs) are employed at different levels, to evaluate, repair, maintain, and rehabilitate different pavements. For optimal maintenance planning, especially for rehabilitation measures based on technical and economic criteria, it is important to consider the results of bearing capacity measurements (Jansen 2009). Therefore, the deflection basin determined in Non-Destructive Deflection Testing is commonly used to evaluate the structural conditions of surface pavement. Deflection measurements at defined load pulses with a falling weight deflectometer (FWD) are a method for load capacity measurements that has been used worldwide since the 1960s (Jansen 2009).

FWD is a global and advanced method that measures load capacities and deflection changes in asphalt pavement under load. The physical properties of asphalt pavement can be measured by this device. FWD data is mainly used to estimate the capacity of layers of asphalt pavement. In other words, it is used to determine if a road surface is overloaded or not.

Even though FWD is a popular device for the planning of maintenance and rehabilitation measures of roads and widely used in Europe, the US and beyond, the technique has a few considerable downsides. The FWD can only take measurements of a limited number of kilometers of road per day. On a large scale, this method is expensive in terms of time and resources. Moreover, in some cases, data sets obtained from FWD are not sufficient, or they may be limited or missing, which makes the planning of maintenance and rehabilitation measures more error-prone.

Nowadays, an artificial neural network (ANN) is a very effective tool in the domain of pavement maintenance (Nejad and Zakeri 2012, Attoh-Okine 1999). A number of studies has demonstrated that ANNs can be used for prediction in the domain of pavements maintenance. By modelling FWD data, the deflection can be predicted at any arbitrary point on the entire route measured by FWD.

The successful use of ANNs in pavements has been demonstrated in many studies, most of them focused on the backcalculation of the modulus for each pavement layer considering complex material properties in viscoelastic and nonlinear modulus (Kim and Kim 1998, Ceylan 2005, Ceylan et al. 2007, Li and Wang 2019;). Some works also document the success of ANN in directly (without backcalculation) predicting pavement responses (Saltan and Terzi 2007, Wang et al. 2020, Rahimi Nahoujy 2020).

2 FWD TESTING DEVICE

The FWD Primax (Figure 1) is one of the earliest deflectometers and has been commercially available since 1969 in Europe and the United States. In this device, an impact load of 10 to 50 kN is generated by a weight released from various heights. The load is transmitted to the road surface (pavement) via a plate (300 mm diameter). The deflections are measured by ten transducers, one of which is located at the center of the plate, while the rest are located at a distance of 2100 mm from the plate center (Huang 2004).

An operator can perform around 50 measurements per hour using a standard version of the Primax 1500 in Germany. The distance between points is 25 meters, and the applied load is 50 kN. The time history of the impulse load is recorded via a load cell. The impulse load generated by the FWD is used to simulate a wheel overrun and, therefore, should have the same pulse duration as the one resulting from a wheel overrun with a vehicle (pulse duration between 20 and 30 ms). The size of the impulse load should correspond to the wheel load of a truck (Jansen 2009, FGSV 2008).

Geophones record the short-term vertical deformation of the road surface (deflection bowl) in response to the impulse load (FGSV 2008). In this study, one geophone is located in the load center, and the other nine are located at following distances from the load center: 0–200–300–450–600–900–1.200–1.500–1.800 (mm). In addition to FWD measurements, the temperature is measured at any point and the asphalt thickness is measured at all points by the ground penetrating radar (GPR) device.

Figure 1. Primax 1500 FWD.

3 METHODOLOGY

3.1 *ANN*

ANNs are introduced as a useful tool in solving engineering problems with highly nonlinear functional approximations. ANNs have been successfully applied to tasks such as recognition of function approximation, function optimization, prediction, data recovery, automatic control and many other cases.

An ANN consists of a group of artificial neurons, which are the smallest unit of information processing. Every artificial neuron receives inputs, processes them, and finally produces an output signal. Two or more neurons can be combined in the form of a layer; an individual network can be made up of multiple layers. Each layer in the weight matrices network may have its biased vector and output.

An ANN is created with a small group of artificial neurons, receptive to training, aiming at solving complex problems. The grouping of artificial neurons can be described as mapping training, which means the neurons learn to create a mapping between the explanatory variables and the response variables (Hagan et al. 2014). The network displayed in Figure 2 includes input neurons, output neurons, and hidden neurons.

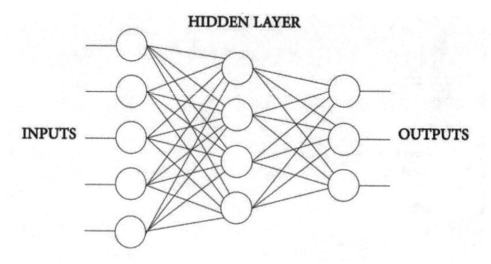

Figure 2. The basic structure of an ANN.

ANNs are programmed or trained in a way that a specific input has a particular target output. Figure 3 demonstrates this situation where the network is modified based on the difference between the target and the output until the output is close enough to the target. To train a network, many input/target pairs should be used, so as to supervise the learning process (Demuth et al. 2001).

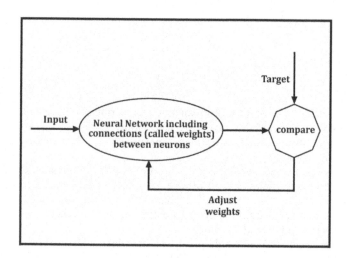

Figure 3. Basic principles of ANNs (Demuth et al. 2001).

3.2 *Multilayer Perceptron (MLP)*

Defined as one of the significant ANNs and common in engineering, the perceptron is considered as one of the most applicable networks. It is a non-recursive network which uses a supervised training algorithm. These networks are able to perform accurately a non-linear mapping by choosing the appropriate number of layers and neural sets. (Hagen et al. 2014, Menhaj 1998).

A MLP neural network model was chosen for the purpose of this study because it can deal with those cases, when not enough FWD deflection data is available. These networks are particularly suited to modeling complex data as provided by FWD due to the capability to learn complex nonlinear behavior.

Usually, a MLP is trained with a backpropagation (BP) algorithm (Principe et al., 2000; Rumelhart et al., 1986). The BP learning algorithm is one of the most important historical developments in neural networks. The MLP modeling with BP algorithm consists of two steps (Hagen et al. 2014, Menhaj 1998):

The route forward: in this route, the input vector is applied on the network, and its effects are transferred via hidden layers to the output layers. The created output vector in the output layer gives the real response of MLP.

The route backward: in this route, the error vectors are reverted from the last layer to the first one. In other words, in this route, the output layer, marks the beginning of the procedure, and the error vector is distributed from output to input, which means that it moves from the last layer to the first one.

The error vector is the difference between the network response (output) and the measured value (target). The error value is calculated in the second step from the output layer and distributed through the network layers throughout the network.

3.3 Training architecture

In this study, an MLP network is used with two hidden layers. The sigmoid function is in the first and second hidden layer and the linear function in the output layer. Further hidden layers are not considered as they complicate the problem and do not produce better results. The goal is to achieve the best result, so, certain modifications are applied to inputs, neurons, and algorithms (training functions) at each iteration. In other words, a neural network is trained for a single output with various inputs, different number of neurons and different algorithms in order to determine the best network among the possible training methods.

The input layer consists of three parameters: measuring point, thickness, and temperature. The output layer has just one parameter: deflection. The model is shown in Figure 4.

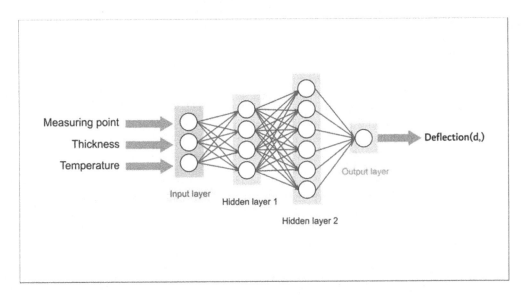

Figure 4. A schematic chart of a MLP.

3.4 Dividing the modeling data

The main capability of ANN is its generalizability. It refers to the possibility of generalizing the results from previous data to new data after training. In fact, the networks are trained in a way that makes them capable of extracting trends and patterns from a specific data set and generalize them to other data. The objective is to ensure the best modeling. Data sets are divided into three sections:

1. Training data: Training data, which refers to data used during the training process, is evident, and is applied as template model. The network is trained using the bulk of data (70% was used in this study) to have the highest possible correspondence between input and output.
2. Validation data: A network might be trained incorrectly, as it tends to "memorize" the data and thus loses the ability to generalize. This is called overtraining (or overfitting). To avoid overtraining, the training must be validated during the process. Thus, another, independent set of data (15% in this study) is tested during the training in order to validate the training process and prevent overtraining from happening. Training will continue if the validation results are positive. Negative results are recorded as "fail". The simultaneous testing and training processes continue, until failed cases reach a certain extent, when the training is stopped. This means that further training would not produce any positive results. In other words, validation results have the ability to stop the training.
3. Testing data: After completion of the training process, testing data is used with independent targets, in order to validate the accuracy of learning. Testing the data is intended to produce the outcome of training. In other words, testing data (15% in this study) provides the final answers to whether the network was trained successfully. Unlike validation data, these data do not have the ability to stop the training and are provided after the training as discussed later.

4 CASE STUDY AND RESULTS

For this study, FWD measuring stations were installed at 20 m intervals and thickness measuring stations (GPR) were placed at 8 m intervals. A part of thickness data produced by the same station is used to draw the matrices for input and output. As a result, 158 data series of input and output were obtained (Rahimi Nahoujy 2020). In this paper, the MLP network is implemented in the MATLAB® programing language and computing environment, where the MLP model is trained with FWD data.

In order to obtain an optimum architecture, different hidden layers, neurons and training algorithms were applied. Each architecture was trained with the same input and output dataset. The models are the outcome of training measuring point, asphalt body temperature, and thickness as input and deflection (d0) as output.

This study finds that the best model has two hidden layers with 4 neurons in the first and 6 in the second hidden layer. The training algorithm used is Conjugate Gradient Backpropagation with Polak-Ribiére Update (CGP). Furthermore, R-value and performance between experimental and simulated data were determined. Figure 5 shows a diagram of performance. This network was trained for 2000 epochs and epoch No. 962 represents the best performance in this network. The network was stopped due to "Maximum epoch reached". The best performance of training, the best performance of validation and the best performance of test are 0.0031, 0.0042 and 0.0050, respectively. In this study mean square error (MSE) was used to indicate for performance. The linear regression for training, validation, testing and the entire network is presented in Figures 6-9.

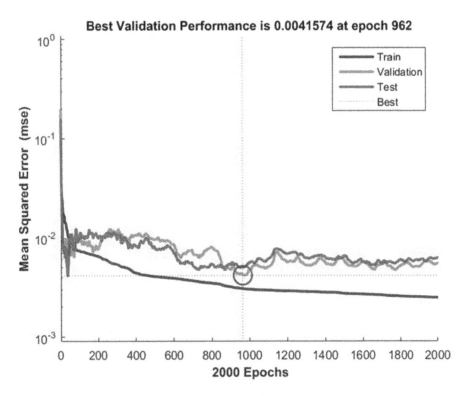

Figure 5. Performance plot for training, validation and test data sets.

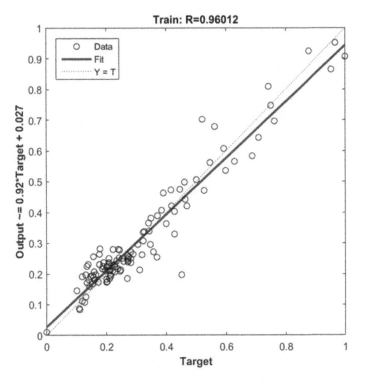

Figure 6. ANN prediction accuracy regression graphs for training data sets.

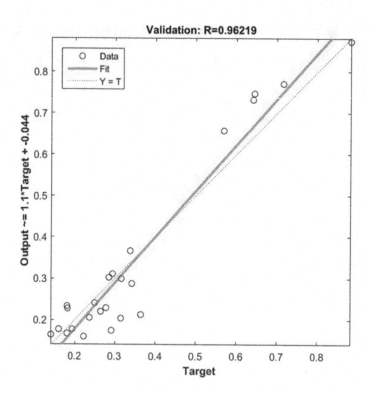

Figure 7. ANN prediction accuracy regression graphs for validation data sets.

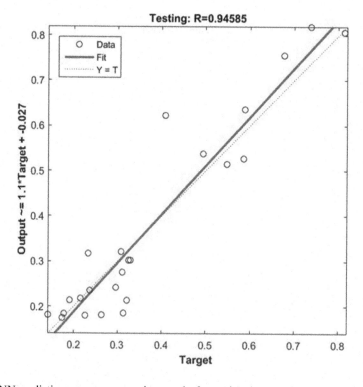

Figure 8. ANN prediction accuracy regression graphs for testing data sets.

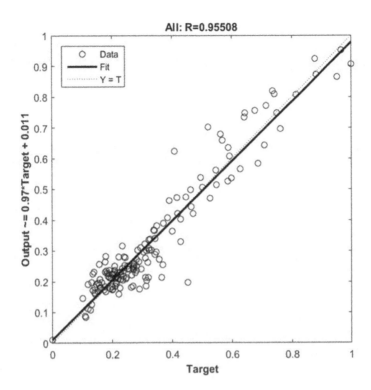

Figure 9. ANN prediction accuracy regression graphs for all data sets.

According to Figures 6-9, the R-value for training is 0.96 with a MSE of 0.0031, indicating an excellent possibility for training the network; the R-value for validation is 0.96 with a MSE of 0.0042. It shows that the model is robust against overtraining, the R-value for test is approx. 0.95 with a MSE of 0.0050. Thus, it is clear that this model has very good generalization ability. The R-value for all of data is approx. 0.96. Therefore, this network is a very good network with powerful generalizability strength.

5 CONCLUSIONS

Deflection value is an essential index in approximating the remaining life of highway flexible pavements. It is measured by FWD in developed countries. The measurement is complicated and time consuming, so that for a short route of highway, a full day for road maintenance works is expected. Moreover, in some cases, data sets are limited, and not enough data can be obtained from FWD.

In this study, based on MLP networks, a model of ANN was developed for calculating deflection basins in flexible pavements. The basis of ANN in any engineering problem solution is to train the input and output they receive. One particular application of ANN in pavement engineering – thanks to its ability for learning complex non-linear behavior – can be investigating the complex deflection basin form to determine the structural capacity of flexible pavements.

In case of insufficient data, a perceptron can be an efficient tool for various sections of surface pavements. A perceptron can accurately do an antilinear mapping by selecting the appropriate number of layers and neural cells. MLP are among the most frequently used and most powerful neural networks when it comes to function approximation problems.

In this study, a feed-forward ANN model was developed for a route. It is based on backpropagation (BP) by a MLP for asphalt pavement. This model was trained with the available data sets, generalizing to other points where data sets were missing.

This study showed that the neural network is able to model the FWD deflections and is characterized by powerful generalizability. After being modeled, the network can provide the hitherto unavailable data for all points between the measuring points with scant error. The best model had a MSE of 0.5 % and a R-value of approx. 95% for the test data.

This method can be used to model many engineering problems caused by a lack of data. It is of high significance for highway agencies in assessing the structural capacity of flexible pavement. Fewer measuring points are necessary because the missing data can be calculated from the equation. As a consequence, this method offers a great potential for the optimization of the traditional measurements: First with regard to measurement costs, and second, it significantly enhances the accuracy of road maintenance planning, solving the problem of limited or missing data sets.

Furthermore, ANN models are a reliable computational tool to solve different complex problems in this field and could also be applied to the prediction of pavement layer moduli. For future research, it might be worth investigating, if ANN-based backcalculation models can be a solution for the analysis of the large number of pavement deflections needed for routine pavement evaluations.

REFERENCES

Attoh-Okine, N. O., 1999. *Analysis of learning rate and momentum term in backpropagation neural network algorithm trained to predict pavement performance*. Advances in Engineering Software. 30:291–302.

Ceylan, H., Gopalakrishnan, K., Guclu, A. 2007. *Advanced approaches to characterizing nonlinear pavement system responses*. Transportation Research Record, 86–94.

Ceylan, H., Guclu, A., Tutumluer, E., Thompson, M. R. (2005). *Backcalculation of full-depth asphalt pavement layer moduli considering nonlinear stress-dependent subgrade behavior*. International Journal of Pavement Engineering, 6(3),171–182.

Demuth, H., Beale, M., 2001. *Neural network toolbox, user guide (version4)*. The MathWorks, Inc.

FGSV, 2008. *Arbeitspapier Tragfähigkeit von Verkehrsflächenbefestigungen Teil B 2.1.Falling Weight Deflectometer (FWD): Gerätebeschreibung, Messdurchführung -Asphaltbauweisen*. FGSV-Verl. Köln, 2008, FGSV 433 B 2.1

Hagan, M. T.; Demuth, H. B.; Beale, M. H.; De Jesus, O., 2014. *Neural network design*. Wrocław. Amazon Fulfillment Poland Sp. z o.o.

Huang, Y. H., 2004. *Pavement design and analysis*. Upper Saddle River, NJ. Pearson/Prentice Hall.

Jansen, D., 2009. *Temperaturkorrektur von mit dem Falling-Weight-Deflectometer gemessenen Deflexionen auf Asphaltbefestigungen*. Dissertation, Institute für Straßenbau und Verkehrswesen, Universität Duisburg-Essen, Schriften-reihe Heft 2, Essen, Germany.

Kim, Y., Kim, Y. R. 1998. *Prediction of layer moduli from falling weight deflectometer and surface wave measurements using artificial neural network*. Transportation Research Record, 1639, 53–61.

Li, M. Y., Wang, H. 2019. *Development of ANN-GA program for backcalculation of pavement moduli under FWD loading with viscoelastic and nonlinear parameters*. International Journal of Pavement Engineering, 20(4),490–498.

Menhaj, M. B., 1998. *Fundamentals of neural networks*. Computational intelligence. No. 1.

Nejad, F. M., Zakeri, H., 2012. *The hybrid method and its application to smart pavement management*. Metaheuristics Water Geotech Transp Eng, 439 H.

Principe, J. C., Euliano, N. R., Lefebvre, W. C., 2000. *Neural and adaptive systems: fundamentals through simulations (Vol. 672)*. New York: Wiley.

Rahimi Nahoujy, M. 2020. *An Artificial Neural Network approach to model and predict asphalt deflections as a complement to experimental measurements by Falling Weight Deflectometer*. Doctoral dissertation, Ruhr-Universität Bochum, Germany.

Rumelhart, D. E.; Hinton, G. E.; Williams, R. J., 1986. *Learning internal representations by error propagation*. In: Parallel distributed processing: explorations in the microstructure of cognition, vol. 1, pp. 318–362. MIT Press.

Saltan, M., Terzi, S. 2008. *Modeling deflection basin using artificial neural networks with cross-validation technique in backcalculating flexible pavement layer moduli*. Advances in Engineering Software, 39 (7),588–592.

Wang, H., Xie, P., Ji, R., Gagnon, J. 2020. *Prediction of airfield pavement responses from surface deflections: comparison between the traditional backcalculation approach and the ANN model*. Road Materials and Pavement Design, 1–16.

Analysis and modelling of the main causes of unsatisfactory quality of transportation infrastructures

F.G. Praticò, A. Astolfi & R. Fedele
DIIES Department, University Mediterranea of Reggio Calabria, Reggio Calabria, Italy

ABSTRACT: Construction process affects the quality of a transportation infrastructure, the development of communities, and the same sustainability of the built environment. Acceptance procedures and regulations are in place, but the resulting quality is sometimes unsatisfactory. This calls for a careful analysis. Based on the above, this study aims at focusing on the key mechanisms and factors that may affect this process and at providing a useful algorithm. The problem has been modelled as a function of one dependent factor, i.e., the profit, and five independent factors, i.e., contract price, pay adjustment, construction costs, bribery, and inefficiency. Each factor has been deeply analyzed, considering main causes of contractor mismanagement, the costs related to a road infrastructure (construction, maintenance and rehabilitation), the methods for deriving penalties, and possible differences emerging from different classes of methods (i.e., empirical and statistical). Results highlight that there are no rational and legal justifications for having unsatisfactory quality of works. The only reasons refer to maliciously and/or inefficiency-related mismanagement, which implies unacceptable buyer's risks, and a paradigm shift towards inconsistency between contractor and citizens' targets. The model here set up is original and can benefit researchers, practitioners, and agencies in assessing the main causes of poor quality works and in setting up effective countermeasures.

Keywords: Mismanagement, pay adjustment, poor quality, quality characteristic, quality measure

1 INTRODUCTION

Among all the possible causes and conditions of mismanagement, corruption, incompetent or dishonest management of processes (in terms of construction costs reduction due to the use of low-quality materials), and unsatisfactory controls/acceptance procedures are very relevant (Burati et al., 2002a; Bandiera et al., 2009; Praticò et al., 2012, 2013, 2015).

According to literature (Burati et al., 2002; Bandiera et al., 2009; Praticò et al., 2012, 2013, 2015), a contractor's profit mainly depends on (cf. Figure 1A) contractor's expenses (e.g., construction costs), contract price, and pay factors (based on the acceptance plans and monitoring method used, cf. (Praticò and Astolfi, 2017; Fedele et al., 2018). Furthermore, contractor's expectations refer to profit optimization. This latter may derive from having higher pay off, lower penalties, or lower costs (cf., Figure 1B).

Despite the fact that detailed technical specifications and laws are in place, the quality of roads is often unsatisfactory (Howard, 2012), and the motivations behind this study refer to the need to analyze and point out the primary causes that may lead to poor quality work.

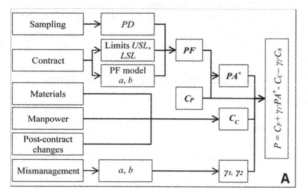 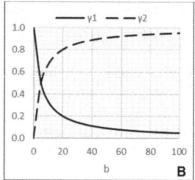

Figure 1. A) From causes to effects; B) Bribery impact.

Symbols. *PD*: percent defective. *LSL*: lower specification limit. *USL*: upper specification limit. *PF*: Payment Factor. *a* is the accept/reject factor, *b* is the bribery level. *PA**: generalized pay adjustment. C_P: contract price. C_C: construction cost. C_b: bribe cost (corruption); γ_1 and γ_2: mismanagement factors that affect contractor profit (cf. Equation 4) *P*: contractor profit (before taxes).

Consequently, the main objective is to set up a model (herein called P-model) to investigate, study in depth, and quantify the main factors that may cause poor quality work, and to provide recommendations. In order to achieve the objective mentioned above, the following tasks were carried out: Task 1: Set up of the P-model (cf. section 2). This task focused on setting up the new model. Task 2: Simplified model implementation (cf. section 3). A simple application was presented and discussed. Task 3: Detailed model implementation (cf. section 3). In this section, in order to apply the new P-model herein set up, its input PA was derived according to different methods based on the literature.

2 MODELLING (TASK 1)

The P-model aims at estimating the dependent variable "contractor's profit", *P*, as a function of several independent variables that, directly or indirectly, affect the road pavement construction process.

The analysis of the literature (in section 1) suggests that contractor profit may be affected by mismanagement. In turn, this latter may depend on the bribery level and/or the inefficiency level (Bandiera et al., 2009).

Let us suppose that the bribery level, *b* (0-100), and the inefficiency level, *i* (0-100), impact the profit, *P*, through the following factors (cf. Figure 1B):

$$\gamma_1 = \frac{\varepsilon}{\varepsilon + i + b} \text{ with } 0 \leq \gamma_1 < 1 \quad (1)$$

$$\gamma_2 = \frac{b}{\varepsilon + b} \text{ with } 0 \leq \gamma_2 < 1 \quad (2)$$

where $\varepsilon > 0$ is a factor to calibrate (e.g., $\varepsilon = 5$), and γ_1 and γ_2 are assumed to be rounded to the first decimal. If there is no bribery ($b = 0$) and no inefficiency ($i = 0$), then the first factor tends to one and the second to zero. In the opposite conditions (high bribery level and inefficiency), the first factor tends to zero while the second tends to one.

In order to derive the profit, *P*, and the profit-to-contract price ratio, P/C_P, the following expressions are herein set up:

$$P = C_P - C_C + \gamma_1 \cdot PA^* - \gamma_2 \cdot C_b \tag{3}$$

$$\frac{P}{C_P} = \frac{C_P - C_C}{C_P} + \gamma_1 \cdot \frac{PA^*}{C_P} - \gamma_2 \cdot \frac{C_b}{C_P} \tag{4}$$

where P is the profit [€], C_P is the contract price [€], C_C is the cost of construction [€], γ_1 is a factor which takes into account the mismanagement level ($\gamma_1 = 1$: correct process; $\gamma_1 \to 0$ total misleading in the acceptance process), PA^* is the generalized pay adjustment [€], γ_2 is a factor which depends on the bribery level ($\gamma_2 = 0$, no bribery; $\gamma_2 \to 1$ full state of corruption), C_b is the bribe cost (due to corruption) [€]. The generalized pay adjustment, PA^*, is herein defined as:

$$PA^* = a \cdot PA + (-1+a) \cdot \left(C'_C + C_{MIL} - PA'\right) \tag{5}$$

where a is the accept/reject factor (= 1 acceptance; = 0 reject), PA is the pay adjustment [€], C'_C is the cost of reconstruction (in case of reject of the construction, i.e. when $a = 0$) [€], C_{MIL} is the cost for milling- and landfill-related activities (in case of reject of the construction, i.e. when $a = 0$) [€], PA' is the possible pay adjustment after reject and reconstruction [€].

Note that: 1) If the construction is acceptable ($a = 1$), then $PA^* = PA$. 2) If construction is rejected ($a = 0$), then PA^* depends on the costs C_{MIL} and C'_C, and on PA'.

Finally, the Payment Factor, PF, is given by:

$$PF = 1 + PA\% = 1 + \frac{PA^*}{C_P} \tag{6}$$

PA is negative if it relates to penalties and PF is positive and ≤1. Note that the concept of profit equals the concept of surplus, after all expenses (pavement cost) are subtracted from revenue, i.e., contract price plus penalties (Muench and Mahoney, 2001; Willoughby and Mahoney, 2007). Importantly, it differs from the net income, which is the contractor profit after all expenses, including interests and taxes, are subtracted from revenue. Let g be the percentage surplus of C_P (contract price) with respect to C_{CAD} (expected, as-design value of C_C):

$$g = \frac{C_P - C_{CAD}}{C_{CAD}} \tag{7}$$

where g is expected to be around 10%.

The rationale behind the proposed model (Equations 3-7) is that: 1) The profit depends on costs (first term, C_C), penalty-mismanagement interaction (terms with γ_1 and γ_2), and contract price, C_P. 2) To study the reasons of poor/good quality works attention must be focused on how P can be increased through the modification of the terms above (particularly C_C, γ_1, and γ_2).

In practice, if the quality measure (i.e., the tool chosen to quantify quality, for example, the percentage within limits, PWL) is lower than the reject quality level (RQL, e.g., 50), then the profit is approximately given by: $P = C_P - 2C_C - C_{MIL}$ (corresponding to $PA' = 0$, $C_C \approx C_C'$, $a = 0$, $C_b = 0$, $\gamma_1 \to 1$ and $\gamma_2 = 0$). If the quality measure is higher than RQL ($a = 1$), then the profit is given by: $P = -C_C + PA + C_P$ (with to $PA' = 0$, $C_C \approx C_C'$, $C_b = 0$, $\gamma_1 = 1$ and $\gamma_2 = 1$). Ideally, the following situations can be given (see Table 1).

Importantly, the factor γ_1, herein introduced, pertains to buyer's and seller's risks. Buyer's risk (β, risk of accepting a rejectable quality level material) and seller's risk (α, risk of rejecting an acceptable quality level material) affect the average value of γ_1. Large samples size would be needed, i.e., 10 to 20 or more (Burati et al., 2002b), to minimize buyer's risk β. As a matter of fact, if controls (sample size) are maliciously minimized, γ_1 tends to zero (as above) and contractor's profit may be maximized, whatever the quality

Table 1. Parameters and results of the P-model considering the main possible situations.

Situation	Parameters	Result	Eq.
Simple acceptance	$a = 1$	$PA^* = PA = 0$	(8)
	$i \rightarrow 0$	$PF = 1 + \frac{PA^*}{C_P} = 1$	(9)
	$b = 0$	$P = C_P - C_C$	(10)
	$\gamma_1 \rightarrow 1$	$\frac{P}{C_P} \cong g$	(11)
	$\gamma_2 = 0$		
	$PA = 0$		
	$C_C \approx C_{CAD}$		
Acceptance with bonus	$a = 1$	$PA^* = PA$	(12)
	$i = 0$	$PF > 1$	(13)
	$b = 0$	$P = -C_C + C_P \cdot \left(1 + \frac{PA}{C_P}\right)$	(14)
	$\gamma_1 = 1$	$\frac{P}{C_P} = \frac{-C_C}{C_P} + \left(1 + \frac{PA}{C_P}\right)$	(15)
	$\gamma_2 = 0$		
	$PA > 0$		
Acceptance with penalty	$a = 1$	$PA^* = PA$	(16)
	$i \rightarrow 1$	$PF = 1 + PA < 1$	(17)
	$b = 0$	$P = C_P - C_C + \gamma_1 \cdot PA$	(18)
	$\gamma_1 \rightarrow 1$	$\frac{P}{C_P} = \frac{-C_C}{C_P} + \left(1 + \frac{PA}{C_P}\right)$	(19)
	$\gamma_2 = 0$		
	$PA < 0$		
Reject	$a = 0$	$PA^* = -C_C' - C_{MIL} + PA' \cong -C_C' - C_{MIL}$	(20)
	$i = 1$	$PF = 1 + \frac{-C_C' - C_{MIL} + PA'}{C_P} \cong 1 - \frac{C_C' + C_{MIL}}{C_P} < 1$	(21)
	$b = 0$	$P = C_P - C_C - C_C' - C_{MIL} + PA' \cong C_P - 2C_C - C_{MIL}$	(22)
	$\gamma_1 \rightarrow 1$	$\frac{P}{C_P} = 1 - \frac{C_C + C_C' + C_{MIL} - PA'}{C_P} \cong 1 - \frac{C_C + C_C' + C_{MIL}}{C_P}$	(23)
	$\gamma_2 = 0$		
		Note that C_C' (reconstruction) may be considered the lowest construction cost that complies with $PA \geq 0$.	
Total Mismanagement	$a = 1, i = 1$	In case of total mismanagement (PA^* is entirely not given) $PF > 1$	(24)
	$b = 1$		
	$\gamma_1 \approx 0, \gamma_2 \approx 1$	$P = C_P - C_C - C_b$	(25)

measures are. Finally, note that if $\gamma_1 = 0$ this does not imply that $P > 0$. This latter inequality depends on being $C_P > C_C + C_b$, which may be hindered by: 1) Having C_P very low due an excessive price cut. 2) Having C_P very low due to a design out-of-date. 3) Having excessive C_C. 4) Having C_b too high.

Figure 1B above illustrates how b affects γ_1, γ_2, and consequently P/C_P ($C_b = 0.05 \cdot C_P$; $\varepsilon = 5$; $i = 0$, $C_C = 8$, $C_P = 10$).

3 RESULTS: MODEL IMPLEMENTATION AND VALIDATION (TASKS 2 AND 3)

3.1 *Simplified model implementation (Task 2)*

In this section the P-model, defined above by the Equations 3-7, was implemented considering a hypothetical and simplified Pay Adjustment, *PA*, based on the percentage between limits, *PWL*, according to a simple *PF* formula (*PF*=55+0.5×*PWL*, cf. (Burati et al., 2002b; Praticò, 2007).

Table 2 reports the results of this task, which was carried out considering the five situations described in Table 1, where the first four cases are without mismanagement/bribery (b=0), while the fifth case refers to the total mismanagement (b=100). The first case refers to the acceptance with bonus (b=0), where the percentage within limits is *PWL*=100% (very high

quality) and there is a small bonus (this does not apply to the Italian laws, where bonus is not given).
- The second case refers to the acceptance without where neither penalties nor bonus are given ($b=0$), and the construction is almost perfect ($PWL=90\%$).
- The third case refers to the acceptance with penalty ($b=0$), where an appreciable part of the construction is defective, 30%, with $PWL=70\%$. In this case a penalty is given.
- The fourth case refers to the reject vase ($b=0$), where the quality of the construction is not acceptable, and the reject and reconstruction is enforced to the contractor. This implies a considerable loss of money for the contractor.
- In principle, in terms of construction quality, the fifth case is the same as the fourth one (very low quality, rejection needed), but due to mismanagement (bribery, $b=100$), the agency does not enforce the "reject and reconstruct". This could be due to the fact that the agency does not investigate properly about the quality level and does not quantify this occurrence ($PWL=40\%$). In this case, the Reject with total mismanagement ($b=100$) is given. Importantly, the profit is very high.

Table 2. Results of the simplified implementation of the P-model.

	Acceptance with bonus ($b=0$)	Simple acceptance ($b=0$)	Acceptance with penalty ($b=0$)	Reject ($b=0$)	Reject with total mismanagement in a reject condition ($b=100$)
PWL	100	90	70	40	40
C_C [€/m²]	10.42	10.00	9.17	7.92	7.92
PA [€/m²]	0.625	0	-1.25	0	0
Cb [€/m²]	0	0	0	0	0.625
a [-]	1	1	1	0	1
PA^* [€/m²]	0.63	0.00	-1.25	-12.00	-12.00
γ_1 [-]	1	1	1	1	0
γ_2 [-]	0	0	0	0	1
PF [-]	1.05	1.00	0.90	R	1.00
P [€/m²]	**2.7**	**2.5**	**2.1**	**-7.4**	**4.0**
P/C_P [-]	22%	20%	17%	-59%	32%
g [-]	25%	25%	25%	25%	25%

Hypotheses and symbols. Calibration factor=ε=5. Inefficiency=i=0. As-design (re)construction cost=C_{CAD}=Cc'=10 €/m². Reconstruction pay adjustment=PA'=0 €/m². Milling cost=C_{MIL}=2 €/m². Contract price=C_P=12.5 €/m². R=reject. PWL=percent within limits. C_C=construction cost. PA=pay adjustment. Cb=bribe cost. PF=pay factor. a=accept/reject factor. PA^*=generalized pay factor. γ_1 and γ_2=mismanagement factors. P=profit. g= percentage surplus.

3.2 Detailed model implementation (Task 3)

In this section, the P-model set up above (cf. equations 3-7) was implemented considering several models taken from the literature to derive PF and PA. It is important to underline that usually each authority has its own method for calculating penalties and PA, which are often a function of measurable quality characteristics (e.g., thickness).

For this reason, the analysis described in this section was carried out by referring to two quality characteristics (i.e., thickness and air void content) and involving a number of methods for penalty estimation, through the following steps:

- Derivation of the effective construction cost (C_C) per investigated layer. This cost is a function of the value of the quality characteristic (e.g., measured characteristic, AV_M, Burati et al., 2004).
- Definition of quality measure (e.g., Percentage Within Limits, PWL).

- Definition of the Acceptable Quality Limit, AQL, and Reject Quality Limit, RQL, for the quality measure.
- Classification of the construction as (1) Acceptable, (2) Acceptable with penalties, and (3) Rejectable.
- Derivation and application of the penalty (e.g., derived through ANAS, 2016, or through Burati et al., 2004).
- Derivation of the actual Payment Factor (PF).
- Derivation of the profit, P.

The methods below were applied to two different pavement courses, i.e., open-graded friction courses, OGFC, and dense-graded friction courses, DGFC. Measured thickness (t_M) and measured air voids content (AV_M) were used as quality characteristics (input parameters).

Results of the P-model are presented below in terms of PF (cf. Equation 6) and profit-to-contract price ratio (P/C_P, cf. Equation 4), using to the following methods to derive PA:

1) ANAS's method (AN) (ANAS, 2016).
2) Burati's method (BN) (Burati et al., 2002b).
3) Elyamany-Attia's method (E&A)(Elyamany, 2013).
4) CIRS's method (CIRS) (Ministero dei Lavori Pubblici, n.d.).
5) Calgary's method (CAL)(Calgary, 2012).
6) Pellinen's method (PM) (Pellinen et al., 2005).

Note that 1) Burati's method was indicated with BN because of the fact that the normal probability density function, N, was used. 2) The method provided by (Pellinen et al., 2005) was adapted to the common as-design values of asphalt concretes, and is herein termed PM.

Figure 2 illustrates how (PF, Equation 6) and profit (P, Equation 4) depend on quality characteristics (i.e., thickness and air void content), based on different methods (e.g., Burati's method), and on sampling variance. In particular, Figures 2A-2F refer to nonconformities of measured thickness (t_M; see Figures 2A, 2C and 2E) and air void content (AV_M; see Figures 2B, 2D and 2F) for OGFCs, while Figures 2G and 2H refer to the same nonconformities for DGFCs. Importantly, AV_M variations are due to roller passes.

For each method, Figure 2A shows the Payment Factor (PF, y-axis) as a function of the measured thickness (t_M, x-axis) of an OGFC, with a standard deviation of 5 mm. In addition, the Reject Quality Limit (RQL) is marked with a circle. For example, for PF_{BN}, $RQL = 50$ and corresponds to an average thickness of 41 mm. It can be noted that the maximum PF is obtained when the Burati's method is used, while the minimum PF is given by the Calgary's method.

Figure 2B shows the Payment Factor (PF) as a function of AV_M of an OGFC with a standard deviation of 0.5%. In this case, the maximum PFs are provided by the Burati's and the Elyamany-Attia's methods (i.e., PF_{BN} and $PF_{E\&A}$, respectively), while the minimum PF is given by the CIRS's method (i.e., PF_{CIRS}).

Figure 2C shows the profit-to-contract price ratio (P/C_P) as a function of the measured thickness (t_M) of an OGFC with a standard deviation of 5 mm ($\gamma_1 = 1$, $\gamma_2 = 0$). It should be noted that the maximum profit is obtained using Burati's method (i.e, P_{BN}), and Pellinen's method (i.e., P_{PM}). On the contrary, the minimum profit is reached when the Calgary's method is applied (i.e, P_{CAL}). Note that if $\gamma_1 = 0$ (mismanagement) and $C_C > C_P$, then P/C_P monotonically decreases (if thickness increases; cf. Figure 2C). This behavior is critical and may explain a tendency to produce low-quality infrastructures which, in turn, may impact life line performance (Praticò et al., 2012; Marcianò et al., 2015), and how efficient is a nation in providing public services (Bandiera et al., 2009). It seems critical to point out that if $\gamma_1 = 1$ there is a win-win situation (contractor's and dwellers interest superpose, and prefer zone I in Figure 2E), while if $\gamma_1 = 0$ dwellers and contractor's targets are opposite. In Figure 2C, additionally, the P/C_P curve for $\gamma_1 = 0$ and $\gamma_2 = 0$ is represented. It refers to the case in which the C_C corresponding to the as-design thickness is lower than C_P. This implies that the lower the thickness is, the higher the P/C_P ratio becomes. The case represented refers to $C_C = 0.8C_P$ (cf., "mismanagement").

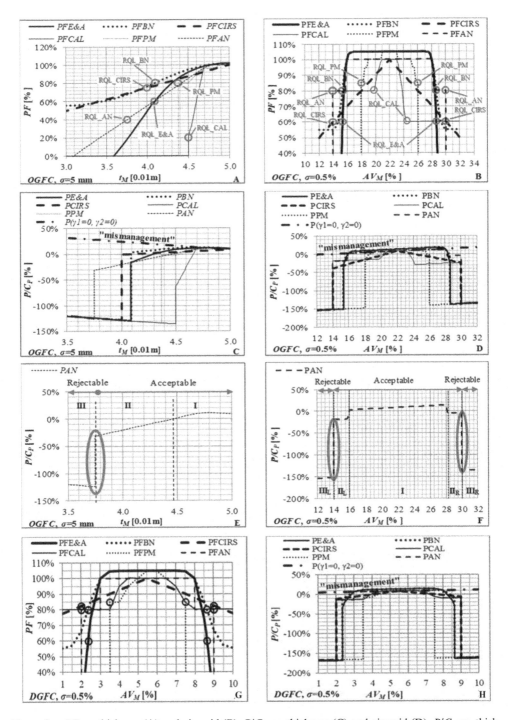

Figure 2. *PF* vs. thickness (A) and air void (B), *P/C$_P$* vs. thickness (C) and air void (D), *P/C$_P$* vs. thickness (E) and air void (F) applying the ANAS's method, P_{AN}, for OGFC, and *PF* vs. air voids (G) and *P/C$_P$* vs. air voids (H) for DGFC.

Symbols. t_M: measured thickness, [cm]; AV_M: measured air voids, [%]; σ: standard deviation, [cm]; C_P: contract price, [€/m^2]; *PFn*: Payment Factor, [%], and *Pn*: profit, [€/m^2], with n = E&A, BN, CIRS, CAL, PM, AN, which stand for the methods: Elyamany-Attia (2013), Burati (2003, normal distribution), CIRS, The City of Calgary (2012), modified Pellinen (2005) and ANAS (2010), respectively; $P(\gamma_1=0, \gamma_2=0)$: P in case of "mismanagement" [€/m^2] derived using the P-model.

Figure 2D shows P/C_P ratio as a function of AV_M of an OGFC reference road pavement with a standard deviation of 0.5%. The behavior of the methods is similar in this case, except for the Pellinen's and the Calgary's methods (i.e., P_{PM} and P_{CAL}, respectively). The highest profits can be obtained using the Burati's and the Elyamany-Attia's methods (i.e., P_{BN} and $P_{E\&A}$, respectively). The lowest profit is given by the Calgary's method (i.e., P_{CAL}). Note that each PF (or P/C_P) curve may have the following domains (Figure 2E): 1) A first domain, I, in which $P > 0$, and there may be the absolute maximum of P/C_P (e.g., 45-50 mm); 2) A second domain, II, (e.g., 38-45 mm) in which $P < 0$ but the layer is acceptable and the higher is the thickness, the higher is the profit; 3) A third domain, III, in which $P < 0$, the layer is rejectable, but the lower is the thickness, the higher is the profit.

Figure 2F shows that in the case of AV_M (with $\gamma_1 = 1$, $\gamma_2 = 0$), the three zones almost duplicate (I, II$_L$, III$_L$ and II$_R$, III$_R$, where L stands for left and R for right). Importantly, it may happen that $\gamma_1 = 0$ (mismanagement), for example, because of $C_C > C_P$, and the contractor may decide, for example, to reduce the filler content, having a higher AV (right hand of Figure 2F). Again, this would imply a paradigm shift (zone I to the right side of the plot instead that in the middle).

Figures 2G and 2H show PF and P/C_P as a function of AV_M of a DGFC with a standard deviation of 0.5%. Also in this case, the maximum profits are gained applying the Burati's and the Elyamany-Attia's methods (i.e., P_{BN} and $P_{E\&A}$, respectively), whereas the minimum profit is given by the ANAS's method (i.e., P_{AN}). The mismanagement curve ($\gamma_2 > 0$, Figure 2H) points out two concepts 1) Bribery may imply a positive P/C_P for a wide range of AV_M; 2) Situations in which high AV_M are caused by local scarcity of fillers (sandy soils), and absence of countermeasures can be easily "tolerated" in case of irregularities in public procurements ($\gamma_1 = 0$). Finally, it seems relevant to observe that the relationship between construction costs and AV_M depends on the reason of AV_M variations.

Figure 3 illustrates how the causes of AV_M variation can affect the relationship above. In particular, the reduction of filler percentage (i.e., P200) causes a quasi-linear variation of costs with AV_M (the lower filler percentage is, the higher AV_M is, and the lower C_C is). For roller passes (N), a lower number of roller passes, implies lower C_C and generally higher AV_M. Note that, in each plot, curves intersect at the design point (AV_M, C_C), i.e. 22.0%, 9.81 €/m² (OGFC, Figure 3A), and 5.5%, 7.34 €/m² (DGFC, Figure 3B).

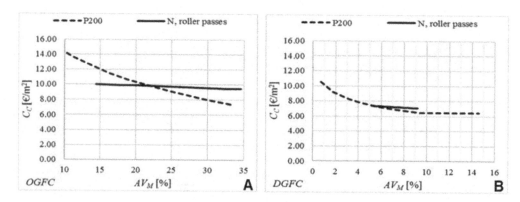

Figure 3. AV_M vs. construction cost for OGFC (A), and DGFC (B). Symbols. C_C: construction cost; AV_M: measured air voids, [%]; DGFC: dense-graded friction courses; OGFC: open graded friction courses; P200: passing the 0.075 mm (No. 200) sieve; N: number of roller's passes.

4 CONCLUSIONS AND RECOMMENDATIONS

In this study, the primary causes that lead to an unsatisfactory quality of transportation infrastructures were analyzed and an innovative model was setup and implemented to deruve the profit. The limitations of the study refer to the lack of data to validate appropriately the method and appear hardly to overcome because of the critical context. This notwithstanding, data available are consistent with the model which provides a realistic tool to approach the problem and try to realistically solve it. The following conclusions can be drawn:

1. Data collected and analyses demonstrate that if quality assurance controls are really carried out, contractor best option still remains to improve process performance and materials quality in order to be enough close to targets and, consequently, to achieve a full compensation (Payment Factor, $PF \geq 1$).
2. Indeed, under the hypotheses that acceptance procedures are properly carried out, contractors may optimize their profit if they reduce the variability of product characteristics, and if they better target values (quality characteristics and quality measures) that are close to the limits.
3. In contrast, there are no rational justifications for having quality measures far from being acceptable (unsatisfactory quality of works). The only justification refers to maliciously (bribery level, $b \gg 0$, and bribe costs, $C_b > 0$) and/or inefficiency-related (inefficiency level, $i \gg 0$) mismanagement in sampling plans ($\gamma_1 \to 0$), which implies unacceptable buyer's risks, and a paradigm shift towards inconsistency between contractor and citizens' targets. This point is very crucial and two basic questions here emerge that call for high-level solutions: A) What are the reasons to move towards "mismanagement" (e.g., very low contract prices)?. B) how can contractor mismanagement strategy be (un)successful?
4. To this end, attention should be focused on conflicts of interest between who has to control (e.g., Director) and who is controlled (e.g., Contractor), as per the abovementioned factors γ_1 and γ_2).

Based on the conclusions reported above, the following recommendations can be listed:

- Controlling that contractor and supervisor do not have conflicts of interest.
- Carrying out proper sampling.
- Controlling that supervisors have an outstanding technical and deontological level.
- Having contract specifications very clear, very detailed, and with bonuses: a major push towards a win-win situation for stakeholders: citizens, contractor, and agency.
- Minimizing the mismanagement risks by involving third parties (quality assurance).

Even though results show that contractor can maximize the profit paying attention on product characteristic and limits, future work will be carried out to: 1) Demonstrate that innovative methods for the assessment and the monitoring of the road pavement characteristics can improve the estimation of the penalties. 2) Find and compare other methods for the estimation of the penalties to define the most convenient for both the contractor and the agency.

REFERENCES

ANAS, 2016. Capitolato Speciale di Appalto - Norme Tecniche per l'esecuzione del contratto - Parte 2.
Bandiera, O., Prat, A., Valletti, T., 2009. Active and passive waste in government spending: Evidence from a policy experiment. American Economic Review 99, 1278–1308. https://doi.org/10.1257/aer.99.4.1278
Burati, J.L., Weed, R.M., Hughes, C.S., Hill, H.S., Kopac, A., 2002a. Optimal Procedures for Quality Assurance Specifications.
Burati, J.L., Weed, R.M., Hughes, C.S., Hill, H.S., Kopac, A., 2002b. Optimal Procedures for Quality Assurance Specifications.
Calgary, C. of, 2012. Standard Specifications - Roads Construction, City.
Elyamany, A.H., 2013. Using MEPDG to Develop Rational Pay Factor for Hot Mix Asphalt Construction. IOSR Journal of Mechanical and Civil Engineering 10, 30–37. https://doi.org/10.9790/1684-1043037
Fedele, R., Merenda, M., Praticò, F.G., Carotenuto, R., Della Corte, F.G., 2018. Energy harvesting for IoT road monitoring systems. Instrumentation Mesure Metrologie 17, 605–623. https://doi.org/10.3166/I2M.17.605-623

Howard, C., 2012. Three Ways Contractors Still Cheat Each Other. Construction Management.
Ministero dei Lavori Pubblici, n.d. CIRS (Centro sperimentale Interuniversitario di Ricerca Stradale). Norme tecniche di tipo prestazionale.
Muench, S.T., Mahoney, J.P., 2001. A Quantification and Evaluation of WSDOT's Hot Mix Asphalt Concrete Statistical Acceptance Specification, Research Project Agreement T9903.2, Task A5 Asphalt Concrete Specifications A.
Pellinen, T.K., Weiss, J., Kuczek, T., Dauksas, G., 2005. Comparison of Various INDOT Testing Methods and Procedures to Quantify Variability in Measured Bituminous and Concrete Properties. https://doi.org/10.5703/1288284313159
Praticò, F.G., 2007. Quality and timeliness in highway construction contracts: A new acceptance model based on both mechanical and surface performance of flexible pavements. Construction Management and Economics 25, 305–313.
Praticò, F.G., Astolfi, A., 2017. A new and simplified approach to assess the pavement surface micro- and macrotexture. Construction and Building Materials 148, 476–483.
Praticò, F.G., Vaiana, R., Gallelli, V., 2012. Transport and traffic management by micro simulation models: Operational use and performance of roundabouts. In: C. A. Brebbia, J. W. S. Longhurst, W. P. (Ed.), WIT Transactions on the Built Environment. 383–394. https://doi.org/10.2495/UT120331
Praticò, F.G., Vaiana, R., Giunta, M., 2013. Pavement Sustainability: Permeable Wearing Courses by Recycling Porous European Mixes. Journal of Architectural Engineering 19, 186–192. https://doi.org/10.1061/(asce)ae.1943-5568.0000127
Praticò, F.G., Vaiana, R., Iuele, T., 2015. Macrotexture modeling and experimental validation for pavement surface treatments. Construction and Building Materials 95, 658–666.
Willoughby, K., Mahoney, J.P., 2007. An Assessment of WSDOT's Hot-Mix Asphalt Quality Control and Assurance Requirements.

Influence of acceleration and deceleration of Super Heavy Loads (SHLs) on the service life of pavement structures

Ali Morovatdar
Graduate Research Assistant, Department of Civil engineering, University of Texas at El Paso, Texas, USA

Reza S. Ashtiani
Associate Professor, Department of Civil engineering, University of Texas at El Paso, Texas, USA

ABSTRACT: Super Heavy Load (SHL) vehicles typically consist of heavy axles that impart substantial damages on transportation facilities such highway pavements and bridges. The taxing loading conditions, when combined with acceleration and deceleration forces from SHL vehicles, results in substantial interfacial shear stresses that manifests itself in premature failure of transportation facilities. This was the motivation for our research team to explore the detrimental effects of the acceleration and deceleration on the service life of roadways. To achieve this objective, our research team, initially deployed portable Weight-in-Motion (WIM) devices to ten sites with high frequency of SHLs in overload corridors of east and southeast Texas. Subsequently, nondestructive field tests such as Falling Weight Deflectometer (FWD) and Ground Penetrating Radar (GPR) were performed during summer and winter months for the back-calculation of the layer moduli in ten sites. The field observations and measurements were in turn incorporated in a three-dimensional (3D) finite element code for characterization of the influence of the acceleration and deceleration on the mechanical responses of the pavement structures. Then, a series of scenarios consisted of different patterns of acceleration, and deceleration were simulated in the 3D finite element software for ten sites and the results were contrasted with the steady rolling conditions. The sensitivity analyses of the representative pavements sections in this study showed that the deceleration or braking forces from the SHL vehicles resulted in the development of significant shear stresses in surface layers of thin structures. The stresses imparted by sudden changes in the velocity of super heavy vehicles can potentially jeopardize the longevity of the pavements structures and result in premature failure of transportation facilities. Therefore, analysis of acceleration and deceleration of the SHL vehicles should be an integral component in risk management studies of overload corridors.

Keywords: Super Heavy Load (SHL), Overload Corridors, Pavement Service Life, Acceleration and Deceleration Forces, Finite Element Method

1 INTRODUCTION

Recent traffic trends and permit issuance records indicate significant demands for operation of the Super Heavy Load (SHL) vehicles in the energy sector and overload corridors of Texas. This is primarily due to the recovery of the state's economy, as well as drastic improvements in freight transportation in overweight corridors. Essentially, movement of the SHL vehicles is

DOI: 10.1201/9781003222880-31

attributed to a variety of economic activities such as energy production, freight, and transportation of heavy machinery and equipment such as well servicing units, power transformers, and electric generators. Despite facilitating the movement of heavy, large, and non-divisible loads, SHL vehicles typically consist of multi-axle trailer units that weigh several times of the permissible weight limits set forth by regulatory agencies. Operation of these non-conventional vehicles with heavy tires and complex axle arrangements, has been a major source of damages for pavement structures in the network.

The detrimental impact of SHLs on transportation infrastructure is even more pronounced with consideration of the acceleration and deceleration forces applied to the roadway structures. Evidently, this is an ongoing statewide challenge for design agencies across the nation and worldwide, with limited precedence in the literature. Therefore, there is an urgent need to properly assess the distresses imparted by SHL vehicles at multiple loading scenarios, such as sudden acceleration and deceleration, with consideration of seasonal effects on material properties, and type of the pavement facility in the impacted networks. The results can potentially provide insights on the Pavement Life Reduction (PLR) studies, and provide means to improve pavement design protocols.

Several researchers studied different methodologies to evaluate the service life of pavements subjected to SHL vehicle operations. Dong and Huang (2013) evaluated the detrimental impacts of SHLs on pavements in Tennessee through field measurements of the pavement responses. Based on field measurements of the pavement surface deflection, the authors concluded that the evaluated SHLs did not cause considerable deformation on the studied flexible pavement. Chen et al. (2013), using numerical simulations, investigated the consumption of pavement service life due to movement of a SHL vehicle in Louisiana. The researchers converted the SHL axles to equivalent number of standard axles to characterize the loading conditions in their analysis. The authors in turn used this information to estimate the costs associated with the pavement damages. In a recent study, Hajj et al. (2018) developed a mechanistic framework to investigate the impacts of the SHLs on pavements on a case-by-case basis. The researchers used records of issued permits as the primary source to characterize the traffic distributions, and loading conditions attributed to SHLs. In the numerical simulation phase, the researchers considered uniform distribution of contact stresses over a circular loaded area.

Several researchers investigated the influence of acceleration and deceleration of the conventional vehicles on the tire-pavement contact stresses (Wang et al., 2012, and Hu et al., 2017, Satvati et al., 2021). Based on the numerical simulations, the authors reported that decelerating (braking) vehicles can result in higher contact stresses and therefore higher pavement responses under sudden braking conditions.

The majority of the previous research studies for predicting the pavement service life either rely on the general observations and measurements made in the field, or are based on limited sections and data points. The simplifying assumptions such as the use of permit records in lieu of field data collection, limitation of type of pavement facilities in the study, overlooking the influence of seasonal variation material properties, and unique characteristics of the pavement structure in each location can potentially jeopardize the accuracy and reliability of the damage quantification and remaining life analyses of pavement facilities. Another anomaly persistent in the literature pertains to unrealistic simulation of the tire-pavement contact stresses using uniformly distributed load, rather than considering the non-uniform distribution of the contact stresses. Relying on such simplifying assumptions can be detrimental to the accuracy of the analysis of structural impacts of SHL vehicles with demanding loading conditions. In addition to the elaborated shortcomings, acceleration and deceleration of the non-conventional SHLs with taxing loading conditions, as well as their impacts on the pavement service life is often overlooked in the literature.

The highlighted issues and concerns were the motivations for our research team to devise a protocol study to assess the influence of acceleration and deceleration of SHL vehicles on the imparted damages and loss of service life of pavements for several Farm-to-Market (FM) roads, State Highways (SHs), and US Highways in ten overload corridors. The developed mechanistic framework, accounts for the demanding loading conditions attributed to SHLs, site-specific Axle Load Spectra (ALS) databases, various acceleration/deceleration patterns, realistic tire-pavement interactions, and unique features of pavement facilities in the network.

2 RESEARCH METHODOLOGY

Figure 1 shows the flowchart of the proposed procedure for the mechanistic characterization of the loss of pavement service life due to acceleration and deceleration of the SHL vehicles. As illustrated in the figure, the proposed analysis framework consisted of three main segments including: (1) characterization of the traffic loading conditions and pavement material properties from field testing, (2) quantification of damages imparted on the pavements using 3D FE modeling, and (3) prediction of the service life of pavements.

To characterize the input parameters, initially, the pavement layer properties were obtained from analyzing the Ground Penetrating Radar (GPR) and Falling Weight Deflectometer (FWD) field data. The pavement design plans were further instrumental in validating or supplementing the information on the material properties, and layer configurations. Subsequently, the authors deployed Portable Weigh-In-Motion (P-WIM) devices to collect site-specific ALS during summer and winter months, for accurate characterization of traffic in ten representative sites. GVWs, vehicle classifications, axle load distributions, axle weights, axle configurations, and wheel loads were the most relevant traffic information that were extracted from the ALS databases. The authors further developed a procedure for routine measurements of tire pressure and tire footprint in the calibration process of P-WIM devices to achieve an accurate account of the loading conditions. Additionally, the research team conducted an extensive data mining on the permit records and SHL vehicle plans to complement the information on the axle and tire characteristics of the SHLs operating in Texas overload corridors.

Subsequently, the site-specific pavement layers properties and the traffic loading information were in turn incorporated into a 3D FE program for determination of the critical pavement responses under the SHL tires at various rolling conditions, i.e., steady rolling, acceleration, and deceleration (braking). The calculated pavement responses were further incorporated into the mechanistic procedure for pavement damage quantification described in Morovatdar et al. (2020a) to determine the corresponding Equivalent Axle Load Factors (EALFs) tailored towards the specific SHL axle, roadway type, and season of the year.

As stated earlier, the last stage of the analysis pertains to characterization of the influence of tire rolling condition of the SHLs on the pavement service life. To accomplish this, the authors initially incorporated the projected Equivalent Single Axle Load (ESAL) values over a 20-year design life for three evaluated loading scenarios, i.e., ESAL $_{(steady\ rolling)}$, ESAL $_{(acceleration)}$, and ESAL $_{(deceleration)}$, to characterize the traffic in numerical simulations for further comparison purposes. The cumulative 18-kip (8.16 tons) ESAL values were calculated from Equation 1 as:

$$ESAL = \sum_{i=1}^{m} (EALF)_i n_i \qquad (1)$$

where, $(EALF)_i$ represents the calculated EALF value for i^{th}-axle; and n_i is the projected number of passes of ith-axle load group during the design period. As shown in Equation 1, pavement damage equivalency factors, calculated at the second stage of the analysis, are integral components for accurate characterization of the accumulated pavement damages. The number of axle load repetitions were also derived from the site-specific ALS databases. Then, the EALF values for each axle type and axle weight were multiplied by the projected number of load repetitions to calculate the projected ESAL values considering the design life of pavements.

Ultimately, the research team incorporated the ESAL values into the ME pavement design software for comparative analysis of service of life of pavements for the aforementioned three sets of traffic data. Climatic data from an adjacent weather station, as well as performance criteria limits, were also assigned using the agency defined thresholds. The analysis results were contrasted with the steady rolling conditions which serve as the benchmark in this study. Therefore, the difference between the service life analyses represents the reduction of the service life associated with the acceleration/deceleration of the SHL vehicles. Additionally, "*pavement life reduction*" index was calculated to provide a quantitative measure of the severity of the distresses imparted by acceleration/deceleration of the SHL vehicles in the network as:

$$\text{Pavement Life Reduction (PLR) Index} = \frac{PL_{steady\ rolling} - PL_{acceleration/deceleration}}{PL_{steady\ rolling}} \times 100 \quad (2)$$

where, $PL_{steady\ rolling}$, $PL_{acceleration}$, and $PL_{deceleration}$ represent the expected pavement life considering the steady rolling, acceleration, and deceleration conditions of the SHL tires, respectively.

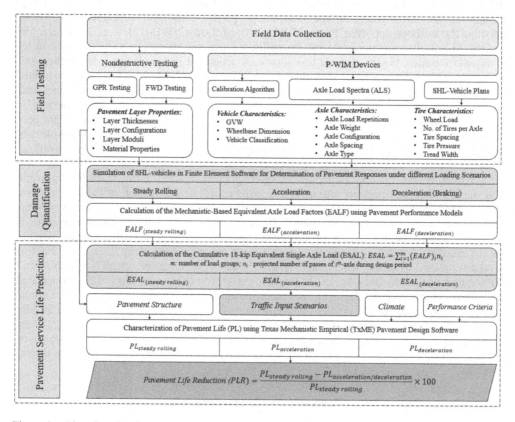

Figure 1. Flowchart for the proposed mechanistic approach for the quantification of the pavement life reduction (PLR) due to acceleration and deceleration of SHL vehicles.

2.1 Field data collection

The site-specific information was collected for ten representative sites strategically distributed throughout the overload corridors in Texas including Corpus Christi, Laredo, Yoakum, and San Antonio Districts. The authors deployed GPR and FWD to the field for the determination of the layer profile and back-calculation of layer moduli in representative pavement sections, as shown in Figures 2(a) and 2(b). Analyzing the GPR data validated by pavement design plans showed that representative pavement sections in FM roadways consisted of less robust pavement profile with thin structure, compared to more robust pavement structures of SH and US highways (Morovatdar et al., 2019, 2020b, 2020c). The authors also developed field testing plans for FWD testing in two different seasons (summer and winter), to properly evaluate the effect of environmental parameters on the back-calculated layer modulus of pavement layers.

The P-WIM devices were also deployed in selected FM, SH, and US roadways to collect the site-specific traffic information. The P-WIM units were temporarily installed at each location and were left to continuously record traffic information for at least two weeks in each site. The process was repeated for both summer and winter times to capture the seasonal effect of traffic variations. Figure 2(c) shows the piezo-electric sensors inserted into specialized pocket tapes, as well as the data acquisition system, as principal components of the P-WIM units deployed for characterization of the ALS in the field.

  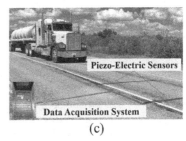

(a)   (b)   (c)

Figure 2. Data collection effort (a) GPR testing, (b) FWD testing, (c) P-WIM deployment.

The analysis of the ALS indicated high number of SHL operations in the surveyed network. Table 1 shows the SHL axle ranges for different axle group types, based on field data. Legal and the maximum permissible weight limits for trucks operating in Texas are also provided for comparison purposes. Axle weights heavier than legal limits are primarily considered as overweight, while the axles that exceed the maximum allowable weight limits are essentially regarded as super heavy loads. As evidenced in Table 1, the maximum recorded axle loads substantially deviate from the maximum permissible limits. The heaviest axle load was found to be 63.5 tons for quad axles, as recorded during the data collection in SH 123 in Corpus Christi. Relevant information on extensive field data collection efforts, and the synthesized findings attributed to all ten representative sites, are provided in Ashtiani et al. (2019).

Table 1. Super Heavy Loads Collected by P-WIM Devices.

Axle Type	Axle Weight, *lb. (tons)*		
	Legal Weight Limit (OW Threshold)	Maximum Permissible (SHL Threshold)	Maximum Recorded Axle Weights in Field Trials
Single Axle	20,000 (9.1)	25,000 (11.3)	42,000 (19.1)
Tandem Axle	34,000 (15.4)	46,000 (20.8)	120,000 (54.4)
Tridem Axle	42,000 (19.1)	60,000 (27.2)	114,000 (51.7)
Quad Axle	50,000 (22.7)	70,000 (31.8)	140,000 (63.5)

2.2 Finite element modeling

In this study, ABAQUS FE program was used for determination of the SHL-induced pavement responses under various tire rolling conditions, with considerations of the viscoelastic nature of Asphalt Concrete (AC) layer, complex traffic loading conditions, and tire-pavement interactions. Subsequent sections provide the relevant information on the 3D FE models developed in this study for simulation of the pavement structures, material behavior models, SHL tires at acceleration/deceleration conditions, and tire-pavement interactions. Detailed information on input parameters including material models, material properties, and traffic characteristics are provided in Ashtiani et al. (2019).

2.2.1 Simulation of the pavement structure

Based on the nondestructive testing results, pavement structures consisting of AC, base, and subgrade layers were modeled in the software. The site-specific information on the pavement layer thicknesses, layer configurations, as well as the layer moduli were incorporated into the developed FE code to simulate the structural characteristics of the representative pavement sections. The viscoelastic properties of the AC surface layers were also incorporated in the analysis to account for the influence of temperature and frequency of loading in this study.

2.2.2 AC viscoelastic behavior modeling

The structural capacity of multi-layer pavement structures is not monolithic under different loading frequencies and different temperatures (Muzenski et al. 2020, Salimi et al. 2019; 2021). This is primarily attributed to the time- and temperature-dependent behavior of the viscoelastic AC layers. In this study, the viscoelastic properties of the AC surface layers were incorporated in the analysis to account for the influence of temperature and frequency of loading. Hence, the generalized Maxwell model was used to characterize the viscoelastic properties of the AC layers through FE numerical simulations in this study, as presented in the following equations:

$$s = \int_{-\infty}^{t} 2G(t-\tau)\frac{de}{d\tau}d\tau \qquad (3)$$

$$p = \int_{-\infty}^{t} K(t-\tau)\frac{d(tr[\varepsilon])}{d\tau}d\tau \qquad (4)$$

where s is deviatoric stress, e is deviatoric strain, p is volumetric stress, $tr[\varepsilon]$ is trace of volumetric strain, t is relaxation time, and K and G are the bulk and shear moduli of AC layer, respectively. Bulk (K) and shear (G) moduli of AC were obtained from laboratory dynamic modulus tests. The Prony series were then used to calculate the corresponding modulus values in time domain to satisfy the FE software requirements, as indicated through Equations 5 and 6:

$$G(t) = G_0\left[1 - \sum_{i=1}^{n} G_i\left(1 - e^{-t/\tau_i}\right)\right] \qquad (5)$$

$$K(t) = K_0\left[1 - \sum_{i=1}^{n} K_i\left(1 - e^{-t/\tau_i}\right)\right] \qquad (6)$$

where G_0 and K_0 are instantaneous shear and volumetric elastic moduli; and G_i, K_i, and τ_i are the Prony series parameters. The information on the dynamic modulus tests, as well as the Prony series parameters, were obtained from a relevant study conducted by Hu et al. (2017).

2.2.3 Simulation of the SHL tire

The tire element, including tire ribs and grooves, was explicitly modeled with consideration of the characteristics and tread patterns of SHL-vehicle tires operating in Texas overload corridors. Figure 3 illustrates the cross-sectional and 3D views of the modelled tire. Based on the information extracted from the review of the SHL vehicle plans and permit records in Texas, it was found that the SHL tire's height ranged from 28 in. to 35 in. (71 cm to 88 cm), while tread width of the tires varied between 6 and 10 in. (15 cm and 25 cm), depending on the tire configuration and axle assembly of different SHL vehicles. The elastic material properties of the tire element were also defined based on the relevant information provided in the literature (Wang et al., 2012).

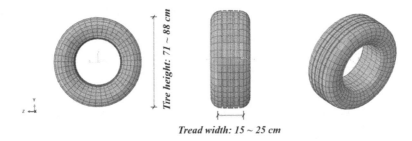

Figure 3. Simulation of the SHL tire using 3D FE models.

2.2.4 *Tire-pavement contact stresses*

Realistic simulation of the tire-pavement contact stresses is of greatest importance for accurate assessment of the SHL-induced pavement responses. Figure 4 provides an example associated with the simulated tire-pavement contact stresses under a SHL tandem-axle with 120 kips (54.4 tons) axle weight and 15 kips (6.8 tons) load magnitude on each individual tire. As shown in Figure 4, the simulated tire element in this study properly captured the non-uniform distribution of the tire-pavement contact stresses under each individual tire rib. The characterized convex-shape distribution pattern of contact stresses along the tire contact length is consistent with the experimental measurements and observations made by previous researchers (Douglas et al., 2000; Wang and Al-Qadi 2009). Based on the contract stress distributions provided in Figure 4, the maximum calculated contact stress was 180 psi (1,241 kPa) within the middle tire ribs; while, the average value was calculated as 125 psi (861 kPa). This average value for contact stresses is in line with our research team field measurements of tire pressure in Texas (Ashtiani et al., 2019).

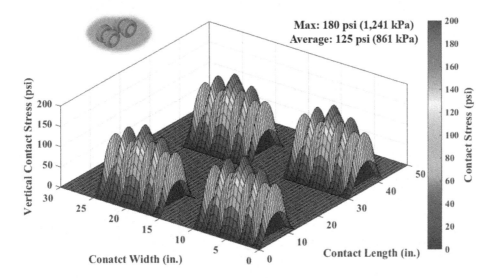

Figure 4. Non-uniform tire-pavement contact stresses under a 120-kips (54.4 tons) SHL tandem-axle.

It should be also noted that a friction coefficient of 0.5, as recommended by Wang et al. (2012), was incorporated into the numerical simulations to account for the frictional effects at the tire-pavement interface. To accomplish this, the penalty method, as well as the Coulomb friction law were used to simulate the normal and tangential interactions, respectively.

2.2.5 Various tire rolling conditions

The authors used relevant user subroutines and dynamic analysis offered in ABAQUS to simulate the tire rolling procedure. Translational speed (v), as well as the corresponding rotational speed (ω), were assigned to the SHL-tires to simulate the steady rolling condition. The kinematic coupling constraint method was also used to allow movement of the tires along traffic direction. Furthermore, translational acceleration (a) and rotational acceleration (a) were separately defined to account for the additional driving/braking torques applied on the SHL tire under acceleration/deceleration conditions. Relatively low vehicle speed (i.e., $v = 32$ kph, $\omega = 11.7$ rad/s) to account for slow-moving nature of SHLs, and different acceleration/deceleration levels (i.e., a = 0.6 m/s^2, 1.5 m/s^2, 3.5 m/s^2; a = 1.2 rad/s^2, 2.9 rad/s^2, 5.8 rad/s^2) were incorporated into the comprehensive sensitivity analyses conducted in this study.

3 FIELD DATA ANALYSIS AND RESULTS

3.1 Influence of SHL acceleration/deceleration on damage factors

Pavement damage factors, namely EALFs, allow for the quantification of the pavement damages per pass relative to a standard 18-kip single axle. The EALF values attributed to the various tire rolling conditions (i.e., steady rolling, acceleration, and deceleration) were contrasted with each other for comparison purposes in Figure 5. The results are specifically tailored towards the SHL tandem-axle load group in State Highways. As evidenced in the plot, in all axle weights, decelerating (braking) tires had the highest damage factors, followed by accelerating and ultimately steady rolling tires. This is primarily attributed to the fact that when a tire is under deceleration/acceleration conditions, the resulting longitudinal tire-pavement contact stresses significantly increase due to the frictional interactions at the tire-pavement interface. Such excessive contact stresses lead to development of the significant shear stresses in the pavement structure, which in turn translates into higher shear rutting in pavements. Consequently, damage assessment and service life analysis protocols should properly account for deceleration and acceleration of the SHLs and their detrimental impacts on the longevity of pavement structures. It is also noted that similar EALF trends were observed from the numerical simulations for various axle types and roadways evaluated in this research study.

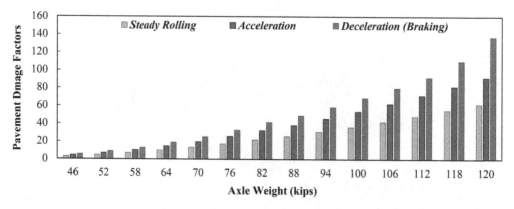

Figure 5. Pavement damage factors for various tire rolling conditions of SHL tandem-axles.

3.2 Influence of SHL acceleration/deceleration on pavement service life

Figure 6 provides the results pertaining to one of the heavily trafficked sections, SH123 in Corpus Christi, as an example to showcase the analysis protocol developed in this study for mechanistic quantification of the loss of pavement service life. The results are based on frequent application of SHL vehicles as captured by P-WIM devices. The plot shows the rutting performance over a 20-year design period for SH123, considering three evaluated tire rolling scenarios. Based on the internal distress calculation models in mechanistic analysis, and steady-rolling based damage factors, it takes 168 months for SH123 to develop 0.5 in. (12.7 mm) of rut depth. However, if the acceleration-based damage factors were incorporated into the analysis framework, it takes 135 months to develop 0.5 in. (12.7 mm) of cumulative rut depth. In other terms, inclusion of acceleration patterns of SHL vehicles in SH123 can essentially consume (168-135 = 33) months of the service life of this pavement section. Additionally, deceleration considerations of the SHL vehicles in SH123 can result in 63 months reduction in its pavement service life. Consequently, the analysis of the loss of pavement service life for SH123 indicated that acceleration and deceleration of the SHL vehicles operating in this roadway can impart significant PLRs as high as 20%, and 38%, respectively.

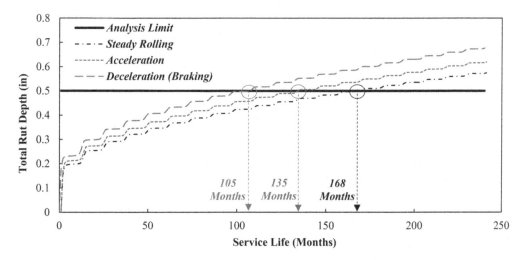

Figure 6. Pavement life reduction for SH123 due to acceleration and deceleration of SHLs.

3.3 Pavement life reduction for various roadway types

Figure 7 illustrates the results associated with the pavement life reduction for various roadway types (i.e., FM, SH, and US highways) due to acceleration and deceleration of SHL vehicles. According to the sensitivity results of multiple loading scenarios, FM roadways with less robust pavement structures were found to be more sensitive to the acceleration and deceleration of SHL vehicles, compared to the representative pavement sections in SH and US highways with higher structural capacity. This highlighted the importance of the pavement profile on the analysis of the acceleration and deceleration of the SHL vehicles.

The results illustrated in Figure 7 also underscored the significance of sudden variations in SHL vehicle speed. According to the results provided in Figure 7, frequent acceleration of SHL vehicles in US, SH, and FM roadways can potentially consume 18%, 20%, and 26%, respectively, of service life of their pavement sections; while, deceleration of such vehicles can result in PLRs as high as 32%, 38%, and 42% for US, SH, and FM roadways, respectively. Therefore, pavement structures servicing overload corridors, particularly at intersections or stop–go sections with significant frequencies of braking, can potentially experience excessive distresses due to absence of a protocol to account for horizontal and tangential forces exerted during acceleration and deceleration scenarios during the pavement design process.

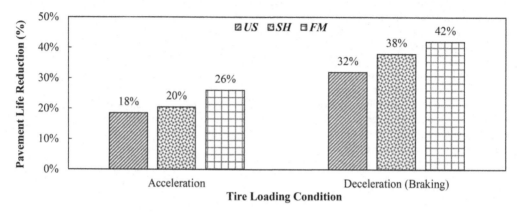

Figure 7. PLR due to acceleration/deceleration of SHLs for different roadway types.

4 SUMMARY AND CONCLUSIONS

In this study, a mechanistic framework was developed for characterization of the loss of pavement service life due to acceleration and deceleration of the SHL vehicles. The developed framework accounts for the taxing stress paths imparted by SHLs, tire-pavement interactions, unique features of transportation systems, and the environmental factors for accurate assessment of the pavement service life. The numerical simulation results in this study underscored the significance of acceleration and deceleration loading scenarios for accurate assessment of SHLs in overload corridors. The sensitivity analysis provided in this study showed that accelerating and decelerating heavy vehicles can impart substantially higher damages on the pavement structures, compared to the same vehicle travelling at constant speed. This is primarily attributed to the dynamic nature of the analysis, and the added horizontal and tangential forces associated with speeding and braking actions.

Further mechanistic analysis of the pavement service life in this study showed that deceleration of the SHL vehicles results in higher levels of PLR, compared to the acceleration of these vehicles under similar conditions. Loss of pavement service life was more substantial in FM roadways with less robust pavement structures, compared to SH and US highways. The post-processed results for the representative pavement sections in this study revealed that acceleration and deceleration of the SHL vehicles operating in FM roadways can impart PLR as high as 26% and 42%, respectively.

The results of this study highlighted the significance of acceleration/deceleration of SHL vehicles and their impact on the accumulated damages imparted on the pavement sections. Therefore, pavement structures, particularly at intersections, junctions, or stop–go sections, can potentially experience excessive distresses due to absence of a protocol to account for the detrimental impacts of acceleration and deceleration scenarios during the pavement design process. Consequently, analysis of acceleration and deceleration of SHL vehicles should be an integral component in risk management studies of pavement facilities servicing the overload corridors.

REFERENCES

Ashtiani, R. S., Morovatdar, A., Licon, C., Tirado, C., Gonzales, J., and Rocha, S., 2019. *Characterization and Quantification of Traffic Load Spectra in Texas Overweight Corridors and Energy Sector Zones.* No. FHWA/TX-19/0-6965-1.

Chen, X., Lambert, J. R., Tsai, C., and Zhang, Z., 2013. *Evaluation of Superheavy Load Movement on Flexible Pavements.* International Journal of Pavement Engineering, 14(5), 440–448.

Dong, Q., and Huang, B., 2013. *Field Measurement of Pavement Responses under Super Heavy Load.*

Douglas, R.A., Woodward, W.D.H., and Woodside, A.R., 2000. *Road Contact Stresses and Forces under Tires with Low Inflation Pressure.* Canadian Journal of Civil Engineering, 27, 1248–1258.

Hajj, E. Y., Siddharthan, R. V., Nabizadeh, H., Elfass, S., Nimeri, M., Kazemi, S. F., … and Piratheepan, M., 2018. *Analysis Procedures for Evaluating Superheavy Load Movement on Flexible Pavement.* Report No. FHWA-HRT-18-049, FHWA, Washington, DC.

Hu, X., Faruk, A. N., Zhang, J., Souliman, M. I., and Walubita, L. F., 2017. *Effects of Tire Inclination (Turning Traffic) and Dynamic Loading on the Pavement Stress–Strain Responses using 3-D Finite Element Modeling.* Pavement Research and Technology, 10 (4).

Morovatdar, A., Ashtiani, R. S., Licon, C., and Tirado, C., 2019. *Development of a Mechanistic Approach to Quantify Pavement Damage using Axle Load Spectra from South Texas Overload Corridors.* In Geo-Structural Aspects of Pavements, Railways, and Airfields Conference (GAP 2019), Colorado Springs, CO, USA.

Morovatdar, A., Ashtiani, R. S., Licon, C., Tirado, C., and Mahmoud, E., 2020a. *Novel Framework for the Quantification of Pavement Damages in the Overload Corridors.* Transportation Research Record 2674, no. 8, 179–191.

Morovatdar, A., Ashtiani, R. S., and Licon Jr, C., 2020b. *Development of a mechanistic framework to predict pavement service life using axle load spectra from Texas overload corridors.* In International Conference on Transportation and Development 2020 (pp. 114–126). Reston, VA: American Society of Civil Engineers.

Morovatdar, A., & Ashtiani, R. S. (2020c). Evaluation of pavement service life reduction in overload corridors. In Advances in Materials and Pavement Performance Prediction II: Contributions to the 2nd International Conference on Advances in Materials and Pavement Performance Prediction (AM3P 2020), 27-29 May, 2020, TX, USA (p. 211).

Muzenski, S., Flores-Vivian, I., Farahi, B., & Sobolev, K. (2020). Towards ultrahigh performance concrete produced with aluminum oxide nanofibers and reduced quantities of silica fume. Nanomaterials, 10(11), 2291.

Salimi, K., & Ghazavi, M. (2019). Soil reinforcement and slope stabilisation using recycled waste plastic sheets. Geomechanics and Geoengineering, 1–12.

Salimi, K., Cerato, A. B., Vahedifard, F., & Miller, G. A. (2021). Tensile Strength of Compacted Clays during Desiccation under Elevated Temperatures. Geotechnical Testing Journal, 44(4).

Satvati, S., Nahvi, A., Cetin, B., Ashlock, J. C., Jahren, C. T., & Ceylan, H. (2021). Performance-based economic analysis to find the sustainable aggregate option for a granular roadway. Transportation Geotechnics, 26, 100410.

Wang, H. and Al-Qadi, I.L., 2009. *Combined Effect of Moving Wheel Loading and Three-Dimensional Contact Stresses on Perpetual Pavement Responses.* Transportation research record, No. 2095. Washington, DC: TRB, 53–61.

Wang, H., Al-Qadi, I. L., and Stanciulescu, I., 2012. *Simulation of Tyre–Pavement Interaction for Predicting Contact Stresses at Static and Various Rolling Conditions.* International Journal of Pavement Engineering, 13(4), 310–321.

Analytic analysis of a grid-reinforced asphalt concrete overlay

J. Nielsen & K. Olsen
S&P Reinforcement Nordic ApS, Odder, Denmark

A. Skar & E. Levenberg
Department of Civil Engineering, Technical University of Denmark, Kgs. Lyngby, Denmark

ABSTRACT: This paper presented an analytic investigation of pavement systems subjected to mill-and-overlay treatments - including grid reinforcement in-between the new asphalt concrete (AC) overlay and the underlying (existing) cracked and aged AC. The investigation was based on an updated version of the classic layered elastic theory capable of handling a fragmented layer. Such a layer mechanically replicates a multi-cracked AC offering considerable vertical stiffness alongside low bending rigidity. A thin high-modulus layer represented the reinforcing grid, fully bonded to the abutting AC layers. Three mill-and-overlay cases and an additional reference case were investigated for a pavement system under the loading of a dual-tire configuration. The cases differed by the milling depth (thin, medium, and thick), and by the inclusion or exclusion of a reinforcing grid. Key responses in the structure and subgrade, commonly associated with different pavement distress, were calculated and compared across the different cases. The analysis suggests that a reinforcing grid can potentially reduce bottom-up cracking and permanent deformation within the AC overlay for the medium and thick mill-and-overlay cases. For the thin mill-and-overlay case, the analysis suggests that top-down cracking is the expected distress mechanism. In this context, the inclusion of a reinforcing grid seemed to be ineffective. Finally, it is found that adding reinforcement to any of the mill-and-overlay cases produces only a marginal effect on key responses linked to the development of permanent deformation deeper in the structure and subgrade.

Keywords: Pavement analysis, mill-and-overlay, reinforcement, layered elastic theory, fragmented layer formulation

1 INTRODUCTION

Pavement systems serve as the economic and social foundation of every country, with asphalt pavements being the most common type (EAPA and NAPA 2011). These systems continually deteriorate under traffic loadings in combination with weather effects and therefore require regular maintenance. With the increase in traffic loadings (e.g., heavier trucks, platooning), and with climate change effects (i.e., more extreme weather situations, sea rise, prolonged rain periods, higher moisture contents in the unbound layers), maintenance activities often prioritise the improvement of load-carrying capacity (i.e., the time until major repair work is needed) – and not just damage repair and return to pristine conditions. One common maintenance treatment in asphalt pavements is mill-and-overlay (Correia and Bueno 2011). This treatment applies to the full-width of the pavement or the full-width of a lane, and consists of partial-depth milling of the existing –

DOI: 10.1201/9781003222880-32

aged and damaged – asphalt concrete (AC), and then repaving with new AC. When the paved thickness has (nominally) the same thickness as the milled depth, the original load carrying capacity may be approached; when the paved thickness is larger, the overall structural thickness is increased, and the pavement's load-carrying capacity may be improved. Mill-and-overlay is also often employed to address functional distress, such as poor skid resistance and excessive unevenness.

After a mill-and-overlay, pavement systems consist of new AC that is supported by an aged and damaged AC layer. From a mechanistic standpoint, an aged and damaged AC layer – especially if severely cracked – exhibits little bending rigidity and therefore sub-optimal support to the new overlay. A possible approach to tackle this situation is to introduce grid reinforcement at the interface between the existing (aged) AC and the new overlay. Asphalt grid reinforcements are commonly made of carbon-, glass-, or polymer-fibres which are geometrically arranged in a thin bituminous-coated mesh, and thus offer high in-plane stiffness and tensile strength (Zofka et al. 2017).

The engineering analysis of asphalt pavement systems is currently based on layered elastic theory (LET), wherein each layer is considered a continuum. When it comes to including the effects of densely cracked AC layers in LET, it is a common approach to assign them reduced elastic properties (Mamlouk et al. 1990; Baltzer et al. 2017), even though AC stiffens with age (Bell et al. 1994; Harvey and Tsai 1997; NASEM 2007; Baek et al. 2012). Such approach originates from continuum damage mechanics, where properties are degraded isotropically to represent randomly oriented cracks (Lemaitre and Desmorat 2005). Due to the nature of traffic loadings, in combination with environmental effects, cracking in AC has a preferred vertical orientation – with layers gradually transforming under service into a fragmented state, e.g., alligator cracking and block cracking. Thus, from a mechanistic standpoint, the isotropic modulus reduction approach results in under-estimation of the vertical rigidity and over-estimation of the bending rigidity.

To address this shortcoming, Levenberg and Skar (2020) recently proposed an analytic formulation (validated against finite element solutions) for a fragmented layer (FL) that can be incorporated into LET. The idea is to model the FL as a Winkler spring-bed with Pasternak-type shear layer(s) for introducing some interaction between the Winkler springs. The spring-bed stiffness, characterised by k, represents the vertical rigidity of the FL; the Pasternak layer(s), characterised by G, produce some bending rigidity for the FL. An additional horizontal spring-bed, characterised by k_h, is included in the FL model to represent the ability to transfer parallel-oriented (horizontal) shear stresses.

As for analysing the effects of grid-reinforcement, there is currently no accepted engineering approach. Models based on the Finite Element Method (e.g., Taherkhani and Jalali, 2017), are computationally expensive and mandate a large number of input parameters. Such an approach is unlikely to gain acceptance among practitioners. Nielsen et al. (2020) proposed to treat a reinforcing grid as a thin layer within the LET framework, characterised by: (i) Young's modulus, (ii) Poisson's Ratio, (iii) an effective thickness, and (iv) conditions of bonding with the adjoining AC layers. From this simple consideration, representing a grid reinforcement could be achieved by relatively few input parameters, without adding extra computational demand to the framework.

The objective of the current work is to analyse and generate initial intuition on the expected effects of installing an interlayer reinforcing grid as part of a mill-and-overlay maintenance treatment. This is pursued synthetically (i.e. utilizing a computational model and not measurements from the field), by considering a layered elastic system representing a traditional asphalt pavement consisting of AC layer, unbound granular base layer, and soil subgrade. A FL is employed to describe an existing (damaged and aged) AC after milling, and a thin high-modulus layer is employed to represent the grid-reinforcement. The paper investigates several cases for a dual-tire assembly loading that differ by the milling depth and by the inclusion or exclusion of an interlayer grid reinforcement. In each case, key responses commonly linked to performance in pavement design are calculated and presented in contour plots; response peaks are subsequently identified and compared. A summary of findings and a short discussion are offered at the end.

2 CASES FOR ANALYSIS

Consider a three-layered pavement system with a 150 mm thick aged and cracked AC layer. This layer rests on an unbound granular base and subbase with a combined thickness of 500 mm. The structure is supported by a subgrade soil extending to a large depth. This pavement is maintained by a mill-and-overlay treatment considering four cases which differ by the milling depth and overlay thickness – see Figure 1. Common to all cases is that the milling depth is equal to the overlay thickness, i.e., returning to the original surface elevation. In Case (a) 30 mm are milled and repaved, in Case (b) 60 mm are milled and repaved, in Case (c) 90 mm are milled and repaved, and in Case (d) the entire aged AC thickness of 150 mm is milled and repaved.

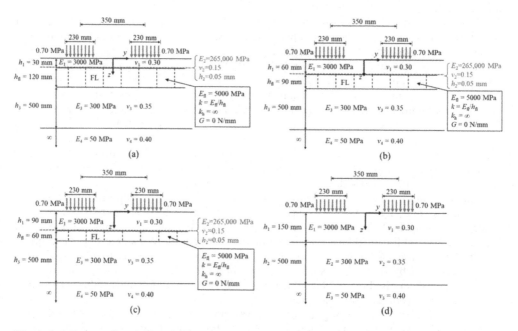

Figure 1. Cross-sectional view of four pavements for model analysis: (a) 30 mm mill-and-overlay, (b) 60 mm mill-and-overlay, (c) 90 mm mill-and-overlay, and (d) 150 mm mill-and-overlay.

The pavements in Figure 1 are modelled within the LET framework. A Cartesian coordinate system is considered with its origin located at the surface, and the z-axis is oriented towards the subgrade. The model layers are parallel to the x-y plane and extend to infinity. The top layer, representing the AC overlay, has a modulus of $E_1 = 3000$ MPa and Poisson's Ratio $v_1 = 0.30$. Any existing (aged) AC is treated as a FL with a Winkler spring-bed stiffness (k) that is equal to a modulus of $E_{fl} = 5000$ MPa divided by the FL thickness h_{fl}, a horizontal spring-bed stiffness k_h of infinity, and a Pasternak layer providing zero resistance to shear deformation (i.e., $G = 0$). The latter assumption corresponds to a densely cracked AC layer with fragments that do not interact. The choice of 5000 MPa to represent an aged AC modulus (and not 3000 MPa as done for the AC overlay) aims to embody age stiffening. The unbound granular materials are represented by a single layer, with a modulus of $E_3 = 300$ MPa and a Poisson's Ratio $v_3 = 0.35$. The subgrade soil is treated as a semi-infinite medium with $E_4 = 50$ MPa and $v_4 = 0.40$.

As can be seen in Figure 1, a thin layer representing grid-reinforcement is included for the first three mill-and-overlay cases. This layer resides in-between the top layer and the FL and is assumed to be fully bonded to both. The properties of this layer are directly based on an existing grid product – S&P Carbophalt® G 200/200 (S&P 2020). This grid consists of carbon fibres characterised by a modulus of 265 GPa and Poisson's Ratio of 0.15. The carbon fibres are arranged in square openings, 15 mm × 15 mm in size, with an average cross-sectional area per unit width of the grid of about 50 mm^2/m. Thus, the grid characteristics are smeared based on the volume of fibres per unit area to correspond to an layer thickness of $h_2 = 0.05$ mm, modulus of $E_2 = 265{,}000$ MPa, and Poisson's Ratio $v_2 = 0.15$. In actuality, the grid is almost 1 mm thick because of a bituminous material that coats the fibres. This coating contributes to the bonding with the adjacent AC layers; its contribution to the elastic properties and tensile strength of the grid is considered negligible.

Finally, as shown in Figure 1, all four cases are loaded by two circular areas representing the tire-pavement contact due to a dual-tire assembly. The centres of the circles are located along the y-axis, on each side of the coordinate origin with equal offset. The centre-to-centre spacing is 350 mm, and the diameter is 230 mm. Both areas are uniformly loaded by vertical stress with an intensity of 0.7 MPa. The above-described characteristics are based on the Danish pavement design guidelines for a standard single-axle load of 120 kN (Baltzer et al. 2017).

3 ANALYSIS AND RESULTS

In this section, the first three mill-and-overlay cases in Figures 1a–1c are analysed. The analysis is repeated twice – without and with the inclusion of grid reinforcement. Because Case (d) consists of full removal and repaving of the AC layer, it serves as a reference. Assuming that damage and ageing effects are only confined to the AC layer, Case (d) also represents the pristine mechanical condition of the pavement system. Ultimately, the reinforced cases, unreinforced cases, along with the reference case amount to a total of seven different pavement systems. Calculations were done utilising the LET code ALVA (Skar and Andersen 2020; Skar et al. 2020) that was extended to include the FL formulation from Levenberg and Skar (2020). ALVA is an open-source LET kernel code based on the numerical computing package MATLAB. Soil mechanics sign convention is followed, wherein compressive stresses and strains are positive.

Four key responses are investigated, namely: (i) horizontal strain oriented along the axle direction in the AC overlay ε_y; (ii) von Mises stress in the AC overlay σ_{vm}; (iii) vertical stress in the aggregate base layer σ_z^b; and (iv) vertical stress in the upper 500 mm of the subgrade σ_z^{sg}. The choice to focus on these responses is related to their correlation with cracking and rutting distresses. Specifically, horizontal tensile strains in the AC are closely associated with crack initiation. The von Mises stress in the AC, and the intensity of vertical compressive stresses in the unbound layers and subgrade soil, are all linked to the accumulation of permanent deformation in these materials (Oeser and Möller 2004; Baltzer et al. 2017).

The analysis commenced with a graphical investigation of key response distributions. This was done in the y-z plane for $x = 0$ and for y in the range of ± 600 mm (z is variable). In this context, Figures 2 and 3 present colour-coded contour plots for Case (a) and Case (c) respectively (see Figure 1); these represent a thin overlay situation and a thick overlay situation. Each figure includes four pairs of heat maps, each representing a different key response – one for an unreinforced pavement system (left-hand side) and another for a reinforced pavement system (right-hand side). Peak values of key responses for all four cases are included in Table 1. Note that the colour-coding differs in-between heat map pairs as well as in-between the two figures. To prepare the plots, calculations were performed over a 2 mm × 2 mm grid in the y-z plane.

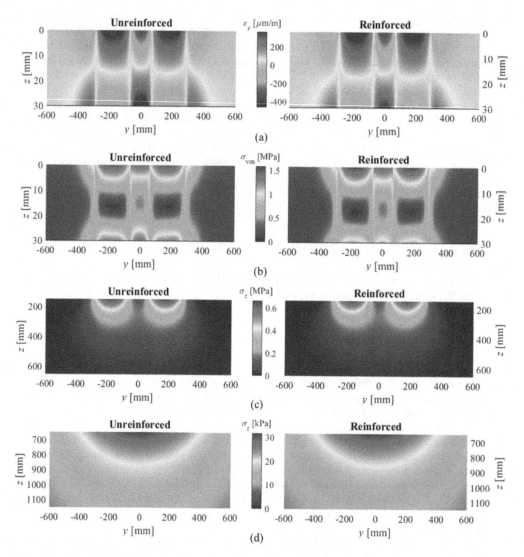

Figure 2. Distributions of the sought key responses in Case (a) of Figure 1: (a) horizontal strain in AC overlay, (b) Von Mises stress in AC overlay, (c) vertical stress in the aggregate base layer, and (d) vertical stress in the subgrade.

Figure 2a presents contour plots of the horizontal strain in y-direction within the thin AC overlay, i.e., for the range of $z = 0$ to $z = 30$ mm. In both plots, it can be observed that under the two loading areas, compressive horizontal strains occur at the AC surface while tensile strains occur at the AC bottom. Outside the loading areas, the strain changes sign from compression to tension at the top, and from tension to compression at the bottom. The peak tensile strain is located in-between the two loading areas. While the two contour plots appear visually similar, the introduction of reinforcement causes a slight reduction in the compressive and tensile strain magnitudes at the bottom of the AC overlay, and a slight increase of strains at the AC surface. Figure 2b presents contour plots of the von Mises stress within the thin AC overlay. In both plots, it can be observed that two peak stress values occur at the top and bottom – directly under and in-between the two loading areas. Visually, the stress peaks are of slightly larger magnitude at the top. The reinforcement appears to generate a reduction in the von Mises stress at the bottom and a slight increase at the top (compared to the unreinforced situation). Figure 2c presents contour plots of the vertical stress in the aggregate base layer within the range of $z = 150$ mm to $z = 650$ mm. In both plots, it can be seen that stresses are only compressive (positive) with two peaks occurring at the surface – directly under the two loading areas. Visually, the stress distributions appear similar for both the unreinforced and reinforced situations. Lastly, Figure 2d presents contour plots of the vertical stress within the subgrade for the depth range $z = 650$ mm (i.e., formation level) to $z = 1150$ mm. In both plots, it can be observed that stresses are only compressive with one peak located at the subgrade surface in-between the two loading areas. Visually, the unreinforced and reinforced situations appear identical.

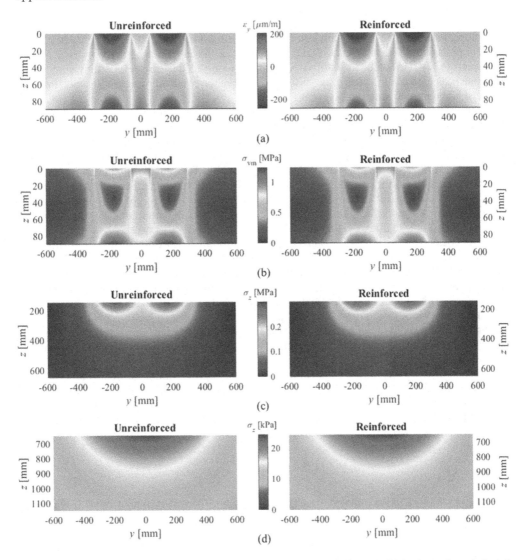

Figure 3. Distributions of the sought key responses in Case (c) of Figure 1: (a) horizontal strain in AC overlay, (b) Von Mises stress in AC overlay, (c) vertical stress in aggregate base layer, and (d) vertical stress in subgrade.

Figure 3a presents contours plots of the horizontal strain in the y-direction within the thick AC overlay, i.e., for the range of $z = 0$ to $z = 90$ mm. In both plots, it can be observed that under the two loading areas, compressive horizontal strains occur at the AC surface while tensile strains occur at the AC bottom. Outside the loading areas, the strain changes sign from compression to tension at the top, and from tension to compression at the bottom. Two peak tensile strains are identified, each located under one of the two loading areas at the AC bottom. While the two contour plots appear visually similar, the introduction of reinforcement causes a slight reduction in the compressive and tensile strain magnitudes at the bottom of the AC overlay, and a slight increase of strains at the AC surface. Figure 3b presents contour plots of the von Mises stress within the thick AC overlay. In both plots, it can be observed that two peak stress values occur at the bottom – under the two loading areas. The reinforcement appears to generate a reduction in the peak von Mises stress (compared to the unreinforced situation). Figure 3c presents contour plots of the vertical stress in the aggregate base layer within the range of $z = 150$ mm to $z = 650$ mm. In both plots, it can be seen that stresses are only compressive (positive) with two peaks occurring at the base's surface – under the two loading areas. Visually, the stress distributions appear similar for both the unreinforced and reinforced situations. Lastly, Figure 3d presents contour plots of the vertical stress within the subgrade for the depth range $z = 650$ mm (i.e., formation level) to $z = 1150$ mm. In both plots, it can be observed that stresses are only compressive with one peak located at the subgrade surface in-between the two loading areas. Visually, the unreinforced and reinforced situations appear identical.

Contrasting the contour plots of Figure 2 and Figure 3 (i.e., thin overlay vs. thick overlay), several observations can be made. First, concerning the horizontal tensile strain in the AC layer: in the thin overlay case (Figure 2a) the peak value is located at the surface, in-between the two loading areas (for both situations without and with reinforcement); in the thick overlay case (Figure 3a) there are two peaks, both located at the bottom of the AC, each under a loading area (for both situations without and with reinforcement). Second, concerning the von Mises stresses in the AC layer: in the thin overlay case (Figure 2b) there are two peaks, each located at the surface under a loading area (for both situations without and with reinforcement); in the thick overlay case (Figure 3b), there are also two peaks – however, they are both located at the bottom – each under a loading area (for both situations without and with reinforcement). Third, concerning the vertical stresses in the aggregate base layer: in the thin overlay case (Figure 2c) the stress fields at the base surface are highly concentrated under each of the loading areas (for both situations without and with reinforcement); in the thick overlay case (Figure 3c) the stress fields at the base surface are more smeared in-between the loading areas (for both situations without and with reinforcement). Lastly, concerning the vertical stresses at the top of subgrade; stress fields appear similar when contrasting the thin overlay case (Figure 2d) against the thick overlay case (Figure 3d), regardless of the inclusion or exclusion of reinforcement.

The next analysis step utilised the contour plots in Figures 2 and 3, as well as contour plots for Cases (b) and (d) (which are not presented) to identify response peaks across the different mill-and-overlay cases in Figure 1 (with and without reinforcement). The results are summarised in Table 1 for a total of seven situations. The considered peaks are: (i) horizontal tensile (negative) strain in AC overlay $\varepsilon_{y,min}$, (ii) von Mises stress in AC overlay $\sigma_{vm,max}$, (iii) vertical stress in the aggregate base layer $\sigma^b_{z,max}$, and (iv) vertical stress in the subgrade $\sigma^{sg}_{z,max}$.

Table 1. Peak values of key responses for all four cases in Figure 1 without reinforcement (w/o) and with reinforcement (w).

	Case (a)		Case (b)		Case (c)		Case (d)
	w/o	w	w/o	w	w/o	w	w/o
$\varepsilon_{y,min}$ [μm/m]	-355	-364	-260	-222	-206	-181	-134
$\sigma_{vm,max}$ [MPa]	1.51	1.58	1.49	1.31	1.24	1.11	0.83
$\sigma^b_{z,max}$ [MPa]	0.66	0.65	0.43	0.42	0.30	0.29	0.17
$\sigma^{sg}_{z,max}$ [kPA]	31.7	31.5	27.7	27.7	24.3	24.2	18.6

First, with respect to the peak tensile strain in the AC overlay: in the thin overlay case, i.e., Case (a), including reinforcement leads to a marginal 2% increase compared to the unreinforced situation; in the thicker overlay cases, i.e., Case (b) and Case (c), including reinforcement lead to a 15% and 12% reduction (respectively) compared to the unreinforced situation. Second, with respect to the peak von Mises stress in the AC overlay: in the thin overlay case, i.e., Case (a), including reinforcement leads to a marginal 4% increase compared to the unreinforced situation; in the thicker overlay cases, i.e., Case (b) and Case (c), including reinforcement lead to an 11% and 10% reduction (respectively) compared to the unreinforced situation. Third, with respect to the peak vertical stress in the aggregate base layer, including reinforcement has a marginal reduction in Case (a), Case (b) and Case (c) compared to the unreinforced situation (respective reduction of 2%, 4%, and 3%). Fourth, with respect to the peak vertical stress on top of the subgrade, including reinforcement has a negligible effect in Case (a), Case (b) and Case (c) compared to the unreinforced situation. Lastly, the table shows that all four peak responses are lowest for Case (d).

4 CONCLUSION

The objective of this work was to provide initial intuition on the effects of introducing interlayer reinforcing grid into a mill-and-overlay maintenance treatment. The analysis was based on a three-layered pavement system subjected to surface loading of a dual-tire assembly, consisting of an AC layer supported by an aggregate base layer resting on a subgrade soil extending to a large depth. The analysis utilised the LET framework to synthetically investigate three mill-and-overlay cases and a reference case. A FL was applied to represent the fully damaged and aged AC layer that was retained in the structure after milling. The reinforcing grid, introduced in-between the FL and the AC overlay, was modelled as a thin high-modulus layer. In each case, horizontal strains and von Mises stresses in the AC overlay, along with vertical stresses in the aggregate base and subgrade were simulated and presented in contour plots for selected cases. Based on these simulations, peaks were identified and contrasted across the different cases.

From this investigation it was found that: (i) for a thick overlay case, the peak tensile strain was located at the bottom – implying bottom-up cracking mode. Adding reinforcement reduced the peak horizontal tensile strain and peak von Mises stress – suggesting potential reduction in cracking and permanent deformation within the overlay; (ii) for a thin overlay case, the peak tensile strain was located at the surface – implying top-down cracking mode. Adding reinforcement marginally increased the peak tensile strain and peak von Mises stress (possibly due to a slight downward shift in the neutral bending axis of the overlay) – suggesting that the reinforcement is ineffective; and (iii) for both thick and thin overlay cases, adding reinforcement has marginal positive effect on the peak vertical stress in the aggregate base layer, and negligible effect on the peak vertical stress in the subgrade.

In studies dealing with grid-reinforced granular base layers, it has been shown that the presence of the grid affects the medium properties in its vicinity – within a so-called transition zone (Luo et al. 2017). This is due to the stress-dependent nonlinear nature of geo-materials, and was not considered in the current analysis. One approach to account for the transition zone within LET framework is to assign an effective (fictitious) modulus to the grid and retain its physical smeared thickness. Such an approach was suggested in the work of Kutay et al. (2020) for grid-reinforced granular base layers, where the modulus within the transition zone was increased. In this context, the grid modulus utilised in this research can be considered as a lower limit value, given that it only represents the actual reinforcement modulus and does not account for the transition zone. In future studies, it is planned to characterise the effective grid modulus based on full-scale experimental investigations. Furthermore, it is anticipated that the effective modulus would appear to be rate- and temperature-sensitive, corresponding to the mechanical nature of AC mixtures.

REFERENCES

Baek, C., Underwood, B. S., Kim, Y. R., 2012. *Effects of Oxidative Aging on Asphalt Mixture Properties.* Transportation Research Record, Vol. 2296(1), pp. 77–85. https://doi.org/10.3141/2296-08

Baltzer, S., Tønnesen, P., Gleerup, S., Holst, M. L., & Thorup, C, 2017. *Design of pavements and reinforcement layers.* Vejregler – Danish Road Directorate, Ministry of Transport, Building and Housing.

Bell, C. A., Wieder, A. J., and Fellin., M.J., 1994. *Laboratory aging of asphalt-aggregate mixtures: field validation.* Washington, DC: Strategic Highway Research Program, National Research Council.

Correia, N. d. S and Bueno, B.d. S., 2011. *Effect of bituminous impregnation on nonwoven geotextiles tensile and permeability properties.* Geotextiles and Geomembranes, vol. 29, no. 2, pp. 92–101. https://doi.org/10.1016/j.geotexmem.2010.10.004

EAPA and NAPA, 2011. *The asphalt paving industry a global perspective.* Third Edition

Harvey, J., and B. W. Tsai., 1997. *Long-Term Oven Aging Effects on Fatigue and Initial Stiffness of Asphalt Concrete.* Transportation Research Record. Vol. 1590(1), pp. 89–98.

Kutay, M. E., Hasnat, M., and Levenberg, E., 2020. *Layered Nonlinear Cross Anisotropic Model for Pavements with Geogrids.* Submitted for proceedings of Advances in Materials and Pavement Performance Prediction, Aug 3–7.

Lemaitre, J. and Desmorat, R., 2005. *Engineering Damage Mechanics.* Springer-Verlag Berlin Heidelberg. https://doi.org/10.1007/b13888210

Levenberg, E. and Skar, A., 2020. *Analytic pavement modeling with a fragmented layer.* International Journal of Pavement Engineering, 1–13. https://doi.org/10.1080/10298436.2020.1790559

Luo, R., Gu, F., Luo, X., Lytton, R. L., Hajj, E. Y., Siddharthan, R. v., Elfass, S., Piratheepan, M., Pournoman, S., 2017. *Quantifying the Influence of Geosynthetics on Pavement Performance.* Transportation Research Board. https://doi.org/10.17226/24841

Mamlouk, M., Zaniewski, J., Houston, W. and Houston, S., 1990. *Overlay design method for flexible pavements in Arizona.* Transportation research record, Vol. 1286, pp. 112–122.

National Academies of Sciences, Engineering, and Medicine (NASEM), 2007. *Environmental Effects in Pavement Mix and Structural Design Systems.* Washington, DC: The National Academies Press. https://doi.org/10.17226/23244

Nielsen, J., Olsen, K., Levenberg, L., and Skar, A., 2020. *Fleksible belægninger med asfaltarmering - Mekanisk beregningsmodel.* Trafik og Veje - April, pp.15–18. (In Danish)

Oeser, M., and Möller, B, 2004. *3D constitutive model for asphalt pavements.* International Journal of Pavement Engineering, 5(3), 153–161. https://doi.org/10.1080/10298430412331314281

S&P (2020), *S&P Carbophalt® G 200/200 Pre-bituminised asphalt reinforcement.* https://www.sp-reinforcement.eu/sites/default/files/field_product_col_doc_file/tds_carbophalt_g_200200_eu_en_0.pdf (date accessed October 7, 2020).

Skar, A. and Andersen, S., 2020. *ALVA: An adaptive MATLAB package for layered viscoelastic analysis.* Journal of Open Source Software, Vol. 5(52), 2548. https://doi.org/10.21105/joss.02548

Skar, A., Andersen, S., and Nielsen, J., 2020. *Adaptive Layered Viscoelastic Analysis (ALVA).* Technical University of Denmark. Software. https://doi.org/10.11583/DTU.12387305

Taherkhani, H. and Jalali, M., 2017. *Investigating the performance of geosynthetic-reinforced asphaltic pavement under various axle loads using finite-element method.* Road Materials and Pavement Design, Vol. 18(S5), pp. 1200–1217. https://doi.org/10.1080/14680629.2016.1201525

Zofka, A., Maliszewski, M., and Maliszewska, D., 2017. *Glass and carbon geogrid reinforcement of asphalt mixtures.* Road Materials and Pavement Design, Vol. 18(S1), pp. 471–490. https://doi.org/10.1080/14680629.2016.1266775

Calibration of fatigue performance model for cement stabilized flexible pavements

E. Rodriguez & R. Ashtiani
Department of Civil Engineering, The University of Texas at El Paso, Texas, USA

ABSTRACT: Chemical stabilization of reclaimed and marginal materials with cement has proven to be a viable option for sustainable design and construction of flexible pavements. The primary role of the treated base layers is to provide a robust platform that withstands complex stress paths imparted by moving traffic loads. In addition to considerations of load-related distresses in the structural design process of pavements with cement treated layers, the selection of appropriate cement content in the laboratory mixture design greatly influences the service life of pavements. For instance, excessive cement content in base mixes can potentially result in shrinkage cracking that will further propagate to the surface layers. The primary focus of this study was to develop and calibrate a new generation of fatigue performance model with considerations of the traffic-induced stresses, as well as the shrinkage parameters in the laboratory mixture design of cement treated layers. To achieve this objective, a comprehensive laboratory experiment design was developed to characterize the tensile and shrinkage behavior of cement stabilized virgin and reclaimed materials. Subsequently, a logical performance model was proposed and further calibrated using 51 pavement sections. Ultimately, the performance of the developed model was tested on a new set of data for cross-validation purposes. The proposed calibrated fatigue performance model with the inclusion of IDT and shrinkage cracking characteristics of the cement treated materials provides a practical approach for the design of pavements with cement stabilized foundations.

Keywords: Cement stabilization, Recycled concrete aggregate, Full depth reclamation, Fatigue performance model, Model Calibration

1 INTRODUCTION

The application of calcium-based treatment agents such as cement, lime, fly ash, among others, is a sustainable means to improve the mechanical properties of pavement foundations with reclaimed materials. Improved orthogonal load-bearing capacity and enhanced durability of cement treated layers are the primary motivations for pavement engineers to include stabilization strategies in their toolbox for the design of new roads and the rehabilitation of existing pavement structures. The use of an adequate amount of cement during the design of cement stabilized mixes will considerably influence the structural and functional performance of a pavement structure. For example, stabilized granular base mixes with excessive amounts of cement are prone to develop shrinkage cracking that will propagate to the surface layers and ultimately jeopardize the longevity of the pavement structure.

Even though the mechanical properties of cement treated materials have been well studied, very limited research has been conducted to relate such properties to the performance of pavement structures (Mohammad et al., 2000, Saxena et al., 2010, Molenaar & Pu, 2008; Wu et al., 2011; Li et al., 2019). The latest pavement design methodologies, such as The Mechanistic-Empirical Pavement Design Guide or MEPDG (ARA & NCHRP, 2004) and the Texas Mechanistic-Empirical Flexible Pavement Design System or TxME (Hu et al., 2014), are based on performance models to predict fatigue cracking and permanent deformation of pavement structures with cement stabilized layers. Unfortunately, the fatigue performance model incorporated in current versions of the MEPDG and the TxME for cement treated layers have proven to be impractical and not yet calibrated.

To overcome this issue, several researchers have developed and calibrated new performance models applicable to cement stabilized materials. Wu et al. (2011) developed a finite element model to simulate the permanent deformation behavior of cement treated materials in flexible pavements. The proposed model was calibrated using surface permanent deformation data from six accelerated pavement test sections constructed in Louisiana. The authors reported good agreement between the predicted and measured rut depths in two low-volume roads. In a relevant study, Li et al. (2019) developed and calibrated a bottom-up and top-down fatigue model for cement stabilized layers in flexible pavements, with considerations of variations of modulus due to erosion, as well as freeze-thaw and wet-dry cycles. The researchers calibrated the performance model based on field-measured modulus values obtained from Falling Weight Deflectometer (FWD) backcalculated data. Saxena et al. (2010) provided a discussion on the shortcomings of the characterization method and performance models in the MEPDG. The authors postulated that the MEPDG fatigue model does not adequately address the effect of shrinkage cracking. Therefore, they recommended the incorporation of a shrinkage distress model for comprehensive modeling of cement stabilized materials in the MEPDG.

The current version of the material models incorporated in the MEPDG overlooks the significance of shrinkage cracking. The primary criterion in the current model is the ratio of the load-induced tensile stresses and modulus of rupture of materials. Ashtiani and Tarin (2016) outlined the theoretical and practical issues regarding the flexural bending beam test for lightly and heavily cement treated base layers. The researchers performed a series of numerical simulations of the test considering different material properties and stress paths. Furthermore, the authors evaluated the mechanical behavior of 570 specimens as part of the laboratory experiment design. Through extensive numerical modeling complemented by laboratory testing, the authors showed in some permutations, that nearly 35% of the cross-section of the beams stayed in compression in unique bending loading conditions. Therefore, the traditional equations with assumptions of linear distribution of the tensile stresses along the transverse direction of the beam greatly overestimate the tensile capacity of cement treated base materials. In addition to theoretical concerns, the authors outlined practical issues, such as non-uniformity of compaction in 15.24x15.24x50.8 cm prismatic beams, and disintegration of large beams during demolding, handling, and transportation to the test set up for lightly cement stabilized base and subgrade soils. Therefore, the authors provided a modification to the indirect diametrical tensile (IDT) test for an efficient and repeatable alternative for determining the tensile strength properties of cement treated base materials in the laboratory.

The primary objective pursued in this study was to develop and calibrate a novel performance model for cement stabilized base layers. The model incorporates IDT strength, as well as both load-induced and shrinkage cracking behavior of cement treated foundations. To accomplish this objective, a comprehensive laboratory experiment design coupled with numerical simulations were developed to characterize the tensile and shrinkage behavior of cement stabilized virgin and reclaimed materials. Subsequently, a logical performance model was proposed, calibrated, and further validated using 64 pavement sections lasting in service for 85 years, on average. 80% of pavement sections (51 sections) were used to calibrate the model while the remaining 20% (13 sections) were used for cross-validation purposes.

2 METHODOLOGY

Figure 1 shows the methodology for the development and calibration of the predictive model for the estimation of the fatigue performance of cement stabilized virgin and reclaimed base materials in flexible pavement structures. The process includes three major areas that will be discussed in detail in the upcoming sections: Development of Fatigue Performance Model, Field Data Collection and Analysis, and Field Calibration and Validation Algorithm. Relevant databases were collected based on laboratory test results, finite element simulations, and field testing for each subcategory illustrated in Figure 1.

In this study, a total of 120 specimens, considering replicates were fabricated and subjected to various laboratory tests. The Unconfined Compressive Strength (UCS) test, static and dynamic IDT strength tests, submaximal modulus tests, and the coefficient of thermal expansion (COTE) or shrinkage test were incorporated in the laboratory experimental design to characterize compressive and tensile strengths, shrinkage behavior, resilient properties, and permanent deformation potential of cement stabilized materials. Four different aggregate base materials including crushed limestone and siliceous gravel, as virgin aggregates, and two sources of reclaimed materials including Recycled Concrete Aggregate (RCA) and blends of Full Depth Reclamation (FDR) were integrated in the experiment design. Three levels of cement contents, 2%, 3%, and 4%, were added to the mixtures to evaluate the contribution of the calcium-based treatment agents in lightly stabilized and heavily stabilized permutations.

Finally, a total of 64 pavement sections were used in the calibration and cross-validation process of the newly developed fatigue performance model for cement treated base layers.

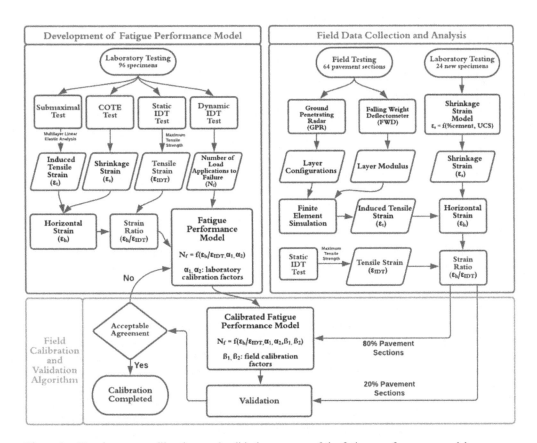

Figure 1. Development, calibration, and validation process of the fatigue performance model.

3 DEVELOPMENT OF FATIGUE PERFORMANCE MODEL

The fatigue performance model developed in this study is based on the strain-ratio concept for the prediction of the fatigue life of cement treated materials. In establishing the fatigue performance model, traffic-induced strain, shrinkage strain, and failure strain in the IDT test were incorporated in the model to provide a holistic view of field conditions. The strain ratio ($\varepsilon_h/\varepsilon_{IDT}$) was computed based on strain values equivalent to the horizontal strain (ε_h), as the summation of tensile strain induced by traffic loads (ε_t) from finite element simulations and shrinkage strain from laboratory tests, as well as the strain that corresponds to the maximum tensile strength of materials (ε_{IDT}).

Strain values that corresponded to the ε_{IDT} and shrinkage strain (ε_s) were obtained in the laboratory from the static IDT test and shrinkage test, respectively. Multilayer linear elastic analysis was conducted to compute ε_t at the bottom of the cement treated layer for site-specific traffic conditions, pavement layer configurations, layer thicknesses, and modulus values. The moduli for cement treated layers were obtained experimentally after conducting the submaximal modulus test on cement stabilized virgin and reclaimed materials in the laboratory.

The selected form of the fatigue performance model was based on the relationship between the strain ratio and the number of load applications to failure (N_f). The latter parameter was obtained experimentally from the dynamic IDT test. The rationale to relate $\varepsilon_h/\varepsilon_{IDT}$ to N_f was to take into consideration the notable effect of shrinkage cracking on the bottom-up tensile-fatigue life of cement stabilized base layers. Figure 2 provides the relationship between N_f and the strain ratio for cement treated virgin aggregates and cement treated reclaimed materials.

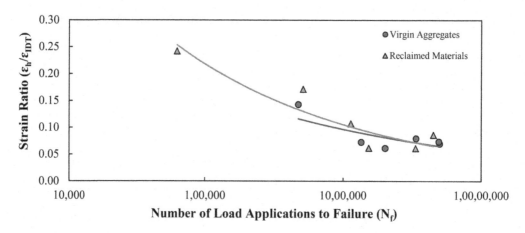

Figure 2. Relationship between the number of load applications to failure (N_f) and strain ratio ($\varepsilon_h/\varepsilon_{IDT}$) for cement stabilized virgin aggregates and reclaimed materials.

As depicted in Figure 2, the descending nature of the trendlines suggests that permutations with higher shrinkage strains and taxing loading scenarios manifested by higher tensile strains exhibit lower fatigue life compared to other counterparts. Furthermore, the slope of the best fit lines in the fatigue failure and strain ratio plots are appreciably different for the two groups of materials tested in this study. This underscores the influence of the type of virgin and reclaimed material to partake in hydration and strength reactions, thus resulting in mixtures with different fatigue performances. Equation 1 shows the general form of the fatigue performance model for cement stabilized materials in this study.

$$ln(N_f) = \frac{\beta_1 x \alpha_1 - (\varepsilon_h/\varepsilon_{IDT})}{\beta_2 x \alpha_2} \qquad (1)$$

Where: N_f is the number of load repetitions to failure, $\varepsilon_h/\varepsilon_{IDT}$ is the strain ratio, α_1 and α_2 are laboratory calibration factors, and β_1 and β_2 are field calibration factors.

The dynamic IDT results were further used for the determination of two sets of laboratory shift factors, α_1 and α_2 for virgin and reclaimed materials. The authors believe it is prudent to separate the databases for virgin and reclaimed materials as their mechanical behavior was observed to be drastically different in the laboratory. Based on the curve fitting procedure adopted in this study, the laboratory shift factors for virgin aggregates, α_1 and α_2, were 0.4561 and 0.026, respectively. For reclaimed materials, the model parameters α_1 and α_2, were 0.7163 and 0.043, respectively. Section 5 of this paper provides the procedure for the estimation of the field calibration parameters, β_1 and β_2, values.

4 FIELD DATA COLLECTION AND ANALYSIS

4.1 Field data collection

A comprehensive database of cross-sectional characteristics and material properties using nondestructive testing equipment, such as FWD and ground penetrating radar (GPR), was compiled from 64 pavement sections. The evaluated pavement sections covered a broad range of site characteristics. The authors combined the field database with available databases from published relevant literature. The combined dataset was incorporated into the calibration algorithm and cross-validation process of the fatigue performance model. The database required for executing the calibration and validation of the model included the location of the pavement sections, layers thickness (based on design blueprints and GPR results), back-calculated FWD modulus, materials properties of stabilized base layers (i.e., UCS, maximum dry density, optimum moisture content), and type of material for each layer.

Upon finalizing the pavement feature database, the next step corresponded to the determination of the traffic-induced tensile strains using finite element simulations.

4.2 Pavement response calculations by Finite Element (FE) method

To simulate field conditions and calculate critical pavement responses, the authors used the FE software Abaqus to model the 64 pavement sections. Different layer configurations of pavement structures were simulated in the FE program. Figure 3 (a) exhibits an example of meshing, boundary conditions, geometry of layers, and traffic-induced tensile strain results for a representative pavement section that contains an asphalt concrete surface layer (AC), cement treated base layer (CTB), and the subgrade soil (SG).

Site-specific structural properties of pavements were assigned to each representative field roadway section in the FE model. Since the entire three-dimensional model was symmetric, only a quarter size of the structure was simulated. This led to optimizing the computational efficiency of the analysis. A finer mesh was defined under the wheelpath, while a coarser one was used in regions far from the loading area. Bottom and lateral boundaries of the FE model were appropriately defined to assure a realistic field model.

The traffic loads considered for the simulation of field roadway sections were based on the historic traffic (ranging from 4.8 million to 28.5 million equivalent single axle loads), as well as the most frequent truck classes and tire pressure in the studied network (Ashtiani et al., 2019). FE models were developed based on tandem wheel loads with 42.7 kN, a contact area of 274.2 cm^2, a tire pressure of 779.1 kPa, and a wheel spacing of 30.48 cm. Lastly, pavement structures associated with distinct materials properties and layer configurations were analyzed for specific sites to obtain the traffic-induced tensile strains (ε_t) as displayed in Figure 3 (b).

Upon finalizing the estimation of ε_t at the bottom of the cement stabilized base layer in all pavement sections, the upcoming step was related to the determination of ε_s and ε_{IDT}. These strain values are mainly relevant to calculate ε_h and, consequently, calculate $\varepsilon_h/\varepsilon_{IDT}$ of the evaluated field sections.

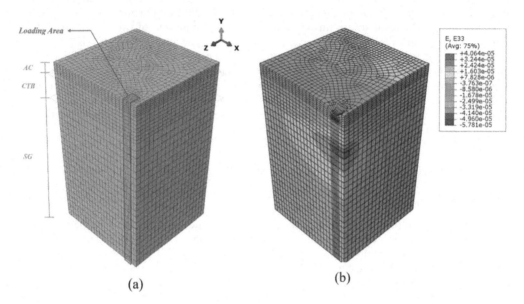

Figure 3. (a) Meshing and pavement layers simulation in Abaqus. (b) Pavement responses contours for induced tensile strain (ε_t).

4.3 Determination of the shrinkage strain the laboratory

Shrinkage strain (ε_s) values for the stabilized base layers of pavement sections in the field were calculated based on a shrinkage model developed in the laboratory suitable for cement treated virgin aggregates and reclaimed materials (Ashtiani et al., 2020). A total of 48 treated specimens with 2, 3, and 4% of cement were prepared and subjected to UCS and COTE tests. Four different aggregate sources including crushed limestone and siliceous gravel, as virgin aggregates, and RCA and FDR, as reclaimed materials, were incorporated in the experiment design to develop the shrinkage model. Multi-variate regression analysis was used to establish robust relationships between shrinkage strain, cement content, and unconfined compressive strength. Equation 2 shows the model for the prediction of shrinkage strain in cement stabilized systems.

$$\varepsilon_s = 0.000187 + 1076.69 \times \frac{C\%^2}{S_C} \qquad (2)$$

Where: ε_s is the shrinkage strain, $C\%$ is the percent of cement content, and S_C is the unconfined compressive strength (kPa).

The prediction model developed by Ashtiani et al. (2020) provided a reasonably good estimate of the shrinkage strain of cement stabilized base materials. Statistical results showed good agreement between the predicted and laboratory-measured shrinkage strain after seven days of temperature cycles.

Upon completing the calculation of shrinkage strain (ε_s) of field roadway sections, the horizontal strain (ε_h) was determined based on the summation of the induced tensile strain (ε_t) and the shrinkage strain (ε_s) to consider the impact of shrinkage behavior on the cement stabilized base layers. The further analysis consisted of calculating the strain ratios of the cement treated base layers from field pavement sections. Therefore, it deemed necessary to determine the strains that corresponded to the maximum tensile strength of materials (ε_{IDT}).

Figure 4 Shows the relationship between the maximum tensile strength and its corresponding strain obtained for different types of materials with different stabilizer percentages. As evidenced in Figure 4, there is a narrow range for the variation of shrinkage strain for different types of materials.

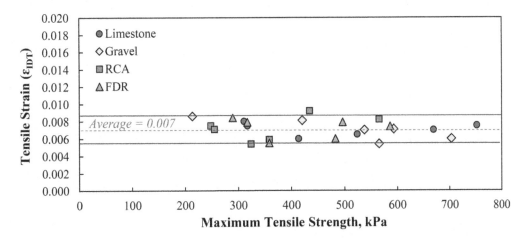

Figure 4. Relationship between tensile strain (ε_{IDT}) and maximum tensile strength.

For each cement stabilized material considered in this study, the authors calculated the coefficient of variation (CV) and range of ε_{IDT} as illustrated in Figure 5. As observed in the plot, the average strain values corresponding to the maximum tensile strength for virgin aggregates and reclaimed materials were 0.0070 and 0.0072, respectively. The statistical analysis showed that ε_{IDT} had a coefficient of variation lower than 20% in all cases. This implies that the scatter of the data points revolves around the average value.

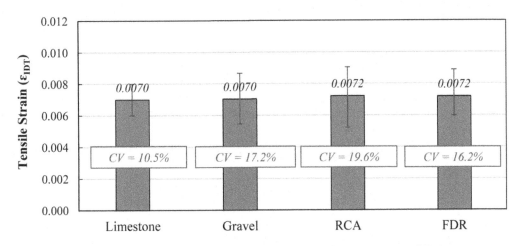

Figure 5. Tensile strain corresponding to the maximum tensile strength (ε_{IDT}) for different cement stabilized materials.

5 FIELD CALIBRATION PARAMETERS AND VALIDATION ALGORITHM

To provide credibility to the structural design process of pavements with cement treated base layers, the study incorporated a field calibration process for the novel fatigue performance model. The split-sample approach was used to verify the accuracy of the prediction model during this calibration effort. Approximately 80% of the sections (51 field pavement sections) were randomly selected for field calibration purposes and 20% of the sections (13 field pavement sections) were selected for cross-validation.

Laboratory calibration factors (i.e., α_1 and α_2) of the fatigue performance model were adjusted to improve precision using the representative field database consisted of 51 pavement sections. The predicted strain ratios in the laboratory were compared to the strain ratios obtained from pavement sections based on the collected field database for stabilized virgin aggregates and stabilized reclaimed materials. More details on the parameters included in the database are provided in Ashtiani et al. (2020). Figure 6 presents the initial relationship between predicted and measured strain ratios before calibration.

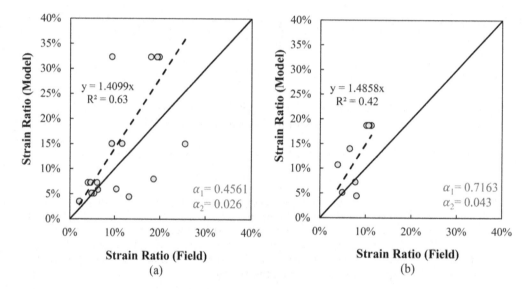

Figure 6. Predicted (model) versus measured (field) strain ratios before calibration for (a) stabilized virgin base layers and (b) stabilized reclaimed base layers.

Field calibration factors (i.e., β_1 and β_2) were introduced in the fatigue performance model to eliminate the bias and reduce the standard error between the predicted and the measured strain ratios. Figure 7 shows the final relationship between predicted and measured strain ratios considering calibration factors. As observed in the plot, bias was eliminated and the precision was improved for the developed fatigue performance model. Based on the comparison of strain ratios, 83% of predicted strain ratios had a difference lower than 5% when compared to the measured strain ratios for stabilized virgin base layers. Similarly, 80% of predicted strain ratios had a difference lower than 5% when compared to the measured strain ratios for stabilized reclaimed base layers.

Laboratory and field calibration factors of the model for different cement stabilized materials are tabulated in Table 1. The results clearly show improvements in the predicted values after incorporation of the field calibration values.

Table 1. Laboratory and field calibration factors of the fatigue performance model.

Fatigue Performance Model	α_1	α_2	β_1	β_2	R	R^2
Model for Virgin Aggregates	0.4561	0.026	1.40	1.36	0.85	0.73
Model for Reclaimed Materials	0.7163	0.043	1.16	1.09	0.81	0.65

As stated earlier, the field calibrated model was cross-validated using an independent set of data (20% of the field dataset) to evaluate the predictions. The validation results are shown in Figure 7. The close scatter of the validation data cloud around the equality line is an indication of the capability of the model to provide a reasonable agreement between the prediction and field conditions. Additionally, the maximum differences between predicted and measured strain ratios were 6.7% and 5.0% for virgin and reclaimed materials, respectively.

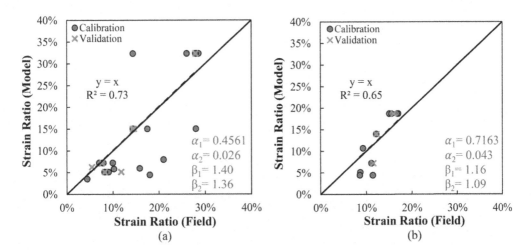

Figure 7. Predicted (model) versus measured (field) strain ratios after calibration for (a) stabilized virgin base layers and (b) stabilized reclaimed base layers.

6 CONCLUSIONS

The focus of this research effort was to develop a new fatigue performance model for cement treated virgin and reclaimed base materials incorporating load-induced, as well as shrinkage cracking behavior of cement stabilized base layers. To achieve this objective, multiple tests including the UCS test, static and dynamic IDT tests, submaximal modulus tests, and shrinkage tests were incorporated in the laboratory experiment design to characterize compressive and tensile strengths, shrinkage behavior, resilient properties, and permanent deformation potential of cement stabilized materials. A total of 120 treated specimens were prepared with four different granular material sources, namely crushed limestone and siliceous gravel, as virgin aggregates, and RCA and FDR materials, as reclaimed aggregates. All four aggregate sources were stabilized with 2%, 3%, and 4% cement. Therefore, the model incorporates both lightly stabilized systems, as well as heavily stabilized aggregate matrices. The primary divergence from existing fatigue performance models in this study was the departure from using modulus of rupture and incorporation of the IDT test and the shrinkage strain for the estimation of fatigue life of pavements with cement treated base layers.

The study also aimed to calibrate and cross-validate the proposed fatigue performance model based on field performance data. This objective was achieved by developing a comprehensive catalog of cross-sectional features and material properties of 64 pavement sections based on non-destructive field testing, such as FWD and GPR, in combination with available databases from published pertinent literature. The fatigue performance model was calibrated for 51 pavement sections with different cross-sectional characteristics and material properties to enhance the generalization of the model. The precision of the fatigue performance model was evaluated on a new set of 13 pavement sections for cross-validation purposes. The statistical results show the superiority of the newly developed model to better represent the fatigue performance of pavements with cement stabilized base layers.

REFERENCES

Applied Research Associates, and Inc. National Cooperative Highway Research Program, 2004. *Guide for Mechanistic-Empirical Design for New and Rehabilitated Pavement Structures*. Final Report, NCHRP 1-37A. Transportation Research Board, National Research Council.

Ashtiani, R.S. and Tarin, J., 2016. *Testing Procedures for Long Life Heavy Duty Stabilized Bases*. Report No. FHWA/TX-06/0-5223-2, The University of Texas at El Paso's Center for Transportation Infrastructure Systems, Austin, Texas.

Ashtiani, R.S., Morovatdar, A., Licon, C., Tirado, C., Gonzales, J., and Rocha, S., 2019. *Characterization and Quantification of Traffic Load Spectra in Texas Overweight Corridors and Energy Sector Zones*. Report No. FHWA/TX-19/0-6965-1, The University of Texas at El Paso's Center for Transportation Infrastructure Systems, Austin, Texas.

Ashtiani, R.S., Rashidi, M., Rodriguez, E., Ordaz, M., Cruz H., Garay, G., and Rocha, S., 2020. *Establishing Best Practices for Construction and Design of Cement Treated Materials*. Report No. FHWA/TX-19/0-6949-0, The University of Texas at El Paso's Center for Transportation Infrastructure Systems, Austin, Texas.

Hu, S., Zhou, F., and Scullion, T., 2014. *Development of Texas Mechanistic-Empirical Flexible Pavement Design System (TxME)*. Report No. FHWA/TX-14/0-6622-2, Performed in cooperation with the Texas Department of Transportation and the Federal Highway Administration, College Station, Texas.

Li, X., Wang, J., Wen, H., and Muhunthan, B., 2019. *Field Calibration of Fatigue Models of Cementitiously Stabilized Pavement Materials for Use in the Mechanistic-Empirical Pavement Design Guide*. Transportation Research Record, No. 2673, pp. 427–435.

Mohammad, L.N., Raghavandra, A., and Huang, B., 2000. *Laboratory Performance Evaluation of Cement-Stabilized Soil Base Mixtures*. Transportation Research Record, No. 1721, pp. 19–28.

Molenaar, A.A.A., and Pu, B., 2008. *Prediction of Fatigue Cracking in Cement Treated Base Courses*. Pavement Cracking: Mechanisms, Modeling, Detection, Testing, and Case Histories, Taylor & Francis Group.

Saxena, P., Tompkins, D., Khazanovich, L., and Balbo, J.T., 2010. *Evaluation of Characterization and Performance Modeling of Cementitiously Stabilized Layers in the Mechanistic-Empirical Pavement Design Guide*. Transportation Research Record, No. 2186, pp. 111–119.

Wu, Z., Chen, X., Yang, X., and Zhang, Z., 2011. *Finite Element Model for Rutting Prediction of Flexible Pavement with Cementitiously Stabilized Base-Subbase*. Transportation Research Record, No. 2226, pp. 104–110.

Case study

Stone mastic asphalt as an ungrooved runway surface: Case study on Emerald airport

G. White
School of Science and Engineering, University of the Sunshine Coast, Sippy Downs, Queensland, Australia

H. Dutney
Airport Engineer, GHD Pty Ltd, Brisbane, Queensland, Australia

ABSTRACT: The pavements at Emerald airport, located in central western Queensland, were resurfaced in 2019. The taxiway and apron were surfaced with dense graded asphalt (DGA) and the runway was surfaced with stone mastic asphalt (SMA). The project was a pilot project to demonstrate ungrooved SMA is a viable alternate to grooved DGA for runway surfacing in Australia. The mixture designs indicated that the SMA is expected to perform in the field similarly to the DGA, with regard to fracture, deformation and moisture damage resistance. The construction trial and surface construction confirmed that the surface texture exceeded the regulatory requirements, as did the wetted surface friction values. Furthermore, the aircraft skid resistance compliance was achieved immediately following construction, which is a significant advantage over grooved DGA, which usually does not achieve compliance until the grooves are sawn some 4-8 week later. The production and construction results demonstrated the ability of the continuous drum mixing plant to achieve adequate SMA production consistence, comparable to that expected for DGA. However, it is recommended that the asphalt production binder content acceptance tolerance be increased for SMA, to reflect the more variable nature of the test results due to unavoidable contamination of the tools by the high bituminous binder content associated with SMA. The project also highlighted the need to monitor the SMA construction to minimise the risk of isolated bitumen-rich spots. Overall, the project was successful and it is recommended that other airports consider ungrooved SMA as an alternate to grooved DGA in the future. Ongoing monitoring of the SMA surface at Emerald airport is required to determine the performance with age, compared to otherwise similar DGA in the same environment.

Keywords: Stone mastic, Airport, Runway, Ungrooved, Surface

1 INTRODUCTION

Emerald airport is located approximately 5 km south of the town of Emerald in the Central Highlands region of Queensland, Australia. The airport provides medical and general support to the township, which is approximately a 10-hour drive from the state capital, Brisbane. The airport services approximately eight public transport flights per day, including regional turboprop and jet aircraft.

The main runway at Emerald airport is nominally 30 m wide and approximately 1,900 m long, with turning nodes at each end. Two perpendicular taxiways connect the runway to the parking apron, which provides six permanent parking bays (Figure 1). In 2018, the

DOI: 10.1201/9781003222880-34

runway surface was old and severely eroded (Figure 2) with some isolated depressions near the two runway ends. In 2019 the runway was resurfaced, including isolated patching of depressed pavement areas, unpaved flank regrading, airfield lighting modifications and re-linemarking. One of the taxiways and approximately half of the parking apron were also resurfaced, but the secondary runway and the associated taxiway and apron for light aircraft were not.

Figure 1. Emerald airport, the area of study outlined in red and the area of construction trial filled in yellow (SMA) and green (DGA).

Figure 2. Typical runway surface prior to 2019 resurfacing.

The resurfacing work was originally planned to use the grooved dense graded asphalt (DGA), which has been normal in Australia since the 1990s, designed, produced and constructed according to the Australian performance-related airport asphalt specification (AAPA 2018). However, recent Australian research on the introduction of stone mastic asphalt (SMA) as an ungrooved runway surface (Jamieson 2019) prompted the runway to be surfaced with ungrooved SMA, as a pilot or demonstration project. Factors that made Emerald airport well suitable to being a pilot SMA runway surfacing included:

- local availability of aggregate sources that met the slightly stricter requirements of airport SMA, compared to aggregate requirements for use in airport DGA,
- relatively good access windows in which to perform the resurfacing works, compared to busier major city airports,
- the semi-arid environment which means that the runway at Emerald airport would only be grooved to meet the regulatory requirements for aircraft skid resistance, rather than an actual underlying desire for a grooved runway surface.

Initially, only the rectangular 30 m by 1,900 m of the runway was intended to be surfaced with SMA. The turning node widenings at each runway end, the taxiway and aircraft parking apron were all intended to be surfaced with ungrooved DGA because aircraft skid resistance requirements do not apply to these areas. Furthermore, underlying patching and shape correction was intended to be performed using DGA. However, as detailed later, a range of issues resulted in more of the work being completed with SMA than originally intended. However, this case study is focussed primarily on the 30 m by 1,900 m rectangular portion of the runway surfaced with generally 40-60 mm of SMA over a variable thickness of shape correcting DGA.

Being the first significant runway in Australia to be resurfaced with ungrooved SMA, there was considerable uncertainty regarding the outcomes. Practitioners and airport managers questioned whether the SMA would:

- be porous, allowing water to enter the surface layer,
- perform as well as dense graded asphalt,
- have a comparable lifespan to otherwise similar DGA,
- achieve the surface texture and wetted friction levels required by regulations,
- be able to be produced reliably in typically available mobile drum production plants, and
- cost significantly more than grooved DGA.

This paper presents the resurfacing of Emerald airport's runway as a case study on the use of SMA as an ungrooved runway surface. The aim is to answer the various questions raised by practitioners regarding the viability of SMA as an ungrooved runway surface, in order to enable SMA to be considered for other regional airport runway resurfacing works in Australia. The mixture design is compared to the specification requirements and the construction planning is outlined, along with the execution and monitoring of the construction trial. The asphalt production and surface construction quality assurance testing results are then compared to the specification requirements, before the surface texture and wetted friction results are presented and compared to regulatory aircraft skid resistance requirements. Finally, the cost of the SMA resurfacing is compared to the cost of the grooved DGA alternate.

2 BACKGROUND

2.1 *Airport asphalt surface performance*

In 2017, Australia developed performance requirements for DGA used as an airport pavement surface (Table 1). A performance-related specification was subsequently developed (AAPA 2018). The specification retains the traditional DGA mixture volumetrics and composition, but allows the mixture designer to select or develop any bituminous binder to achieve the specified asphalt performance properties (Table 2). This approach reflects the absence of a reliable durability test, meaning that the traditional volumetric composition continued to be the primary basis on which reasonable weather-related durability is controlled. However, deformation resistance, fracture resistance and moisture-related durability are now controlled by performance-related laboratory testing at the mixture design stage. Surface friction and testing are more complex and these are discussed further below.

Table 1. Summary of airport asphalt performance requirements (White 2018).

Physical requirement	Protects against	Level of importance
Deformation resistance	Groove closure Rutting Shearing/shoving	High
Fracture resistance	Top down cracking Fatigue cracking	Moderate
Surface friction and texture	Skid resistance Compliance requirement	High
Durability	Pavement generated FOD Resistance to moisture damage	Moderate

Table 2. Performance-related airport asphalt requirements (AAPA 2018).

Test Property	Test Method	Requirement
Indirect Tensile Strength Ratio (TSR)	AG:PT/T232	Not less than 80%
Wheel Tracking Test (10,000 passes at 60°C)	AG:PT/T231	Not more than 2.0 mm
Fatigue life (at 20°C and 200 μm)	AG:PT/T274	Not less than 500,000 cycles to 50% of initial flexural stiffness

In 2019 the Australian performance-related specification for dense graded airport asphalt was expanded to include the requirements of SMA (Jamieson & White 2019). The incorporation of SMA into the specification required:

- minor changes to aggregate property requirements for reliable SMA production,
- adding SMA volumetric composition requirements for mixture design,
- adding volumetric surface texture and binder drain down tests to SMA mixture design, and
- adding the testing and reporting of surface texture during construction.

Despite these changes, the performance requirements (Table 2), asphalt production tolerances and surface construction processes, field air voids content limits and general quality requirements were kept the same for SMA as they were for DGA. A draft of the updated specification was used for the resurfacing of the runway at Emerald airport, with minor changes to the specification made as a result of the lessons learnt from the project.

2.2 *Aircraft skid resistance*

The physical requirements of runways are specified by the International Civil Aviation Organization (ICAO) through a document commonly known as Annex 14 (ICAO 2013). In Australia, these requirements are implemented by the Civil Aviation Safety Authority (CASA) through their Manual of Standards Part 139 (MOS 139) (CASA 2019). Regarding aircraft skid resistance, MOS 139 requires all runways to be constructed and maintained throughout their lifecycle to provide either:

- surface macro-texture exceeding 1 mm, or
- friction levels that exceed the minimums recommended by ICAO, or
- sawn grooves perpendicular to the runway centreline.

Surface macro-texture is commonly measured by volumetric sand patch or estimated by a laser scanner (White et al. 2019). Friction levels must be measured by continuous friction measuring equipment (CFME) immediately behind the application of a nominal 1 mm film of water (FAA 1997). Grooves are traditionally 6 mm deep and 6 mm wide, spaced 38 mm from centre to centre, but trapezoidal grooves of equivalent volumetric density (Zuzelo 2014) are becoming popular in some countries, including Australia.

Grooved DGA relies on the presence of grooves to achieve adequate aircraft skid resistance. In contrast, SMA relies on surface macro-texture for regulatory compliance. Either way, CFME survey results are commonly used to verify wet friction levels and to assist in the management of friction over the lifetime of an asphalt surface (White & Azevedo 2022). Because it is not grooved, SMA must reliably provide a surface macro-texture not less than 1.0 mm, and/or return wetted friction levels that exceed the maintenance interventions levels recommended by ICAO.

The performance-related airport asphalt specification (AAPA 2018) did not have friction or surface texture requirements for DGA, except to test and report the wetted friction on completion of the resurfacing work. This reflects the inability of an asphalt mixture designer to significantly control the friction, the inability to test wetted friction at high speeds until at least 500 m of the surface is constructed, and the general reliance on post-construction grooving to meet the regulatory requirement for friction. When SMA is used, the primary means of achieving regulatory compliance is surface macro-texture. Therefore, volumetric surface texture was added to the mixture design requirements, as well as to the construction trial process. Surface texture was also added to the construction quality control testing, as a report only parameter, in addition to the wetted friction levels on completion of the construction.

2.3 *Dense graded versus stone mastic asphalt*

As stated above, airport asphalt in Australia and many other countries is predominantly DGA designed according to the Marshall method (White 1985) with samples compacted by 75 blows of a Marshall hammer. Runway asphalt generally has a 14 mm maximum nominal aggregate size. In contrast, SMA is designed using 50 blows of a Marshall hammer, to prevent aggregate particle crushing, has a gap graded coarse aggregate skeleton, a higher bituminous binder content and a higher fine aggregate content. The coarse aggregate skeleton provides stone-on-stone contact to provide high deformation resistance, while the high mastic (combination of binder and fine aggregate) content fills the additional voids in the coarse aggregate skeleton and produces a non-porous mixture that has high resistance to fracture and ageing because of the high bituminous binder content. Table 3 and Figure 3 detail the volumetric properties specified in the preliminary airport DGA/SMA specification. The laboratory asphalt mixture performance requirements are the same for both SMA and DGA, as detailed in Table 2.

In practice, Marshall designed airport DGA usually contains 0.5-1.5% hydrated lime as an active filler. The higher fine content associated with SMA would require more than 4% hydrated lime, which would excessively stiffen the mixture. Therefore, a blend of hydrated lime and crushed limestone is usually used for SMA production. Furthermore, the wider SMA envelope associated with the 4.75-9.5 mm sieves is required because SMA is generally produced from fewer aggregate fractions, resulting in less control of the overall grading, as well as the comparatively lower portions retained on the larger sieves, meaning that changes in the 14 mm and 10 mm aggregate fraction gradings has a greater impact on the overall grading on these sieves.

Table 3. Volumetric properties of SMA and DGA.

Property	Units	Test method	Dense graded	Stone mastic
Maximum size	mm	N/A	14	14
VMA	% (by volume)	AS/NZS 2891.8	13-17	17-20
VFB*	% (by volume)	AS/NZS 2891.8	70-80	75-85
Air voids	% (by volume)	AS/NZS 2891.8	3.5-4.5	3.0-4.0
Binder content*	% (by mass)	AS/NZS 2891.8	5.2-5.8	6.2-6.8
Binder type	N/A	Selected by designer	Polymer modified	Polymer modified
Surface texture	mm	Sand patch	Not applicable	>1.0
Binder drain down	%	AG:PT/T235	Not applicable	<0.3
Layer thickness	mm	For construction planning	40-80	45-60

VMA = voids in the mineral aggregate, VFB is the voids filled with binder. * denotes properties not explicitly specified but are effectively fixed by other specification requirements.

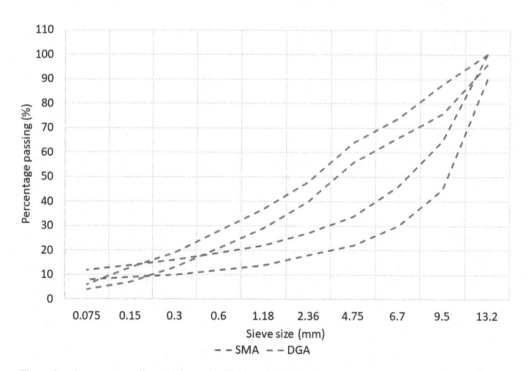

Figure 3. Aggregate grading envelopes for SMA and DGA.

3 CASE STUDY

3.1 *Mixture design*

The volumetric composition and performance properties of the 14 mm sized SMA mixture were designed to meet the requirements of the Australian airport asphalt specification, as shown in Table 4 and Figure 4, which also shows the 14 mm sized

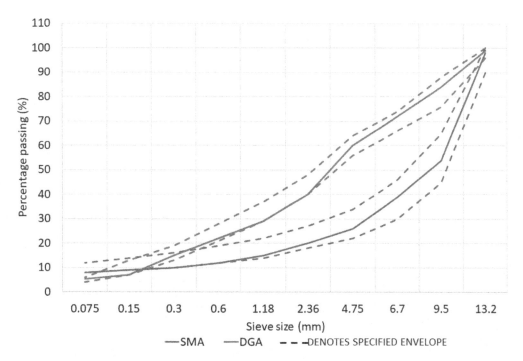

Figure 4. Designed aggregate grading for SMA and DGA.

DGA parameters used for apron/taxiway resurfacing, patching and the runway shape correction layer, as a reference. Both the SMA and DGA mixtures used the same bituminous binder, which was an elastomeric (approximately 7% SBS) polymer modified binder grade known in Australia as A10E (Austroads 2014). Both mixtures also used the same coarse aggregate and fine aggregate sources. The only exception being the SMA included 3% crushed lime stone, as well as 1% hydrated lime. The same hydrated lime was included in the DGA, but the limestone was not included and is not a normal component in airport DGA asphalt mixtures (White 2018). The requirement for crushed lime stone is expensive and logistically complex and is discussed in detail later. Cellulose fibres were added at 0.3% (by mass) to reduce binder drain down risk.

The SMA and DGA generally showed similar performance in the laboratory, including comparable moisture resistance (TSR), deformation resistance (wheel tracking) and cracking resistance (fatigue life). However, the SMA had a significantly lower resilient modulus and indirect tensile strength, as well as a significantly higher Marshall Flow. These results all reflect the higher binder content associated with the SMA, 6.4% compared to 5.4% for DGA. The Marshall Flows were higher than traditionally required for both mixtures, but this reflects the highly elastomeric nature of the A10E binder and a range of 3-5 mm is normally expected for 14 mm sized airport DGA produced with A10E.

The mixture design surface texture was measured on the samples of asphalt compacted using a laboratory slab compactor to a target air voids content of 5%, for the Cooper's wheel tracking test. The result was 1.5 mm, which exceeds the 1.0 mm minimum required by

Table 4. Designed volumetric and performance properties of SMA and DGA.

Property	Units	Test method	SMA	DGA
VMA	% (by volume)	AS/NZS 2891.8	18.9 (17-20)	14.4 (13-17)
VFB	% (by volume)	AS/NZS 2891.8	78	76
Air voids	% (by volume)	AS/NZS 2891.8	3.9 (3.0-4.0)	4.2 (3.5-4.5)
Binder content	% (by mass)	AS/NZS 2891.8	6.4	5.4
Stability	kN	AS/NZS 2891.5	9.2	13.2
Flow	mm	AS/NZS 2891.5	6.2	3.7
TSR	%	AG:PT/T232	89 (>80)	87 (>80)
Tensile strength	MPa	AG:PT/T232	522	1036
Wheel tracking	mm	AG:PT/T231	1.7 (<2.0)	1.7 (<2.0)
Fatigue life	%	AG:PT/T274	17 (<50)	18 (<50)
Resilient modulus	MPa	AS/NZ 2891.13.1	1,210	1,870

regulations for an ungrooved surface. Although surface texture was not measured in the laboratory for the DGA, experience consistently indicates a range of 0.4-0.6 mm for 14 mm dense graded airport asphalt (AAA 2017). Furthermore, the binder drain down result was 0.13%, which is below the maximum limit of 0.15%.

Overall, the SMA mixture design achieved the volumetric, performance, surface texture and binder drain down requirements and was approved for use on the runway. Subsequent effort focussed on the construction planning and verifying that the mixture design was achieved in the on-site asphalt production plant.

Values in brackets are specified limits for mixture design purposes, where applicable

3.2 *Construction planning*

Airport specifications in Australia have generally allowed 14 mm sized DGA to be paved in single layer thicknesses within the range 40-80 mm. The performance-related specification reflects the same requirement for surface layers, but relaxes this for correction layers and patches, to allow works to be performed efficiently in limited night-time work windows (AAPA 2018). Despite this, it is common practice to design and plan construction using a more limited range of layer thicknesses. Consistent with other projects, the Emerald airport minimum designed level lift was 38 mm, providing a minimum DGA planned minimum layer thickness of 45 mm, after 7 mm deep texturing to remove the existing surface grooves. In contrast, the SMA specification limits the single layer thickness to 45-55 mm (Jamieson & White 2019). This reflects the perception that SMA is generally harder to construct without compromising the achieved compaction and surface finish. However, where shape correction is required to optimise the finished surface geometry, the reduced surface layer thickness flexibility reduces the efficiency of the construction.

The Emerald airport runway resurfacing work included significant geometric shape correction, which is typical of all existing airport resurfacing projects, particularly those in regional locations. In addition, the last 300 m of pavement at each of the runway ends was found to have less structural capacity than the intermediary portion of the runway, so additional asphalt thickness was planned in those areas. Overall, the increase in surface level, from the existing surface to the new surface, was planned to range from 9 mm to 152 mm, with an average lift of 71 mm. After 7 mm texturing and deeper profiling to achieve the minimum 45 mm thickness was taken into account, the new asphalt thicknesses ranged from 45 mm to 159 mm, with an average of 78 mm. Approximately 50% of the asphalt thicknesses were between 55 mm and 80 mm (Figure 5).

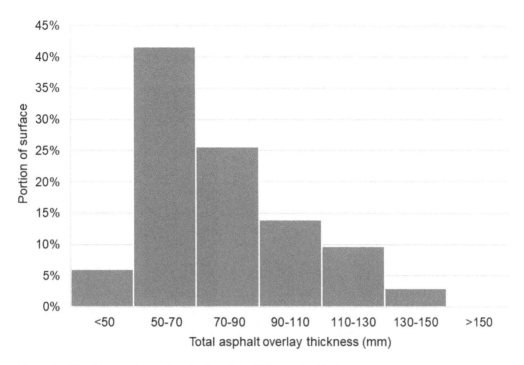

Figure 5. Total (correction and surface) asphalt thickness distribution.

If the job was performed using DGA, the 80 mm maximum layer thickness would have required 38% of the runway area to be pre-corrected prior to the 45-80 mm surface layer being constructed. However, the lower (55 mm) maximum layer thickness for SMA meant that 86% of the runway required correction prior to surfacing with 45-55 mm thick SMA. Furthermore, the distribution of correction thickness changed from a relatively uniform spread from 5 mm to 55 mm for the DGA option (Figure 6) to being dominated by thinner 5 mm to 25 mm of correction for the 45-55 mm SMA option (Figure 6). This issue was further exacerbated when the minimum thickness of 14 mm sized DGA, as a correction layer (30 mm) was taken into account. Significantly deeper profiling would be required to achieve a minimum 30 mm thickness of DGA correction layer, even where the target correction lift was only 5-25 mm.

As detailed below, the construction trial included SMA thicknesses from 40 mm to 65 mm. Based on the positive results, the SMA thickness limits were revised to 45-65 mm. This reduced the area of the runway requiring correction from 86% to 62% of the total runway area. However, the correction thickness was still dominated by 5 mm to 15 mm lifts (Figure 6) meaning that significant profiling was still required. To further reduce this requirement, a 10 mm sized DGA mixture was designed. The 10 mm sized DGA was allowed to be constructed in layers 20-40 mm thick, reducing the profiling and associated correction layer volume of asphalt significantly.

It is noted that the alternate approaches to minimising the correction layer area and thickness were also considered. These included constructing all the DGA first and then the SMA. However, this would leave unacceptable steps in the runway between work periods and this option was not viable. Feathering out the correction layer to effectively zero thickness was also considered. This approach is generally only permitted on runways where the surface layer is constructed over the correction layer in the same work period, so that aircraft never traffic

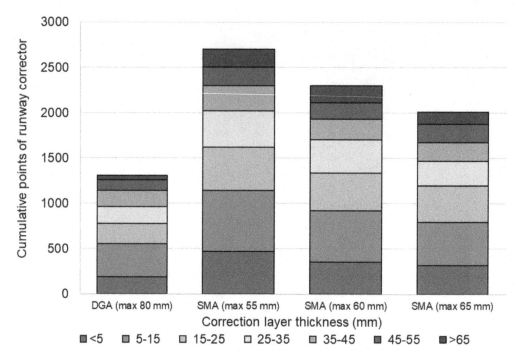

Figure 6. Correction layer thickness distribution for various surface layer thicknesses.

the feathered edge of the corrector, which would introduce the risk of loose stones ravelling out and damaging aircraft engines and undercarriage. Consequently, this approach was only viable if the SMA surface layer was constructed in the same work period as the DGA correction layer. However, the time required to change aggregate storage bins and clean out the plant and equipment, to avoid contamination of the SMA with residual DGA, was considered prohibitive and this option was abandoned at the contractor's discretion, despite it being technically feasible.

The effect of the reduced layer thickness range associated with SMA surfaces was the single biggest issue encountered on the Emerald airport runway overlay. In part, this was a function of the required surface level lifts and associated shape correction. However, the challenge associated with SMA as an ungrooved runway surface, where shape correction is required, was highlighted. Consequently, it was concluded that SMA is more viable where either two layers of asphalt surface are proposed over the full runway length and width, allowing all correction to be performed in the lower (DGA) layer, or where surface correction requirements are minimal, so that the correction can be performed within the thickness limits of a single SMA layer.

3.3 *Construction trial*

As shown in Figure 1 the construction trial was located on the secondary runway. This allowed wetted friction testing to be performed prior to the surface SMA layer construction commencing. The friction testing results are detailed later. The construction trial included a -100 m long section of DGA as well as a 100 m long section of SMA. Both sections were two paver runs wide, totaling 7.5 m.

In addition to the normal construction verification processes, the construction trial allowed a relaxation of the SMA thickness to be tested. The SMA trial was nominally 40-65 mm thick. There was no difference in the surface texture, visual appearance, air voids, level control or joint condition noticed throughout the variable thickness SMA trial section. Consequently, the SMA single layer thickness limits were relaxed from 45-55 mm to 45-65 mm, significantly reducing the area of correction layer required and the volume of existing surface profiling required to accommodate the minimum correction later thickness. The trial also allowed the nuclear density gauge to be calibrated to both the DGA and SMA mixtures, which is critical when non-destructive (gauge-based) density testing is adopted for construction acceptance (White & Alrashidi 2021).

Apart from the normal production and construction compliance verification, the primary function of the construction trial was to allow the owners of the airport to inspect, evaluate and otherwise test the SMA surface in a less critical area of the airfield, before it was confirmed for the main runway surfacing. The correction layer construction and the apron/taxiway work allowed a period of four weeks between construction of the trial sections and commencement of the main runway SMA surface layer construction. During this period the trial area was visually assessed, coarse aggregate was attempted to be mechanically dislodged, the volumetric macro-texture was measured and an ICAO-approved self-wetting CFME survey was performed one day and seventeen days after the SMA trial was constructed.

All production and construction compliance requirements were achieved, the surface was found to be consistent and the coarse aggregate could not be mechanically dislodged. The wetted friction values exceeded the ICAO maintenance intervention values required at both 65 km/hr and 95 km/hr test speeds (Figure 7) and the surface macro-texture exceeded the regulatory minimum average 1 mm (Table 5). The DGA trial results are included for reference. The average friction values reduced slightly from day 1 to day 17, which reflected the reduced tackiness of the surface immediately after construction. The surface texture results were generally lower than the 1.5 mm reported during the mixture design, but were acceptable.

Overall, the construction trial was a success, confirming all production and construction processes and validating the SMA surface macro-texture and wetted friction values met the regulatory requirements for an ungrooved runway surface. The trial provided a sound benchmark for the asphalt production and surface construction on the main runway.

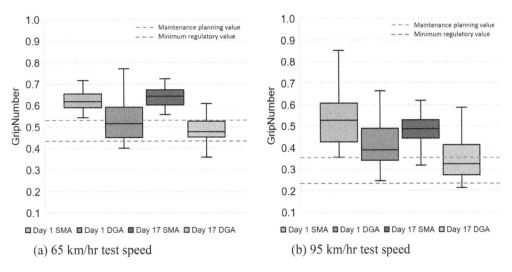

Figure 7. Wetted surface friction results of construction trial SMA and DGA.

Table 5. Statistics of construction trial SMA macro-texture.

Statistics	SMA value (mm)	DGA value (mm)
Minimum	0.94	0.46
Average	1.10	0.51
Maximum	1.23	0.62
Count of results	10	10

3.4 Asphalt production

One of the concerns expressed by practitioners is the ability of continuous drum mixing plants, which dominate in Australia, to consistently achieve the tight asphalt production requirements normally associated with airport surfacing. Although the asphalt production tolerances are usually achieved for DGA, the majority of SMA aggregate is provided by the 14 mm aggregate fraction, meaning that the produced asphalt grading is highly reliant on the consistency of the 14 mm aggregate fraction grading. Despite this concern, the currently specified asphalt production tolerances for SMA are no different to those for DGA.

The asphalt production results were compared to the specified asphalt production tolerances (Table 6). The magnitude of a 95% confidence interval was also calculated based on the production data. All of the aggregate grading 95% confidence internals were smaller than the tolerance width, meaning that 95% of the results were within the specified tolerance. The only SMA production properties with a 95% confidence interval larger than the production tolerance were the binder content, as well as the mathematically linked air void content. However, this was not the result of SMA production. Unlike the aggregate grading, which is fundamentally different for SMA, binder addition and mixing is identical for both mixture types. Rather than being a production issue, the higher variability associated with binder content is a testing issue. The higher fines and binder content required for SMA means that the tools used in the on-site laboratory become covered in excess mastic. The tools must be manually cleaned and the mastic manually split and added to the four sub-samples of loose asphalt that are used for the various production quality assurance tests. This unavoidably affects the subsequently measured binder content. Consequently, a larger production tolerance for binder content may be justified for SMA, compared to DGA, primarily to reflect this challenge associated with the sampling, splitting and testing of SMA. Because the air voids are calculated for the same samples as the binder content is tested, the variable binder content also affects the air voids results.

3.5 Surface construction

The SMA was monitored during the works and has been inspected periodically since its completion. Based on feedback from the paving crew managers and the visual inspections, generally the SMA:

- was no more difficult to pave, compact and finish than the DGA produced as part of the same project,
- produced a tight and generally consistent surface finish with the expected surface texture (Figure 8),
- had tight and almost undetectable hot construction joints between paving lanes, and
- had well matched and finished cold construction joints between work periods.

Table 6. Asphalt production variability compared to tolerances.

Property	Average	Std. Dev.	CoV	95% CI	Tolerance
VMA (%)	19.2	0.4	2.0%	1.6	4.0
Air voids (%)	3.8	0.8	21.9%	**3.2**	3.0
Binder content (%)	6.4	0.3	4.3%	**1.2**	0.6
Stability (kN)	8.9	0.9	10.6%	3.6	6.0
Flow (mm)	6.3	0.5	8.6%	1.8	2.0
Percentage passing (%) the sieve sized (mm)					
13.2	93	2.2	2.3%	9	12
9.5	61	2.4	3.9%	10	12
6.7	37	2.5	6.7%	10	12
4.75	30	2.3	7.5%	9	12
2.36	21	1.4	6.9%	6	10
1.18	15	1.1	7.3%	4	10
0.6	12	0.9	7.4%	4	8
0.3	9.9	0.8	7.7%	3.2	6
0.1	8.4	0.7	8.0%	2.8	4.0
0.075	6.8	0.6	8.1%	2.4	3.0

VMA = voids in the mineral aggregate. Std. Dev. is the standard deviation. CoV is the coefficient of variability. 95% CI is equal to four times the Std. Dev. and represents the range of values, centred on the average, expected to included 95.4% of the results. Tolerance is two times the specified tolerance, which is specified as a ± value deviation from the target. Values in red identify where the 95% CI was not smaller than the tolerance.

The only blemish associated with the SMA was three isolated bitumen-rich spots, such as the example in Figure 9. These small areas were all located within the first paving run of the respective work periods, meaning they were associated with the first SMA produced that evening. The SMA had been stored in the truck too long and some of the binder had drained down to the bottom to become a localised area of excessive binder that subsequently bled to the surface during rolling and created the bitumen-rich spot. This was subsequently minimised by reducing the time between the first SMA production and the commencement of paving. The three areas totalled 4 m^2 (<0.01% of the runway surface area) and although the surface texture was reduced to below the regulatory 1 mm minimum, the air voids remained within the specified limits and the three isolated areas were deemed to be of no practical importance.

The primary properties tested for surface construction are the in-field air voids content and surface texture, with the wetted friction measured on completion of the physical work. The surface texture and wetted friction are discussed below. The in-field air voids content results all met the specified limits, both on the joints and away from the joints, based on both individual results and the average value for each work period (Table 7) with just one exception. One individual result (out of 165) away from the joints fell below the 2% minimum, at 1.5%. This one-off was considered an anomaly.

Overall, the surface construction met all expectations with the only exception being the minor and isolated bitumen-rich areas, commonly known as fatty spots, which were subsequently avoided by ensuring that the asphalt was produced and paved in a timeframe that prevented the binder draining to the bottom of the cartage trucks prior to the commencement of paving each work period.

3.6 *Aircraft skid resistance*

Despite the positive results from the construction trial, aircraft skid resistance was continually monitored through the construction by volumetric surface macro-texture testing and was

Figure 8. Typical SMA surface finish.

verified on completion by wetted friction testing using the same CFME used on the construction trial area.

The surface texture results ranged from 1.0 mm to 1.7 mm, with an average of 1.3 mm (Figure 10). This was consistent with the construction trial area and reliably exceeded the 1.0 mm minimum value, without exceeding this regulatory requirement excessively.

The runway friction testing results were averaged over 100 m lengths of the runway, which is normal practice for runway friction surveys. Because runways are tested at four offsets and two directions, each 100 m section is actually tested eight times. The combined eight 100 m run result values are shown in Figure 11 (65 km/hr) and Figure 12 (95 km/hr). It was noted that the 100 m section at each end of the runway had higher variability and significantly higher friction values, at both test speeds, compared to the rest of the runway length. This reflects the unreliability of the results as the test device was accelerated to the target speed and these results are generally excluded from any compliance evaluation. Even excluding the 100 m at each runway end, the results all exceeded the regulatory minimum and maintenance intervention levels (Table 8) meaning that the runway was immediately compliant with regulatory aircraft skid resistance requirements, within a week of resurfacing. This contrasts with grooved DGA, which generally requires grooving to achieve compliance, with the grooving processes commonly delayed until 4-8 weeks after resurfacing, to allow the asphalt to cure as the initial volatiles are lost (AAA 2017).

Some practitioners have expressed concern that the high level of surface texture will allow water to enter the mix and cause moisture damage. The mixture design testing (Table 4) indicated a high resistance to moisture damage, with a TSR of 89%, compared

Figure 9. Example of isolated bitumen-rich area of SMA.

Table 7. In-field air voids content (%) results.

Statistic	Away from joints		On the joints	
	Individual	Average	Individual	Average
Minimum	1.5	2.7	2.5	4.8
Average	3.9	3.9	5.6	5.6
Maximum	5.7	4.5	8.2	6.5
Std. Dev.	0.97	0.49	1.37	0.66
Count of results	165	17	64	16
Specified limits	2.0-6.0	3.0-5.0	2.0-9.0	3.0-8.0

to the DGA mixture results of 87%. However, to demonstrate the impermeable nature of the SMA, bulk material was sampled during SMA production and was tested for permeability using a falling head test on samples prepared in a gyratory compactor with different target air void contents. The results showed that the SMA was impermeable at air void contents below approximately 7% (Figure 13). Because the field air void contents ranged from 1.5% to 5.7%, the surface was considered impermeable, except at the

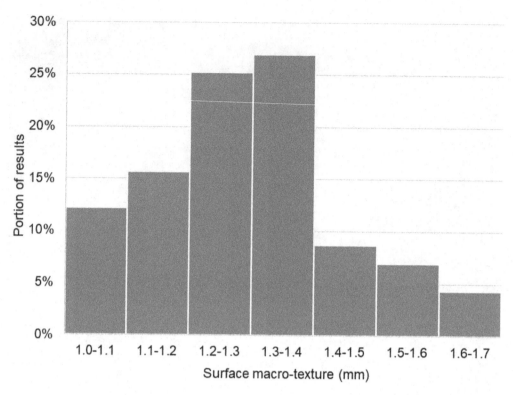

Figure 10. Distribution of SMA surface macro-texture results.

joints. That is, despite the high surface texture, the SMA does not allow water to drain through interconnected air voids and the SMA is consequently no more susceptible to moisture damage that DGA.

3.7 *Cost comparison*

To allow Emerald airport to consider their options, both grooved DGA and ungrooved SMA options were tendered. The tendered costs were $8.9 M (grooved DGA) and $9.5 M (ungrooved SMA). The difference reflects the omission of grooving but the higher cost of SMA associated with the higher binder content, the inclusion of fibres to reduce binder drain down and the reduced productivity associated with the increased area required to be corrected prior to constructing the surface layer. In contrast to the additional cost associated with the SMA option, SMA is well established as having a greater expected service life, evidenced by research from the USA (Yin & West 2018) and experience at Cairns airport (Jamieson & White 2020). Based on an expected four year longer service life associated with the SMA, the ungrooved SMA has an annualised depreciation rate that is 25% lower than grooved DGA (Table 9). It is also shown that even if the ungrooved SMA service life is only one year longer than for the grooved DGA, the annualised depreciation is still lower for ungrooved SMA than for grooved DGA. The Emerald airport runway surface must be monitored over its life-cycle and its condition compared to the adjacent DGA to determine the actual increase in surface life.

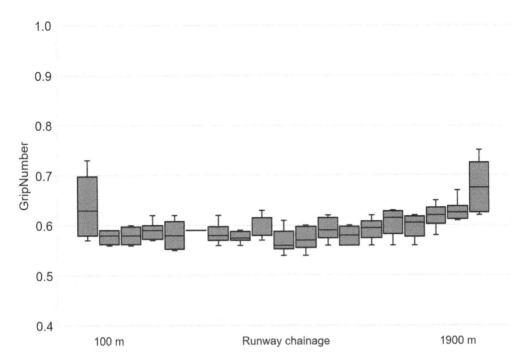

Figure 11. 65 km/hr SMA wetted friction values.

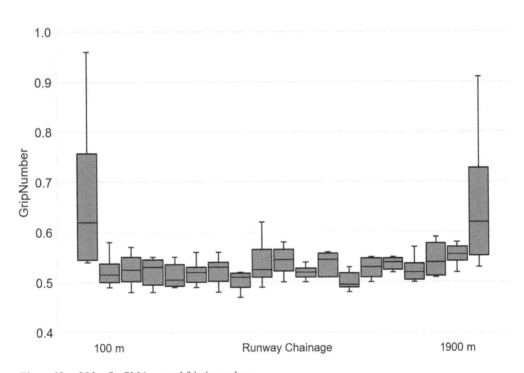

Figure 12. 95 km/hr SMA wetted friction values.

Table 8. Runway CFMD friction levels six days after resurfacing.

Statistic	65 km/hr test speed		95 km/hr test speed	
	Full length	Excluding ends	Full length	Excluding ends
Minimum	0.54	0.54	0.47	0.47
Average	0.60	0.58	0.54	0.53
Maximum	0.75	0.67	0.96	0.62
Std. Dev.	0.04	0.02	0.06	0.03
Count of results	152	136	152	136
Regulatory minimum	0.43		0.24	
Maintenance intervention	0.53		0.36	

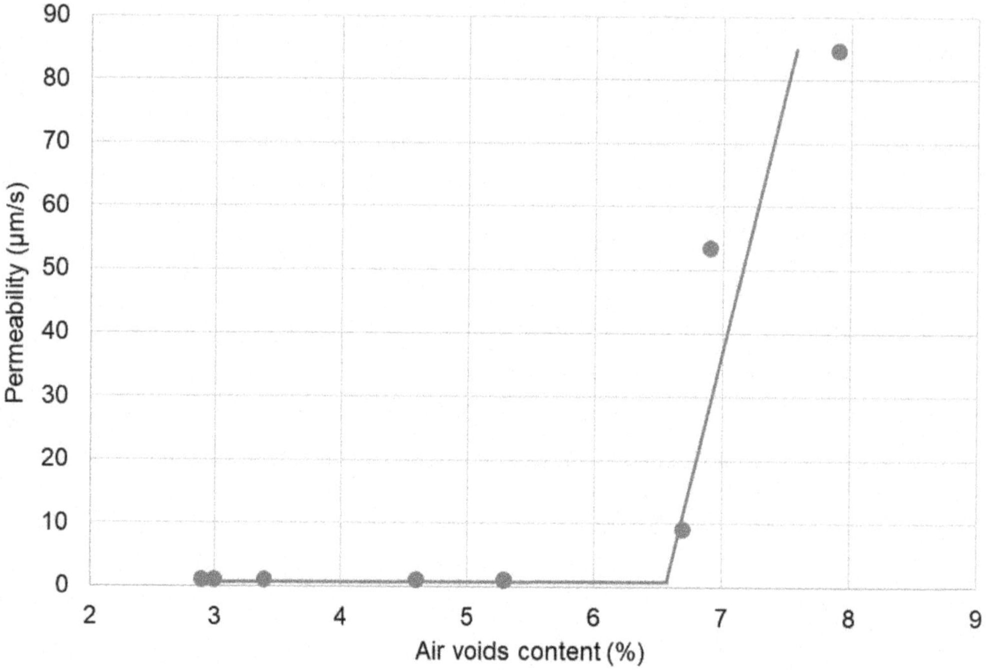

Figure 13. SMA permeability increase with air voids content.

Table 9. Annualised surface depreciation for different service lives.

Service life	Ungrooved SMA	Grooved DGA	SMA to ten year DGA
10 years	$0.93 k/annum	$0.89 k/annum	104%
11 years	$0.85 k/annum	-	95%
12 years	$0.78 k/annum	-	87%
13 years	$0.72 k/annum	-	80%
14 years	$0.68 k/annum	-	75%
15 years	$0.68 k/annum	-	70%

4 CONCLUSIONS

The pilot project at Emerald airport has achieved the aim of demonstrating ungrooved SMA as a viable alternate to grooved DGA for runway surfacing. The mixture designs indicated that the SMA is expected to perform in the field similarly to DGA, with regard to fracture, deformation and moisture damage resistance. The construction trial confirmed that the surface texture exceeded the regulatory requirements, as did the wetted surface friction values. Furthermore, the aircraft skid resistance compliance was achieved immediately following construction, which is a significant advantage over grooved DGA, which usually does not achieve compliance until the grooves are sawn some 4-8 weeks later. The production and construction results similarly confirmed the aircraft skid resistance requirements were met, as well as the ability of the continuous drum mixing plant to achieve adequate SMA production consistence, comparable to that expected for DGA. However, it is recommended that the asphalt production binder content acceptance tolerance be increased for SMA, to reflect the more variable nature of the test resulting from unavoidable contamination of the tools by the high bituminous binder content. The Emerald airport project also highlighted the need to monitor the SMA construction to minimise the risk of isolated bitumen-rich spots. Overall, the project was considered a successful pilot and it is recommended that other airports consider ungrooved SMA as an alternate to grooved DGA in the future. Ongoing monitoring of the Emerald Airport SMA surface is required to determine the performance with age, compared to the otherwise similar DGA in the same environment.

REFERENCES

AAA 2017, *Airfield Pavement Essential*, Airport Practice Note 12, Australian Airports Association, Canberra, Australian Capital Territory, Australia, April.

AAPA 2018, *Performance-based Airport Asphalt Model Specification*, Australian Asphalt Pavement Association, Melbourne, Victoria, Australia, ver. 1.0, January.

Austroads 2014, *Specification Framework for Polymer Modified Binders*, Report AGPT/T190, Austroads, February.

CASA 2019, *Manual of Standards: Part 139 – Aerodromes*, Civil Aviation Safety Authority, Canberra, Australian Capital Territory, Australia, 5 September.

FAA 1997, *Measurement Construction and Maintenance of Skid-Resistant Airport Pavement Surfaces*, Advisory Circular 150/5320-12C, Federal Aviation Administration of the United States of America.

ICAO 2013, *Aerodrome Design and Operations*. Annex 14, Volume 1, to the Conventions on International Civil Aviation, International Civil Aviation Organization, Montreal, Canada, February.

Jamieson, S 2019, *Stone Mastic Asphalt for Australian Airport Asphalt*, A thesis submitted for the award of Master of Science in Civil Engineering, University of the Sunshine Coast, Sippy Downs, Queensland, Australia, 17 June.

Jamieson, S & White, G 2020, 'Review of stone mastic asphalt as a high performance ungrooved runway surfacing', *Road Materials and Pavement Design*, vol. 21, no. 4, pp. 886–905.

Jamieson, S & White, G 2019, 'Developing a performance-based specification for stone mastic asphalt as an ungrooved runway surface', *International Airfield and Highway Pavements Conference*, Chicago, Illinois, USA, 21-24 July.

White, G 2018, 'State of the Art: Asphalt for Airport Pavement Surfacing', *International Journal of Pavement Research and Technology*, vol. 11, no. 1, pp. 77–98.

White, G & Alrashidi, F 2021, 'Field evaluations of nuclear and non-nuclear gauges as alternates to destructive coring for airport asphalt density testing', *International Symposium on Frontiers of Road and Airport Engineering*, Delft, Netherlands, 12-14 July.

White, G & Azevedo, R 2022, 'Managing the evolution of early-life runway surface friction', *11th International Conference on Managing Pavement Assets*, Chicago, Illinois, USA, 7-10 June, article-in-press.

White, G, Ward, C & Jamieson, S 2019, 'Field evaluation of a handheld laser meter for pavement surface macro texture measurement', *International Journal of Pavement Engineering*, article-in-press, doi.org/10.1080/10298436.2019.1654103.

White, TD 1985, 'Marshall procedures for design and quality control of asphalt mixture', *Proceedings Asphalt Pavement Technology*, no. 54, pp. 265–285.

Yin, F & West, R 2018, 'Performance and life-cycle cost benefits of stone matrix asphalt', *1st International Conference on Stone Matrix Asphalt*, Atlanta, Georgia, USA, 5-7 November.

Zuzelo 2014, 'The Benefits of Runway Grooving', Presented to the *Airfield Engineering and Maintenance Summit*, Furama Riverfront, Singapore, 25-28 March.

Central Highlands Regional Council gave permission for the publication of this paper and Boral Asphalt supported its preparation. These critical contributions by the asset owner and the contractor are greatly appreciated and gratefully acknowledged.

Field investigation of the deterioration of flexible polymer modified pavements: A case study in northern New England

M. Elshaer & C. Decarlo
US Army Corps of Engineers Engineering Research and Development Center, Cold Regions Research and Engineering Laboratory, Hanover, USA

ABSTRACT: The durability of flexible pavements in cold regions is a challenge due to the impact of environmental conditions such as extremely low temperatures and freeze-thaw cycling. In an attempt to tackle this issue, several asphalt modifiers have been evaluated as a potential solution to resist degradation and reduction of service life. Among these modifiers, polymer modification has a strong track record in improving the resistance of asphalt mixtures to various distresses typical in cold climates, both loading and environmental driven. The objective of this study is to investigate the structural deterioration of **solid, pelletized** polymer-modified asphalt pavement sections over the course of one year. Four test sections were constructed in Northern New Hampshire with different polymer dosage rates; control (with no polymer), 2.5% polymer, 5% polymer, and 7.5% polymer by weight of asphalt binder. To assess the structural capacity and performance of the pavement sections, Falling weight deflectometer (FWD) testing was conducted at each test over the course of one year after construction. A comprehensive analysis was performed to scrutinize the measured FWD deflection data for each test section. Backcalculation analysis was conducted to estimate the stiffness of pavement layers and track the asphalt layer's structural deterioration over time for each pavement section. The field investigation results showed that the modified polymer sections exhibited less degradation over time compared to the control section. In addition, the results suggested that polymer modification to the AC layer results in structural benefits to the pavement structure.

Keywords: Polymer Modification, Flexible Pavement, FWD, Backcalculation, Cold Regions, Deterioration

1 INTRODUCTION

Asphalt pavements in cold climates experience a significant amount of distress, including both traffic and environment driven, throughout their service lives. Historically, many transportation agencies have sought to improve the mechanical performance of their asphalt mixtures through the use of additives. Among the many additives used, polymer modification is one of the most popular forms of additives used in modern asphalt construction due to its ability to improve the overall materials' high-temperature properties without compromising low-temperature behavior. Polymer modification is beneficial in cold climates as it allows the use of a soft base binder to prevent thermal cracking while maintaining significant amounts of stiffness at high temperatures to reduce susceptibility to rutting.

Multiple studies have documented the performance benefits of using polymer modification (Bates and Worch 1987, Bahia et al 1997, Roque et al 2004). Newman (2003) investigated the effectiveness of using styrene-butadiene-styrene and styrene-butadiene-rubber (SBS and SBR) in airfield mixtures. Results from bending beam fatigue testing indicated a significant increase in fatigue cycles to failure for both polymers. Additionally, Isaacson and Zheng (1998) investigated five different polymer-modified materials in three asphalt mix designs using the thermal stress restrained specimen test. Results showed that while the effectiveness of the additives heavily depended on the specific polymer, aging levels, and mixture volumetrics, all of the polymer additives improved the mixture's resistance to low-temperature thermal cracking.

While polymer additives have been shown to clearly improve asphalt mixture performance, they are not without drawbacks, especially when considered in construction inexpedient and remote environments. One of the most prominent challenges is storage and handling of the material, where specialized heating and storage tanks are required. If improperly handled, the polymer network in the asphalt binder will be disturbed, significantly reducing the performance gains of the material (Wegan 2001). While these are commonly available in areas with routine asphalt construction, this equipment is typically not available in remote and expedient environments where the use of temporary, small-scale production plants are more common.

One potential way to address this problem is the use of pelletized polymer additives (Iterchimica 2008). Pelletized polymers are added in as solids directly into the pugmill or mixing drum, requiring minimal modifications to existing plant architecture. Being able to store the additives in dry, bulk form allows them to always be available and ready for quick deployment in remote and expedient construction. While laboratory results with pelletized polymer-modified mixtures have shown promising results (Azam et al 2019), there is a lack of field studies that have been conducted analyzing the performance of the material, particularly in cold climates. Therefore, this pelletized polymer is needed to be investigated in the field to determine the performance of polymer-modified mixes over time.

With this knowledge gap in mind, the overall **objectives** of this study are as follows:

- Construct field sections of pelletized polymer modified asphalt in a cold climate and monitor the performance of those sections over time.
- Evaluate the effect of pelletized polymer dosage on-field performance.
- Gain an understanding of the deterioration of pelletized polymer modified mixtures over the course of a winter season as compared to conventional asphalt mixtures.

2 MATERIALS, CONSTRUCTION, AND TESTING PROGRAM

2.1 *Materials*

To satisfy the objectives of this study, four asphalt mixtures were produced and placed in the field. For this study, a virgin, 12.5mm NMAS, dense-graded Superpave surface mixture with a PG64-22 binder was selected as the control mixture. This control mixture was then modified with three levels of the solid polymer by weight of asphalt binder: 2.5%, 5.0%, and 7.5% for a total of four asphalt mixtures. The four mixtures were produced at a batch plant using a dry mix process where the pelletized polymer additive was introduced into the pugmill with the heated aggregates per the additive manufacturers' recommendations. It should be noted that this is in contrast to typical polymer modification where the additives are in the asphalt binder. The pelletized polymers were allowed to mix with aggregate heated to 170-180°C for 30 seconds in the pugmill before the asphalt binder was introduced into the mixture where another 30 seconds of mixing was allowed to thoroughly coat the aggregates. Table 1 shows the design gradation and volumetrics of the control mixture design as well as the gradation and volumetrics of the as-produced mixtures that were used for field construction and testing. ASTM D2172 and ASTM D6925 were followed to determine the gradation and volumetric properties of the mixtures.

Table 1. Production quality control results.

Mixture		Design Gradation	Control QC[a] Test	2.5% Polymer QC Test	5.0% Polymer QC Test	7.5% Polymer QC Test
Percent Passing (%)	3/4 in./19.0 mm	100.0	100.0	100.0	100.0	100.0
	1/2 in./12.5 mm	96.8	95.8	96.0	95.9	96.0
	3/8 in./9.5 mm	83.6	84.2	83.6	85.9	85.3
	#4/4.75 mm	55.1	52.1	50.4	54.6	50.1
	#8/2.36 mm	35.0	36.5	33.6	39.8	34.8
	#16/1.18 mm	24.4	27.6	24.8	30.1	25.5
	#30/0.60 mm	17.8	19.0	17.1	19.9	17.0
	#50/0.30 mm	10.1	10.1	9.1	9.6	9.7
	#100/0.15 mm	6.0	4.8	4.3	3.8	5.0
	#200/0.075 mm	3.0	2.8	2.4	2.4	3.0
Binder Content (%)		5.80	5.79	5.92	5.89	6.12
Air Voids at 50 Gyrations (%)		4.0	4.3	3.3	4.2	3.2
Maximum Specific Gravity		2.492	2.497	2.482	2.487	2.485
Voids in Mineral Aggregate (%)[b]		14.5	14.8	14.6	15	14.6
Voids Filled with Asphalt (%)[c]		73.8	70.9	77.2	72.4	77.8
Dust-to-Binder Ratio		0.68	0.63	0.51	0.52	0.63

[a] Quality Control.
[b] Minimum Superpave voids in mineral aggregate for 12.5 mm NMAS mix is 14.0.
[c] Allowable Superpave voids filled with asphalt range for 12.5 mm NMAS mix is 70–80.

2.2 Field sections

As part of this study, four field test sections were constructed to evaluate the polymer-modified mixture's performance in a cold climate. The site chosen for this work was a low-volume road serving mostly residential and some truck traffic in northern New England, where the average winter low temperature is -27°C with 30-60 annual freeze-thaw cycles. The test sections were designed as a 50mm overlay over an existing pavement. The existing pavement is a conventional asphalt pavement (12.5 NMAS with PG64-22 binder) with a 75 mm asphalt surface and 200 mm full-depth reclamation (FDR) A-1-a base material, all of which is constructed on a A-2-4 silty sand subgrade. The most prominent distress in the existing pavement was moderate severity transverse cracks spaced approximately every 9 m.

The test sections used in this study were laid out as shown in Figure 1. Four sections were set up as pairs in the adjacent lane of the low-volume road. The target length for each test section was set as 152.4 meters with 15-meter transition zones at the beginning, end, and between the test sections to facilitate paving operations and allow smooth transitions between test sections. Prior to paving of the test sections, the existing transverse cracks were sealed

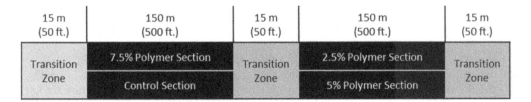

Figure 1. Test sections layout.

with a mastic-based crack sealant and a RS-1 rapid setting emulsion tack coat was applied at a residual rate of 0.22 L/m².

During the construction of the test sections, the density of the overlay layer was monitored using an electronic density guage. After construction, cores were taken to determine the as-built thickness of the overlay sections and to verify the density measurements. The final densities for the four sections were 94%, and the average overlay thickness were found to be 53 mm for the control section, 51 for 5% polymer section, 42 mm for the 2.5% polymer section, and 43 mm for the 7.5% polymer section.

In general, the construction of the test sections was without significant challenges. No differences were noticed in paving and compacting the modified layers with respect to the control. One observation worth noting was after the breakdown rolling, small check cracks were observed in all four sections. Although inconsistent in their amount and location, check cracking was observed in all four sections. After finish rolling and initial trafficking, the check cracking was not visible and no obvious deterioration due to the check cracking has been observed to this point.

2.3 *Testing program*

The test sections were visited periodically after the construction in June 2019 to assess the field condition of the control overlay section and three-solid-polymer-modified overlay sections. The field investigations of the four asphalt overlay sections involved the assessment of the structural capacity using Falling Weight Deflectometer (FWD) test at different locations. A Dynatest 8000 FWD setup with a 300 mm diameter loading plate was used to measure the deflection of the pavement surface on the mid lane and outer wheel path in the direction of travel following LTPP protocol (Schmalzer 2006). FWD tests were conducted every 7 m for about 152 m in length for all four sections. FWD tests were conducted three times; 2 months, 5 months, and 12 months after construction at each section to investigate the structural integrity and deterioration over time.

To enhance the interpretation of FWD deflection, the impact of temperature, moisture content, traffic loads, and pavement structure were taken into account. The subgrade layer was investigated by collecting soil samples using a hand auger and sent to the laboratory to determine the soil material properties and the soil moisture content during the time of conducting the FWD test. The air and surface temperatures were measured through the sensors installed in the FWD device. Four holes were drilled at different depths (17 mm, 25 mm, 50 mm, and 76 mm from the pavement surface) and filled with oil to measure the asphalt gradient temperature every 30 minutes during the time of conducting the FWD test. To evaluate the structural capacity of four overlay sections and determine the deteriorations over time, comprehensive methods were carried out as follows:

2.4 *Maximum deflection (D_0)*

The maximum pavement deflection at the load plate reflects the strength of the overall pavement structure. The maximum deflections for each test point were normalized to 40 kN and then corrected to a reference temperature of 20°C, (Kim et al. 1995), using the measured mid-depth asphalt temperature:

$$D_{68} = D_T \times \left[10^{\alpha(68-T)}\right] \quad (1)$$

where D_{68} = the adjusted deflection (in.) to the reference temperature of 20°C (68°F), D_T = the deflection (in.) measured at temperature T (°F), $\alpha = 3.67 \times 10^{-4} \times t^{1.4635}$ for wheelpaths, t = the thickness of the AC layer (in.), and T = the AC layer mid-depth temperature (°F) at the time of FWD testing.

2.5 Backcalculation analysis

The FWD deflection values were backcalculated using Evaluation of Layer Moduli and Overlay Design (ELMOD) software to determine the stiffness of the pavement layers. ELMOD performs layer elastic analysis to iteratively adjust the seed modulus for each layer to match the predicted and measured deflections within some tolerable error.

To start the backcalclation analysis, pavement layer thickness and seed moduli of pavement layers should be defined. The thickness of the pavement layers of four sections was determined from the field cores conducted during the construction of the four sections. The asphalt layer was treated in the backcalculation analysis as two individual layers; the top layer is a thin asphalt overlay with a seed moduli of 3447 MPa at 20°C determined based on the dynamic modulus master curve conducted in the laboratory. The second layer is the old existing asphalt layer with a seed moduli of 1034 MPa at 20°C and treated as a fixed layer in the backcalculation analysis (Pierce et al. 2017). This gives a better insight into the comparison of the stiffness of the overlay layer among the four sections. Several attempts were made to minimize the associated error resulted from the thin overlay in the backcalculated analysis to 1-3%. For the base layer, the seed modulus value for the FDR layer was assumed to be 413 MPa (AASHTO 2008). For the subgrade layer, the seed modulus of the subgrade at the time of the FWD measurements was determined from the measured moisture content. The soil samples from four depths (400, 475, 550, and 600 mm from the pavement surface) were collected each time during the FWD test and sent to the laboratory to measure the moisture contents following ASTM D (2019a) 2216-19. Witczak's equation, presented in equation (2), was used to determine the resilient modulus at the time of the FWD measurements based on the measured moisture content. This resilient modulus value was considered to be the seed modulus value of the subgrade in the backcalculation analysis

$$\log \frac{Mr}{Mr_{opt}} = a + \frac{b-a}{1 + \exp[\ln \frac{-b}{a} + k_m(S - S_{opt})]} \quad (2)$$

where Mr = the resilient modulus at any degree of saturation; Mr/Mr_{opt} = the resilient modulus ratio; Mr_{opt} = the resilient modulus at a reference condition; a = the minimum of log Mr/Mr_{opt}; b = the maximum of log Mr/Mr_{opt}; k_m = the regression parameter, and $S-S_{opt}$ = the variation in degree of saturation expressed in decimals.

The backcalculation analysis utilized the FWD deflections measured in the outer wheel path at the last drop of the 40 kN load. The outer wheel path was selected because pavement distress usually occurs in the wheel path. The measured mid-depth temperature of the asphalt overlay layer was considered as the effective temperature to represent the asphalt layer temperature at the time of the FWD measurement. The backcalculated moduli of the asphalt overlay layer were then corrected to the standard temperature of 20°C using the asphalt temperature adjustment factor (ATAF) following the model presented in Lukanen et al. 2000. The first time of FWD measurement was considered to be the reference point to which the overlay backaclualted moduli ratio was calculated. The ratio of the backcalculated moduli of the overlay layer at different times to the first time of measurements was computed to investigate the stiffness degradation over time.

2.6 Modified Structural Number (SNC)

The structural number represents the effective structural strength of the existing pavement. The modified structural number (SNC) based on the FWD maximum deflection (D_0) for asphalt pavement (equation 3) was calculated to determine the structural capacity of four sections. SNC was calculated based on the normalized maximum deflection to 40 kN and corrected to a reference temperature of 20°C for reasonable comparison among four sections:

$$\text{SNC} = 3.2 \times D_0^{-0.63} \tag{3}$$

where SNC = the modified structural number and D_0 = the maximum deflection (mm).

The ratio of the SNC at different times to the first time of measurements was computed to investigate the structural capacity deterioration over time.

3 RESULTS AND DISCUSSIONS

3.1 *Maximum deflection (D_0)*

Figure 2 shows the average normalized maximum deflection to 40 kN load and corrected to 20°C with the associated standard deviation in the outer wheel path for each section at different time of measurements. As the figure shows, the sections with polymer-modified asphalt showed relatively lower deflection values compared to the control section at each time of measurements. This suggests that the polymer slightly improved the structural capacity of the sections regardless of the dosage rate. This figure also depicts that the polymer dosage had a smaller impact on the deflection values. The section modified with 5% polymer showed relatively the lowest deflection values at each time of measurement.

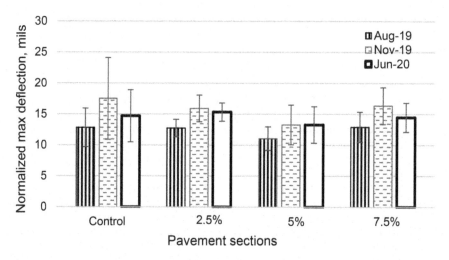

Figure 2. Normalized maximum deflection for all pavement sections.

Figure 3 shows the ratio of normalized maximum deflection at different time after construction to normalized maximum deflection right after construction. The largest ratio was for the control section, where there is a 15-37% increase in the deflection after one year of construction. On the other hand, there is an average of 12-20% increase in the deflection after one year of construction for the polymer modified sections. The polymer dosage had a slight impact on the rate of change in deflection. This figure also depicts that the percent of change in deflection for the control is higher than those in polymer-modified sections. This indicates that the pelletized-polymers could have the potential to lower the degree of deterioration the pavement will experience over time.

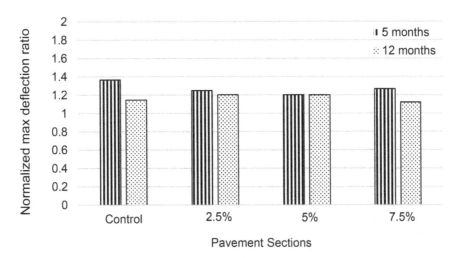

Figure 3. Ratio of seasonal normalized maximum deflection for all pavement sections.

3.2 *Backcalculation analysis*

Figure 4 shows the ratio of the AC overlay moduli at a different time of measurements to AC overlay moduli right after construction. This figure illustrates that the rate of deterioration in the control section was the greatest among all other sections after 5 months of construction, where the control section showed a 70% decrease in the AC overlay moduli. On the other hand, the sections modified with polymer shows a different trend. As the figure shows, the polymer dosage had a significant impact on the rate of deterioration. The section that was modified with 2.5% and 7.5% polymer exhibited no deterioration in the AC overlay moduli. The sections that were modified with 5% showed a 15% drop in the AC overlay moduli. After one winter freeze-thaw season and one year after construction, 5% and 7.5% polymer sections showed similar performance which there is a 35% decrease in the AC overlay moduli while the control and 2.5% polymer sections showed no change in the AC overlay moduli. This indicates that the section modified with 2.5% polymer may lower the rate of deterioration over time. These mixed results in the figure could be due to the viscoelastic behavior of the asphalt overlay layer, where the backcalculation approach presented in this study wasn't able to distinguish the accurate influence on the thin AC overlay. This also could be due to the contribution of the high moisture content accumulated in the base layer after the spring-thaw time, which adversely impacted the stiffness of remaining layers.

3.3 *Modified Structural Number (SNC)*

Figure 5 shows the ratio of the modified structural number (SNC) for all four overlay sections over one year. The results show that there is a slight impact from the pavement section on the change in the SNC. The results also revealed that the dosage of the polymer has a smaller impact on change in the structural capacity of pavements. A 9-18% reduction of the structural capacity for control section and 7-13% reduction for polymer sections are observed over one year after construction.

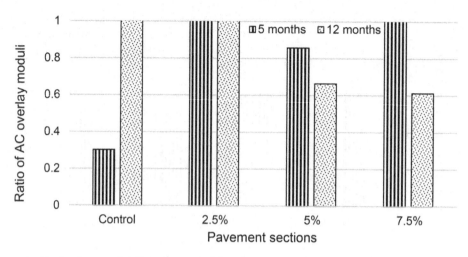

Figure 4. Ratio of seasonal AC overlay moduli for all pavement sections.

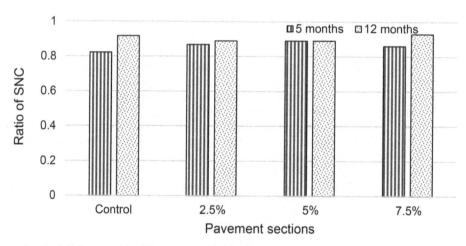

Figure 5. Ratio of seasonal SNC for all pavement sections.

4 CONCLUSION

A control section and three pelletize-polymer modified sections were constructed in northern New England to better understand the deterioration rate of pelletized-polymer-modified sections over time. The asphalt mixture for the control section was produced in the plant using a Superpave mix design with a PG 64-22 binder. The asphalt mixtures for the three other sections were modified using pelletize polymer, which was added during the plant production using the dry mix procedure. No issue was observed during the production and construction of solid polymer-modified sections. FWD test was conducted shortly and periodically over one year after construction on the control and three polymer sections to preliminary investigate the structural capacity of these four sections and determine the deterioration rate over one year of service (two summer seasons and one winter season). Three different analysis methods were conducted to better analyze the FWD deflection data. The findings and conclusions can be summarized as follows:

- The results from the maximum deflection showed that the polymer-modified sections showed less degradation over one year of service. It is concluded that the addition of polymer may slightly improve the structural capacity of the pavement structure.
- The backcalcualtion analysis showed that the climate season and the dosage of polymer had an impact on stiffness degradation over time. Before the winter season, the control section exhibited the largest degradation after 5 months of service among all other polymer sections. The sections modified with 2.5% and 7.5% polymer exhibited no degradation and a slight degradation was observed from 5% polymer section. After one winter season, the control and 2.5% polymer sections exhibited no degradation. Though, 2.5% and 5% polymer sections exhibited degradation of almost 35%. These mixed results between before and after winter season could be due to the seasonal variations of unbound layers' moisture and the limitation of the backcalculation analysis to differentiate the impact of AC overlay.
- The structural strength analysis using the structural number showed that there is a similar degradation over time among control and three polymer sections. The rate of polymer dosage had a smaller impact on the change in the structural capacity over time.

Therefore, it can be concluded that this type of pelletized polymer has the potential to improve the structural capacity and lower the degradation the pavement will experience over time. However, further extensive field investigation is still needed over time to display a definitive conclusion about this polymer type.

ACKNOWLEDGEMENTS

The use of trade, product, or firm names in this document is for descriptive purposes only and does not imply endorsement by the U.S. Government. The tests described and the resulting data presented herein, unless otherwise noted, are based upon work supported by the US Army ERDC under PE 0603119a, Project T26 "Military Engineering Applied Research", Task "Innovative Construction Materials for the Arctic". Permission was granted by the Director, Cold Regions Research and Engineering Laboratory to publish this information. This project was supported in part by an appointment to the Research Participation Program at the US Army Cold Regions Research and Engineering Laboratory (CRREL), US Army ERDC, U.S. Department of Defense (DoD), administered by the Oak Ridge Institute for Science and Education through an interagency agreement between the U.S. Department of Energy and DoD. The findings of this paper are not to be construed as an official Department of the Army position unless so designated by other authorized documents.

REFERENCES

Azam, A. M., S. M. El-Badawy, and R. M. Alabasse. 2019. "Evaluation of Asphalt Mixtures Modified with Polymer and Wax." *Innovative Infrastructure Solutions 4* (1): 43.
Bates R, Worch R. Engineering Brief No. 39, *Styrene–butadiene rubber latex modified asphalt*. Federal Aviation Administration, Washington, DC, 1987.
H.U. Bahia, D. Perdomo, P. Turner. *Applicability of Superpave binder testing protocols to modified binders*. Transportation Research Record, 1586 (1997), pp. 16–23.
Iterchimica 2008. *Road and Airfield uses of SUPERPLAST Modified Hot mix Asphalt*.
Isacsson, U., and H. Zeng. 1998. Low-Temperature Cracking of Polymer-Modified Asphalt. *Materials and Structures* 31 (1): 58–63.
Kim, Y. R., B. O. Hibbs, and Y. C. Lee. 1995. "Temperature Correction of Deflections and Backcalculated Asphalt Concrete Moduli." *Transportation Research Record 1473*, 55–62. Washington, DC: Transportation Research Board, National Research Council.
Lukanen, E. O., R. Stubstad, and R. Briggs. 2000. *Temperature Predictions and Adjustment Factors for Asphalt Pavement*. FHWA-RD-98-085. McLean, VA: Federal Highway Administration.
Newman, J. K. 2003. "Flexural Beam Fatigue Properties of Airfield Asphalt Mixtures Containing Styrene-Butadiene Based Polymer Modifiers." In Proceedings of the *Sixth International RILEM Symposium on Performance Testing and Evaluation of Bituminous Materials*, 357–363. Bagneux, France: RILEM Publications SARL.

Pierce, L. M., J. E. Bruinsma, K. D. Smith, M. J. Wade, K. Chatti, and J. Vandenbossche. 2017. *Using Falling Weight Deflectometer Data with Mechanistic-Empirical Design and Analysis, Volume III: Guidelines for Deflection Testing, Analysis, and Interpretation.* FHWA-HRT-16-011. McLean, VA: U.S. Federal Highway Administration.

Roque R, Birgisson B, Tia M, Kim B, Cui Z. *Guidelines for the use of modifiers in Superpave mixtures: Executive summary and volume 1 of 3 volumes: Evaluation of SBS modifier.* State Job 99052793. Florida Department of Transportation, Tallahassee, FL, 2004.

Schmalzer, P. N. 2006. *Long-Term Pavement Performance Program Manual for Falling Weight Deflectometer Measurements.* FHWA-HRT-06-132. McLean, VA: U.S. Federal Highway Administration, Office of Infrastructure Research and Development.

Wegan, V. 2001. *Effect of Design Parameters on Polymer Modified Bituminous Mixtures.* Copenhagen, Denmark: Danish Road Directorate.

NDT and APT

Traffic speed deflectometer measurements at the Aurora instrumented road test site

C.P. Nielsen
Greenwood Engineering A/S, Copenhagen, Denmark

P. Varin & T. Saarenketo
Roadscanners Ltd, Rovaniemi, Finland

P. Kolisoja
Tampere University, Tampere, Finland

ABSTRACT: The results from a comprehensive set of validation measurements of the Traffic Speed Deflectometer (TSD) are presented. The measurements took place at the Aurora instrumented road test site in Muonio, Finnish Lapland, at a range of different driving speeds and road temperatures. The ability of the TSD to accurately measure pavement surface response, was validated by comparing the TSD measurements to measurements by in-situ displacement transducers. In addition, a linear viscoelastic back-calculation algorithm was applied to the TSD measurements to produce estimates of stresses and strains inside the pavement structure. These estimates were compared to measurements by in-situ transducers at the Aurora test site. The tests revealed an excellent agreement between the surface responses measured by the two systems, and a good agreement between the strains predicted from TSD measurements and the strains measured by in-situ transducers. As expected, the back-calculated moduli of the asphalt were found to increase with driving speed and decrease with temperature.

Keywords: Traffic Speed Deflectometer, TSD, Aurora, Structural condition, Instrumented pavement

1 INTRODUCTION

The Traffic Speed Deflectometer (TSD) sees increased use throughout the world for network level monitoring of pavement structural condition. In recent years, improved sensor technology and new analysis techniques have pushed the boundaries for what is possible with TSD measurements. Whereas the earliest TSDs were mainly suitable as screening devices, the newest generation of TSD is able to characterize structural condition at high spatial resolution and deliver high level results such as back-calculated elastic moduli and pavement strains (Nielsen 2019).

The ability of the TSD to characterize broad structural condition has earlier been validated by comparisons with multi-depth deflectometer (Velarde et al. 2016; Kannemeyer, Lategan, and Mckellar 2014), FWD (Manoharan et al. 2018; Roberts et al. 2014), and Benkelman beam measurements. However, the improved capabilities of the newest TSDs make it relevant to engage in a more detailed and ambitious program of validation. In this paper, we present results from a comprehensive measurement campaign taking place at the Aurora instrumented road test site in

Northern Finland. The Aurora site consists of two extensively instrumented road sections with transducers for measuring surface displacement and surface acceleration in vertical direction, AC layer horizontal strain, base layer vertical strain, and vertical stress in the base layer (Kolisoja et al. 2019). The comparisons between the TSD measurements and the transducer measurements take place in two stages. In the first stage, the ability of the TSD to measure the surface response accurately, is validated by a direct comparison of the surface measurements from both systems. The second stage focuses on derived values, and thus constitute a test of the combined measurement and data analysis system. In this stage, back-calculated pavement strains are compared with the strains measured in-situ by the strain transducers. To assess the importance of driving speed and temperature, the measurements took place at a range of driving speeds and at different times of day.

2 MEASUREMENT DESCRIPTION

Two measurement systems were used in this study to provide independent readings of the pavement response. The Traffic Speed Deflectometer is a nondestructive pavement evaluation device, which uses laser Doppler vibrometers to measure the pavement response to a moving 10 tonne axle. The Doppler lasers measure the deflection velocity of the pavement surface at several points on the line passing between the center of the truck's twin tires. Combining this with a measurement of the driving speed, the slope of the deflection basin is recovered. See Refs. (Krarup et al. 2006; Nielsen 2019) for more information.

The Aurora instrumented road test site is part of an open testing ecosystem of intelligent transport and infrastructure solutions launched by the Finnish Transport Infrastructure Agency (FTIA). It includes two road sections extensively instrumented for monitoring both structural responses and condition of these road sections. The Aurora 1 site is located on a stiff subgrade soil, where an old road structure with good bearing capacity was strengthened by laying a new 50 mm overlay on top of the old pavement leading to 120 mm of bound layers. The Aurora 2 site is located in softer subgrade with deformation problems in the old pavement structure. For this reason, the top part of the old pavement structure was mixmilled and a bound base and AC wearing course, altogether 90 mm, was paved on top of the mix-milled base. A special feature in the base course material at the Aurora test site is that it has very high suction properties. The transducers at the Aurora test site operate with a sampling rate of 1 kHz and they allow us to probe the true pavement behavior, both at the surface and inside the pavement structure. To reduce noise, the transducer outputs were filtered with a forward-backward third order Butterworth filter. The cutoff frequency for the filter was chosen to correspond to a wavelength of 40 cm at the driving speed of the TSD.

The measurements took place on the 6th and 7th of July 2020 at the Aurora instrumented road test site in Finnish Lapland. Over the course of the measurements the air temperature ranged between 12.5 °C and 20.5 °C, and the road surface temperature ranged between 13 °C and 24 °C. To test the effect of driving path, 11 measurements were made with a constant driving speed of 80 km/h and with varying distances to the road side. To test the effect of driving speed, 3 additional measurements were made with each of the driving speeds 10 km/h, 20 km/h, 40 km/h, and 80 km/h. These measurements were made in the evening on the 6th with a road surface temperature around 23 °C. In the morning the next day, another 3 measurements were made with each of the driving speeds 20 km/h, 40 km/h, and 80 km/h. This time with a road surface temperature around 14.5 °C. Also, a single measurement with a driving speed of 5 km/h was made. The approximate lateral position of the TSD was determined using vertical accelerometers located with 10 cm spacing across the driving path. The last bit of alignment was done by correlating the back-calculated TSD results with the transducer measurements.

3 COMPARING SURFACE DISPLACEMENT MEASUREMENTS

The Aurora 2 site has three surface displacement transducers mounted at different lateral positions on the road. By differentiating the displacements with respect to time, the deflection velocities can be found. This enables a direct comparison between the deflection velocities measured by the TSD and the deflection velocities measured by the in-situ displacement transducers. This comparison is, however, dependent on the driving path of the TSD; only for the cases where the center of the twin tires passes directly over a displacement transducer, should we expect a good agreement between the two systems. In Figure 1 two examples of deflection velocities measured by the TSD and the displacement transducers are shown. By using the TSD driving speed v_0, the Aurora recording times are converted to positions $x = v_0 t$. This allows for a direct comparison of the two systems on the same axis. Here, and in the remainder of the paper, the reported TSD measurements were averaged over a distance of 2 meters. The black dots indicate the velocities measured by the TSD, and the dashed colored lines show the velocities measured by the displacement transducers. Each line is labeled by the transducer's lateral distance to the path measured by the TSD. It is seen that the displacement transducers close to the TSD path have an excellent agreement with the velocities measured by the TSD.

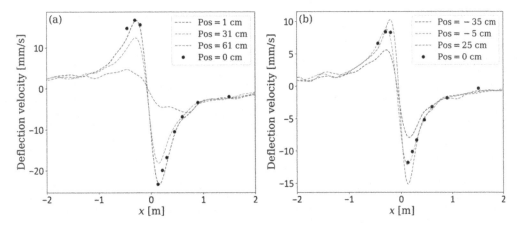

Figure 1. Deflection velocities measured by the TSD (black dots) and measured by the displacement transducers (dashed lines). (a): measurement from day one with a driving speed of 80 km/h. (b): measurement from day two with driving speed of 20 km/h.

4 BACK-CALCULATION

The TSD measurements were processed using a linear viscoelastic back-calculation algorithm developed specifically for the TSD. By taking material viscoelasticity and damping into account the algorithm is able to capture the asymmetric nature of the deflection basin under a moving load. The outputs of the algorithm are back-calculated layer moduli, as well as displacements, stresses and strains at any point in the pavement structure. See Ref. (Nielsen 2019) for details.

The layer thicknesses used in the back-calculation were based on GPR measurements, probing with a sounding rod, and visual inspection of the upper layers. The layer thicknesses are shown in *Table 1*. For the purposes of back-calculation the stiff moraine layer is implemented as bedrock.

Table 1. Pavement layers at the two Aurora sites.

Aurora 1	Aurora 2
12 cm asphalt concrete (AC)	9 cm asphalt concrete (AC)
20 cm base course	25 cm base course
30 cm sub-base course	30 cm sub-base course
50 cm coarse grained material	336 cm sandy material and subgrade
Stiff moraine	Stiff moraine

For each measurement point, the back-calculation algorithm provides values for the complex modulus in each of the four pavement layers. The modulus of the AC layer is of particular interest, since this is expected to change with driving speed and temperature. In Figure 2 the real part of the back-calculated AC modulus is plotted versus driving speed at the Aurora 1 and Aurora 2 sites. The red set of measurements were made in the evening of day one, with a road surface temperature of 23 °C ($\pm$ 1 °C), and the blue set of measurements were made in the morning of day two, with a road surface temperature of 14.5 °C ($\pm$ 1 °C). It is seen that the back-calculated moduli increase with driving speed and decrease with temperature. This agrees with standard AC behavior, and supports our assumption that the back-calculation algorithm finds the true AC layer moduli.

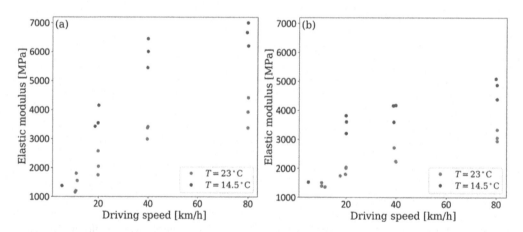

Figure 2. Back-calculated elastic moduli of the AC layer plotted versus driving speed for two different temperatures at the Aurora 1 site (a) and the Aurora 2 site (b).

4.1 Back-calculated field values

The back-calculated deflection velocities, strains, and stresses can be directly compared to the values measured by the various transducers. In the following, back-calculated and measured field values are shown for two of the measurements at Aurora 2. The first measurement was made on day one with a driving speed of 80 km/h and the other was made on day two with a driving speed of 20 km/h.

In Figure 3 examples of back-calculated and measured deflection velocities are shown for the two measurements at Aurora 2. The black dots show the deflection velocities measured at the centerline of the twin tires, the dashed lines show the velocities measured by the transducers, and the full lines show the back-calculated deflection velocities evaluated at the position of the transducers. Each line is labeled by the transducer's lateral distance to the path measured by the TSD. In Figure 3 (b) the deflection velocity obtained from the transducer with the green dashed line is seen to exceed the deflection velocity back-calculated from TSD measurements. The reason for this is probably that the measurement is made under one of the tires, and therefore the details of the tire footprint become important. In the current implementation, the back-calculation algorithm treats each tire as a circular load of radius 10 cm. For most purposes this is sufficiently realistic to yield good results, but it is expected to lead to discrepancies very close to or under the tire.

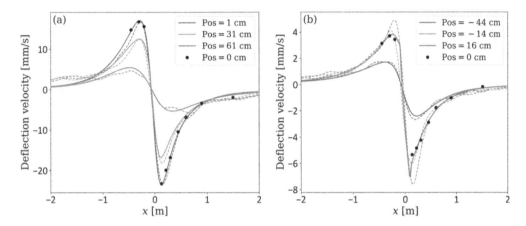

Figure 3. Deflection velocities measured by the TSD (black dots), back-calculated from TSD measurements (full lines), and measured by the displacement transducers (dashed lines). (a): measurement from day one with a driving speed of 80 km/h. (b): measurement from day two with driving speed of 20 km/h.

In Figure 4 examples of back-calculated and measured longitudinal strains in the bottom of the AC layer are shown. There is seen to be good agreement between the back-calculated and measured strains. In Figure 5 examples of back-calculated and measured transversal strains in the bottom of the AC layer are shown. The agreement between measured and back-calculated strains is quite good in Figure 5 (b), but in Figure 5 (a) there are significant differences between the predicted and measured strains. The measurements in Figure 5 (a) are made close to or under the tire, so the observed discrepancies could be due to the assumption of a circular tire footprint. In Figure 6 examples of back-calculated and measured vertical strains in the base layer are shown. There is seen to be good agreement between the predicted and measured values. In Figure 7 examples of back-calculated and measured vertical pressures in the base layer are shown. To improve readability of the figure, only results from 4 of the 8 installed pressure transducers are plotted. These sensors were located at a depth of 280 mm (blue and orange lines) and 180 mm (green and red lines). The agreement between the predicted and measured values is seen to be quite poor, with some of the pressures differing by more than a factor of two. In light of the good agreement between the base layer strains in Figure 6, it is somewhat surprising to observe so large discrepancies in the base layer stresses. This issue is discussed further in the next section.

4.2 Comparison of peak values

In the preceding section the back-calculated and measured field values were compared in detail for two of the measurements. In order to take all 34 measurements into account, a more

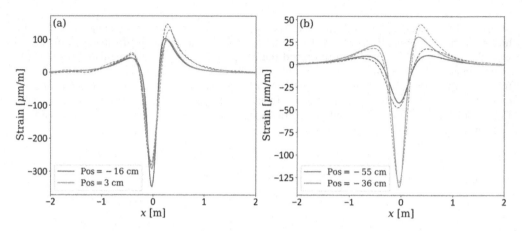

Figure 4. Longitudinal strains in bottom of AC layer back-calculated from TSD measurements (full lines) and measured by strain gauges (dashed lines). (a): measurement from day one with a driving speed of 80 km/h. (b): measurement from day two with driving speed of 20 km/h.

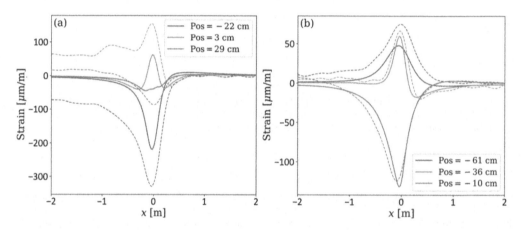

Figure 5. Transversal strains in bottom of AC layer back-calculated from TSD measurements (full lines) and measured by strain gauges (dashed lines). (a): measurement from day one with a driving speed of 80 km/h. (b): measurement from day two with driving speed of 20 km/h.

compact way of comparing data is applied in this section. For each measurement, the peak values of the measured fields are extracted and plotted versus the corresponding peak values of the back-calculated fields. For the deflection velocity, both the positive and the negative peaks are plotted.

In Figure 8 (a) the peak deflection velocities are plotted together with the line of equality. There is seen to be an excellent agreement between the peak velocities measured by the two systems. In Figure 8 (b) the peak longitudinal strains in the bottom of the AC layer are plotted. The relation between the TSD strains and the Aurora strains is better described by a line with slope 0.85 (grey) than by the line with slope 1 (black). This scale error is likely caused by small deviations in the AC layer thickness or sensor depth. For instance, a difference in AC layer thickness of just 13 mm is enough to explain the observed scale error.

In Figure 9 (a) the peak transversal strains in the bottom of the AC layer are plotted. Some amount of scatter around the line of equality is observed, but in general the agreement between the two systems is reasonably good. In Figure 9 (b) the peak vertical strains in the

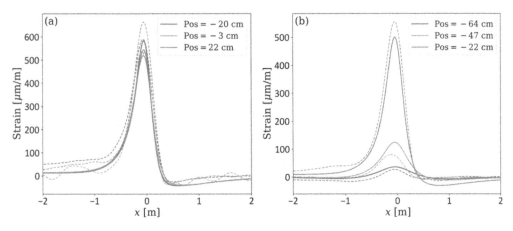

Figure 6. Vertical strains in base layer back-calculated from TSD measurements (full lines) and measured by strain transducers (dashed lines). (a): measurement from day one with a driving speed of 80 km/h. (b): measurement from day two with driving speed of 20 km/h.

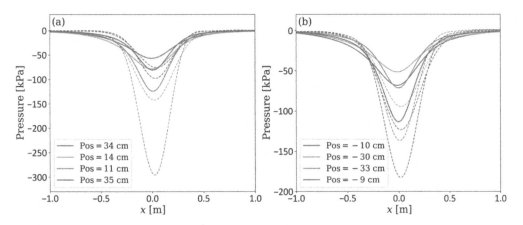

Figure 7. Vertical pressure in base layer back-calculated from TSD measurements (full lines) and measured by pressure transducers (dashed lines). (a): measurement from day one with a driving speed of 80 km/h. (b): measurement from day two with driving speed of 20 km/h.

base layer are plotted. At low strains, the TSD strains are seen to be larger than the corresponding Aurora strains. In general, the agreement between the systems is, however, quite good.

In Figure 10 (a) the peak vertical pressures in the base layer are plotted. The data points are seen to fall far from the line of equality, and in general the agreement between the two systems is poor. This is consistent with the finding in Figure 7, where the detailed pressure distributions were considered. As was previously noted, it is surprising to observe so large discrepancies in the stresses, when all the strain components exhibit a good agreement between the two systems. The likely reason for this discrepancy is nonlinear behavior of the unbound layers, something which is not accounted for in the linear viscoelastic pavement model used to back-calculate the TSD measurements. The simplest model for unbound layer nonlinearity is the so-called "k-theta" model, originally suggested by Brown and Pell (1967). It states that the resilient modulus M_r depends on the sum of principal stresses θ as,

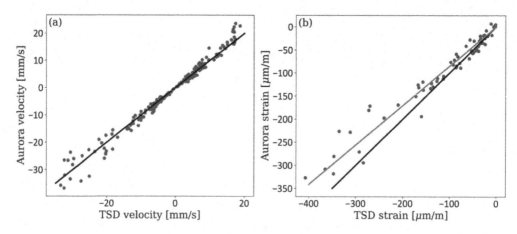

Figure 8. (a): Aurora peak deflection velocities plotted versus the TSD peak deflection velocities. The line of equality is shown in black. (b): Aurora peak longitudinal strains plotted versus TSD peak longitudinal strains. The line of equality is shown in black and a line with slope 0.85 is plotted in grey.

$$M_\mathrm{r} = k_1 P_\mathrm{a} \left(\frac{\theta}{P_\mathrm{a}}\right)^{k_2} \qquad (1)$$

where k_1 and k_2 are material parameters, and the atmospheric pressure P_a is used here to make Equation 1 dimensionally consistent. Since the back-calculated strains have a good agreement with the measured strains, it should be possible to recover the pressure $\sigma_{zz}^{\mathrm{Aurora}}$ from the resilient modulus and the back-calculated strain E_{zz}^{TSD},

$$\sigma_{zz}^{\mathrm{Aurora}} \propto M_\mathrm{r}\varepsilon_{zz}^{\mathrm{TSD}} \propto \theta^{k_2}\sigma_{zz}^{\mathrm{TSD}} \propto \left(\sigma_{zz}^{\mathrm{TSD}}\right)^{1+k_2} \qquad (2)$$

where the last expression makes the simplifying assumption that the sum of principal stresses θ is proportional to $\sigma_{zz}^{\mathrm{TSD}}$. In Figure 10 (b) the Aurora pressure is plotted versus the back-calculated pressure in a log-log plot. It is seen that the points are well approximated by a power law with exponent 1.5 (red line). This corresponds to a k_2 value of 0.5, which agrees with the values reported in the literature (Kolisoja 1997). It is noted, however, that for high pressures, the data points in Figure 10 (b) start to deviate from the simple power law. This implies that the k-theta model is not able to describe the nonlinear unbound material behavior correctly at high deviator stress levels.

The arguments given above indicate, that there is a considerable difference between the subgrade modulus determined by the linear viscoelastic back-calculation and the true resilient modulus of the subgrade. Nevertheless, the estimated and measured subgrade strains are seen to agree reasonably well. The reason for this is, that the subgrade strain has to be consistent with the measured surface deflection. So even though the subgrade modulus is poorly estimated, the estimated subgrade strain cannot deviate too much from the true value. This is in line with the findings in (Nielsen 2020), where it was shown that good estimates of the pavement strains can be obtained from knowledge of the deflection basin, even when the elastic moduli are not known.

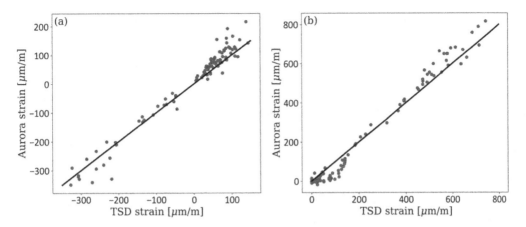

Figure 9. (a): Aurora peak transversal strains plotted versus TSD peak transversal strains. The line of equality is shown in black. (b): Aurora peak vertical base course strains plotted versus TSD peak vertical base course strains. The line of equality is shown in black.

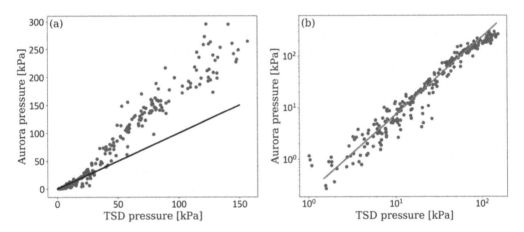

Figure 10. (a): Aurora peak vertical pressures plotted versus the TSD peak vertical pressures. The line of equality is shown in black. (b): Same as (a), but plotted with logarithmic axes. The red line is a power law with exponent 1.5.

5 CONCLUSION

The ability of the TSD to assess pavement response under a moving load, was evaluated by comparison with in-situ transducers at the Aurora instrumented road test site. The pavement deflection velocities measured by the TSD were found to have an excellent agreement with the deflection velocities measured by the in-situ displacement transducers. Estimates of stresses and strains inside the pavement structure were obtained by applying a linear viscoelastic back-calculation algorithm to the TSD measurements. There was found to be a good agreement between measured and back-calculated horizontal strains in the bottom of the AC layer and vertical strains in the bottom of the base layer. In contrast, there was a poor agreement between measured and back-calculated pressures in the base layer. This was found to be due to stress hardening of the base layer material, something which was not accounted for in the linear viscoelastic pavement model. In agreement with standard rheological models for asphalt, the back-calculated moduli of the AC layer were found to increase with driving speed and decrease with temperature.

ACKNOWLEDGEMENTS

The authors would like to thank Nuutti Vuorimies and Antti Akkanen from Tampere University for valuable assistance during the measurement campaign.

REFERENCES

Brown, S. F., and P. S. Pell. 1967. *An Experimental Investigation of the Stresses, Strains and Deflections in a Layered Pavement Structure Subjected to Dynamic Loads*, In Proceedings of the 2nd International Conference on Structural Design of Asphalt Pavements, 487–504.

Kannemeyer, L, W Lategan, and A Mckellar. 2014. *Verification of Traffic Speed Deflectometer Measurements Using Instrumented Pavements in South Africa*, 1–29.

Kolisoja, P. 1997. *Resilient Deformation Characteristics of Granular Materials*, Tampere University of Technology.

Kolisoja, P, N Vuorimies, A Kurki, and T Saarenketo. 2019. *Open Structural Monitoring Data from Two Extensively Instrumented Road Sections – Case Aurora*, In Proceedings of the XVII ECSMGE–2019

Krarup, J., S. Rasmussen, L. Aagaard, and P. G. Hjorth. 2006. *Output from the Greenwood Traffic Speed Deflectometer*, In Paper Presented to 22nd ARRB Conference. Canberra, Australia.

Manoharan, S., G. Chai, S. Chowdhury, and A. Golding. 2018. *A Study of the Structural Performance of Flexible Pavements Using Traffic Speed Deflectometer*, Journal of Testing and Evaluation 46 (3): 1280–89.

Nielsen, C.P. 2019. *Visco-Elastic Back-Calculation of Traffic Speed Deflectometer Measurements*. Transportation Research Record.

Nielsen, C.P. 2020. *Deriving Pavement Deflection Indices from Layered Elastic Theory*. Transportation Research Record.

Roberts, Jon, Ulysses Ai, Tyrone Toole, and Tim Martin. 2014. *Traffic Speed Deflectometer Data Review and Lessons Learnt*.

Velarde, J. A., S. Rocha, S. Nazarian, G. Rada, S. Thyagarajan, and R. V. Siddharthan. 2016. *Use of Embedded Sensors to Evaluate Performance of Traffic Speed Deflection Devices*, Journal of Testing and Evaluation 45 (4): 1316–25.

Unbound pavement materials' response to varying groundwater table analysed by falling weight deflectometer

M. Fladvad
Road Technology, Norwegian Public Roads Administration, Trondheim, Norway
Department of Geoscience and Petroleum, NTNU – Norwegian University of Science and Technology, Trondheim, Norway

S. Erlingsson
Pavement Technology, Swedish National Road and Transport Research Institute (VTI), Linköping, Sweden
Faculty of Civil and Environmental Engineering, University of Iceland, Reykjavik, Iceland

ABSTRACT: Expected climate changes will in many areas represent a shift towards increased precipitation and more intense rainfall events. This may lead to increased moisture within road structures and possible overloading of road drainage systems. Pavement design methods must therefore be able to predict the behaviour of pavement materials at increased moisture levels. An instrumented accelerated pavement test (APT) has been conducted on two thin flexible pavement structures with coarse-grained unbound base course and subbase materials using a heavy vehicle simulator (HVS). The two pavement structures were identical except for the particle size distribution (PSD) of the subbase materials, where one had a dense 0/90 mm curve with a controlled fines content, and the other had an open-graded 22/90 mm curve. The APT was conducted using constant dual wheel loading, and three different groundwater levels were induced in order to change the moisture content in the structures. Falling weight deflectometer (FWD) measurements were conducted at each groundwater level during the APT. Additional FWD measurements were conducted as the groundwater was lowered after the APT loading was finished. The moisture content in the unbound materials was continuously measured throughout the test. The analysis is focussed on the response of the unbound aggregate layers to varying moisture levels in the pavement structures. Analysis results show how the dense- and open-graded materials respond to the moving groundwater table, and how this affects the deflection of the full structures.

Keywords: Falling weight deflectometer, groundwater table, unbound granular materials, large-size aggregates, accelerated pavement testing

1 INTRODUCTION

Climate changes are expected to result in increased precipitation and more intense rainfall events in many regions (Seneviratne et al., 2012). Such changes in environmental conditions may lead to increased moisture in unbound pavement materials. Overloading of road drainage systems is another potential effect, which may lead to saturation of pavement materials.

Pavement saturation during flooding is one of the key deterioration processes that result in degradation of pavement materials (Lu et al., 2018). When unbound materials are saturated, pavement structures observe a significant loss of structural capacity (Elshaer et al., 2019). As

DOI: 10.1201/9781003222880-37

the moisture content increases, the friction between aggregate particles becomes lower, and the resistance to differential particle deformation is reduced, leading to a reduced resilient modulus of unbound aggregates (Lekarp and Dawson, 1998; ARA Inc., 2004; Erlingsson, 2010).

The moisture content in pavement materials is affected by both groundwater level, drainage conditions and moisture sensitivity in the material itself and underlying layers. Hence, testing individual materials will not give a full overview of the behaviour of a pavement structure in changing moisture conditions. The influence of moisture on pavement structures has been investigated successfully using heavy vehicle simulator (HVS) in research by e.g. Erlingsson (2010), Saevarsdottir & Erlingsson (2013) and Camacho-Garita et al. (2020).

Falling weight deflectometer (FWD) is a non-destructive test commonly used for pavement response evaluation. Surface deflections resulting from a dynamic load is registered by the deflectometer, and layer moduli is backcalculated using a mechanistic model (Chen et al., 1999; Marecos et al., 2017; Cafiso et al., 2020).

In the Nordic countries, large-size (upper sieve size D ≥ 90 mm) unbound pavement materials are commonly used in pavement structures (Fladvad et al., 2017). Materials where D > 90 mm are not covered by the European standards for construction aggregates (EN 13242, 2007; EN 13285, 2018). The impact of increased moisture in large-size unbound pavement materials is not well understood, as the particle size makes laboratory methods for measuring moisture content unsuitable. Open-graded materials can be used to ensure drainage of pavement structures. Such materials may be difficult to analyse with traditional pavement design methods, because of their lack of an optimum moisture content. Degree of saturation has been found to be a valuable descriptor of the moisture conditions in pavement materials (ARA Inc., 2004; Tatsuoka & Correia, 2018), but cannot be applied for draining materials.

To better account for expected climate changes in pavement design, experimental evaluation of large-size pavement materials' response to changing moisture content is necessary. The aim of the present research was to investigate:

1. How moisture content in pavement materials change as groundwater levels vary
2. How unbound large-size pavement materials respond to moisture changes in terms of stiffness measured by FWD.

This paper focusses on FWD measurements conducted during an accelerated pavement test (APT). Response modelling and permanent deformation modelling based on stress and strain registrations from the same APT are reported by Fladvad & Erlingsson (2021a; 2021b).

2 MATERIALS AND METHODS

2.1 *Pavement structures*

Two thin flexible pavement structures were tested in an APT using an HVS. The structures were constructed in a concrete test pit which was 5 m wide, 15 m long, and 3 m deep. The pavement structures had a total thickness of just below 0.6 m, and the remaining 2.4 m test pit depth was filled with a silty sand subgrade. Layer thicknesses for both structures are shown in Figure 1.

Both pavement structures were constructed from asphalt concrete (AC) surface and binder layers over a 0/32 mm unbound base layer. The subbase materials differed between the structures, where one structure had a wide-graded 0/90 mm crushed rock subbase with a controlled fines content, whereas the other structure had an open-graded 22/90 mm crushed rock subbase. Particle size distribution (PSD) curves for the unbound pavement materials are shown in Figure 2.

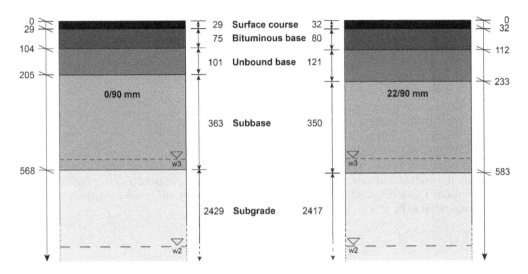

Figure 1. Outline of pavement structures. Depth below surface and layer thicknesses in mm. Blue dashed lines represent groundwater levels for phases w2 and w3.

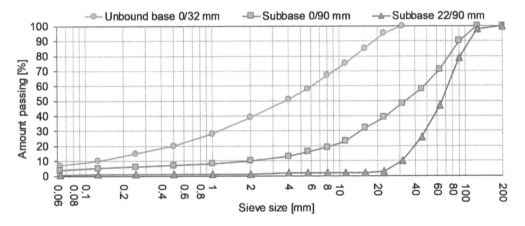

Figure 2. Particle size distribution curves for unbound base and subbase materials.

2.2 Groundwater variations

The pavement structures were subjected to three different groundwater levels in succession. First, the groundwater table (GWT) was located at great depth below the surface, leaving the pavement materials and subgrade in a natural moist state (phase w1). Next, in phase w2, water was added to the test pit, resulting in a constant GWT 30 cm below the subbase layers. This GWT level corresponds to a functioning drainage system located 30 cm below the pavement structure. In phase w3, GWT was additionally raised to about 5 cm into the subbase layers. This GWT level corresponds to an overloaded drainage system. The GWT levels from phase w2 and w3 are illustrated in Figure 1.

During phases w1-w3, the pavement structures were subjected to accelerated traffic loads. After the traffic loading finished in phase w3, GWT was lowered in steps to the levels from phases w2 and w1.

2.3 Falling weight deflectometer measurements

FWD measurements were conducted in each groundwater phase. The measurements were conducted at the following times:

w1.1: Before accelerated traffic loading started
w2.1: After traffic loading in phase w2 was finished
w3.1: Before traffic loading in phase w3 started
w3.2: After traffic loading in phase w3 was finished
w2.2: When GWT was reduced to the level from phase w2
w3.2: When GWT was reduced to the level from phase w1

The measurements were conducted at ambient temperature, varying from 3.8 to 15.5°C. Knowing the asphalt concrete stiffness $E_{T_{ref}}$ at a reference temperature T_{ref}, the stiffness E_T at a temperature T can be calculated from Eq. (1). The reference stiffness $E_{T_{ref}}$ used in the calculations was 6500 MPa at T_{ref} 10°C, and b was assumed 0.065.

$$E_T = E_{T_{ref}} e^{-b(T-T_{ref})} \tag{1}$$

2.4 Accelerated traffic

The pavement structures were subjected to accelerated traffic load from an HVS. The traffic was applied as bidirectional dual-wheel loading with 60 kN dual-wheel load, corresponding to a 120 kN axle load. The tyre pressure was 800 kPa, and the HVS kept a constant speed of 12 km/h. During traffic loading, the temperature in the heavy vehicle simulator was kept constant at 10°C using a climate chamber.

The accelerated traffic amounted to 1 233 000 load repetitions, divided between 550 000 load repetitions in phase w1, 385 000 load repetitions in phase w2 and 298 000 load repetitions in phase w3. The total accelerated traffic load corresponds to 2.55 million 10-tonne standard axles, which again may correspond to 20 years of an annual average daily traffic (AADT) of 3000 vehicles, assuming 10 % heavy vehicles.

3 RESULTS AND DISCUSSION

3.1 Moisture content

The volumetric water content in the unbound base, subbase and subgrade registered at the time of FWD measurements is summarised in Table 1. The Sb registrations represent the average from two moisture sensors, while the UB and Sg moisture was registered by one sensor in each layer. The subgrade moisture is common to both structures, as both were built on the same silty sand subgrade.

The moisture registrations show that the two subbase materials, as expected, have very different moisture contents. The open-graded 22/90 mm subbase has a very low and nearly constant water content around 2 %, varying by only 0.5 % through all groundwater phases. The well-graded 0/90 mm subbase maintains a moisture content around 7.4-8.4 %.

The moisture content in the unbound base reaches a maximum of 10.0 % in both structures, but this level was reached much earlier in the 0/90 mm structure.

For the subgrade, water content varies from 12.8 % in the naturally moist state (w1), to 26-28 % in the fully saturated state when GWT is located above the moisture sensor.

3.2 Falling weight deflectometer

Table 2 shows the stiffnesses backcalculated from measured deflection at all groundwater levels. In the backcalculation, the AC stiffness is adjusted to a surface temperature of 10°C.

Table 1. Volumetric water content [%] registered at the time of FWD measurements. UB – Unbound base, Sb – Subbase, Sg – Subgrade.

Measurement ID	0/90 mm structure			22/90 mm structure		
	UB	Sb	Sg	UB	Sb	Sg
w1.1	7.7	7.4	12.8	7.1	1.8	12.8
w2.1	8.6	7.8	26.6	7.2	1.9	26.6
w3.1	8.7	7.9	26.6	7.1	2.0	26.6
w3.2	10.0	8.4	27.7	10.0	2.3	27.7
w2.2	9.3	7.9	27.5	9.7	2.0	27.5
w1.2	9.0	7.6	16.5	9.6	1.9	16.5

The backcalculation was conducted using the ERAPave software (Erlingsson and Ahmed, 2013). The subgrade is divided into two layers, where the boundary between the layers is the GWT level in phase w2. The subbase is divided into three layers, where the lowest layer is below the GWT in phase w3.

The backcalculated deflection results are compared in Figure 3. The results are based on 50 kN falling weight load.

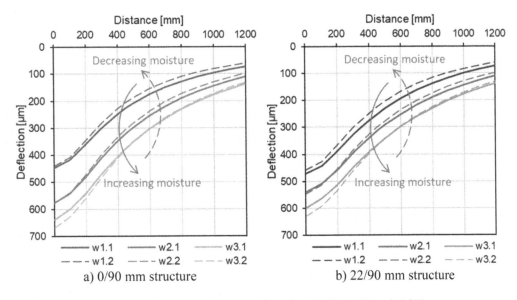

a) 0/90 mm structure   b) 22/90 mm structure

Figure 3. Backcalculated deflection, temperature adjusted to 10 °C. FWD load 50 kN.

As expected, increased GWT and moisture content resulted in increased deflection and reduced stiffness of the pavement materials. This tendency is reversed when GWT is lowered. Although the 0/90 mm structure originally shows lower deflection and higher subbase stiffness, this structure is more affected by the moisture changes and shows higher deflection than the 22/90 mm structure in phase w2 and w3. The increase in deflection in the load centre is summarised in Table 3. The difference is most

Table 2. Backcalculated stiffnesses for pavement materials.

a) 0/90 mm structure

Layers	Thickness [mm]	Stiffness [MPa]					
		w1.1	w2.1	w3.1	w3.2	w2.2	w1.2
Surface and binder course	104	6500	6500	6500	5671	5671	5671
Unbound base	101	250	225	225	203	214	225
Subbase 1	163	220	157	157	149	157	220
Subbase 2	150	220	143	143	129	143	220
Subbase 3	50	220	129	129	116	129	220
Subgrade 1	300	82	66	44	46	70	98
Subgrade 2	2132	82	53	44	46	60	98

b) 22/90 mm structure

Layers	Thickness [mm]	Stiffness [MPa]					
		w1.1	w2.1	w3.1	w3.2	w2.2	w1.2
Surface and binder course	112	6500	6500	6500	5671	5671	5671
Unbound base	121	250	250	250	203	214	225
Subbase 1	150	143	143	143	135	143	154
Subbase 2	150	143	143	143	135	143	154
Subbase 3	50	143	143	143	135	143	154
Subgrade 1	300	82	66	44	46	70	98
Subgrade 2	2117	82	53	44	46	60	98

prominent from phase w1 to w2, where the increase is about double for the 0/90 mm structure compared to the 22/90 mm structure.

Table 3. Increase of deflection in load centre, compared to phase w1.

	0/90 mm structure	22/90 mm structure
w2	29 %	15 %
w3	49 %	34 %

The 0/90 mm subbase material regains its original stiffness from measurement w1.1 when the GWT is lowered to the original state in w1.2. The 22/90 mm subbase material, on the other hand, shows increased stiffness in w1.2 compared to w1.1, indicating stiffening and stabilisation due to traffic helped by moisture.

Neither subbase material showed any change in stiffness between w2.1 and w.3.1. This is likely an effect of too short time between the GWT raise and the FWD measurement, as the moisture content had not yet increased in these layers. At the time of measurement w3.2, moisture content was increased in the 0/90 mm subbase, and stiffness was decreased accordingly.

The differences in stiffness for the unbound base between the two structures in w2.1 and w3.1 can be explained by the difference in moisture content. The stiffness of the unbound base decreases more quickly in the 0/90 mm structure, corresponding to the quicker moisture increase seen in Table 1. This development is likely due to the limited ability of the open-graded subbase to transport moisture from the GWT to the unbound base. After long time, in measurement w3.2, the moisture content in the unbound base is the same for both structures.

The backcalculation analysis revealed a reduction of the AC stiffnesses between measurement w3.1 and w3.2. The AC stiffness reduction is likely an effect of the accelerated traffic, not the GWT variations.

For all three GW phases, the subgrade stiffness is higher in the second measurement. The increased stiffness shows that the APT has resulted in post-compaction of the subgrade.

The differences in deflection and stiffness from w3.2 to w2.2 and w1.2 are undisturbed by traffic, as these measurements were conducted after the traffic loading was finished. The results clearly show how the lowered GWT reduces deflection and increases the stiffness in all unbound layers, including the subgrade.

Tensile strain at the bottom of the AC layers caused by the FWD load was calculated from the backcalculated stiffnesses using ERAPave (Table 4). In the beginning of the test (w1.1), the tensile strain was similar in both structures, as they have the same stiffness in AC and unbound base. The maximum tensile strain was calculated in measurement w3.2, where the moisture content was highest.

Table 4. Calculated tensile strain ε_t from FWD load at the bottom of AC layers.

Measurement ID	Tensile strain [μstrain]	
	0/90 mm structure	22/90 mm structure
w1.1	224	224
w2.1	248	226
w3.1	251	229
w3.2	280	262
w2.2	270	253
w1.2	246	246

The allowable number of load repetitions N_f to cause fatigue cracking is related to the tensile strain ε_t at the bottom of AC layers through Eq. (2) (Huang, 2012).

$$N_f = f_1(\varepsilon_t)^{-f_2} \qquad (2)$$

To quantify the consequences of the increased ε_t caused by increased moisture content in the pavement, N_f for phase w1 and w3 can be compared. Assuming $f_2 = 4$, ε_t from w3.2 reduces N_f to 41 % compared to w1.1 for the 0/90 mm structure, and 53 % for the 22/90 mm structure.

For the 0/90 mm structure, when surface deflection increases by 49 %, the tensile strain under the AC increases by 25 %, and the fatigue life time is reduced to only 41 % of the original. Similarly for the 22/90 mm structure; surface deflection increases by 34 %, tensile strain increases by 17 % and the fatigue life time is reduced to 53 % of the original. All these indicators show that the 0/90 mm structure is more affected by the raised GWT than the 22/90 mm structure.

3.3 *Particle Size Distribution (PSD)*

Figure 4 displays PSD curves for the subbase materials before and after accelerated pavement testing. Post testing samples were collected separately from the upper and lower parts of the subbase layers. Each sample weighed minimum 80 kg.

For the 0/90 mm subbase, amount of material increased in the size from 1–20 mm, and decreased from 40–125 mm. For the 22/90 mm subbase, amount of material increased from 20–45 mm, and decreased from 63–125 mm. Both materials showed an increase in fines < 0.063 mm.

Both subbase materials show degradation in the form of abrasion and fragmentation as a result of the construction process and traffic loading. However, the FWD results show that the structural capacity of the pavement layers has not been reduced; the 22/90 mm even show increased stiffness in measurement w1.2 compared to w1.1.

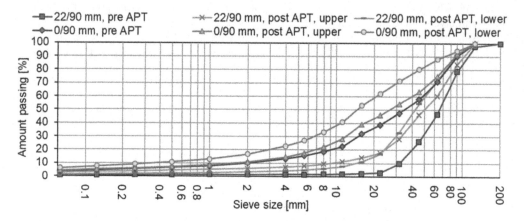

Figure 4. Particle size distribution curves before and after accelerated pavement testing.

4 CONCLUSIONS

Two flexible pavement structures were evaluated using FWD as GWT beneath the structures moved and induced changes in the moisture content of the pavement materials. Differences in the response of unbound pavement materials to changing moisture conditions were successfully measured by FWD.

The bearing capacity of the pavement structures were affected by the changing GWT beneath and in the pavement structure. For the well-graded structure, the moisture content increased in the unbound base even when GWT was raised only to 30 cm below the subbase, showing how the moisture development in pavement materials are affected by material choice in adjoining layers.

Using an open-graded subbase material makes the structure less affected by changes in the GWT:

– Increasing GWT from great depth to the design drainage level increased deflection by 15 % for the open-graded structure, and 29 % for the well-graded structure.
– GWT corresponding to an overloaded drainage system led to a 34 % increase in deflection for the open-graded structure, and a 49 % increase in the well-graded structure.
– Although the open-graded structure had lower stiffness at the beginning of the test, this structure was less affected by the groundwater changes and showed a 7.7 % increase in subbase stiffness after the APT.
– The fatigue life time is substantially reduced in both structures as GWT is raised, but the 0/90 mm structure is most heavily affected.

Both structures show less deflection after 1.23 million load repetitions than in the measurement conducted before the accelerated traffic started, although the moisture content is higher in all layers in the final measurement. This shows that both structures were able to withstand substantial traffic load without degradation of their bearing capacity even though the PSD curves show some degradation of the aggregate particles in the subbase layers.

Pavement and drainage design must account for increased precipitation and more intense rainfall events in order to limit damage caused by increased moisture in pavement materials.

ACKNOWLEDGEMENTS

The research presented in this paper was financed by the Norwegian Public Roads Administration. The testing was conducted at the Swedish National Road and Transport Research Institute. The authors would like to thank the construction company Veidekke for supplying the subbase materials for the APT.

REFERENCES

ARA Inc. (2004) Guide for the Mechanistic-Empirical Design of New and Rehabilitated Pavement Structures, Final report, NCHRP 1-37A. Washington DC, USA. Available at: http://onlinepubs.trb.org/onlinepubs/archive/mepdg/guide.htm.

Cafiso, S., Di Graziano, A., Fedele, R., Marchetta, V., & Praticò, F. (2020). 'Sensor-based pavement diagnostic using acoustic signature for moduli estimation'. International Journal of Pavement Research and Technology, 13(6), 573–580. doi: 10.1007/s42947-020-6007-4

Camacho-Garita, E. et al. (2020) 'Effect of Moisture on Full-Scale Pavement Distress', Journal of Testing and Evaluation, 48(1), p. 20180902. doi: 10.1520/JTE20180902.

Chen, J., Hossain, M. and Latorella, T.M. (1999). 'Use of Falling Weight Deflectometer and Dynamic Cone Penetrometer in Pavement Evaluation', Transportation Research Record, 1655(1):145–151. doi:10.3141/1655-19

Elshaer, M., Ghayoomi, M. and Daniel, J. S. (2019) 'Impact of subsurface water on structural performance of inundated flexible pavements', International Journal of Pavement Engineering, 20(8), pp. 947–957. doi: 10.1080/10298436.2017.1366767.

EN 13242 (2007) Aggregates for unbound and hydraulically bound materials for use in civil engineering work and road construction EN 13242:2002+A1:2007. Brussels, Belgium: European Committee for Standardization.

EN 13285 (2018) Unbound mixtures - Specifications EN 13285:2018. Brussels, Belgium: European Committee for Standardization.

Erlingsson, S. (2010) 'Impact of Water on the Response and Performance of a Pavement Structure in an Accelerated Test', Road Materials and Pavement Design, 11(4), pp. 863–880. doi: 10.1080/14680629.2010.9690310.

Erlingsson, S. and Ahmed, A. W. (2013) 'Fast layered elastic response program for the analysis of flexible pavement structures', Road Materials and Pavement Design, 14(1), pp. 196–210. doi: 10.1080/14680629.2012.757558.

Fladvad, M., Aurstad, J. and Wigum, B. J. (2017) 'Comparison of practice for aggregate use in road construction—results from an international survey', in Loizos, A., Al-Qadi, I., and Scarpas, T. (eds) Bearing Capacity of Roads, Railways and Airfields. Athens, Greece: CRC Press, pp. 563–570. doi: 10.1201/9781315100333-74.

Fladvad, M. and Erlingsson, S. (2021a) 'Modelling the response of large-size subbase materials tested under varying moisture conditions in a heavy vehicle simulator', Road Materials and Pavement Design. doi: 10.1080/14680629.2021.1883462.

Fladvad, M. and Erlingsson, S. (2021b) 'Permanent deformation modelling of large-size unbound pavement materials tested in a heavy vehicle simulator under different moisture conditions', Road Materials and Pavement Design. doi: 10.1080/14680629.2021.1883464.

Huang, Y. H. (2012) Pavement Analysis and Design. 2nd edn. Pearson Education Inc.

Lekarp, F. and Dawson, A. R. (1998) 'Modelling permanent deformation behaviour of unbound granular materials', Construction and Building Materials, 12(1), pp. 9–18. doi: 10.1016/S0950-0618(97)00078-0.

Lu, D., Tighe, S. L. and Xie, W.-C. (2018) 'Impact of flood hazards on pavement performance', International Journal of Pavement Engineering, pp. 1–7. doi: 10.1080/10298436.2018.1508844.

Marecos, V., Fontul, S., de Lurdes Antunes, M., & Solla, M. (2017). 'Evaluation of a highway pavement using non-destructive tests: Falling Weight Deflectometer and Ground Penetrating Radar'. Construction and Building Materials, 154, 1164–1172. doi: 10.1016/j.conbuildmat.2017.07.034

Saevarsdottir, T. and Erlingsson, S. (2013) 'Water impact on the behaviour of flexible pavement structures in an accelerated test', Road Materials and Pavement Design. Taylor & Francis, 14(2), pp. 256–277. doi: 10.1080/14680629.2013.779308.

Seneviratne, S. et al. (2012) 'Changes in climate extremes and their impacts on the natural physical environment', in Field, C. B. et al. (eds) Managing the Risk of Extreme Events and Disasters to Advance Climate Change Adaptation. A Special Report of Working Groups I and II of the Intergovernmental Panel on ClimateChange (IPCC). Cambridge, UK, and New York, NY, USA: Cambridge University Press, pp. 109–230. Available at: https://www.ipcc.ch/site/assets/uploads/2018/03/SREX-Chap3_FINAL-1.pdf.

Tatsuoka, F., & Correia, A. G. (2018). 'Importance of controlling the degree of saturation in soil compaction linked to soil structure design'. Transportation Geotechnics, 17, 3–23. doi: 10.1016/j.trgeo.2018.06.004

FWD quality assurance in Germany

D. Jansen
Federal Highway Research Institute – BASt, Bergisch Gladbach, Germany

ABSTRACT: The importance of bearing capacity measurements for structural assessment is increasing worldwide. The establishment of quality assurance procedures is therefore all the more important. As a rule, these include self-monitoring by the operator and calibration of the measuring equipment at the manufacturer's premises or at calibration centers. A complete quality assurance includes, in addition to the inspection and calibration of individual devices and device components, the participation in comparative investigations. For the Falling Weight Deflectometer (FWD), comparative measurements were established in Germany on the basis of international existing comparative tests and guidelines. The measurements take place annually on the premises of the Federal Highway Research Institute (BASt) according to a fixed schedule. In addition to checking the comparability of the deflection measurement, the repeatability and the temperature measuring equipment and the FWD setups are also checked. Furthermore, the annual event enables the important exchange between the FWD operators, which in turn is also useful for quality assurance.

Keywords: FWD, bearing capacity, comparison, quality

1 INTRODUCTION

The first Falling Weight Deflectometers (FWD) were put into operation in Germany in the early 1990s. The scientific introduction took place in 1996 (Straube, E., Beckedahl, H., & Huertgen, H., 1996). A series of working papers on the planning, execution and evaluation of bearing capacity measurements were and are being prepared by the bearing capacity committee of the German Research Association for Roads and Transport (FGSV).

In the FWD related working paper (FGSV, 2008) recommendations are given for internal and external monitoring of the measuring equipment following the recommendations of COST Action 336 (European Commission, 2005), which are mainly the device calibration and check up by the manufacture and repeatability tests by the operator/owner in his own responsibility. Also, the execution of interlaboratory tests is recommended: "A central test centre will be set up to check the repeatability and reproducibility of test results of the FWD (interlaboratory tests) used in Germany." The establishment of such a central test center has not been carried out for a long time. However, the results of comparable international interlaboratory comparisons and comparative measurements organized from time to time by individual groups show that the permanent organization of such comparative measurements is an essential element of quality assurance.

The Federal Highway Research Institute (BASt), in its role as quality assurance agency for a number of measurement procedures used in the road sector and as a client for FWD measurements within the framework of research projects, has had a great interest in establishing comparative measurements for FWD in Germany. Since such comparative measurements have been established internationally for many years, the experience gained from these projects was used in the planning of the comparative measurements. In particular, the experience gained in the Netherlands was used (Van Gurp, 2013), which was also included in the report on COST Action 336 (European Commission, 2005). In 2015 the first comparative measurements as a start of new series have been conducted at BASt premises.

Figure 1. Participants of 2019 event.

2 HISTORY OF COMPARATIVE EVENTS

As said the first of a new series of comparative event as has been performed in late 2015 followed by a second event in late 2016. At the request of the participants, the event then was moved to spring, i.e. virtually shortly before the start of the measuring season. With this the third event has been performed in early 2018, followed by the fourth in early 2019 (Figure 1). The 2020 event has been cancelled a view days before due to the worldwide COVID-19 epidemic. A new event was planned for March 2021 but will also be postponed to 2022.

Table 1. History of comparative events.

Nr.	Year	FWD/HWD	Participants
1	2015	4x Sweco 1x Dynatest	5x Germany
2	2016	4x Sweco 2x Dynatest	5x Germany
3	2018	6x Sweco 2x Dynatest	7x Germany 1x Belgium
4	2019	8x Sweco 6x Dynatest	7x Germany 1x Belgium 1x Netherlands 2x Denmark 1x Lithuania 1x Estonia
-	2020 cancelled	8x Sweco 10x Dynatest	8x Germany 1x Belgium 3x Netherlands 2x Denmark 3x Poland 1x Switzerland

With the years the number of participating companies increased. The event is open to everybody; therefore, the event has turned into an international event. Table 1 gives an overview of the events and participants.

Since the evaluation and assessment of the comparative measurements are essentially based on statistics, the highest possible number of participants has a quality-enhancing effect on the event.

3 TEST SETUP

The design of the experimental programme is closely aligned with the European consensus elaborated in COST Action 336. The experimental program pursues the following core points:

– Examination of repeatability (each FWD separately),
– Examination of comparability (comparison between the FWDs, also known as reproducibility),
– Examination of the load pulse by evaluation of the time history data as well as
– Investigation of the precision of temperature measurements.

3.1 *General setup*

In order to act as independently as possible from the aspects of traffic safety, only measuring points on the BASt site will be considered. To a certain extent, this can ensure that repeated measurements are carried out in the coming years in the same situation. A total of 21 measurement points are available per measurement round. Two measuring points are located on the model road in test hall 9, 15 measuring points on the open area of the vehicle technical test facility (FTVA) and its access road as well as four further measuring points on the visitor parking lot. Figure 2 shows a map of the situation. Table 2 shows the with the help of GPR measurements determined asphalt thicknesses for area of interest. Figure 3 shows the mean deflections (load centre deflection and deflection at 1.800 mm from load centre) of all selected test points. These data are only to understand the full test setup and will not be used for the interpretation of results.

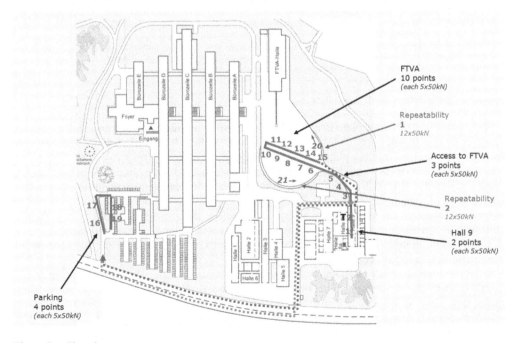

Figure 2. Situation map.

Table 2. Asphalt thickness at measuring points.

measuring point nr.	location	thickness of asphalt [cm]
1	hall 9	ca. 26 cm (on 15 cm hydraulic bound base)
2	hall 9	ca. 12 cm
3-5	FTVA	ca. 30 cm
6-15	FTVA	ca. 18 cm [1]
16	parking	ca. 20 cm [1]
17	parking	ca. 17 cm [1]
18	parking	ca. 17 cm [1]
19	parking	ca. 20 cm [1]
20-21	FTVA	ca. 30 cm

[1] from GPR

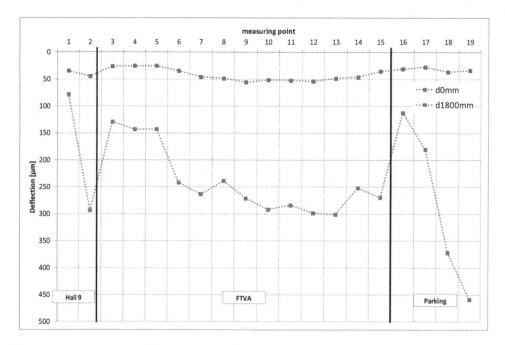

Figure 3. Mean (reference) deflections of test points.

The position of the load plate is marked beforehand at all measuring points with a circle, diameter 40 cm. In addition, cones are set up, at which the left side of the vehicle has to be driven closely past, so that the alignment of the geophone beam is comparable.

In general, the FWDs should be equipped in accordance with the FGSV Working Paper Bearing Capacity, Part B.2 (FGSV, 2008) and brought into operational readiness as for the performance of regular measurements. These and additional requirements are given in Table 3 and will be given to the participants a few weeks before their arrival. In addition, a device audit will be carried out in which questions are asked about the model type and number, year of construction, number of geophones and their positions, number and positions of temperature recordings and other equipment features.

After the arrival of all participants, a short presentation of the process planning is be given by the test supervisor. The different test setups, the position of the measuring points and general information are presented and discussed. Each participant receives a folder with the compilation of all information and a documentation sheet (logbook) for each measurement round, which also clearly showed the measurement sequences and the associated test setups.

Table 3. General equipment requirements.

Geophone positions	0 – 200 – 300 – 450 – 600 – 900 – 1.200 – 1.500 – 1.800 mm
Load pulse	target load = 50 kN/load pulse duration between 25 and 30 ms
Time history	recording should be activated
Filtering	Peak smoothing and Cut-Off frequency at 60 Hz activated
GPS	if available activated, used only for plausibility checks
Temperatures	air and surface temperature (if available) plus asphalt temperatur probe equipment
System time	will be synchronized with the time of the test supervisor

The order of the FWDs among each other is determined, if possible, on the basis of existing experience in such a way that slow FWDs line up at the back. If even more FWDs participate, it is planned to introduce a staggered start at different measuring points in order to optimize the timing of the sequence.

At each event two measurement rounds are carried out, one in the morning and one in the afternoon. After each round, the operators are asked to export their results to a pre-agreed file format (txt, fwd or f25 format) and hand them over to the test leaders. Before the second test round starts the preliminary assessed data of the first round is presented to all participants. Based on this the operators have the chance to do modifications at their system if needed. Normally, however, no changes are made.

The detailed assessment is done shortly after the event. The results are summarized in a report and together with a certificate of participation given to the participants. In order to safeguard the economic interests of the FWD operators, the results of the comparison measurements are shown anonymously. The respective FWD operators are informed of the coding for their FWD with the notification of the results.

3.2 *Repeatability*

According to (FGSV, 2008) and (European Commission, 2005), a distinction must be made between short- and long-term repeatability tests. In the short-term repeatability test, the load plate remains on the ground, whereas in the long-term repeatability test, the measuring point is approached several times (at longer intervals). The object of the comparative investigations is the examination of the short-term repeatability.

Two measuring points are determined to check the repeatability. At these measuring points 12 load strokes with 50 kN each (target value) are carried out. The load plate must not be lifted. Although only the last ten load strokes are evaluated, all load strokes are to be stored.

3.3 *Comparability*

To check comparability (reproducibility), 19 measuring points are selected. At each of these measuring points, five load strokes of 50 kN each (target value) are performed. The load plate must not be lifted. Although only the last four load strokes were evaluated, all load strokes are to be stored.

3.4 *Examination of load pulse*

In order to investigate the load pulse, it is required to record the time history data at all measuring points.

3.5 *Temperature measurements*

Each FWD operator is asked to use all available measuring instruments for temperature measurement. Typically, these are:

– air temperature (mounted on the FWD),
– surface temperature (mounted on the FWD or hand-held unit) and
– asphalt body temperature (mounted on the FWD or hand-held unit).

For the adjustment of the measuring instruments for asphalt body temperature measurement, a controlled temperature water bath is set up at one of the measuring points. The water bath temperature is set low in the morning and high in the afternoon. As a reference, the continuous measurement is carried out with a calibrated external temperature logger.

In addition, temperature loggers are installed at two measuring points to record the surface temperature.

4 EVALUATION

In the following the evaluation methodology and results are shown and discussed. Exemplarily, the results of one event each are presented.

4.1 *Repeatability*

Repeatability (here short-term repeatability, see above) occurs when an FWD at the same measuring point without lifting the load plate, operated by the same person, produces reproducible results with several load drops. To evaluate the repeatability, the last 10 load drops are each normalized to 50 kN. The mean value and the standard deviation of the 10 load drops are then determined. Repeatability is given if the standard deviation is less than or equal to 2 µm or less than or equal to the sum of 1.0 µm and 0.75 % of the mean value (whichever is greater) (European Commission, 2005) (Van Gurp, 2013).

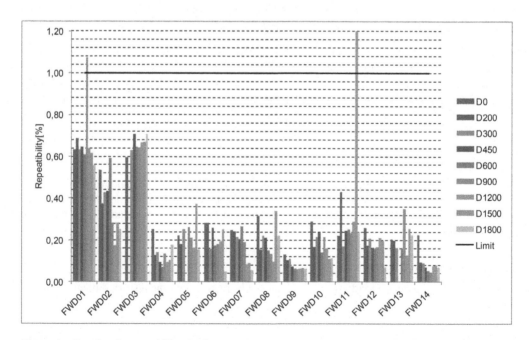

Figure 4. Results of repeatability check.

Figure 4 represents the relative comparison of the results of a repeatability test, where the value 1.0 represents the relative limit (the smaller the value, the better). The missing data bars are due to different arrangements of the geophones. Usually this criterion is very good fulfilled by all FWDs. Deviations from this usually led to the replacement of a geophone identified as defective and a self-check by the operator/owner afterwards.

4.2 *Comparability*

Comparability (reproducibility) exists when different FWDs, operated by different persons, produce comparable measurement results at identical measurement points. The measurements must also be carried out under identical conditions. Since the FWD visit each measuring point per measuring round in a short period of time, identical conditions are to be assumed. A cross-check will be done based on the recorded surface temperature.

The basis for a comparative assessment is the determination of a requirement level. In the case of comparative investigations, this is a 'comparison bowl' per measuring point. The comparison bowl is formed from the measured values of all FWDs and must be determined anew for each comparative study due to changing conditions. It should be noted that the comparison bowl is not an absolute reference. The following procedure was used to determine the comparison bowl and for the subsequent evaluation (European Commission, 2005) (Van Gurp, 2013).:

1. normalization of the load impact 2 to 5 to 50 kN
2. calculation of the comparison bowl by averaging the load normalized deflections per geophone position of all FWDs, whereby the highest and lowest values are excluded
3. calculation of the standard deviation of the comparison bowl from the load normalized deflections per geophone position of all FWDs, whereby the highest and lowest values are excluded
4. calculation of the mean values of the load normalized deflections per geophone position and per FWD
5. check for each FWD and each geophone position whether the mean value (from 4.) lies within the range of the comparison bowl (from 2.) ± the standard deviation (from 3.)
6. calculation of correction factors for each geophone position from the ratio of the mean value (from 4.) and the comparison bowl (from 2.)
7. calculation of correction factors for each FWD from the mean of the correction factors (from 6.)

Figure 5 Shows a typical distribution of correction factors. Again, most times these criteria are fulfilled or lead to repair, new calibration or part replacement.

4.3 *Load impulse*

The pulse durations for most FWDs range from 25 to 30 ms. Pulse durations of up to 35 ms have been observed sometimes and can sometimes be connected to noticeable results at the other quality checks. If necessary the operators are recommended to check and adjust the pulse durations. At the first events the load impulse, i.e. the representation of the shape of the load impulse has been plotted and the energy input as an integral over time has been calculated. Figure 6 and 7 show some typical results. It is under discussion if the so given information gives a benefit to the comparison measurements since the time history assessment needs a lot of effort. However, the raw data are available and can be evaluated on request.

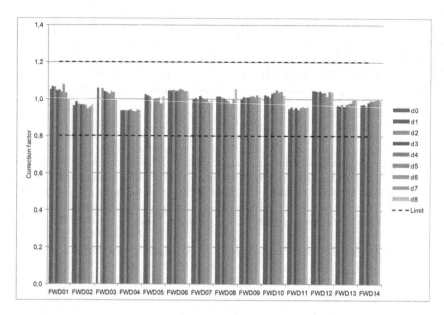

Figure 5. Calculated correction factors for each geophone of each FWD.

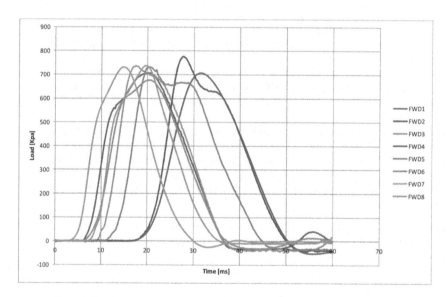

Figure 6. Load impulse history of several FWDs.

4.4 *Temperatures*

Calibration of the temperature sensors is usually not included in the annual FWD calibration services. Due to the high temperature dependence of the asphalt, the correct determination of the temperatures is very important. It was assumed that the measurement of temperatures is relatively simple and accurate. However, the experience of the last years shows that the correct measurement of temperatures is a big problem.

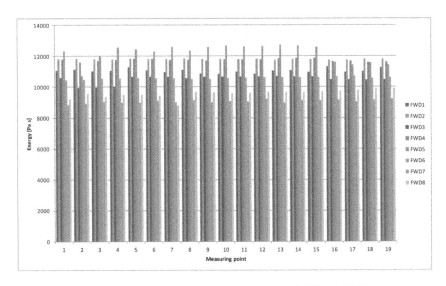

Figure 7. Calculated load impulse input as an integral over time of different FWDs.

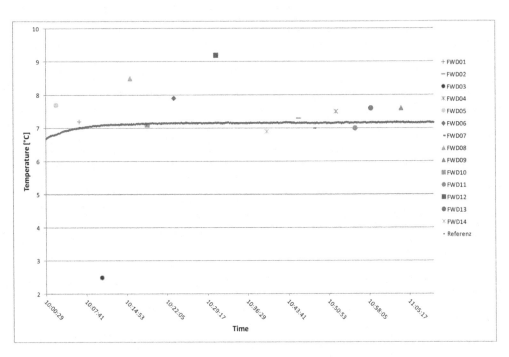

Figure 8. Comparison of the temperature measurement by FWD in a water bath (sensor for the asphalt body temperature measurement) and data logger (reference).

Both the alignment of some sensors in the water bath, as well as the comparison of the measured values of different FWD among themselves and in comparison to data of the meteorological station existing on the BASt premises show partly large and unsystematic deviations (Figure 8 and 9). Causes of errors could not be clarified conclusively so far. The reasons are assumed to be the installation positions of the

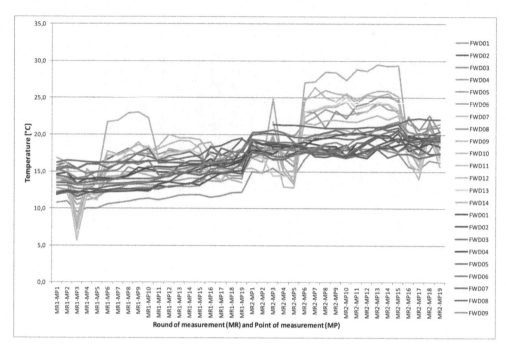

Figure 9. Comparative representation of the measured surface temperatures.

temperature sensors (e.g. near the exhaust system of in-built FWDs) and also the partly very short contact times during surface temperature measurement with contact thermometers.

5 SUMMARY

Bearing capacity measurements with the FWD must be carried out in a quality assured manner. This is the only way to ensure that a valid basis for decisions on maintenance management is available. The execution and evaluation of comparative measurements offers the possibility to test the entire measuring system under real conditions, i.e. also with the usual influences by the operator.

The results of the events carried out so far have shown that this method can clearly identify faults in the measurement vehicles. The results also show that the FWD method is a relatively robust method. Attention should be paid in the future to improve the temperature measurement. For this purpose, the support of the manufacturers is required.

The events carried out so far have met with great interest. The event has meanwhile established itself internationally and is usually held annually. On the client side, participation is increasingly desired or even demanded. In addition to the technical aspects, it has also become clear that the exchange of information between operators is an important element of quality assurance. For this purpose, the event offers a good platform in addition to the regular user group meetings, for example.

REFERENCES

European Commission, 2005. *European Cooperation in the Field of Scientific and Technical Research, COST 336 „Use of Falling Weight Deflectometers in Pavement Evaluation"*, Final Report, Brussels, Belgium.

FGSV, 2008. *Arbeitspapier Tragfähigkeit, Teil B2.1 „Falling-Weight-Deflectometer (FWD): Beschreibung, Messdurchführung – Asphaltbauweisen"*, FGSV-Verlag, Köln, Germany.

Straube, E., Beckedahl, H. & Huertgen, H., 1996. *Begleitende Forschung zur Einführung des Falling-Weight-Deflectometer (FWD) in der Bundesrepublik Deutschland*, Straßenbau und Straßenverkehrstechnik, Bonn, Germany.

Van Gurp, C., 2013. *FWD Correlation Trial 2013*. CROW report D13-05, Ede, Netherlands.

Identifying weak joints in jointed concrete pavements from TSD measurements by basis pursuit

M. Scavone, S.W. Katicha & G.W. Flintsch
Center for Sustainable and Resilient Transportation Infrastructure, Virginia Tech Transportation Institute. Blacksburg, USA

ABSTRACT: Weak joints are the source of most performance problems in jointed concrete pavements (JPCP) and composite pavements created by overlaying a JPCP. Until recently, it has been prohibitively time-consuming to evaluate the joints in a road network, and the evaluation of joint load transfer efficiency (LTE) has been restricted to specific project-level applications. This is changing with the advent of the Traffic Speed Deflectometer (TSD), a device that can collect structural information data at a 1 m resolution while moving at traffic speed. The 1 m resolution is adequate to capture the response of the joints to the applied load, but it results in a higher noise level compared to data collected at lower resolutions (e.g. at the typical 10 m). In this paper, we propose the use of Basis Pursuit (BP) to extract meaningful information about the joints' condition from the (noisy) TSD measurements. Weak joints are modeled as spikes (Dirac basis) in the measurements while the remaining features in the measurements are modeled by a wavelet basis. Combining the two bases (Dirac basis and wavelet basis) results in multiple possible representations of the collected measurements: we essentially have twice the number of unknowns than the number of equations to determine these unknowns. BP seeks a representation with a small number of elements from the two bases. Because weak joints are best represented by spikes, BP denoising results in mostly selecting the spikes at the weak joints location. These identified spikes provide a list of weak joints that should be further investigated (e.g. by performing LTE testing) or should receive priority in repair. We present examples of BP using simulated data and actual TSD collected data.

Keywords: TSD, concrete, joints, denoising, BP

1 INTRODUCTION

A major source of distress in jointed pavement structures, like jointed Portland concrete pavements (JPCP) and overlays of these, are deteriorated load-carrying joints. Deteriorated (or "weak") joints are those that lost their capacity to transfer the traffic loads from one slab to the adjacent ones – their Load Transfer Efficiency (LTE) is low (Alavi et al., 2008; Pierce et al., 2003). A jointed pavement may develop surface pathologies following the loss of LTE at its joints: the excessive movement of the slabs contributes to the pumping of the fine materials of the foundation layer, which in turn leads to joint faulting and slab cracking out of the loss of support (Huang, 2004; Delatte, 2014). Besides, if the concrete pavement is overlaid with asphalt, reflection cracking will occur on the surface layer at the location of the weak joints (Huang, 2004).

Thus, knowledge of the LTE of the joints of an in-service pavement is crucial both at the project level (input to the design of overlays or reinforcements (AASHTO, 2015)) and at the

DOI: 10.1201/9781003222880-39

network-wide management level, where structural performance data is most advisable for proper resource allocation (Haas et al., 2015).

At the project level, the Falling Weight Deflectometer (FWD) has been the device of choice to perform LTE testing (Alavi et al., 2008; Pierce et al., 2003; AASHTO 2015; Haas et al., 2015). However, network-wide LTE testing with FWD devices is unfeasible because of the device's stop-and-go nature, which makes it unsuitable for large-scale surveys – Flintsch et al (2013) point out that an FWD can cover no more than 180 tests/workday, which translates to 18 lane-miles at a 0.1-mile resolution. if translated to LTE testing of a jointed pavement where the joints are spaced 15 feet, and two tests are required at each location, the production rate of a single FWD drops to roughly 400m (0.25 miles) per day. Thus, a timely periodic network-wide survey would require a large fleet of FWDs operating simultaneously on in-service roads, which is impracticable both in terms of operational costs and from a safety standpoint (Flintsch et al., 2013; Katicha et al., 2013; Haas et al., 2015; Rada et al., 2016).

Traffic-speed deflection devices (TSDDs) overcome these limitations (European Commission, 1997; Arora et al, 2006; Flintsch et al., 2013; Rada et al.; 2016) by surveying the deflection of the pavement surface (or the velocity at which the pavement deflects) non-stop and in response to the device's own weight as it travels along the test pavement. Several prototypes were built over the years (Arora et al., 2006; Flintsch et al., 2013; Rada et al., 2016; Andersen et al., 2017), but only two types of devices are currently operational: the Rapid Pavement Tester (RAPTOR) and the Traffic Speed Deflectometer (TSD) – the widely-researched Rolling Weight Deflectometer (RWD) (Steele et al., 2020) has been decommissioned as of early 2020. Plenty of published literature exists on the use of these devices for network-wide assessments, some key examples are Flintsch et al. (2013); Rada et al. (2016). In general, comparisons of these devices against traditional deflection devices like the FWD and deflectographs returned good agreement between traffic speed devices and their static counterparts, provided that the contrasts are done in terms of comparable physical magnitudes (Katicha et al., 2014; Flintsch et al., 2013). However, little published material exists on using these devices to survey networks of jointed pavements (plain concrete pavements and overlays of these). A possible reason for that is that so far they are deemed not suitable for rigid pavements for the little deflection produced under its load – in fact, the Australian Standard on the TSD for deflection surveying (Austroads, 2016) specifically advises against using this device on rigid pavements. Yet, in one of the few published case studies of rigid pavement surveying with a TSD (Flintsch et al., 2013), it was found that the TSD would be suitable as a screening tool, that is, a device to perform a first 'scan' of the entire network to detect locations with structural deficiencies worth of further investigation and/or immediate repair. Katicha et al. (2013; 2014; 2016) provide a framework to denoise the dense TSD data [1-meter resolution] and extract meaningful features for further analysis.

2 OBJECTIVES

In this paper, we present the use of Basis Pursuit [BP] (Chen et al., 2001) on dense TSD deflection slope measurements to remove unwanted noise and extract the pulse responses (peaks in deflection) produced by weak locations, such as weak joints in jointed pavement networks. We aim at decomposing the TSD measurements into a combination of continuous components plus pulse responses due to the weak spots. BP is an objective, reproducible, and computationally fast methodology to achieve this task, even with large input datasets. Thus, it is suitable as a framework to easily detect weak spots within a pavement segment that was surveyed at a high resolution.

The remainder of this paper is divided into three sections. Firstly, we briefly present BP as a signal processing technique, relate it to the *Least Absolute Shrinkage and Selection Operator* [LASSO] (Tibshirani, 1996; Chen et al.,2001; Hastie et al., 2009), and discuss practical issues related to the selection of its hyper-parameters. Following, we present a case of signal denoising with BP using a simulated dataset. Finally, we present the results of BP denoising of actual TSD signals from different backgrounds: an experimental test track and an actual survey of overlaid jointed pavements. This paper contains an abridged version of a recently published article (Scavone et al., 2021), yet the intention of this paper is not to dive too much into the mathematical and computational detail behind BP, but actually to focus on the applicability as a denoising technique for TSD measurements, providing an example of an actual TSD survey on composite jointed pavements that we denoised with BP.

3 A BRIEF OVERVIEW OF BASIS PURSUIT

Basis Pursuit [BP] (Chen et al., 2001) is a convex optimization procedure that seeks a representation of a given signal y with n values using the fewest components from a given signal dictionary (a collection of elementary signals forming a basis or set of bases), that is, a sparse representation of y. The size of the dictionary would not affect the decomposition process *per se*, BP can be applied to decompose y over an over-complete dictionary (a dictionary with more than n elementary signals), a problem with many possible solutions, and BP would return a decomposition based on few non-zero components (a *sparse* decomposition).

In practical terms, such over-complete dictionaries may be warranted, as a sparse representation of the signal y may be achieved using sub-sets of components from multiple bases. For instance, the deflection measurements from a TSD assumed as a composition of a continuous component plus pulse responses could be described *sparsely* as a combination of Dirac signals (pulses) plus a combination of elementary continuous signals. Meanwhile, restricting to single-base decomposition (either only continuous signal base or Dirac base only) is likely to lead to non-sparse representations – a remarkable example of this is the Gibbs Phenomenon.

For this application, we are modeling the pulse responses as Dirac signals and the continuous component as a combination of wavelets because most measured signals admit a sparse representation in the wavelet domain, which in turn enhances the removal of noise, and because computations involving wavelets are very fast. Each basis has n components, thus our over-complete dictionary has $2n$ components. Given a TSD signal as input, BP would return a sparse decomposition of it over the two bases: the continuous component would represent any smooth variation in deflection over the surveyed pavement while the Dirac component would contain the pulse responses produced by the weak spots, the feature we attempt to extract in order to locate them.

The BP optimization problem for a generic signal y can be mathematically formulated as:

$$min \|\alpha\|_1 \text{ subject to } y = \phi\alpha$$

In the equation above, Φ is an n-by-$2n$ matrix containing the signal dictionary, and α is the vector of coefficients to reconstruct y as a combination of the columns of Φ. Alternatively, when the ultimate goal is to both remove the noise present in the signal y and simultaneously sparsely decompose y, BP can be formulated as a particular LASSO regression [or L_1 regularization] problem (Tibshirani, 1996; Chen et al., 2001; Hastie et al., 2009):

$$\min_\alpha = \frac{1}{2}\|y - \phi\alpha\|_2^2 + \lambda\|\alpha\|_1 \text{ where}: \ y = z + \sigma\in \text{ and } \in\sim N(0,1)$$

In the particular case of TSD signals, z is the unknown *true* pavement deflection response, and y is the noisy measurements collected with the TSD, σ is the standard deviation of the noise in the TSD signal.

LASSO regression problems have a known solution (Tibshirani, 1996; Chen et al., 2001; Friedman et al., 2010), which obviously depends on the hyper-parameter λ – for one-dimension variables, it is the soft-thresholding function with parameter λ. Friedman et al.'s scheme to solve the LASSO problem is numeric and consists of applying the soft-thresholding function coordinate-wise at each descent step – at each iteration, the coordinates of α are updated one at a time. Furthermore, the fact that the matrix Φ can be thought of as a stack of orthonormal columns (as is the case with BP) is advantageous: the cyclic coordinate descent can be done in *batch-mode*, instead of updating the coordinates of α one by one, all the coordinates that relate to a given stack of orthonormal columns can be updated simultaneously. In the BP denoising case, where $\Phi = [\Phi_1, \Phi_2]$ and $\alpha = [\alpha_1, \alpha_2]$, where Φ_1 is the Dirac base (and α_1 is its parameters' vector) and Φ_2 is the wavelet base (related to the vector α_2), a single iteration of the batch coordinate descent first updates α_1 while leaving α_2 constant, and secondly proceeds with α_2 without altering α_1 until the next iteration step. Refer to Scavone et al. (in press) for more details on the procedure.

Two major trends exist on the choice of the hyper-parameter λ, these are grounded on a bias-variance trade-off decision. One possible choice for λ is the so-called *Universal Threshold* (Donoho and Johnstone, 1994). The *Universal Threshold* is a value of λ high enough that retains only those components of the noisy input signal that have a high probability of being in the true signal. However, using a high threshold may both ignore real signal features whose amplitude is smaller than λ while at the same time will dampen the recovered signal excessively towards zero. Alternatively, λ could be chosen such that the BP output signal resembles the unknown signal z the most – it minimizes the error between z and $\Phi\alpha$. Computationally, however, the procedure may require time, as the BP decomposition must be applied using all candidate values of λ – yet anyway λ is bounded by zero and the *Universal Threshold*, which narrows the search effort. Since z is unknown (and so the fit error cannot be directly calculated), Stein's SURE estimate (Stein, 1981; Tibshirani, 1996) must be used. The formula for SURE that corresponds to optimization problems like BP is given by Tibshirani and Taylor (2012) and reproduced in Scavone et al. (2021).

4 BASIS PURSUIT APPLIED TO SIMULATED DATA

In this section, we provide a demonstration of the capabilities of BP as a denoising and feature extraction tool by reconstructing a simulated signal. The signal (a size-2048 vector) is a composition of a sinusoidal component with amplitude 2, centered around zero, and wavelength of 400 plus a discontinuous signal composed of peaks of amplitude 7 every 50 units (Figure 1). We added random noise to the simulated signal (normally distributed, with mean zero and variance 1), and attempted to reconstruct the true signal as a composition of pulses and wavelets. In this demonstration, we reconstructed the noisy signal by both BP (L_1 regularization) and Ridge regression (L_2 regularization). Figure 1 portrays the true and noisy signals, and Figure 2 presents the reconstruction results for both regularization techniques for the value of λ that minimizes the reconstruction error. Moreover, Figure 3 shows a plot of the coefficients that correspond to both bases according to both signal recovery methods.

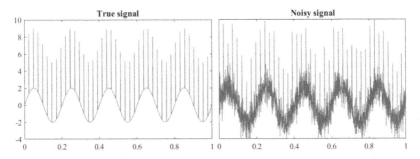

Figure 1. Simulated signal crafted for the BP application example. Left: generated signal, right: signal with added random noise.

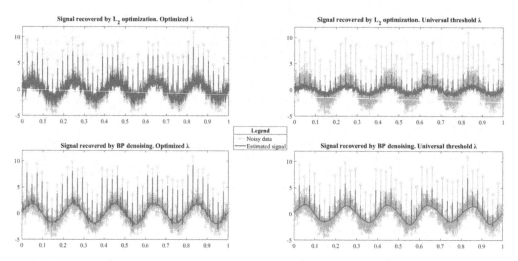

Figure 2. Simulated signal reconstruction results. Left: BP [L_1 regularization], right: L_2 regularization.

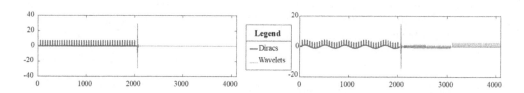

Figure 3. Components from Dirac and Wavelet basis used in signal reconstruction. Left: BP [L_1 regularization], right: L_2 regularization.

The graphical results presented above highlight the evident advantage of BP [L_1 regularization] over L_2 regularization (Hastie et al., 2009): the BP-recovered signal is composed of fewer non-zero components than the L_2-recovered signal, which eases interpretation, and the BP reconstruction also boasts smaller mean square error (MSE) than the reconstruction by L_2 regularization. Besides, BP outperforms L_2-regularization in terms of components from the two bases used to reconstruct the *true* signal – the MSE plummets to a constant, close-to-zero value after about 60 components are used, whereas the L_2 regularization requires roughly as many components as the signal length (about 2000 components) to achieve the same MSE value (Figure 4).

5 BASIS PURSUIT APPLIED TO REAL TSD DATA

In Scavone et al. (in press), it is demonstrated that the TSD is capable of detecting the pulse response from structurally deficient joints in jointed pavements and that BP can extract such responses from a noisy TSD signal. In this chapter, we further elaborate on this subject, providing an example of how this technique could be applied in practice when only one TSD measurement is available.

In July 2019, a TSD surveyed Interstate 66 in the Washington, DC metropolitan area, as part of the network-wide TSD surveying program from the ongoing FHWA TPF-5-385 project. The selected pavements are the eastbound and westbound segments of I-66 between mile markers [MM] 52 and 58, whose pavements are non-overlaid jointed concrete. The spacing between transverse joints is around 15 feet. The TSD operator [ARRB Inc.] provided georeferenced deflection slope and deflection bowl measurements for all segments at a 1-meter

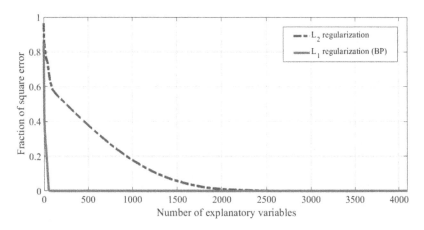

Figure 4. Decay of reconstruction error versus the number of non-zero components used in the reconstruction. Comparison for L_1 and L_2 regularization.

resolution, only one pass of the TSD over each segment was recorded. The TSD data was segmented as two subsections for each direction of travel (namely sections between MM 52-55 and 55-58), so, for the sake of analysis, four signals were processed independently. Each deflection measurement is tied to a picture of the pavement surface taken at that exact location using a downward-facing camera mounted on the TSD.

Signal denoising by BP requires the standard deviation of the measurement noise to be known. Yet, since the TSD ran only once over each segment, the signal's noise standard deviation can not be computed directly (as would happen when repeated measurements are available, like in the case study presented in Scavone et al. (2021). Instead, it has been approximated using the robust estimator presented in Katicha et al. (2015):

$$\hat{\sigma}_{noise} = 1.4826 \times median(x - median(x))$$

A Matlab implementation of BP that processes each segment separately and extracts the recovered pulse components and their geographical coordinates was written for this case study. Processing the four segments required roughly 3 minutes on an average-spec personal computer. Figure 5 below shows the denoised TSD signal and the recovered pulse responses for one of these segments, both using the optimized penalty and the *Universal Threshold*.

The recovered pulse components were exported to a GIS application (QGis 2.18) for easy visualization and interpretation. The map in Figure 6 presents the detected weak spots with an amplitude greater than 300 microns for the EB and WB segments between MM 52 and 54 for the reconstruction with optimized λ Those weak spots also detected by the BP decomposition with the *universal threshold* are highlighted with a circle. The 300 microns threshold used in the map was selected purposefully as it is roughly equal to the product between the *Universal Threshold* [about 4.20] and the TSD signal noise's standard deviation [about 64 microns], any recovered pulse with an amplitude lower than this threshold has chances of being a false positive, and any pulse with an amplitude higher than 300 microns has a high probability of being a *true* pulse representative of a location with a structural deficiency, thus worth highlighting for investigatory purposes.

Figure 7 Presents pictures of the pavement surface at or near high-probability weak spots, the point number corresponds to the point tags in Figure 6. Cases a) and b) represent actual locations where a pulse response was detected: For case a) the pulse response could have been produced either by the transverse joint or by the circular marking (probably the cut for a coil), while for case b) it is most likely that the (weak) transverse joint shown in the picture

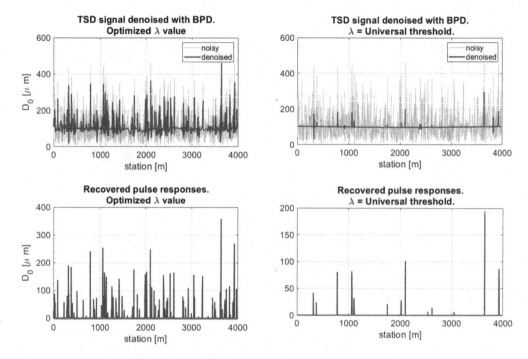

Figure 5. TSD signal recovered by BP I-66 eastbound between MM 52 and MM 55. Left plots: results for optimized penalty parameter λ, right plots: results for *Universal Threshold*.

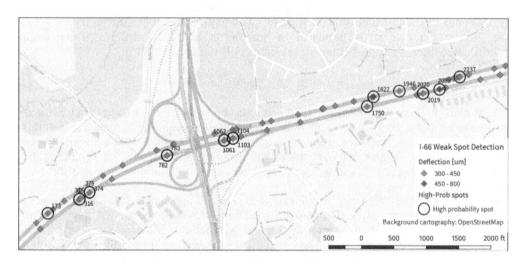

Figure 6. Interstate-66, MM 52-54. Weak spots pulse responses recovered by BPD.

be the culprit – although no surface distress developed in the concrete slabs. Meanwhile, picture c) represents the pavement 10 meters ahead of the weak joint shown in b), a joint in good structural condition for no pulse response was recorded by the TSD.

The remarkable result from this example is that the TSD could flag a potentially weak joint before any surface pathology related to lack of support or low LTE developed in the slab – from the surface, the pavement at locations b) and c) looks in good condition, free of surface defects, yet one location is structurally deficient according to the TSD data. From

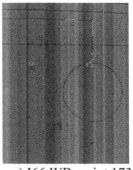  

a) I66 WB, point 173. Joint and cut for coil

b) I66 EB, point 1061. Weak joint

c) I66 EB, ahead of point 1061. Good joint

Figure 7. Pavement surface at selected weak spots. Imagery from the TSD.

a prevention standpoint, this is obviously advantageous as contributes to both rapidly assessing all the joints in the segment and pinpointing those with deficiencies before the lack of LTE takes its toll on the connecting slabs.

6 CONCLUSIONS

The TSD has been originally intended as a screening tool to pinpoint the location of structurally deficient points within a pavement network for investigatory purposes. This paper discusses a data analysis technique (Basis Pursuit, [BP]) to achieve this task in practice: BP analysis of the TSD measurements decomposes such measurements as a combination of elementary signals with an actual meaning or, in other words, that can be interpreted from an engineering perspective. The two examples presented in this paper demonstrate the effectiveness of the TSD to recognize structurally deficient joints within a concrete pavement segment even before the concrete slabs develop distress. The positive impact of such valuable information for preventive management purposes is evident, as it enables the manager to direct early corrective maintenance activities at those locations, saving the structure from developing further damage.

Nonetheless, there is still room for improvement in BP analysis of TSD signals: Reconstruction of signals by BP involves bias/variance trade-off, the penalty parameter λ drives the reconstruction either towards less bias (at the cost of adding false-positive components to the reconstruction, fitting noise) or towards a reconstruction using high-probability components only (but with a damped recovered signal). Improvements to the reconstruction procedure were proposed by Candès et al. (2008) and Javanmard and Montanari (2018). If these lead to unbiased recovered components, further analyses could be made on the structural issues that are causing such pulse responses through back-calculation, as is proposed for instance by Deep et al. (2020). These ramifications are yet to be explored but open draw a promising range of applications for high-speed deflection devices.

ACKNOWLEDGMENTS

The authors would like to express their gratitude towards J. Daleiden and N. Bech from ARRB Inc., for providing the processed TSD data utilized in the case study presented herein, and to B. Diefenderfer from VDOT for his contribution to the I-66 case study.

REFERENCES

Alavi, S.; LeCates, J. F.; Tavares, M. P. (2008): *Falling Weight Deflectometer Usage*. NCHRP Synthesis 381. Transportation Research Board.

American Association of State Highway and Transportation Officials (AASHTO) (2015): *Mechanistic-Empirical Pavement Design Guide. A Manual of Practice* Publication Code MEPDG–2.

Andersen, S., Levenberg, E., & Andersen, M. B. (2017). Inferring pavement layer properties from a moving measurement platform. In *10th International Conference on the Bearing Capacity of Roads, Railways and Airfields* (pp. 675–682). Taylor & Francis.

Arora, J. ., Tandon, V., & Nazarian, S. (2006). *Continuous deflection testing of highways at traffic speeds* (No. FHWA/TX-06/0-4380–1).

Austroads (2016): *Pavement Data Collection with a Traffic Speed Deflectometer (TSD) Device*. Austroads Test Method AG:AM/T017. September 2016.

Candes, E. J., Wakin, M. B., & Boyd, S. P. (2008). Enhancing sparsity by reweighted ℓ_1 minimization. *Journal of Fourier analysis and applications*, *14*(5-6), 877–905.

Chen, S. S., Donoho, D. L., & Saunders, M. A. (2001). Atomic decomposition by basis pursuit. *SIAM Review*, *43*(1), 129–159.

Deep, P., Andersen, M. B., Rasmussen, S., Marradi, A., Thom, N. H., & Presti, D. L. (2020). Simulating Deflection of a Jointed Rigid Pavement Under Rolling Wheel Deflectometer (RAPTOR) Loading. In *Proceedings of the 9th International Conference on Maintenance and Rehabilitation of Pavements—Mairepav9* (pp. 859–870). Springer, Cham.

Delatte, N. J. (2014). Concrete pavement design, construction, and performance. CRC Press.

Donoho, D. L., & Johnstone, J. M. (1994). *Ideal spatial adaptation by wavelet shrinkage*. Biometrika, 81 (3), 425–455.

Flintsch, G. W.; Katicha, S. W.; Bryce, J.; Ferne, B.; Nell, S.; Diefenderfer, B. (2013): *Assessment of Continuous Pavement Deflection Measuring Technologies*. SHRP-2 Report S2-R06F-RW-1. Transportation Research Board, Washington DC.

Friedman, J., Hastie, T., & Tibshirani, R. (2010). Regularization paths for generalized linear models via coordinate descent. *Journal of statistical software*, *33*(1), 1.

Haas, R.; Hudson, W. R.; Falls, L. C. (2015): *Pavement Asset Management*. First Edition. Scrivener Publishing.

Hastie, T., Tibshirani, R., & Friedman, J. (2009). *The elements of statistical learning: data mining, inference, and prediction*. Springer Science & Business Media.

Huang, Y. H. (2004). *Pavement Analysis and Design*. Second Edition. Pearson.

Javanmard, A., & Montanari, A. (2018). Debiasing the Lasso: Optimal sample size for gaussian designs. *The Annals of Statistics*, *46*(6A), 2593–2622.

Katicha, S. W.; Flintsch, G. W.; Ferne, B. (2013): Optimal Averaging and Localized Weak Spot Identification of Traffic Speed Deflectometer Measurements. *Transportation Research Record 2367*, 43–52.

Katicha, S. W.; Flintsch, G. W.; Bryce, J; Ferne, B. (2014): Wavelet Denoising of TSD Deflection Slope Measurements for Improved Pavement Structural Evaluation. *Computer-Aided Civil and Infrastructure Engineering 00*, 1–17.

Katicha, S. W.; Flintsch, G. W.; Ferne, B.; Bryce, J. (2014): Limits of Agreement Method for Comparing TSD and FWD Measurements. *International Journal of Pavement Engineering*, *15*(6), 532–541.

Katicha, S. W., Bryce, J., Flintsch, G., & Ferne, B. (2015). Estimating "True" Variability of Traffic Speed Deflectometer Deflection Slope Measurements. *Journal of Transportation Engineering*, *141*(1), 04014071

Katicha, S. W.; Loulizi, A.; El Khouri, J.; Flintsch, G. W. (2016): Adaptive False Discovery Rate for Wavelet Denoising of Pavement Continuous Deflection Measurements. *Journal of Computing in Civil Engineering*, *31*(2), 04016049-1–10

Pierce, L.; Uhlmeyer, J.; Weston, J.; Lovejoy, J.; Mahoney, J. P. (2003): Ten-Year Performance of Dowel-Bar Retrofit. Application, Performance, and Lessons Learned. *Transportation Research Record 1853*, 83–91

Rada, G.; Nazarian, S.; Visintine, B. A.; Siddhartan, R.; Thyagarajan, S. (2016): *Pavement Structural Evaluation at the Network Level: Final Report*. (FHWA-HRT-15-074). United States Department of Transportation, Federal Highway Administration. September 2016

Scavone, M. A.; Katicha, S. W.; Flinstch, G. W. (2021): Identifying Weak Joints in Jointed Concrete and Composite Pavements from Traffic Speed Deflectometer Measurements by Basis Pursuit. *Journal of Computing in Civil Engineering*. DOI: 10.1061/(ASCE)CP.1943-5487.0000951

Steele, D. A.; Lee, H.; Beckemeyer, C. A. (2020): *Development of the Rolling Wheel Deflectometer (RWD)*. (FHWA-DTFH-61-14-H00019). United States Department of Transportation, Federal Highway Administration. March 2020

Stein, C. M. (1981). Estimation of the mean of a multivariate normal distribution. *The Annals of Statistics*, 1135–1151.

Tibshirani, R. (1996). Regression shrinkage and selection via the Lasso. *Journal of the Royal Statistical Society: Series B (Methodological)*, 58(1), 267–288.

Tibshirani, R. J., & Taylor, J. (2012). Degrees of freedom in Lasso problems. *The Annals of Statistics*, 40(2), 1198–1232.

Assessing the usefulness of the outer-area method in quantifying the structural capacity of full-scale composite pavement sections subjected to accelerated pavement testing

A. Francois, D. Offenbacker & Y. Mehta
Center for Research and Education in Advanced Transportation Engineering Systems, Department of Civil Engineering, Rowan University, Glassboro, New Jersey, USA

ABSTRACT: The focus of this study was to evaluate the overall accuracy of the Outer-Area method in quantifying the structural capacity of full-scale, composite pavement sections that are subjected to APT. Five 9.1 m. long and 3.7 m wide composite field sections were evaluated in this study. All sections contained a similar substructure (i.e., 203 mm, Portland cement concrete base, 406 mm New Jersey I-3 (A-1-a) granular subbase, and 304 mm, compacted subgrade). Section 1 contained a stone matrix asphalt (12.5-SMA) overlay while Section 2 contained a 50.8 mm thick, New Jersey high performance thin overlay (NJHPTO) overlay. Sections 3, 4 and 5 contained an overlay that was a combination of a 25 mm thick, binder rich intermediate course (BRIC) and a 50 mm layer of a Superpave mixture (9.5-ME), SMA, and NJHPTO, respectively. All sections were instrumented with two asphalt strain gauges and subjected to accelerated pavement testing (APT) using a heavy vehicle simulator (HVS). During APT, a 60 kN, dual tire, single axle load was applied to the sections at 8 km/h for 200,000 repetitions. Heavy weight deflectometer (HWD) testing was performed on each test section before and after APT. The collected pavement deflection data from each composite section was analyzed using the outer-Area method to determine the overall structural capacity of the bound pavement layers before and after APT. The actual field performance of the test sections was assessed using the strain data collected during APT. Based on the results of the study it was determined that the Outer-Area method may not be able to accurately quantify the damage full-scale composite pavements experience due to APT.

Keywords: Guidelines, abstract, title, text, figures

1 INTRODUCTION

Heavy weight deflectometer (HWD) testing is a nondestructive method of pavement testing that simulates the deflection of a pavement surface under the action of fast moving traffic load. It is generally used to determine the structural conditions of pavement layers over time. HWD testing involves the application of variable loads on the surface of a pavement via a spring loaded plate. The deformation of the pavement due to the applied loads causes a deflection basin to form within the pavement structure. The pavement response to the applied impulse load is measured by velocity transducers or geophones which, are placed at specific radial distances from the center of the applied load. The measured deflections are then used to compute the estimated layer stiffness through an ill-posed, iterative process referred to as back-calculation (Mehta and Roque 2003). Computer programs are usually used to perform back-calculation of multi-layered systems. These computer programs allow users to input estimated initial or "seed" moduli values for the various layers of a pavement. A subroutine program within the back-calculation software then utilizes finite element or multi-layer analyses to determine an "effective" layer moduli which adjusts for stress-sensitivity and discontinuities (Maestas and Mamlouk 1992).

Most back-calculation software are capable of computing "effective" layer moduli for flexible and rigid pavement systems. However, these software are limited in their ability to analyze HWD data from composite pavement systems: particularly when the pavement system contains thin HMA overlays. As such transportation agencies have developed procedures to forward calculate the apparent stiffness of the uppermost bound layer(s), of pavement systems under an applied surface load. One such procedure is the Outer Area method that was developed by the Federal highway Administration.

2 GOAL

The goal of this study was to evaluate the efficacy of the Outer-Area method in quantifying the structural capacity of full-scale, composite pavement sections that are subjected to APT. In order to accomplish this goal, heavy weight deflectometer testing was conducted on five full-scale, composite pavement sections. The full-scale sections were also subjected to accelerated pavement testing using a heavy vehicle simulator. The HWD data was analyzed using the Outer-Area procedure. The effectiveness of the Outer-Area method in ranking the reduction in structural capacity of the test sections was evaluated by reconciling the analyzed HWD results with the actual damage that occurred in the test sections during to APT.

3 FIELD SECTION DESCRIPTION AND INSTRUMENTATION

Five full-scale, pavement sections were evaluated in this study. These pavement sections were located at Rowan University Accelerated Pavement Testing Facility (RUAPTF). The pavement sections were 9.2 m long by 3.6 m wide. Figure 1 presents a representative pavement structure (i.e., layers) for all five test sections. All sections had the same supporting layers or substructure (i.e., a Portland cement concrete (PCC) base, a granular aggregate subbase, and compacted natural soil as a subgrade). However, different hot mix asphalt (HMA) overlays were utilized on each test section. Section 1, contained a 76.2 mm thick, stone matrix asphalt (SMA) overlay, and section 2 contained a 50.8 mm thick, New Jersey High Performance Thin Overlay (NJHPTO). The overlays on sections 3, 4 and, 5 consisted of a 50.8 mm thick, Superpave mixture (9.5-ME), 50.8 mm thick, SMA mixture, and 50.8 mm thick, 4.75-HPTO mixture, respectively placed on top of a 25.4 mm thick, Binder Rich Intermediate Course (BRIC).

Each test section was instrumented with two H-type strain gauges that were placed 12.7 mm from the bottom of the asphalt overlay. The strain gauges were placed approximately 457 mm apart, at the joint, between the approach and leaving PCC slabs (Figure 2). One of the strain gauges was placed directly beneath the wheel path of one of the loading tires (dual-tire configuration) while the other was placed at the edge of the loading path.

4 TESTING AND EVALUATION PLAN

4.1 *Heavy weight deflectometer testing*

Heavy weight deflectometer testing was performed immediately before and after HVS testing on each test section. The HWD test locations on each full-scale section evaluated in this study is shown in Figure 2. There were six HWD test locations on each test section. Two of these test locations were evaluated in the direction of HVS loading (i.e., TS-1 and TS-2) while the remaining four test locations were evaluated in the opposing direction. The HWD test points were located in two zones on each section as shown in Figure 2. HWD Test Zone 1 consisted of the region loaded by the HVS test wheel while Test Zone 2 consisted of the area outside the HVS loading area. HWD testing was performed in these two specific regions of the test sections in order to evaluate the effect of applied HVS loading on pavement section deterioration. The arrows shown in Figure 2 indicate the position of the geophones used to measure pavement deflections. At each test location on the sections, four drop heights that corresponded to target loads of 29.4 kN, 41.2 kN, 58.8 kN, and 78.4 kN, respectively were utilized. Geophones were spaced at 0, 200 mm, 300 mm, 450 mm, 600 mm, 900 mm, and 1200 mm, respectively from the applied load.

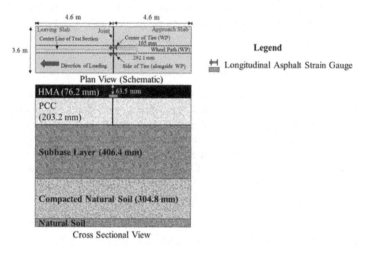

Figure 1. Pavement structure and instrumentation of test sections.

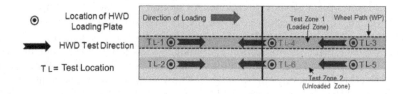

Figure 2. HWD test locations and test zones on each test section.

4.2 *Heavy vehicle simulator testing*

A heavy vehicle simulator (HVS) was used to apply full-scale loading on all three pavement sections evaluated in this study. Each test section was subjected to a 60-kN load using a dual-tire, single-axle wheel configuration. The pressure in both tires was maintained at 758.4 kPa during loading. A total of 200,000 loading passes were applied on each test section at an 8 km/h, loading rate. Data from the embedded strain gauges was collected as HVS loading progressed on the test sections. A data acquisition system was used to collect strain measurements from the gauges during each load application. Strain data was collected at a frequency of 2,000 data points per second. The frequency at which strain measurements were recorded varied based on the stage of APT. Strain data was recorded: every 100 passes between HVS passes 1 and 1000, every 500 passes between HVS passes 1000 and 10,000, every 1000 passes between HVS passes 10,000 and 20,000, every 22,500 passes between HVS passes 20,000 and 50,000, every 10,000 passes between HVS passes 50,000 and 100,000, and every 20,000 between HVS passes 100,000 and 200,000. The measurement of strain data during APT was important because it allowed for the degree of damage accumulation in each HMA overlay to be determined. The measured strain data also facilitated test sections performance comparison. This was essential in order to verify the overall effectiveness of the Outer-Area method in assessing the relative change in structural capacity of the full-scale section due to APT.

5 OVERVIEW OF ANALYSIS PROCEDURES

5.1 *Outer-area method*

The Outer-Area method was adopted in this study to analyze the measured deflections and draw conclusions about the structural capacity of the test sections. This method relies on the computation of a deflection basin curvature index (i.e., the Outer-AREA parameter) that is used together with a composite modulus (E_o) index to forward calculate an approximate modulus for the upper, bound layer of a composite pavement system known as the effective stiffness index (E_s) (Smith et al. 2017). The composite modulus represents an overall modulus for the entire pavement system beneath the loading plate while the Outer-AREA parameter represents a normalized area with respect to the measured deflection basin. Additionally, the Outer-AREA parameter ignores the deflections obtained directly under the applied load. Thus, the Outer-AREA parameter compensates for the magnitude of the imposed HWD load and to minimize the compression effect in the HMA layer (Smith et al. 2017).

The Outer-AREA parameter is a deflection basin shape factor that is related to the ratio between pavement stiffness and the underlying subgrade stiffness. It is strongly influenced by the thickness of a pavement structure and the stiffness of the materials that make up the structure. However, the parameter is inversely related to the mid-depth temperature of the asphalt layer within a pavement system. Therefore, temperature corrections are typically performed on the pavement deflection data in order to minimize the effect of temperature on the Outer-Area parameter (Smith et al. 2017). The effective stiffness index is determined by using Equations 1 to 4.

$$E0 = \frac{(1.5 * a * \sigma_0)}{D_0} \qquad (1)$$

$$\text{Outer} - \text{Area} = 6\left(1 + 2\left(\frac{D3}{D2}\right) + 2\left(\frac{D4}{D2}\right) + 2\left(\frac{D5}{D2}\right) + 2\left(\frac{D6}{D2}\right) + 2\left(\frac{D7}{D2}\right)\right) \qquad (2)$$

$$AF = \left(\frac{k_2 - 1}{k_2 - \left(\frac{\text{Outer-AREA}}{k_1}\right)}\right)^2 \qquad (3)$$

$$Es = Eo * AF * k_3^{(\frac{1}{AF}-1)} \qquad (4)$$

Where: - (E_0) Composite modulus of pavement
- (E_s) Effective stiffness index
- (a) radius of HWD plate
- (σ_0) Peak pressure of HWD impact load under plate
- (D_0) Peak center HWD deflection
- (D_2, D_3, D_4, D_5, D_6, D_7) Measured pavement deflections at geophones
- (AF) Outer-Area factor
- (k1) = 11.037 (Outer-AREA when upper layer stiffness equal that of lower layers)
- (k2) = 3.262 (maximum possible improvement in Outer-AREA (36/11.037))
- (k3) Thickness of upper layer/load plate diameter

5.2 APT strain data analysis procedure

The strain data collected from the embedded strain gauges in the full-scale, test sections were analyzed using the strain data analysis procedure developed by Francois et al. (2019). The strain data analysis procedure consists of four steps which, are illustrated in Figure 3. The first step of the strain data analysis procedure, entails converting the voltage signal recorded by the embedded, strain gauges into actual strain measure ements using calibration factors provided by the manufacturer. This step also involves filtering the strain time history response pulse using a 25-data point, moving average to reduce the number of data points required to accurately capture the strain response at a particular HVS pass.

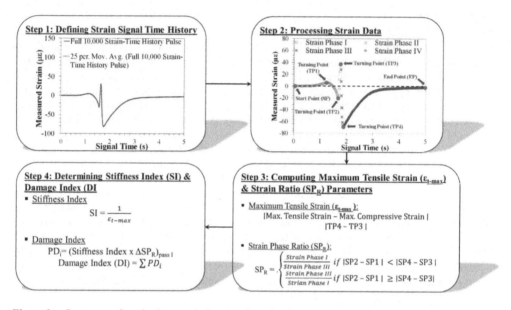

Figure 3. Summary of strain data analysis procedure developed by Francois et al. (2019).

The second step of the procedure involves identifying the critical points on the strain time history response pulse and defining the four phases of this pulse. Strain Phase I reflects the initial compressive strain the overlay experiences at the joint when the load approaches the strain gauge from the approach slab. Strain Phase II captures the tensile strain the HMA overlay experiences when the load was directly on top of the joint. Strain Phase III represents the compressive strain the overlay experiences as the load departs from the joint. Strain Phase IV highlights the gradual increase in strain when the HVS wheel load no longer directly impacts joint (or slab) deflection (Francois et al. 2019). The third step involves computing the maximum tensile strain (ε_t-max) and strain phase ratio (SP_R) parameters based on the strain phases defined in Step 2. The fourth and final step of the strain analysis procedure entails determining the damage index (DI) (Figure 3). The damage index is used to compare the relative damage accumulation in the HMA overlays due to APT.

6 RESULTS AND DISCUSSION

6.1 Analysis of HWD test results

Table 1 presents the Outer-Area parameters that were computed directly from measured pavement deflections at the joint: within the HVS wheel path (TL-4 in Figure 2) and outside the

Table 1. Outer-Area values computed at joint in test sections.

Test Section	HWD Test Location	Outer-Area Parameter		Percent Difference (%)
		Before HVS Testing	After HVS Testing	
Test Section 1	TL-4	37.0	34.6	6.7
(SMA)	TL-6	37.0	36.3	1.9
Test Section 2	TL-4	39.8	31.2	24.2
(NJHPTO)	TL-6	40.8	38.2	6.5
Test Section 3	TL-4	40.4	26.1	43.0
(9.5-ME & BRIC)	TL-6	41.6	31.1	28.9
Test Section 4	TL-4	34.4	30.0	13.7
(SMA & BRIC)	TL-6	33.5	31.7	5.5
Test Section 5	TL-4	30.1	25.7	15.7
(NJHPTO & BRIC)	TL-6	30.5	29.0	5.0

HVS wheel path (TL-6 in Figure 2). The results in Table 1 indicate that Section 3 had the highest Outer Area values before HVS loading followed by Sections 2, 1, 4, and 5, respectively. These results implied that the bound layers of the composite pavement in Section 3 had the highest stiffness of all test sections while the bound layers of Section 5 had the lowest stiffness. This is because higher Outer-AREA values typically reflect higher pavement stiffness. Based on the Outer-Area values computed for TL-4 in each test section, it can be inferred that Section 3 underwent the largest reduction in stiffness due to HVS loading followed by Sections 2, 5, 4, and 1, respectively. These results suggested that more damage accumulation may have occurred in the bound layers of Section 3 during APT while the bound layers in Section 1 may have experienced the least damage accumulation of all test sections.

Table 2 presents the effective stiffness index values that were obtained for all five test sections at the joint in the loaded and unloaded zone of each test section (i. e., HWD test locations TL-4 and TL-6 in Figure 2). These effective stiffness index values are the only ones presented because they quantify the overall stiffness of the composite pavement sections at the joint: where the most critical responses to loading occur. Since TL-4 represented the HWD test location in the loaded zone and TL-6 represented the HWD locations in the unloaded zone, the change in structural capacity of the test sections at the joint due to loading can be compared. The reduction in effective stiffness at TL-4 was generally larger than the effective stiffness reduction seen at TL-6. This trend was expected because TL-4 was located 38.1 mm from the joint in wheel path of the HVS while TL-6 was located 38.1 mm from the joint in the region not loaded by the HVS. Therefore the decrease in structural capacity captured at TL-4 was expected to be more significant than that captured at TL-6. With respect to the relative change in structural capacity of the test sections after HVS loading, it can be observed that Section 3 experienced the largest reduction in effective stiffness followed by Section 1, Section 2, Section 4, and Section 5, respectively. These results suggested that Section 3 experienced the most deterioration (i.e., damage accumulation) during HVS testing while Section 5 experienced the least deterioration (i.e. reduction in stiffness during APT.

6.2 *Analysis of APT test results*

The ε_t-max was computed from all the strain-time history response pulses recorded in the five full-scale sections evaluated in this study. These results are illustrated in Figure 4a. The computed ε_t-max values in all five sections increased (i.e., followed a logarithmic growth trend) as loading passes increased. This trend was expected because the increase in applied loading passes typically amounts to an increase in permanent strain (or damage) within the pavement sections. In addition, given the constant loading applied on top of all five pavement sections (i.e., (60-kN)), the rate of damage accumulation in each section can be compared at a particular loading pass. Based on this fact, the results indicated that the 9.5ME & BRIC overlay experienced the most damage due to APT while 4.75-NJHPTO overlay underwent the least damage due to loading. This is because the rate of increase in ε_t-max in Section 3 were largest followed by Sections 1, 5, 4, and 2 NJHPTO respectively.

Figure 4b Shows the Damage Index values obtained for each test section after each recorded loading pass. The damage index generally increased with loading passes as expected. However, the DI increased at different rates for the different HMA (or sections). By comparing the accumulated damage after applying a certain number of loading passes (for example 160,000 loading passes), it can be observed that Section 4 had the highest rate of accumulated damage followed by Section 5, Section 3, Section 2, and Section 1, respectively.

Table 2. Effective stiffness index computed at joint in test sections.

Test Section	HWD Test Location	Effective Stiffness Index (Es)		Percent Difference (%)
		Before HVSj Testing	After HVS Testing	
Test Section 1 (SMA)	TL-4	125,850	19,690	84.0
	TL-6	23350	16362	30.0
Test Section 2 (NJHPTO)	TL-4	73,095	15,991	78.1
	TL-6	60,981	19,169	68.5
Test Section 3 (9.5-ME & BRIC)	TL-4	64,436	1,226	98.0
	TL-6	21,679	6,933	68.0
Test Section 4 (SMA & BRIC)	TL-4	10,636	2,933	72.5
	TL-6	8,652	4,219	51.2
Test Section 5 (4.75-NJHPTO & BRIC)	TL-4	7,477	4,364	41.6
	TL-6	7,721	2,834	63.3

7 SUMMARY AND CONCLUSIONS

The goal of this study was to evaluate the overall accuracy of the Outer-Area method in assessing the structural capacity of full-scale, composite pavement sections that are subjected to APT. Heavy weight deflectometer and accelerated pavement testing results obtained from five field sections that contained specialty New Jersey overlays were for this assessment. Based on the analyzed APT and HWD results, the following conclusions were drawn:

- Analyzed HWD results showed that Section 3 experienced the largest reduction in effective stiffness followed by Section 1, Section 2, Section 4, and Section 5, respectively.
- Analyzed strain data from full-scale testing indicated that Section 4 had the highest rate of accumulated damage after APT followed by Section 5, Section 3, Section 2, and Section 1, respectively.
- Reduction in structural capacity ranking obtained from the Outer-Area method did not coincide with damage accumulation ranking obtained from the Francois et al. 2009 strain data analysis procedure.
- Outer-Area method may not be able to accurately quantify the damage full-scale composite pavements experience due to APT. However, further research is required as this may appear to be the case due to the limited scope of this study.

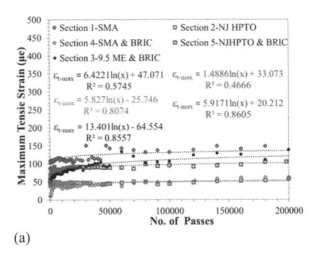

(a)

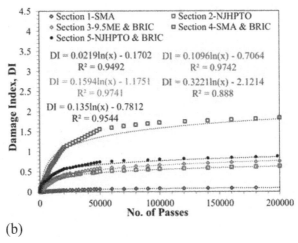

(b)

Figure 4. Summary of strain data analysis procedure developed by Francois et al. 2009.

REFERENCES

Francois, A. Ali, A. and Mehta, Y. 2019. *A proposed approach for processing and analyzing strain data collected in full-scale accelerated pavement testing*. Proceedings of 5th Conference on Smart Monitoring Assessment and Rehabilitation of Civil Structures, Potsdam, Germany.

Maestas, J. and Mamlouk, M. 1992. *Comparison of Pavement Deflection Analysis Methods Using Overlay Design*. Transportation Research Record: Journal of the Transportation Research Board, Vol. 1377, pp.17–21.

Mehta, Y. and Roque, R. 2003. *Evaluation of FWD Data for Determination of Layer Moduli of Pavements*. Journal of Materials in Civil Engineering, Vol. 15, No. 1, pp. 25–31.

Smith, K. Bruinsma, J. Wade, M. Chatti, K. Vandenbossche, J. and Thomas, H. 2017.*Using Falling Weight Deflectometer Data with Mechanistic-Empirical Design and Analysis, Volume I: Final Report*. Federal Highway Administration, Washington D.C., FHWA-HRT–16–009.

Prediction of subgrade soil density using dielectric constant of soils

Ahmed Abdelmawla
Graduate Research Assistant, School of Environmental, Civil, Agricultural, and Mechanical Engineering, University of Georgia, Athens, USA

S.Sonny Kim*
Distinguished Faculty Fellow, School of Environmental, Civil, Agricultural, and Mechanical Engineering, University of Georgia, Athens, USA

ABSTRACT: Subgrade layer plays an important role to support the overlying pavement structural layers. To resist permanent deformation and failure under traffic loadings, subgrade soils are compacted to a required maximum density level during the field construction. This paper presents a simple exponential model to estimate subgrade soils density using Ground Penetrating Radar (GPR). The model was developed based on electromagnetic mixing theory to backcalculate subgrade soils density using GPR measurements. The developed model determines a soil's dielectric constant, considering dielectric and volumetric properties of the three major components of soil: air, water, and solid particles. A series of laboratory tests was conducted on six (6) soil samples at various density levels to validate the estimated densities with the measured ones. The soils density estimated by the new exponential model showed approximately 6% error than soils density measured by sand cone method.

Keywords: GPR, dielectric constant, bulk density, subgrade

1 INTRODUCTION

Subgrade soil is an integral part of the road pavement structure as it provides support to the pavement from beneath. The subgrade soil and its properties are important in the design of pavement structure. The main function of the subgrade is to give adequate support to the pavement structure. Therefore, the subgrade should possess enough stability under adverse climate and loading conditions. The formation of rutting and consequent cracking of pavements are generally attributed due to the poor subgrade condition.

Subgrade soils are typically characterized by their resistance to deformation under load, which can be a measure of strength (stress needed to break a material) or stiffness (the relationship of stress and strain in the elastic range or how well a material is able to snap back to its original shape and size after applying stress). The greater the resistance to deformation a subgrade shows the more the load it can support before reaching a critical deformation.

After scarifying and compacting, subgrade must be tested for compaction and checked to ensure a typical cross section and uniform grade before subsequent courses can be placed. The minimum rate of density testing for untreated subgrade material in place is one test per 1500 linear feet of roadway (full width) according to the Roadway Testing Technician 2018 Study Guide of GDOT. Before pavement Layers are placed on the subgrade, the subgrade must be

---

*Corresponding author

DOI: 10.1201/9781003222880-41

visually checked for soft spots, depressions, etc. Passing compaction tests don't necessarily mean the subgrade is ready for the pavement. Any deficiencies must be corrected prior to placing subsequent layers.

Traditionally, quality control (QC) measures undertaken during the construction of the subgrade often check the moisture content (w) and dry density (γ_d) of the subgrade. Therefore, w and γ_d are the appropriate indicators of level of compaction of the subgrade at the time of construction. In some cases, Dynamic Cone Penetration (DCP), Falling Weight Deflectometer (FWD), or Light Weight Deflectometer (LWD) tests are performed to assess the stiffness of the subgrade during construction. However, these tests do not provide a direct measurement of the resilient modulus and are not commonly performed due to cost and time constraints. Another disadvantage of the traditional QC test methods is the inability of evaluating the entire compacted area. Tests at discrete locations may leave undetected soft spots on the finished subgrade. Poor quality can also have an adverse effect on pavement layers constructed on top of the prepared subgrade (1).

Pavement structure deteriorates as it receives more traffics. Thus, pavement inspection is periodically conducted. Inspection of road condition consists of two steps: First step is the visual inspection to detect cracks' locations and functional evaluation by assessing surface parameters such as roughness and skid resistance. Secondly, a structural evaluation is performed for subsurface layers, to check for decrease in thickness, lack of interlock between different layers, propagation of cracks and loss of subsurface materials (i.e., sinkholes) that might be in progress.

1.1 *Ground Penetrating Radar (GPR)*

Considering time, safety and cost-effectiveness concepts, non-destructive testing (NDT) methods are preferable than costly traditional methods such as coring and destructive testing, those approaches have been used since the beginning of the former century, but they are not accurate all the time and in some cases unsafe for the operators (e.g., on heavily crowded roads).

Usually the GPR is used effectively for several tasks on flexible, semirigid, and rigid pavement include layer thickness assessment, pavement condition evaluation, and estimation of moisture content, asphalt density gauge and compaction. The main GPR applications in tunnel inspection include thickness estimation of lining and backfill grouting, rebar localization and corrosion assessment, localization of reinforcement and cavities in lining, and estimation of moisture content (2).

Currently, GPR is also used for estimating asphalt density, for detecting defects and deteriorations in flexible pavements, and for assessing water content at different levels within the pavement structure. These applications already are at advanced level, but more research is needed; indeed, they are not yet offered/required in a systematic way, in pavement management (2).

Ground Penetrating Radar (GPR) is one of the most efficient NDT methods for subsurface monitoring. GPR is a special type of radar designed to evaluate internal inhomogeneity and predict layer thickness of structures by penetrating the surface with electromagnetic (EM) waves. Among the various types of GPR systems, a pulsed (or impulse) GPR system is the most commercially available and commonly used to evaluate transportation infrastructure (3). The principle of the pulsed systems is that a narrow EM pulse transmitted into the ground is reflected from the interfaces of materials that have distinct dielectric properties. The most common uses of GPR in pavement engineering are to measure pavement layer thicknesses, identify large voids, detect the presence of excess water in a structure, locate underground utilities, and investigate significant delamination between pavement layers (4).

Depending on how the antennae are deployed, GPR systems are classified as air-coupled (or launched) or ground-coupled systems (5,6). Air-coupled antennae are typically mounted 15 cm to 50 cm above the surface. These systems produce a clean radar signal at the pavement surface and allow for high-speed surveys of up to 65 mph. The drawbacks of air-coupled systems are their relative low penetration depth (although this could be overcome by the recent introduction of the 500-MHz air-coupled antenna (7).

The primary material property obtained from GPR surveys is the dielectric constant. The dielectric constant, also known as the relative permittivity, ε_r, of a homogeneous media relates the relative EM velocity in a material to the speed of light in free space, c (8).

$$\varepsilon_r = (c/v)^2 \qquad (1)$$

Where, ε_r: dielectric constant,
v: EM velocity in the material,
c: speed of light in free space of $3*10^8$ m/s.

The use of GPR for soil density measurements has been progressively studied, but the total research devoted to this method is minimal compared with research devoted to other applications for pavement systems. Al-Qadi et al. (9) developed three HMA bulk specific gravity models to predict in-place HMA density using GPR based on the EM mixing theory. In Malaysia, Mardeni et al. (10) developed soil density prediction tool to analyze the effect of soil density with its electrical properties using ground penetrating radar (GPR) principal at frequency range of 1.7 GHz to 2.6 GHz. In Europe, Saarenketo (11) established a method to measure HMA pavement density using GPR. Based on the long-term studies on applying GPR for pavement quality control (QC), an exponential relationship was developed between the surface dielectric constant and the void content. In most of the previous studies, the prediction of HMA density was based on pure statistical analysis of test data. This study applies electromagnetic mixing theory to develop a prediction model for subgrade soils density.

2 RESEARCH OBJECTIVE

Soil is a composite material that consists of solid particles, air, and water. Density of soil depends on the specific gravities and volumetric fractions of its components. In a similar way, the dielectric constant of the soil is a function of the dielectric and volumetric properties of its components. Various EM mixing models are available to predict the dielectric constant based of a mixture on the dielectric constants and volume fractions of its components. Most of these models hypothesize that a mixture is composed of a background material with inclusions of different sizes and shapes.

The objective of this study is to develop mathematical model that can be used to predict subgrade soil dry density, based on its dielectric constant using these EM mixing formulae.

3 MODEL DEVELOPMENT

In accordance with the CRIM mixture theory, the dielectric constant of a homogenous mixture can be estimated by the "power-law" approximation (9,12,13). A widely used class of mixing models is shown in Eq. (2).

$$\varepsilon_{eff}^{\beta} = f\,\varepsilon_1^{\beta} + (1-f)\,\varepsilon_2^{\beta} \qquad (2)$$

Where:

$\varepsilon_{eff}, \varepsilon_i$ and ε_e: are the complex dielectric constants of the mixture, and the two constituents, respectively.

f: the fractional volume of the first constituent (volume of first constituent/total volume).

In this mixing model, a certain power of the permittivity is averaged by volume weights. f is the volume fraction and β is a constant dependent on the mixture's composition and is usually assumed to be 0.5 (13). Assuming the total volume of soils phases is assumed as 1 as seen in Figure 1, Eq (2) can be re-written as Eq (3).

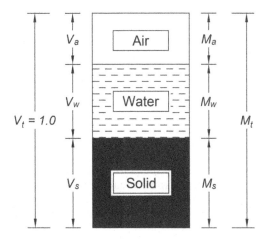

Figure 1. Soil volume and mass composition.

$$\varepsilon_{soil}^{\beta} = V_a \cdot \varepsilon_a^{\beta} + V_w \cdot \varepsilon_w^{\beta} + V_s \cdot \varepsilon_s^{\beta} \qquad (3)$$

Where:
V_a: Air voids volume in soil, V_w: water volume in soil, V_s: solid particles volume in soil,
M_a: Air voids mass = 0, M_w: mass of water, M_s: mass of soilds
ε_{soil}: Soil media dielectric constant, ε_a: Air dielectric constant = 1.0,
ε_w: water dielectric constant = 80-81, ε_s: Soil particles dielectric constant.
From Phase diagram relations, Assuming $V_t = 1.0$

$$\therefore \gamma_d = \frac{GS \cdot \gamma_w \cdot \left(1 - \frac{V_a}{V_t}\right)}{1 + w \cdot GS} = \frac{GS \cdot \gamma_w \cdot (1 - V_a)}{1 + w \cdot GS}, \quad \therefore V_a = 1 - \frac{\gamma_d \cdot (1 + w \cdot GS)}{GS} \qquad (i)$$

$$\therefore \gamma_d = \frac{M_S}{V_t} = M_S, \quad \therefore \gamma_d = M_S \qquad (ii)$$

$$\therefore V_w = w \cdot M_S \qquad (iii)$$

$$V_s = \frac{M_S}{GS} \qquad (iv)$$

$$\varepsilon_{soil}^{0.5} = V_a \cdot \varepsilon_a^{1/\alpha} + V_w \cdot \varepsilon_w^{0.5} + V_s \cdot \varepsilon_s^{0.5} \qquad (v)$$

Substituting i, ii, iii and iv in v

$$\therefore \sqrt{\varepsilon_{soil}} = 1 - \frac{\gamma_d \cdot (1 + w \cdot GS)}{GS} + w \cdot \gamma_d \cdot \sqrt{\varepsilon_w} + \frac{\gamma_d}{GS} \cdot \sqrt{\varepsilon_s}$$

$$\therefore \sqrt{\varepsilon_{soil}} - 1 = \gamma_d \cdot \left[-\frac{1}{GS} - w + w \cdot \sqrt{\varepsilon_w} + \frac{1}{GS} \cdot \sqrt{\varepsilon_s} \right]$$

$$\therefore \sqrt{\varepsilon_{soil}} - 1 = \gamma_d \cdot \left[\frac{1}{GS}(\sqrt{\varepsilon_s} - 1) + w\,(\sqrt{\varepsilon_w} - 1) \right]$$

$$\therefore \gamma_d = \frac{\left(\sqrt{\varepsilon_{soil}} - 1\right)}{\frac{1}{GS}\left(\sqrt{\varepsilon_s} - 1\right) + w\left(\sqrt{\varepsilon_w} - 1\right)} \quad (4)$$

Note that in these equations, the specific gravity of a material is equal to the density of the material divided by the density of water at 4° C (1g/cm³), and therefore is numerically the same as the density of the material in g/cm³.

3.1 Dielectric constant

The dielectric constant (ε) is the ratio of the permittivity of a substance to free space. The value of ε in air is 1 while ε in water is approximately 80. The dielectric permittivity can be determined by measuring radio frequency reflection as it propagates through the soil. Figure 2 shows a schematic diagram of the reflections of a GPR signal from a layered system, while A_o is the amplitude of the surface reflection and A_m is the amplitude of the reflected signal collected over a metal plate placed on the surface.

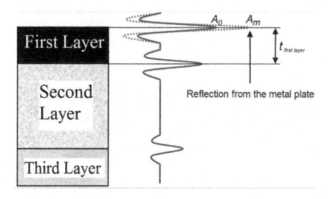

Figure 2. Schematic diagram for reflections from layer interfaces.

Where:
A_o = amplitude of the surface reflection, and
A_m = amplitude of the reflected signal collected over a metal plate placed on the surface.
A_2 = amplitude of the reflected signal collected over second layer surface.
Eq. (5) is used to calculate the thickness (d_i) of any layer:

$$d_i = \frac{ct_i}{2\sqrt{\varepsilon_{r,i}}} \quad (5)$$

Where:
d_i = the thickness of the ith layer
t_i = the EM wave two-way travel time through the ith layer as shown in Figure 2
$\varepsilon_{r,i}$ = dielectric constant of the ith layer,
c = speed of light in free space of $3*10^8$ m/s
The dielectric values of the first and second layers can be computed using the following Eqs. (6) and (7), respectively (14).

$$\varepsilon_1 = \left(\frac{1 + \frac{A_0}{A_m}}{1 - \frac{A_0}{A_m}}\right)^2 \quad (6)$$

$$\varepsilon_2 = \varepsilon_1 \left(\frac{1 - \left(\frac{A_0}{A_m}\right)^2 + \left(\frac{A_2}{A_m}\right)}{1 + \left(\frac{A_0}{A_m}\right)^2 - \left(\frac{A_2}{A_m}\right)} \right)^2 \quad (7)$$

Where
ε_1 = dielectric constant for the first layer,
ε_2 = dielectric constant for the second layer
The dielectric value ε_n for the third layer onwards (n^{th} layer) can be evaluated as Eq. (8):

$$\varepsilon_{r,n} = \varepsilon_{r,n-1} \left(\frac{1 - \left(\frac{A_0}{A_m}\right)^2 + \sum_{i=1}^{n-2} \gamma_i \left(\frac{A_i}{A_m}\right) + \left(\frac{A_n}{A_m}\right)}{1 - \left(\frac{A_0}{A_m}\right)^2 + \sum_{i=1}^{n-2} \gamma_i \left(\frac{A_i}{A_m}\right) - \left(\frac{A_n}{A_m}\right)} \right)^2 \text{ if } n > 2 \quad (8)$$

γ_i is the reflection coefficient at the ith layer interface.

$$\gamma_i = \frac{\sqrt{\varepsilon_{r,i}} - \sqrt{\varepsilon_{r,i+1}}}{\sqrt{\varepsilon_{r,i}} + \sqrt{\varepsilon_{r,i+1}}} \quad (9)$$

Where $\varepsilon_{r,1}$ is the dielectric constant of the top layer. The dielectric constant of the second layer $\varepsilon_{r,2}$ is calculated using Eq. (7) for n equals 1 and using the value of $\varepsilon_{r,1}$ computed in Eq. (6). It should be noted that in the above formulation, the pavement layers are assumed to be homogeneous. Therefore, the dielectric constant of each layer is assumed to be constant in the sense that it does not vary within the layer thickness (7,14).

Topp et al. (15) proposed an equation to estimate the volumetric water content (θ) in relation with dielectric constant of the soil. The water content of soil could be calculated from volumetric water content using Eqs. (10) and (11).

$$\theta = -0.053 + 0.0292(\varepsilon_{soil}) - 5.5 \times 10^{-4}(\varepsilon_{soil})^2 + 4.3 \times 10^{-6}(\varepsilon_{soil})^3 \quad (10)$$

$$W = \theta / \gamma_{bulk} \quad (11)$$

4 MODEL EVALUATION USING LABORATORY TESTING

4.1 *Soil Preparation*

A soil type readily found in North Georgia were used as the subgrade material in this study. The soil index properties for these soils are provided in Table 1. The soil is classified as high plasticity silt (MH) per the Unified Soil Classification System (USCS) or A-7-5 according to the American Association of State Highway and Transportation (AASHTO) classification system.

Table 1. Soil Index Properties.

Specific Gravity	USCS Classification	Plastic Limit	Liquid Limit	Plasticity Index
2.75	MH	37	57	20

Six (6) test specimens were prepared with 1-meter long x 0.7-meter wide each. The soil was mixed with water till reaching desired water content for soil compaction. Then, subgrade soil specimens were constructed with 0.3 m (12 in.) in depth and compacted to reach six (6) different density levels. Table 2 shows the density level for each specimen. It should be noted that specimens, D1 through D5, were compacted with varing compaction efforts while D6 was not compacted at all. Therefore, it is believed that D6 specimen has a higher void ratio compared to the other specimens. After the specimens were prepared, bulk density of each specimen was measured using sand cone tests. The results of sand cone tests are presented in Table 2.

Table 2. Sand cone test results.

Sample	Bulk density (γ_{bulk})		Water content %	Dry density (γ_{dry})	
	(t/m³)	(pcf)		(t/m³)	(pcf)
D1	1.517	94.710	22.49%	1.238	77.3
D2	1.578	98.490	21.05%	1.304	81.4
D3	1.304	81.423	20.43%	1.083	67.6
D4	1.696	105.876	24.43%	1.363	85.1
D5	1.744	108.877	26.60%	1.378	86.0
D6	1.082	67.561	18.69%	0.912	56.9

4.2 GPR data collection

After sand cone tests were conducted, GPR data were collected from six (6) samples using a 2.0-GHz air-coupled antenna. The Data acquisition system was calibrated before each scan in order to validate the signals with a perfect reflection on a metal plate. This calibration also provides some preliminary data for the data processing, with A_m (amplitude of the reflected wave on the metal plate to estimate the dielectric constants for the subsequent layers) in order to calculate the density and water content ratios for the targeted samples by the GPR scan.

5 TEST RESULTS

Figure 3 shows the calibration of the GPR system with metal plate on top of soil specimen. From this calibration process, A_m was found to be 7,942,173. Using the measured GPR amplitude values from each soil specimen, dielectric constans for six (6) soil specimens were calculated using Eq. (7). Table 3 summarizes the calculated dielectric constant for each specimen.

The initial soil bulk density is calculated using Eq. (5) with an assumed gravimetric water content (w) equal to 25%. Volumetric water content (θ) is calculated using Eqs. (11) and (12). With Eq. (12) gravimetric water content (w) is calculated and the error between calculated and estimated gravimetric content is adjusted. Using Generalized Reduced Gradient (GRG) function, the error is minimized and recalculated for the right values through iteration process. Table 4 shows the results of the calculation process.

Table 5 shows the calculated bulk density from the model and measured bulk density from sand cone tests. Generally, the percent error ranged from 4 to 33%. A substantial error with 45% was observed for the specimen with the lowest density.

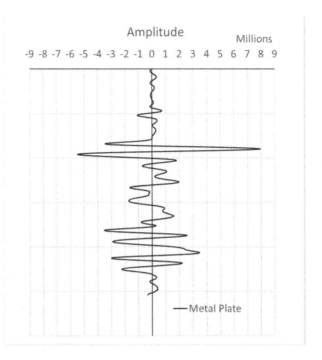

Figure 3. Metal plate calibration result.

Table 3. Dielectric constant results.

Sample ID	Amplitude	ε
D1	5,257,634	24.18
D2	5,318,661	25.55
D3	4,827,136	16.80
D4	5,788,944	40.67
D5	5,740,596	38.63
D6	3,378,338	6.15

Table 4. Proposed model calculation results*.

Sample ID	w % (Measured)	γ_{dry} initial		θ	w % (Calculated)	γ_{dry} Calculated	
		t/m³	pcf			t/m³	pcf
D1	22.5%	1.29	81	39%	24%	1.32	82
D2	21.1%	1.34	84	41%	24%	1.37	85
D3	20.4%	1.02	64	30%	24%	1.04	65
D4	24.4%	1.78	111	51%	23%	1.81	113
D5	26.6%	1.72	108	50%	23%	1.76	110
D6	18.7%	0.49	31	11%	17%	0.50	31

* ε_W = 80 (80-81), ε_s = 15 (5-30); θ = volumetric water content

Abdelmawla and Kim (*16*) proposed that measured density should be validated with baclculated value to reflect the soils physical properties in field. To adjust the error

Table 5. Comprison between Proposed model and Sand cone test results.

Sample ID	γ_{dry} Calc.		Dry density (Sand Cone Test)		Error %
	t/m³	pcf	t/m³	pcf	
D1	1.32	82	1.24	77	7%
D2	1.37	85	1.30	81	5%
D3	1.04	65	1.08	68	4%
D4	1.81	113	1.36	85	33%
D5	1.76	110	1.38	86	28%
D6	0.50	31	0.91	57	45%

between the calculated and measured bulk density, an exponential model was utilized as shown in Eq (12).

$$\gamma_{dry,\ estimated} = \alpha \left(\gamma_{dry,\ calculated} \right)^{\beta} \tag{12}$$

Based on the non-regression analysis, the fitting coefficients of α and β were estimated as 1.051 and 0.591, respectively. The results of the exponential method are shown in Table 6. As a result, Eq. (14) was developed to calculate soil bulk density.

Table 6. Exponential model results.

Sample ID	γ_{dry} Measured.		γ_{dry} from Eq. (12)		Error %
	t/m³	pcf	t/m³	pcf	
D1	1.24	77	1.21	75	2.4%
D2	1.30	81	1.23	76	6.0%
D3	1.08	68	1.11	70	2.9%
D4	1.36	85	1.40	87	2.6%
D5	1.38	86	1.38	86	0.0%
D6	0.91	57	0.91	57	0.0%

6 CONCLUSIONS

In this paper, the bulk density of subgrade soil was estimated using air coupled GPR systems, with antenna operating at 2 GHz central frequencies. An exponential model was developed and tested to estimate the dry density of subgrade soil. The soils density estimated by the newly developed exponential model shows approximately 6% error than soils density measured by sand cone method. The suggested approach provides several potential benefits that may eventually rationalize its use as a method to efficiently and accurately estimate subgrade soil density based on GPR data.

ACKNOWLEDGMENTS

The work presented in this paper is part of a research project (RP 19-21) sponsored by the Georgia Department of Transportation. The contents of this paper reflect the views of the authors, who are solely responsible for the facts and accuracy of the data, opinions, and conclusions presented herein. The contents may not reflect the views of the funding agency or other individuals. The authors would like to acknowledge the financial support provided by the Georgia Department of Transportation for this study.

REFERENCES

1. Barman, M., Nazari, M., Imran, S.A. et al. *Innov. Infrastruct. Solut.* 2016, 1: 23. https://doi.org/10.1007/s41062-016-0020-0
2. Pajewski, L., Fontul, S., & Solla, M. "Chapter 10 - Ground-penetrating radar for the evaluation and monitoring of transport infrastructures". In *Innovation in Near-Surface Geophysics*, 2019, (pp. 341–398). Elsevier Inc. doi: https://doi.org/10.1016/B978-0-12-812429-1.00010-6
3. Al-Qadi, I. L. Radar Testing of Concrete: Where Are We Now. *Presented at 8th International Conference and Exhibition on Structural Faults and Repair*, London, 1999.
4. Advisory Circular: Use of Non-destructive Testing in the Evaluation of Airport Pavements. AC No. 150/5370-11A. FAA, *U.S. Department of Transportation*, 2004.
5. Lahouar, S. Development of Data Analysis Algorithms for Interpretation of Ground Penetrating Radar Data. *PhD dissertation*. Virginia Polytechnic Institute and State University, Blacksburg, 2003.
6. Leng, Z., I. L. Al-Qadi, J. Baek, and S. Lahouar. Selection of Antenna Type and Frequency for Pavement Surveys Using Ground Penetrating Radar (GPR). *Presented at 88th Annual Meeting of the Transportation Research Board*, Washington, D.C., 2009.
7. Al-Qadi, I. L., Leng, Z., Lahouar, S., & Baek, J. In-Place Hot-Mix Asphalt Density Estimation Using Ground-Penetrating Radar. *Transportation Research Record*, *2152*(1). 2010, 19–27. https://doi.org/10.3141/2152-03
8. Al-Qadi, I. L., Samer Lahouar, and Amara Loulizi. Successful Application of Ground-Penetrating Radar for Quality Assurance–Quality Control of New Pavements. Transportation Research Record 1861, Paper No. 03-3512, 2003.
9. Al-Qadi, I. L., and S. M., Riad. Characterization of Portland Cement Concrete: Electromagnetic and Ultrasonic Measurement Techniques. *Report Submitted to National Science Foundation*, Washington, D.C., 1996.
10. R. Mardeni, K. S. Subari and I. S. Shahdan, "Soil density prediction tool using microwave ground penetrating radar, *2013 Asia-Pacific Microwave Conference Proceedings (APMC)*, Seoul, 2013, pp. 579–581. DOI: 10.1109/APMC.2013.6694870
11. Saarenketo, T. Using Ground-Penetrating Radar and Dielectric Prob Measurements in Pavement Density Quality Control. In *Transportatio Research Record 1575*, TRB, National Research Council, Washington, D.C., 1997, pp. 34–41.
12. Birchak J. R., Gardner C. G., Hipp J. E. and Victor J. M., "High dielectric constant microwave probes for sensing soil moisture,"in Proceedings of the IEEE, vol. 62, no. 1, pp. 93–98, Jan. 1974.
13. Sihvola, A. Electromagnetic Mixing Formulas and Applications. *IEEE Publishing, London*, 1999.
14. Tosti, Fabio, Alani, Amir, Benedetto, Andrea, Bianchini Ciampoli, Luca, Brancadoro, Maria Giulia and Pajewski, Lara. A comparative investigation of the pavement layer dielectrics by FDTD modelling and reflection amplitude GPR data. *BCRRA Tenth International Conference on the Bearing Capacity of Roads, Railways and Airfields*, 28–30 June 2017, Athens, Greece. 2017.
15. Topp, G.C., Davis, J.L., & Annan, A.P. Electromagnetic determination of soil water content: Measurements in coaxial transmission lines. *Water Resources Research*, Vol. 16, No. 3, Pages 574–582, June 1980
16. Abdelmawla, A.M. and Kim, S. Application of Ground Penetrating Radar to Estimate Subgrade Soil Density. Infrastructures 2020, 5(2), 12; https://doi.org/10.3390/infrastructures5020012

Traffic speed deflectometer measurements at Copenhagen airport

C.P. Nielsen & K. Jensen
Greenwood Engineering A/S, Copenhagen, Denmark

ABSTRACT: The results from a Traffic Speed Deflectometer (TSD) measurement campaign in Copenhagen airport are presented. The structural condition was measured with high spatial resolution on one runway with flexible pavement and one runway with concrete slabs overlaid with asphalt. The runway with flexible pavement had a high stiffness, and comparisons with HWD measurements showed a good agreement between TSD measurements and HWD measurements. On the runway with asphalt overlaid concrete slabs, the joint movement was quantified based on the TSD measurements. This revealed areas with comparatively high joint movement, which could be susceptible to reflective cracking of the asphalt layer. A total of 189 km runway was measured in four hours.

Keywords: Traffic Speed Deflectometer, TSD, Structural condition, Airport pavement, Runway pavement

1 INTRODUCTION

The structural condition of an airport's runways is of critical importance to the operation of the airport. Unexpected runway breakdowns can lead to substantial economic losses for the airport, and can affect the wider community by disrupting mobility. For these reasons, many airports perform routine Heavy Weight Deflectometer (HWD) surveys of runway structural condition (AC 2016; Pigozzi et al. 2015). The slow measuring speed of the HWD does, however, pose a challenge for its use on airport runways. The runway needs to be closed for an extended period of time while the measurements are done, and time constraints make it impractical to measure along the runway in more than a few lines.

In the road sector, similar challenges have led to the adoption of the Traffic Speed Deflectometer (TSD) for surveying structural condition. The TSD uses an innovative principle based on Doppler lasers to measure the response of the pavement to a 10 tonne axle, while moving at ordinary traffic speed. While the early TSDs were mainly used as screening devices and a supplement to FWD measurements, the newest TSD generations deliver data of a very high quality which can be used for work at the project level. Due to these improvements in measurement accuracy, it is now feasible to perform TSD measurements under the comparatively challenging conditions posed by an airport runway.

In this paper we present results from a TSD measurement campaign in Copenhagen Airport. Two different runways were measured during the campaign; one runway with flexible pavement and one runway with asphalt overlaid concrete slabs. The structural condition of one runway was measured with a high spatial resolution in both the longitudinal and the transverse direction. The other runway was measured with high spatial resolution in the longitudinal direction.

DOI: 10.1201/9781003222880-42

2 THE TRAFFIC SPEED DEFLECTOMETER

The Traffic Speed Deflectometer is a nondestructive pavement evaluation device, which uses laser Doppler vibrometers to measure the pavement response to a moving 10 tonne axle. The Doppler lasers measure the deflection velocity of the pavement surface at several points on the line passing between the center of the truck's twin tires. Combining this with a measurement of the driving speed, the slope of the deflection basin is recovered. The slope of the deflection basin can be used directly as a pavement condition indicator, or a back-calculation algorithm can be applied to find the elastic moduli of the pavement layers. The TSD used in this measurement was instrumented with 11 Doppler lasers; 8 in front of the load and 3 behind the load. See Refs. (Krarup et al. 2006; Nielsen 2019) for more information.

The TSD measurements took place on the 7th of May 2020 at Copenhagen Airport. Over the course of the measurements the air temperature varied between 11.2 °C and 16.0 °C, and the runway surface temperature varied between 14.0 °C and 22.0 °C. The measurements were made with a driving speed close to 80 km/h.

3 MEASUREMENTS ON RUNWAY 04R-22L

Runway 04R-22L is one of the two main runways of Copenhagen airport. The 04R direction is mainly used for aircraft take off, and is responsible for 32 % of the take offs in Copenhagen airport. The 22L direction is mainly used for landings, and is responsible for 60 % of the landings. The runway has a length of 3.3 km and is paved with 350 mm asphalt on top of a 300 mm unbound base layer. The runway was measured from edge to edge (including shoulders) with 48 passes of the TSD.

The performance of the top part of the pavement can be quantified using the surface curvature index SCI_{300}, which is defined as the difference in deflection between the load position and a position 300 mm from the load (Pigozzi et al. 2015). The curvature of the deflection basin is related to the horizontal strain in the bottom of the asphalt layer, which is an indicator of the pavement's susceptibility to fatigue damage. In Figure 1 the SCI_{300} values measured with the TSD are plotted on top of a satellite image of runway 04R-22L. The reported SCI_{300} values are averaged in sections of one meter. The scale of the colorbar is chosen so that it spans the range of most of the values encountered on the runway proper. The SCI_{300} values encountered on the shoulders are significantly higher than the chosen color scale, and the shoulders are therefore shown as yellow in most places. The notable exceptions are the places where taxiways lead into the runway: in these places the SCI_{300} values on the shoulder become comparable to the values on the runway. The SCI_{300} values found on the runway are all very low, in the range 5μm to 25μm, showing that it is a very stiff pavement. Nevertheless, some structure in the reported values is apparent. Around $x = -800$m a large area of comparatively high SCI_{300} is visible on both sides of the runway. These areas were initially thought to coincide with the touch down zone for aircraft landing in the 04R direction. However, only 2 % of the airplanes land in the 04R direction and the designated touch down zone marked on the runway is located near to $x = -1500$m. The reason for these large areas with higher SCI_{300} is therefore not readily apparent. It is probably a combination of usage patterns, small irregularities in the original construction and the maintenance history of the runway which has led to the observed behavior. The same is the case for the other regions of high or low SCI_{300}, which can be observed in Figure 1. It should, however, be stressed that the yellow areas are only weak in comparison. In absolute terms the entire runway is very stiff.

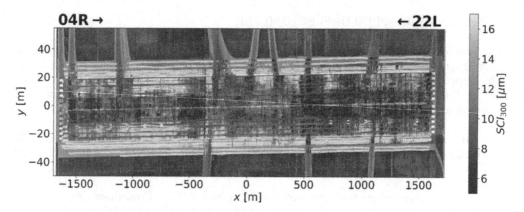

Figure 1. SCI_{300} values measured with the TSD on runway 04R-22L reported in 1 m sections.

3.1 Back-calculation on runway 04R-22L

The SCI_{300} index is convenient for getting a quick overview of the pavement performance. To quantify the pavement performance in more detail, the measurements were processed using a viscoelastic back-calculation algorithm developed specifically for the TSD, see Ref. (Nielsen 2019). The outputs of the back-calculation are the elastic moduli of the pavement layers and the pavement strains evaluated at points of interest in the pavement structure. In the back-calculation a layer thickness of 350 mm was used for the asphalt layer and 300 mm was used for the unbound base layer. The subgrade was assumed to be an infinite half-space. Before applying the back-calculation algorithm, the measurements were averaged in 10 m sections. In Figure 2 the back-calculated modulus of the asphalt layer on runway 04R-22L is plotted. In general, the asphalt is seen to be quite stiff, having elastic moduli in the range from 5000 MPa to 25000 MPa. It is, however, also seen that the behavior is not entirely uniform; within the yellow areas of high modulus there are small blue sections of lower modulus. These blue spots could indicate actual areas of reduced modulus, but on closer inspection it becomes apparent that they are actually areas where the back-calculation fits the data poorly. The reason for these poor fits is that the layer thicknesses used in the back-calculation do not accurately describe the pavement. Most likely, the 350 mm asphalt layer is composed of many different asphalt layers applied over the course of the runway's lifetime. Over time, the layers in the bottom have lost much of their stiffness, whereas the layers in the top still retain a high elastic modulus. This assumption was in fact confirmed by asphalt cores, which had previously been taken by airport maintenance personnel. The cores showed a reduced asphalt quality in the lower parts of the 350 mm asphalt layer. In Figure 3 some examples of TSD measurements and model fits are shown with an assumed top layer thickness of 350 mm (blue) and 150 mm (orange). It is seen that the model with 150 mm captures the measured behavior much better than the model with 350 mm. Nevertheless, to be consistent with the existing HWD back-calculations on the runway, the model with 350 mm asphalt is used in the following.

One of the main motivations for making a back-calculation is that it allows for an evaluation of the critical strains in the pavement structure. In Figure 4 the back-calculated horizontal strain ϵ_{xx} in the bottom of the asphalt layer is plotted. This strain is an important indicator, since it is related to the development of fatigue cracking. It is seen that the back-calculated horizontal strain follows the same pattern as the surface curvature index in Figure 1. It is important to note, that although the issues with the asphalt layer thickness discussed above will have an adverse effect on the quality of the modulus estimates, the strain estimates are still expected to be reasonably good. The reason for this is, that knowledge of the surface deflections puts some bounds on the possible values of the strains. Thus, even if there are some errors in the asphalt modulus, the estimated strain cannot deviate much from the true strain. This point is treated in more detail in (Nielsen 2020), where it is also shown

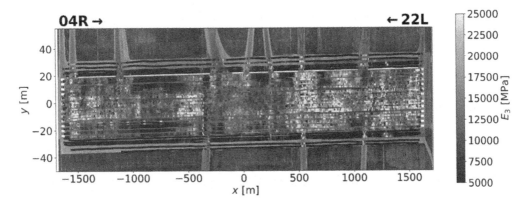

Figure 2. Back-calculated elastic moduli of the asphalt layer on runway 04R-22L.

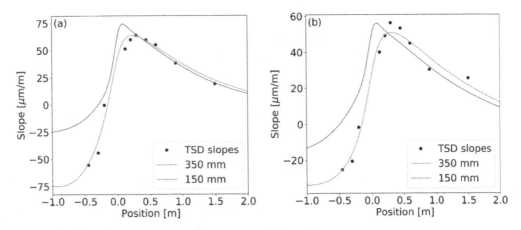

Figure 3. Two examples of measured and fitted slopes. The blue line assumes a top layer thickness of 350 mm and the orange line assumes a top layer thickness of 150 mm.

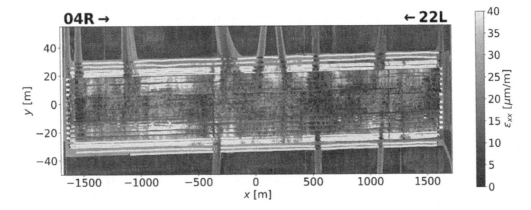

Figure 4. Back-calculated horizontal strain ϵ_{xx} in the bottom of the asphalt layer on runway 04R-22L.

that good estimates for the pavement strains can often be obtained without any knowledge of the elastic moduli.

3.2 Comparison with HWD measurements

The structural condition of runway 04R-22L had previously been measured with an HWD on October 5th 2017. The measurements were made with a spacing of 60 m and at each measurement position three different load levels (100 kN, 150 kN, 220 kN) were used. To avoid disrupting the runway operation, the measurements were made in the night with a runway surface temperature around 8 °C. Comparing the deflections obtained with each load level revealed that the deflections had a close to linear dependence on the load. It should therefore be possible to compare the HWD measurements with the TSD measurements by scaling the measurements with the used load. This also agrees with the findings from sensitivity studies at the National Airport Pavement Test Facility (AC 2016). In Figure 5 the SCI_{300} index measured by the TSD is plotted together with the SCI_{300} index measured by the HWD. The HWD values have been scaled with the ratio of the loads used by the two devices. The SCI_{300} values from the two devices are seen to be comparable, although the patterns do not match completely. In Figure 6 (a) the SCI_{300} values are plotted for the topmost HWD line together with the TSD values for the two lines which fall closest to the HWD line. Again, the HWD values are scaled with the relative loads. The TSD measurements were made close to each other, but they are not coinciding exactly. Therefore, the two TSD measurements deviate a little from each other. The scaled HWD measurements are seen to follow the same pattern as the TSD measurements, and the magnitude of the values have a good agreement with those obtained from the TSD. There are multiple possible explanations for the small discrepancies which are observed between the TSD and HWD results. 1: The HWD and TSD measurements were not made in exactly the same spots, and we should expect some deviation on that account. 2: The HWD measurements were made three years before the TSD measurements, and at a different pavement temperature. 3: The loading levels for the two devices were different, and if the subgrade behaves nonlinearly this could lead to a difference in the responses. This is, however, believed to be a small factor for the pavement at hand, since the HWD measurements revealed a close to linear dependence on the load. 4: The HWD and TSD have different loading mechanisms. The HWD simulates a moving load, by applying an impulsive load of a certain duration. The TSD, on the other hand, measures the pavement response to an actual moving load. Using an impulsive load to simulate a moving load will lead to an error if the pavement behaves viscoelastically. Since the runway pavement has a relatively thick asphalt layer, we would expect viscoelasticity to have a quite significant effect on the pavement behavior. The HWD measurements could therefore be significantly skewed by the used loading mechanism. Overall, however, the measurements from the two devices are seen to match quite well.

In Figure 6 (b) the back-calculated elastic moduli of the asphalt layer based on both devices are plotted. There is seen to be an appreciable difference between the values obtained on the two TSD lines. This is partly due to small differences in driving path, but also, and more significantly, due to the layer thickness assumed in the back-calculation. Nevertheless, the TSD and HWD based moduli are seen to be of similar magnitude and exhibit some of the same patterns.

It is seen that the HWD moduli in Figure 6 (b) contain one point with a very low modulus. The corresponding SCI_{300} point has a value near 100μm and is therefore not visible in Figure 6 (a). The obvious interpretation is that this point is an outlier caused by a measurement error, and that it should be disregarded from the evaluation of the runway. This is, however, only an assumption. Conceivably, there could be a small region at this location, with a much-reduced pavement stiffness. Based only on the sparsely sampled HWD measurements, it is not possible to tell which of these possibilities is true. In contrast, the high spatial resolution of the TSD measurements allows all the stiffness variations to be visualized, and makes it easy to spot whether an outlier has occurred.

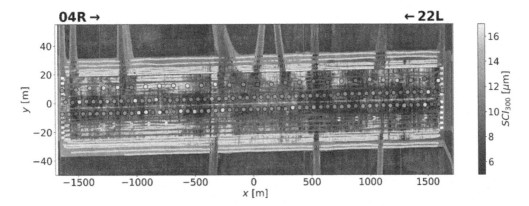

Figure 5. SCI_{300} measured with the TSD (lines) and measured with the HWD (dots). The HWD measurements have been scaled by the relative loads to allow for a direct comparison.

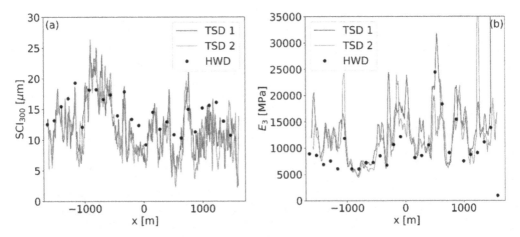

Figure 6. 1D comparison between TSD and HWD measurements. (a): SCI_{300} measured with the TSD (lines, 10 m moving average) and measured with the HWD (dots). The HWD measurements have been scaled by the relative loads to allow for a direct comparison. (b): Back-calculated elastic moduli of the asphalt layer based on the TSD measurements (lines, 40 m moving average) and the HWD measurements (dots).

4 MEASUREMENTS ON RUNWAY 12-30

Runway 12-30 runs orthogonally to the two main runways and is used much less frequently. The 12 direction accounts for 0.9 % of all landings and take offs, and the 30 direction accounts for 4.7 % of the landings. Originally, the runway was composed of concrete slabs, which are still visible on the shoulders and in the 30 end of the runway. On the remaining part of the runway, the concrete slabs have been overlaid with asphalt.

Measuring on a concrete pavement is different from measuring on an ordinary flexible pavement. The concrete slabs are almost always very stiff, and it is therefore less relevant to measure exactly how stiff they are. The pavement structure is, however, weak at the joints of the concrete slabs. This allows for motion of the slabs, and if this motion is too large it may result in erosion of the base layer or reflective cracking of the asphalt overlay. With the TSD it is possible to measure the motion of the concrete slabs, as the loaded axle moves from one slab to the next. This motion is conveniently quantified via the difference in Doppler velocity in front of the load and behind the load $\Delta v = v_+ - v_-$. For the TSD used in this measurement the v_+ laser is located 13 cm in front of the load, and the v_{laser} is located 20 cm behind.

The runway was measured with 6 passes of the TSD and the data reported in 5 cm intervals. In Figure 7 (a) the velocity difference Δv is plotted along one of the measurements. In Figure 7 (b) a zoom of the same measurement is shown. Spikes in Δv are seen at regular intervals, corresponding to the locations of the slab joints. In most of these places the slabs and joints are covered with asphalt, but the joint movement is still clearly visible through the asphalt. In Figure 8 the joint movement is visualized on the surface of the runway. Here, the locations with joint movement have been identified and shown with a dot, and each dot has been colored based to the magnitude of the Δv peak. It is seen that there is only visible joint movement on some parts of the runway, and the magnitude of the joint movement varies significantly from region to region. As an example, the region near $x = 300$m and $y = -10$m contains numerous joints with Δv close to 35mm/s. This indicates that the region could be prone to reflective cracking. Inspecting satellite images of the runway, it was found that there are signs of reflective crack repairs in that area (also faintly visible in Figure 8). As another example, we consider the region between $x = 700$m and $x = 1200$m. Here, the concrete slabs in the middle of the runway were broken into smaller pieces a few years previously. It is seen, that this had the desired effect; there is no joint movement in the middle of the runway. Only the bottom line around $y = -15$m exhibits joint movement. Evidently, this line is outside the region that was reconstructed.

To put the TSD measurements on runway 12-30 in perspective, measuring the load transfer of every joint with an FWD would require one measurement for every five meters. At one minute per measurement and a runway length of 2.7 km, it would take 9 hours to measure one lane with an FWD. With the TSD, each lane was measured in around two and a half minutes.

We note that the Δv index used to quantify the joint movement in this survey, is different from the conventionally used load transfer efficiency (LTE). The two indices are obviously correlated, and in general places with a high Δv will have a low LTE. However, the relation between the two indices is not one-to-one. In an upcoming paper on TSD measurements on concrete pavements, the relation between LTE and TSD based load transfer indices is explored further.

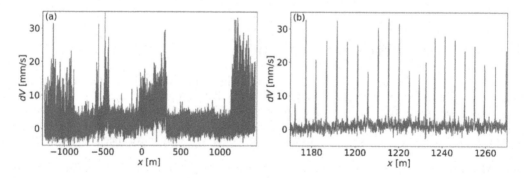

Figure 7. (a): Plot of Δv along one measurement on runway 12-30. (b): Zoom of (a).

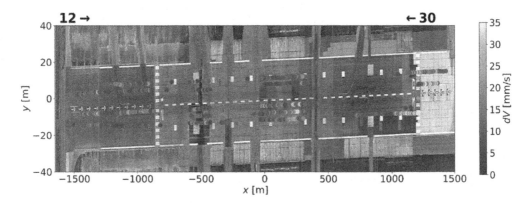

Figure 8. Locations of joint movement on runway 12-30. The color scale shows the magnitude of the joint movement.

5 SUMMARY AND CONCLUSION

The ability of the TSD to measure on airport runways was demonstrated on two runways in Copenhagen airport. The measurements on runway 04R-22L proved that the TSD is able to measure on very stiff flexible pavements, and the TSD proved its ability to consistently discriminate between stiff and very stiff areas of the pavement. The SCI_{300} values measured with the TSD had a good agreement with earlier HWD measurements on the runway. It was, however, found that the layer thicknesses previously used for HWD back-calculation did not represent the pavement structure well. For this reason, the elastic moduli back-calculated based on HWD and TSD did not agree as well as the raw measurements.

The pavement on runway 12-30 was composed of asphalt overlaid concrete slabs, which made it relevant to investigate joint movement rather than pavement stiffness. In large areas of the runway, movement of the slab joints could be detected through the asphalt overlay. Some of these areas could be susceptible to reflective cracks in the asphalt overlay.

In conclusion, the measurements in Copenhagen airport have demonstrated that the TSD is able to measure runway structural condition on two different pavement types, and that it is able to do it rapidly, consistently, and with high spatial resolution.

REFERENCES

AC. 2016. *150/5320-6F. Airport Pavement Design and Evaluation.*
Krarup, J., S. Rasmussen, L. Aagaard, and P. G. Hjorth. 2006. *Output from the Greenwood Traffic Speed Deflectometer*. In *Paper Presented to 22nd ARRB Conference*. Canberra, Australia.
Nielsen, C.P. 2019. *Visco-Elastic Back-Calculation of Traffic Speed Deflectometer Measurements*. Transportation Research Record.
Nielsen, C.P. 2020. *Deriving Pavement Deflection Indices from Layered Elastic Theory*. Transportation Research Record.
Pigozzi, Franco, Silvia Portas, Francesca Maltinti, and Mauro Coni. 2015. *Analysis of Runway Deflectometer Campaign for Implementation on Airport Pavement Management System*. International Journal on Pavement Engineering & Asphalt Technology 15 (2): 11–26.

Field performance assessment versus asphalt pavement design considerations

A. Loizos, K. Gkyrtis, C. Plati & K. Georgouli
Laboratory of Pavement Engineering, National Technical University of Athens (NTUA), Athens, Greece

ABSTRACT: Knowledge of pavement structural condition is desirable throughout a pavement's service life. Although the as-built pavement condition in terms of bearing capacity is most critical for the evolution of its future performance, it still remains challenging for pavement engineers to investigate any dependencies of the field assessment process on the pavement design. The latter becomes even more significant when the assumptions and loading principles used for material characterization during the design, differ from those that can be implemented in the field through the common utilization of Non-Destructive Testing (NDT) systems, including among others, the Falling Weight Deflectometer (FWD). On these grounds, the present research study considers field evaluation challenges of a new asphalt pavement structure of a heavy-duty motorway designed according to a robust international analytical design method. In particular, standardized Asphalt Concrete (AC) materials were assumed for the asphalt base and binder courses, whose stiffness characteristics were determined for the design analysis through the two-point bending test on trapezoidal specimens. Contrariwise, NDT testing was performed in the field and cylindrical cores were also extracted to assess material performance in the laboratory through the uniaxial compression test mode. The analysis for the pavement evaluation follows mechanistic perspectives demonstrating how to address potential limitations of pavement structural evaluation.

Keywords: Asphalt pavements, zero-point assessment, viscoelastic materials, pavement evaluation

1 INTRODUCTION

Pavement condition assessment at any time t throughout a pavement's lifespan is more than desirable, as it enables rational scheduling of maintenance actions (Plati et al. 2020a). In particular, the assessment shortly after construction or else during t=0, which will be hereinafter referred to as "zero-point" assessment, is of utmost significance as its purpose is twofold. First, it provides information on the condition that was achieved during construction (quality control) and second, it serves as historic data (Georgouli and Loizos 2017). While the goal of the former is to define whether the constructed pavement meets the pavement and material design principles, the latter enables the establishment of a reference status based on the as-built pavement condition. This aspect is rather useful for future performance monitoring during a pavement's service life.

For both of the afore-mentioned purposes during pavement evaluation, the major engineering challenge lies upon using tools and analysis theories that lead to reliable results and conclusions in respect to pavement condition. Commonly, data collected through Non-Destructive Testing (NDT) systems, including the Falling Weight Deflectometer (FWD) and the Ground

Penetrating Radar (GPR), is used as input to back-calculate the pavement stiffness profile and proceed with a response analysis that enables pavement life expectancy estimation (Plati et al. 2020b). The Multi-Layered Elastic Theory (MLET) is most often used as the background theory of the previous analysis steps (Marecos et al. 2017, Crook et al. 2012).

However, there is enough evidence that the Asphalt Concrete (AC) materials exhibit visco-elastic behavior, since they are temperature- and loading rate- dependent (Grellet et al. 2012, Chabot et al. 2010). Viscoleastic material characterization can be performed either through laboratory tests on field cores (Gkyrtis et al. 2021, Georgouli et al. 2015) or through the use of algorithms that estimate the AC dynamic modulus E^* (Georgouli and Loizos 2017). No matter the way followed, the use of different approaches to analyze the pavement structure (i.e. through elastic or viscoelastic calculations) can lead to discrepancies on its bearing capacity and life expectancy estimation with a profound effect on maintenance decision-making issues.

On this context, although the as-built pavement condition in terms of bearing capacity is most critical for the evolution of its future performance, it still remains challenging for pavement engineers to investigate any dependencies of the field assessment process on the pavement design. Focusing on the zero-point condition, the latter becomes even more significant because the assumptions and loading principles used for material characterization during the design may differ from those that can be implemented in the field through the use of NDTs (Gkyrtis et al. 2021).

This is the case of the experimental pavement selected in this study for the investigation purposes. In particular, the selected pavement was designed according to an international analytical and robust design method (LCPC-SETRA, 1997). Standardized AC materials were assumed for the asphalt base and binder courses, whose stiffness characteristics were determined for the design analysis through the Two-Point Bending (2PB) test on trapezoidal specimens in the laboratory. The latter implies that the design considerations are not compatible with the cylindrical cores that are usually extracted in the field. Moreover, the FWD load imposed in-situ through a segmented loading plate is not directly comparable with the design loading principles. Such issues make even more complex the zero-point condition assessment in new pavements, since dependencies on the design process may exist. In addition, material characterization in the Linear Visco-Elastic (LVE) region requires core extraction and sophisticated laboratory equipment to obtain the E^*, aspects that usually make it tedious for pavement engineers to consider advanced material characterization in the field.

Nevertheless, pavement engineers are responsible for effectively incorporating pavement mechanistic principles into common engineering issues. The objective of this study is to investigate structural assessment challenges during the zero-point assessment and develop a simplified approach based on NDT and laboratory data in order to reach a reliable overview of the pavement condition. To meet the research aim, a field experiment was organized along an experimental asphalt pavement section, which is part of a heavy-duty motorway and belongs to the trans-European transport network. The analysis was based on response calculations and the emphasis was put on the tensile strains at the bottom of the AC layers.

2 STRUCTURE, MATERIALS AND TESTING

A typical cross section of the trial asphalt pavement is shown in Figure 1. At the time of the experiment, the AC layers included an asphalt binder and asphalt base layer with limestone aggregates. The aggregate gradation is given in Table 1. The asphalt material had a penetration grade ranging from 35-36PEN and a softening point of 55-56°C. Base layer consisted of unbound granular material and the subgrade layer consisted of natural gravel.

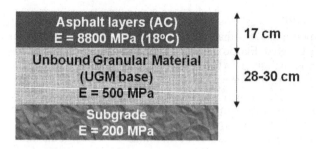

Figure 1. Typical pavement cross-section.

Table 1. Aggregate gradation for AC materials (% passing).

Sieve (mm)	Binder course	Base course
19	100	91
9.5	78	65
4.75	52	44
0.075	4	3.4

In respect to the experiment, a 50kN load was imposed by the FWD on the pavement surface at the outer wheel-path of the heavy-duty traffic lane. Surface deflections were recorded by nine sensors. Temperature measurements in the middle of the AC layers were performed through properly drilled holes within the pavement. At each of the investigated locations (P1-P7), multiple FWD measurements were undertaken at the temperatures of Table 2 in order to obtain a better overview of AC temperature sensitivity. These multiple measurements were taken at different times of the day. GPR surveys and coring yielded granular base and AC layer thicknesses respectively. GPR signal analysis led to thickness prediction errors ranged from less than 1% up to 8% in accordance with previous studies (Plati et al. 2020b).

Table 2. Measured temperatures at the middle of AC layers.

Position	Temperature (°C)
P1	26.7, 31.2, 35.9
P2	27.0, 31.5, 36.2
P3	27.4, 31.8, 36.7
P4	24.0, 39.7
P5	24.0, 39.6
P6	24.0, 39.6
P7	24.0, 39.7

As per the laboratory testing, the dynamic modulus test (AASHTO, 2001) was performed on seven cores extracted in-situ at the same positions of FWD testing. E^* was determined under the Uniaxial Compression (UC) test mode. In particular, thirty combinations of a controlled sinusoidal (haversine) compressive loading were imposed at each core, including five temperatures (4, 15, 20, 25 and 37°C) and six frequencies (25, 10, 5, 1, 0.5 and 0.1Hz). Difficulties in performing the dynamic modulus test at -10 and 54°C led to the selection of the previous temperatures. Similar problems have been also mentioned elsewhere (Bennert and Williams 2009).

3 ANALYSIS CONCEPTUAL FRAMEWORK

Two kinds of response calculations were performed: (a) viscoelastic analysis based on the laboratory determined E^* master curves of field cores, and (b) viscoelastic analysis based on the theoretical E^* master curve of the standardized material assumed in the design. The first case represented the as-built pavement status. For comparison purpose, the second case generated an additional reference strain state adjusted to the field conditions. This implies that the AC characteristics constituted the only factor that could differentiate the pavement condition between the "expected" behavior (in accordance with pavement design considerations) and the "as-built" behavior. The analysis concept is illustrated in Figure 2. The stiffness profile of the underlying layers, pavement geometry, loading conditions and temperature were considered identical in both cases in respect to the in-situ evaluation process. In particular, base and subgrade layers were assumed to consist of elastic materials and their weighted moduli were considered equal to 500 and 200MPa respectively based on FWD results at all locations.

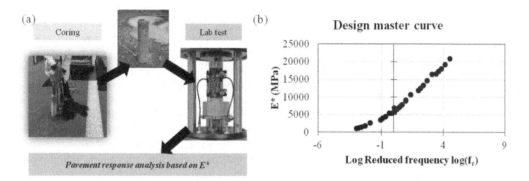

Figure 2. Analysis framework for the (a) "as-built" condition, and the (b) "expected" condition.

The adoption of the strain analysis approach dealt with potential incompatibilities on AC moduli assessment stemming from the different laboratory tests utilized for AC characterization. Furthermore, during the design process, an axle load of 130kN with dual tires was assumed, which generates a strain condition not directly comparable with field related conditions. As such, the developed framework with the FWD loading for both analysis cases can overcome potential obstacles in regards to the dependency of pavement assessment on the design principles.

For both analysis cases (a) and (b), the pavement response analysis was performed in the ViscoRoute software (Chabot et al. 2010), through the calibration of the rheological Huet-Sayegh model. This model consists of two branches including a spring and two parabolic dampers in the first branch and a spring in the second branch (Figure 3).

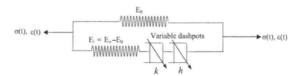

Figure 3. Representation of the Huet-Sayegh model.

Its mathematical formulation is:

$$E(\omega, \theta) = E_0 + \frac{E_\infty - E_0}{1 + \delta(j\omega\tau(\theta))^{-k} + (j\omega\tau(\theta))^{-h}} \quad (1)$$

where E_∞ is the instantaneous elastic modulus (MPa), E_0 is the static elastic modulus (MPa), k and h are dimensionless exponents of the parabolic dampers, and δ is a positive dimensionless coefficient. Also, $\omega=2\pi f$ (f=loading frequency) and θ=temperature. The temperature function $\tau(\theta)$ is defined as per Eq. 2 (A_0, A_1, A_2=thermal coefficients):

$$\tau(\theta) = \exp(A_0 + A_1\theta + A_2\theta^2) \qquad (2)$$

The calibration of Eq. 1 was based on nonlinear least squares regression with the minimization of the Standard-Square Error (SSE), defined as:

$$SSE = \sum_{i=1}^{n} (\log E^*_{lab} - \log E^*_{HS})^2 \qquad (3)$$

where E^*_{lab} is the experimental dynamic modulus (MPa) and E^*_{HS} is the predicted dynamic modulus (MPa). It is noted that the SSE ranged satisfactorily from 0.005-0.028 for all cores, indicating a good adaptability of the Huet-Sayegh model to E^* data. Thereafter, the response calculations were performed for the FWD load level of 50kN (tire pressure 707kPa) considering all of the temperatures enlisted in Table 2 and a moving speed of 80km/h. FWD frequency was estimated equal to 18Hz according to a methodology presented in Crow (1998) and was thereafter transformed to the previous speed according to an equation presented in Mollenhauer et al. (2009).

Finally, considering the integrity destruction in new pavements due to the coring process, the ability of FWD measured deflections to predict the as-built viscoelastic strains was investigated through statistical correlation. The rationale behind this approach was also strengthened because of the highway pressing constraints against lane closures, which is a critical issue for the case of in-service pavements.

4 RESULTS

4.1 Overview of viscoelastic response analysis

The response analysis with the design master curve as input led to an average longitudinal strain $\varepsilon_{x\text{-ref}}$=71.3µm/m and an average transverse strain $\varepsilon_{y\text{-ref}}$=80.4µm/m at the reference temperature of 18°C (assumed in the design), with a coefficient of variation equal to 3.7% and 3.6% respectively. For engineering analysis purposes, the above strains were assumed as the reference strain status for the pavement assessment at the same temperature. As such, in Figure 4, viscoelastic strains based on the E^* of field cores are presented together with the reference strains; the solid line corresponds to the $\varepsilon_{x\text{-ref}}$=71.3µm/m and the dashed line corresponds to the $\varepsilon_{y\text{-ref}}$=80.4µm/m.

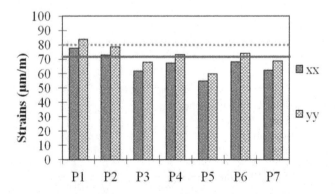

Figure 4. Viscoelastic strains at 18°C.

It can be seen, that 5 out of 7 locations exhibit lower strains at the bottom of AC in respect to the expected strains (solid and dashed lines). The maximum difference in the cases where strains from the as-built condition exceed the reference values (locations P1-P2) is 6.8 and 4.5μm/m for the longitudinal and transverse strains respectively. In order to investigate whether this trend is consistent or not for the whole spectrum of measured temperatures (Table 2), the strain comparison between the expected and the as-built condition was repeated considering the variable field temperatures at each location (Figure 5). It can be seen that for all the measured temperatures, the expected strains exceed the strains calculated considering the as-built material condition.

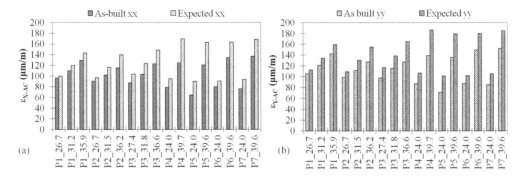

Figure 5. Comparison of viscoelastic strains (a) longitudinally and (b) transversely.

The magnitude of differences varies depending on the analysis temperature, so the impact of AC temperature on pavement evaluation is highlighted. In particular, averaged strains are presented in Figure 6 considering all the coring locations and similar field temperatures. Indeed, the expected strains exhibit a more abrupt change against temperature increase, which may imply that the material placed in-situ was less sensitive to the temperature increase as opposed to the expected performance of the standardized material that was assumed in the design.

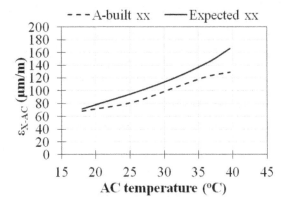

Figure 6. Longitudinal strain sensitivity against temperature.

Additionally, the material sensitivity on temperature is also profound in Figure 7, considering the full longitudinal and transverse strain profiles indicatively for location P4 at the highest measured temperature of 39.7°C. The asymmetry around the loading point in the longitudinal strain profile is less obvious at the as-built strain profile. Similarly, a slight asymmetry is also observed transversely on the expected strain profile around the peak value, which is less visible on the as-built strain profile. The previous remarks indicate again that the

AC behavior of the material placed in-situ is less prone to temperature increase than the behavior of the standardized material considered in the design.

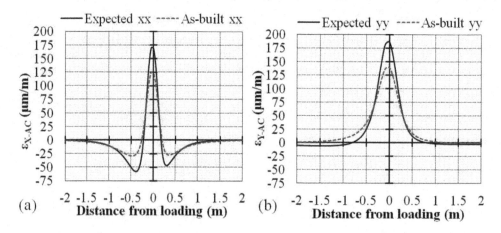

Figure 7. Indicative strain profiles at 39.7°C (a) longitudinally, and (b) transversely.

Overall, the proposed strain analysis approach effectively demonstrated the material behavior differentiation when moving from the design process to the field, implying that coring process might be needed even during the assessment at t=0, since the existence of properly calibrated algorithms for the non-destructive prediction of E^* is usually a matter of concern. In addition, utilizing design E^* data, even at t=0, could result in erroneous estimations of the pavement bearing capacity estimation.

4.2 *Correlations between NDT data and viscoelastic strains*

At this stage, the ability of the FWD measured deflections to act as potential predictors of the viscoelastic strains was investigated. From a linear regression analysis, a promising potential was found, since both longitudinal and transverse strains correlated well with the AC temperature and the Surface Curvature Index (*SCI*), defined as the difference of the surface deflections d_0 and d_{300} ($SCI=d_0-d_{300}$). These deflections are measured at a distance of 0 and 300mm away from the centre of the FWD loading plate. The longitudinal strain variation against AC temperatures and SCI is shown in Figure 8. As can be seen, temperature appears to have better prediction ability than SCI. Nevertheless, it was decided to combine both parameters in order to assess the feasibility of strain estimation.

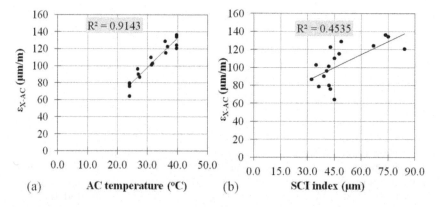

Figure 8. Correlations between (a) strains and AC temperature, and (b) strains and *SCI*.

As such, moving forward with the modelling approach presented in Loizos et al. (2020), the linear regression followed in this study resulted in a typical equation: $\varepsilon_{x\text{-}AC\_predicted} = a \cdot T + b \cdot SCI + c$, where a, b and c are regression constants. The R^2 value of the equation fit was found to be 0.91. The regression constants are not intentionally given, since only data from a unique structure was used, thus the use of the equation cannot be replicated elsewhere. Nevertheless, the presented approach might be proved beneficial in case the coring process needs to limited.

5 CONCLUDING REMARKS

This study demonstrated a simplified approach to perform structural evaluation of a new asphalt pavement by following mechanistic perspectives. The proposed approach could be used to overcome potential dependencies of the pavement evaluation process on pavement design principles. This was the case of the investigated pavement, as the standardized material assumed in the design was theoretically characterized through a test mode that is not compatible with the cylindrical cores that are usually extracted in the field.

Although the experiment took place at t=0, significant differentiation of the AC material was observed during the viscoelastic analysis. The assessment followed a strain comparison concept instead of moduli assessment, since moduli are in general expected to be non-comparable due to the different laboratory methods for the estimation of E^*. Moving forward with the investigation and based on the good variety and availability of NDT data, a practical framework was shown that could assist real-scale pavement evaluation and the related calculations, by integrating mainly NDT data and limited coring data (needed for model calibration) in order to quickly estimate viscoelastic responses.

In other words, pavement assessment at t=0 could be potentially addressed by estimating field viscoelastic strains through NDT data and limited coring (representing the as-built strain status) and simultaneously using the design modulus master curve to generate a reference response status adjusted at field conditions (to reflect the expected strain status according to pavement design).

To conclude, this paper contributed on how to deal with potential pavement structural evaluation limitations during the zero-point assessment and highlighted the significance of this assessment. No matter how well or not a pavement has been designed and constructed, this is expected to be serviceable for a long period beyond its design life. As such, formulating a method for the as-built pavement condition assessment contributes to a database development in order to systematically monitor pavement performance in the future and make a rational scheduling of potential maintenance or rehabilitation actions in the long term. Nevertheless, additional research is needed with a wider material spectrum and NDT data to further assess the feasibility of the proposed approach.

REFERENCES

AASHTO T342-11 (2001) Standard method of test for determining dynamic modulus of HMA. American Association of State Highway and Transportation Officials, Washington.

Bennert, T. and Williams, S.G., 2009. *Precision of AASHTO TP62-07 for use in mechanistic-empirical pavement design guide for flexible pavements.* Transportation Research Record: Journal of the Transportation Research Board, 2127: 115–126.

Chabot, A., Chupin, O., Deloffre, L. and Duhamel, D., 2010. *ViscoRoute 2.0: a tool for the simulation of moving load effects on asphalt pavement.* Road Materials and Pavement Design, 11(2): 227–250.

Crook, A.L., Montgomery, S.R. and Guthrie, W.S., 2012. *Use of falling weight deflectometer data for network-level flexible pavement management.* Transportation Research Record: Journal of the Transportation Research Board, 2304: 75–85.

CROW record, *Deflection profile-not a pitfall anymore. Survey and interpretation methodology-falling weight deflection measurements.* Crow, EDE, Netherlands, 1998.

Georgouli, K. and Loizos, A., 2017. *E* prediction algorithm for pavement quality control assessment.* Loizos et al. (Eds): Proceedings of the 10th International Conference on the Bearing Capacity of

Roads, Railways and Airfields (BCRRA 2017, June 28-30, Athens, Greece), Taylor & Francis group: 799–805.

Georgouli, K., Pomoni, M., Cliatt, B. and Loizos A. 2015. *A simplified approach for the estimation of HMA dynamic modulus for in service pavements.* In Proceedings of the 6th International Conference on Bituminous Mixtures and Pavements (ICONFBMP), June 10-12, 2015, Thessaloniki, Greece, Taylor and Francis Group, pg. 661–670.

Gkyrtis, K., Loizos, A. and Plati, C. 2021. *A mechanistic framework for field response assessment of asphalt pavements.* International Journal of Pavement Research and Technology, 14, 174–185.

Grellet, D., Dore, G., Kerzreho, J.P., Piau, J.M., Chabot, A. and Hornych, P., 2012. *Experimental and theoretical investigation of three dimensional strain occurring near the surface in asphalt concrete layers.* In Proceedings of the 7th Rilem International Conference on Cracking in Pavements, France, SPRINGER: 1017–1027.

LCPC-SETRA, French Design Manual for Pavement Structures: Guide Technique, 1997.

Loizos, A., Gkyrtis, K. and Plati, C., 2020. *Modelling asphalt pavement responses based on field and laboratory data.* Chabot et al. (Eds): Accelerated Pavement Testing to Transport Infrastructure Innovation, LNCE 96: 438–447.

Marecos, V., Fontul. S., Antunes, M.L. and Solla, M., 2017. *Evaluation of a highway pavement using non-destructive tests: Falling Weight Deflectometer and Ground Penetrating Radar.* Construction and Building Materials, 154: 1164–1172.

Mollenhauer, K., Wistuba, M. and Rabe, R., 2009. *Loading Frequency and Fatigue: In situ conditions & Impact on Test Results.* 2nd Workshop on Four Point Bending, Pais (ed.), University of Minho, Portugal, pp. 261–276.

Plati, C., Loizos, A. and Gkyrtis, K. 2020a. *Assessment of modern roadways using non-destructive geophysical surveying techniques.* Surveys in Geophysics, 41, 395–430.

Plati, C., Gkyrtis, K. and Loizos, A., 2020b. *Integrating non-destructive testing data to produce asphalt pavement critical strains.* Nondestructive Testing and Evaluation, https://doi.org/10.1080/10589759.2020.1834555.

Using non-destructive test to validate and calibrate smart sensors for urban pavement monitoring

A. Di Graziano, S. Cafiso & A. Severino
Department of Civil and Architectural Engineering, University of Catania, Italy

F. Praticò, R. Fedele & G. Pellicano
Department of Information Engineering, Infrastructure and Sustainable Energy, University Mediterranean of Reggio Calabria, Italy

ABSTRACT: In the context of smart cities, infrastructures play a strategic role to guarantee sustainability, efficiency, safety, and resiliency. Several solutions can be adopted, but the key factor for the success of the solution selected is its ability of improving the maintenance management process.

Specifically, in order to timely identify pavement needs, early, effective and continuous monitoring is needed to reduce total costs and to extend service life. Over the years several efforts have been made to implement more advanced and effective monitoring systems at ever more contained costs, going from impractical manual and destructive methods through automated in vehicle equipment to the most recent Smart Sensor Network (SSN) embedded into/positioned on the pavement. While traditional systems (GPR and FWD) are currently used in road pavement maintenance where they have shown their reliability and effectiveness, instead smart sensors in pavement maintenance are promising but in the early stage of investigation.

The objective of the presented study is to test a solution that can be used to make smarter the road pavement monitoring. Specifically, this paper details an urban site investigation comprising Ground Penetrating Radar (GPR), Falling Weight Deflectometer (FWD) and vibro-acoustic sensors. GPR and FWD are used for providing reference data of pavement bearing capacity. In the paper, tests and results in selected trial sites are used to identify Strengths, Weaknesses, Opportunities, and Threats (SWOT analysis) of the application of smart vibro-acoustic sensors for the assessment of pavement residual life.

Results show that the method is able to evaluate pavement deterioration (above all in terms of presence and entity of cracks) by means of meaningful features extracted from the vibro-acoustic signatures (acoustic signals) of the road pavement loaded by urban traffic, with the aim of using these data to build innovative performance curves able to improve an urban pavement management system.

Keywords: In-situ measurements, NTD, smart sensors, pavement condition

1 INTRODUCTION

The success of the smart city is strictly related to the ability of transportation infrastructures to guarantee sustainability, efficiency, safety, resiliency, and improve the maintenance management (Praticò, 2001). Specifically, in order to timely identify pavement needs, early, effective, and continuous monitoring is needed, aiming at reducing total costs and at extending service life (Odoki et al., 2015).

Traditional approaches, such as the analysis of pavement samples (cores/slabs), Ground Penetrating Radar (GPR), Falling Weight Deflectometer (FWD), Light Weight Deflectometer (LWD), are currently used in road pavement maintenance. Their reliability and effectiveness is well-known, as well as their potential weaknesses, such as (i) providing punctual measure (lack of effectiveness), (ii) being based on expensive devices or destructive tests (lack of sustainability), (iii) partly or completely implying the closure of the road section (which negatively affects the vehicular traffic and road safety, lack of safety), (iv) using traditional data management (lack of intelligence), and (v) being based on stand-alone devices (lack of connectivity).

Over the years, several efforts have been done to implement more advanced (i.e., effective, intelligent, safe) monitoring systems at lower costs (i.e., sustainable). Several monitoring solutions were proposed from impractical manual and destructive methods, through automated in-vehicle equipment, to the recent Smart Sensor Networks (SSN) embedded or positioned on the pavement (Di Graziano et al. 2020). In more detail, the following classes of solutions were identified together with their advantages and disadvantages:

1) Embedded sensor-based systems (Advantages: Do not affect the drivers' safety; Allow detecting internal conditions of a road section with great accuracy. Disadvantages: Need to be installed during construction or damage the road surface if installed on an existing road. Short life time due to harsh operation conditions. (Zhou et al., 2012; Hasni et al., 2017)).
2) Non-embedded sensor-based systems (Advantages: Non-destructive. Disadvantages: Often allow measuring surface conditions, or deriving internal condition based on surface measurements. (Praticò et al., 2017; Iodice et al., 2020).
3) Mobile systems (Advantages: Do not need installation on site; Able to scan wide road surfaces/sections. Do not interfere with the traffic. Disadvantages: Usually based on expensive devices. Require high computational effort (especially if they use images/videos as input data). Provide results that represent the average condition of a road surface/section (Yi et al., 2015; Cafiso et al., 2019).
4) Stationary systems (Advantages: Provide results (often) more accurate than the ones of mobile systems. Often interfere with the traffic. Disadvantages: Provide results that represent the condition of a point of a road surface/section (Grace, 2015; Fernandes and Pais, 2017).
5) Wireless systems (Advantages: Do not need wire for their connection (energy- and data transmission-related). Allow covering large areas (e.g., using Wireless Sensor Network). Disadvantages: May need widespread system of antennas or satellites. Need fast and secure protocols for data transmission. (Ceylan et al., 2013; Merenda et al., 2020).
6) Wired systems (Advantages: Do not affected by power and connection outages. Disadvantages: Often destructive. Do not allow long distance data transmission. (Yu et al., 2013; Iuele et al., 2019);
7) Self-powered systems (Advantages: Do not need energy sources. Disadvantages: Are in an early stage of development. Allow short time monitoring. Often have low autonomy. (Hasni et al., 2017; Fedele et al., 2018).
8) Traditional data management (Advantages: Based on well-known techniques such as sampling, filtering, and analyzing. Disadvantages: Do not allow automatic extraction of information from the recorded data. (Mounier et al., 2012; Ouma and Hahn, 2017).
9) Smart data management (Advantages: Allow the extraction of meaningful information from the recorded data using machine learning, big data, and artificial intelligence. Allow safe and prompt data transmission/exchange (among things, and between things and users). Disadvantages: Often need big data and interdisciplinary knowledge. Require high computational effort (Ceylan et al., 2014; Praticò et al., 2020).

Despite the advantages promised by the innovative solutions described above, it is really difficult to find applications in real contexts, and often they are in the early stage of investigation.

Based on the above, the main objective of the presented study is to test a solution that can be used to make smarter the road pavement monitoring. In more detail, an urban site investigation was carried out using as source of noise and vibration a FWD and a LWD, and an innovative vibro-acoustic sensor as receiver of the response of the road pavement to well-known loads produced by the sources cited above.

The characterization of the different road sections investigated (damaged and un-damaged) was carried out through a GPR and a FWD, which were used for providing reference data of pavement bearing capacity.

In the paper, tests and results in selected trial sites are used to identify Strengths, Weaknesses, Opportunities, and Threats (SWOT analysis) of the application of smart vibro-acoustic sensors for the assessment of pavement residual life.

Consequently, the remaining part of the paper is organized as follows. Section 2 aims at describing the equipment used in this study to characterize the road sections under investigation. Section 3 aims at presenting the innovative vibro-acoustic sensor proposed in this study. The road trial is included in Section 4, which is followed by Section 5 that presents the main results of the study. Finally, conclusions and future works are followed by the references.

2 MONITORING EQUIPMENTS

2.1 Ground Penetrating Radar (GPR)

The ground penetrating radar is a geophysical radar system with two antennas and receivers used to perform non-destructive investigations of underground characteristics with high resolution and in depth (up to 3.2 m from the surface).

GPR operation principle is based on electromagnetic theory, its functioning consists in sending short electromagnetic pulses into a medium and when pulses achieve an interface they are reflected back partially and collected by the receiving antenna (Figure 1). The reflected energy is displayed in wave-forms and the greatest amplitudes represent the interfaces between layers with distinct dielectric characteristics (Daniels, 2004). Therefore, GPR measures the travel time between the transmission of the energy pulses and its reception. Transmission and reception of radar pulses are performed from one or more antennas that are moved on the investigated medium. The collected data are processed and saved on a control unit, that is also used to generate the necessary pulses for the operation of the antennas.

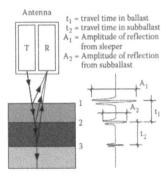

Figure 1. GPR principles (Saarenketo, 2006).

As mentioned GPR can provide a fast, nondestructive measurement technique for evaluating asphalt layer, presence of infrastructural interferences within the pavement, or to assess infrastructural conditions (Goulias et al., 2020).

Thickness of the asphalt layer can be calculated by two methods: two-way travel time or the common midpoint method (Lahouar et al., 2002) when multiple GPR channels are available. Since urban pavement present thin layers, it is often challenging to differentiate the reflection from the top and bottom of the layer. Antennas with a center frequency of 2000 MHz can provide a sufficient resolution to measure a minimum layer thickness less than 2.5 cm (1 in.) with an accuracy of 0.25 cm (0.1 in.) (ASTM D4748). Moreover, when GPR approaches above a pipe, a cable duct or a manhole, the two-way travel time versus travel distance has a parabolic shape (Al-Qadi et.al., 2005), so it is possible to highlight the presence of these types of interferences. The GPR used in the present study uses two antennas with 600 MHz and 2000 MHz frequencies. The K2_FW acquisition software is used to manage the phases of radar acquisition and to review the data acquired directly on site. The default field propagation speed is equal to 10 cm/ns.

2.2 *Falling Weight Deflectometer (FWD)*

The FWD is a non-destructive field test which is designed to simulate deflection of a pavement surface caused by a fast-moving truck. The device simulates the load conditions of a heavy vehicle and estimates the pavement's response by measuring the basin of deflection using several sensors fixed on a beam. The conventional FWD is able to apply loads in the range of 7-140 kN, even if the standard load used for structural pavement analysis is usually 30-50 kN producing about 700 kPa pressure under the load plate. The device allows a variable weight to be dropped from a variable height and the load is applied to the superstructure through a circular loading plate and weights from 50 to 400 Kg. The generated duration of the half sine pulse is typically 30 ms, corresponding to the loading time produced by a truck moving at 40 Km/h. The FWD used in the present study (Figure 2) is equipped with a loading plate of 300 mm diameter, 9 geophones positioned in the direction of the road, a load of 250 kg and 4 standard heights (50-100-200-390 mm) able to produce a stress of 500-1700 kPa (Cafiso and Di Graziano, 2009).

Figure 2. Falling Weight Deflectometer.

3 VIBRO-ACOUSTIC SENSORS

Based on the analysis reported in section 1, an innovative monitoring system was designed and several studies were carried out to set up and improve both the system and the data analysis method (see, e.g., (Fedele and Praticò, 2019; Fedele, 2020; Merenda et al., 2020; Praticò et al., 2020).

In more detail, the proposed solution consists of a system and a method (Figure 3) that aim at monitoring the Structural Health Status (SHS) of road pavements, designed bearing in mind concepts such as innovation, sustainability, efficiency, and intelligence. The solution is based on sensors that are attached on the road surface, i.e., it is a non-destructive test (NDT)-based solution. The solution is innovative because of the fact that it refers to the concept of vibro-acoustic signature. i.e., the vibro-acoustic response of pavement to loads. In more detail: 1) The vibrations are produced by the vehicles (source) that propagate into the road layers. 2) The corresponding variation of air pressure is detected by a microphone placed inside a sound insulating coating.

More precisely, the acoustic receiver consists of sound isolated microphone + external sound card + laptop for recording, processing and analyzing the received data (Figure 3). Importantly: 1) The road pavement is considered as an acoustic filter with a given SHS. 2) The method aims at detecting hidden cracks into the road section (e.g., under the wheel paths), while they are damaging the "acoustic filter" and before they arrive to the road surface. 3) The worsening of the SHS of the road pavement (filter) is recognized using the variation of meaningful features extracted from the recorded signals (vibro-acoustic signatures of the monitored pavement), in such a way as to obtain an efficient and intelligent solution.

Note that the receiver shown in Figure 3 (used in this study) is one of the possible systems that can be used to the purposes mentioned above, and that it is being improved (seeking to obtain a self-powered wireless system) in order to address some of the objectives of the ongoing Italian project USR342-PRIN 2017-2022.

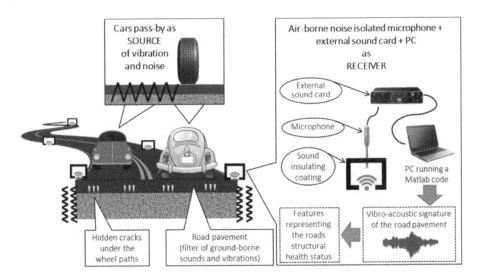

Figure 3. Schematic representation of the monitoring solution proposed in this study.

In this study the following devices were used as receiver (measurement chain): 1) Omnidirectional pre-polarized microphone "Audix TM1" (frequency response: 20-25 kHz ± 2 dB, sensitivity: 6 mV/Pa at 1 kHz, dynamic range: 112 dB), acoustically isolated using a plastic box filled with isolating material and modelling clay (used also to attach the box to the road). 2) External sound card "Roland quad-capture UA-55". 3) Recorder/Analyzer: ASUS ZenBook (model: UX433F, Intel® Core™ i7-8565U, RAM: 16 GB, 64bit x64) running MATLAB codes that allow recording acoustic signal with a sampling frequency of 192 KHz.

4 URBAN ROAD TRIAL TEST

In order to identify the cracked and uncracked areas to be investigated during the urban road trial test and the thickness of asphalt layer in these points, the Ground Penetrating Radar was used. About 300 meters long an urban road in Catania (South Italy) were covered in both directions, for a total of about 600 meters.

Thanks to the K2_FW acquisition software, the thickness of asphalt layer was analysed along the entire path with a step of 1 meter and the presence of any underground services was identified. Only the areas in which the asphalt layer with uniform dense mix was thicker than 15 cm were taken into account (Figure 4).

Figure 4. GPR investigation.

Through a visual inspection, among the areas having a similar thickness greater than 15 cm, four sections were identified, two without surface cracking and two with cracks (Figure 5).

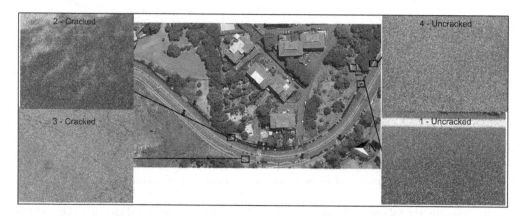

Figure 5. Selection of pavements in different conditions.

In each of the identified pavement areas, the test was carried out by loading the pavement with the FWD following the schema reported in the left part of Figure 6. In particular, once the vibro-acoustic sensor has been positioned on the pavement area under consideration, the tests were carried out as follows:

1. The first three load tests were carried out with the FWD plate in line with the microphone at a distance of 1,5 meters, the second moving the plate one meter back and the third one meter forward along the direction of travel;
2. The fourth loading test was carried out with the LWD positioned in line with the microphone at the same distance as the test with the FWD.

During the tests carried out with the FWD, the mass was dropped 9 times from different heights in order to obtain different loads: i) the first of 56 kN settling, ii) two of 40 kN, iii) two of 56 kN, iv) two of 84 kN and v) two of 120 kN. For each load point, the FWD test was performed twice. The surface temperature was about the same during the overall test.

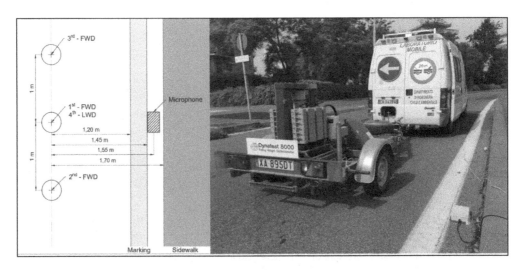

Figure 6. FWD and vibro-acoustic sensor schema of investigation.

5 RESULTS

During the experimental investigation, the FWD was applied as a source of vibration and noise at different loads. For the sake of simplicity, in the following, the load conditions were called FWD1 (=40 kN), FWD2 (=56 kN), FWD3 (=84 kN), and FWD4 (=120 kN). The following Figure 7 shows the deflection basins as the four load varies for un-cracked and cracked pavement referred to sections 1 and 2. In the graphs, the dashed lines show the deflection basins in the cracked sections, while the solid lines refer to the uncracked sections. With 1.X all the FWD tests carried out in the first section (uncracked) are marked, while with 2.X those carried out in the second section (cracked). In both cases, X=1 and X=2 are the load tests carried out in the first point (in axis with the sensor), X=3 and X=4 are the tests carried out one meter before and X=5 and X=6 one meter later in the direction of travel.

Graphically it is already possible to highlight the significant difference in the behaviour of the two sections, a difference which is confirmed by the statistical diversity (t-test 0%) of the averages of the under-plate deflection values (D_0) and the IS_{200} index (D_{200}-D_0) as measure of the bearing capacity of the surface layer, as reported in the Table 1. Similarly, in the different plate positions at the same test section the deflections were similar showing homogeneity of the pavement structure and conditions.

Figure 8 shows how an un-cracked section of the road pavement under investigation responded to the four loads mentioned above. As expected, the amplitude and the length of the road vibro-acoustic signatures undergoes an increase in time domain because of the increase of the applied load (Figure 8). Note that the FWD performed a couple of loads for each load condition, and for this reason the following figure shows a couple of signals (almost coincident; see "signal 1" and "signal 2" in figure).

Subsequently, the road vibro-acoustic signatures were transformed from the time to the frequency domain (Power Spectral Density, PSD, versus Frequency). Four examples of road's

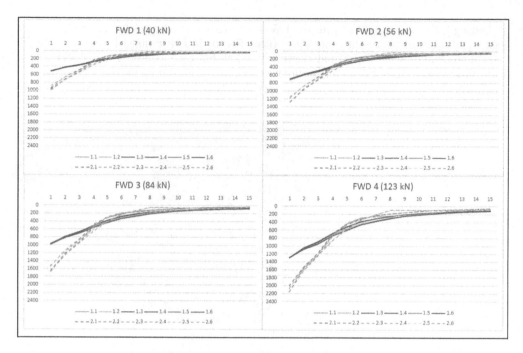

Figure 7. Deflection Basins corresponding to the 4 load levels.

Table 1. FWD data for section 1 and 2.

Section	Stress [kPa]	Force [kN]	D0 [m 10-6]	IS200 [m 10-6]	Section	Stress [kPa]	Force [kN]	D0 [m 10-6]	IS200 [m 10-6]
1-uncracked	545	39	516	69	2-cracked	529	37	948	183
	790	56	704	90		771	54	1222	229
	1181	83	976	126		1155	82	1608	296
	1737	123	1298	170		1696	120	2112	382

vibro-acoustic signature in the frequency domain related to two different SHSs, i.e., uncracked and cracked, are shown in Figure 9.

Note that a 1/3 octave band spectrum was derived from each FWD load for each load condition, and that, as shown in figure, this representation of the road signature (i.e., in the frequency domain) is really useful to define the inappreciable differences among the signals.

From the signatures shown in Figure 9, three features were extracted (the first two ones from the time domain, and the third one from the frequency domain), based on:

1) Δa: difference between the absolute maximum P and the absolute minimum N of the signal amplitudes (arbitrary unit, a.u.).
2) Δt: time delay of N-P (milliseconds).
3) fc: Spectral centroid of the spectrum (PSD vs. frequency) in the frequency range 16–2500 Hz. This is measured in Hz and can be defined as the "center of mass" of a spectrum (Schubert and Wolfe, 2006)). The figures below show the trends of the three aforementioned features as a function of the FWD load.

Based on the features of Figures 10 and 11, it is possible to state that there is room to structure algorithms to use the proposed monitoring solution to detect the occurrence of damages

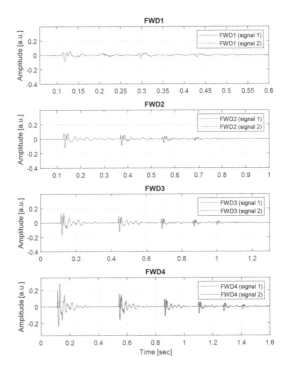

Figure 8. Example of vibro-acoustic responses of an un-cracked (left) and cracked (right) road pavements.

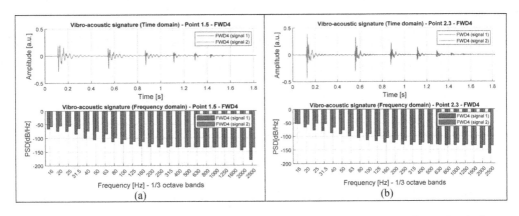

Figure 9. Examples of vibro-acoustic responses of the un-cracked (a) and cracked (b) road under investigation (points 1.5 and 2.3, respectively) and load condition FWD1 (= 120 kN).

in asphalt concrete pavements using meaningful features extracted from the time and the frequency domain of analysis. In more detail:

1) The first feature extracted, Δa, shows that: 1.1) Some peaks of the road signature increase if the road SHS get worse (Figure 10); 1.2) This allows identifying different thresholds between un-cracked and cracked conditions as a function of the load applied (e.g., about 0.6 for 120 kN); 1.3) This trend exhibits (unlike the other features extracted) a clear proportionality with the load applied (i.e., assumes larger values if the load increases); 1.4) This

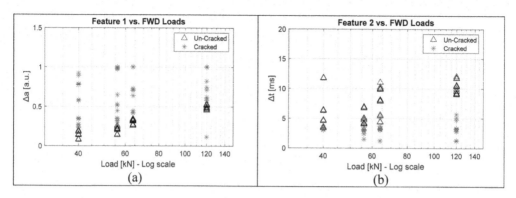

Figure 10. Features extracted in the time domain: (a) Feature 1 (i.e., Peak amplitude variation, Δa), and (b) Feature 2 (Peak delay, Δt) as a function of the FWD load for two road pavement conditions (un-cracked and cracked).

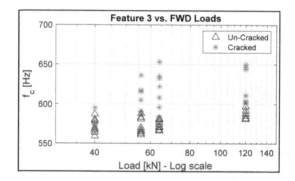

Figure 11. Feature extracted in the frequency domain: Feature 3 (i.e., Spectral centroid, fc) as a function of the FWD load, for 2 road pavement conditions.

shows (like the other features) that the dispersion of the values related to the cracked SHS is greater than that the one related to the un-cracked SHS.
2) The second feature extracted, Δt, shows how the road signatures get shorter (in time) when cracks are present on the road section (Figure 10), because of the reduction of the signals' queue.
3) The third feature extracted, fc, reports how cracks allow the road pavement to act as a mechanical filter that absorbs a part of the signals energy. This latter can be detected in the frequency domain (range of operation 16-2500 Hz) in terms of reduction of the power spectral density, PSD (for selected frequency components). In this case, the damages present into the cracked road section (point 2 above) seem to affect especially the frequencies below 2500 Hz (Figure 9), suggesting that crack lengths are probably higher than the thickness of the bituminous layer. Indeed, low frequencies correspond to high wavelengths and then to a quite high level of damage. This latter consideration could be supported by the fact that higher loads (e.g., greater than 56 kN) allow better discriminating the two SHSs taken into account in this study, because they cause higher stress into the lower layers. In contrast, when a load of 40 KN is used, the Feature 3 appears to lose its ability to detect the SHS.
4) Better insights between un-cracked and cracked condition were obtained using the lowest and the highest loads.

6 CONCLUSIONS AND FUTURE WORKS

Based on the first study, in Table 2 a SWOT analysis of the application of smart vibro-acoustic sensors for the assessment of pavement residual life is proposed to drive the future development of the proposed and similar pavement monitoring systems.

Table 2. SWOT analysis for the application of smart vibro-acoustic sensors.

STRENGHTS *(internal, positive factor)*	WEAKNESSES *(internal, negative factor)*
• Interesting NDT testing comparison results • Reduced costs for a spread urban network pavement management • Reduced number of operators • Application of mathematical and analytical models • Application of a patented NDT monitoring solution (system and method).	• Reliability of the results • Variability of the results • Need of calibration tests for the data acquisition • Need of further development of a patented NDT monitoring solution (system and method) to obtain self-powered wireless sensing nodes
OPPORTUNITIES *(external, positive factor)*	**THREATS** *(external, negative factor)*
• Development of new models for the interpretation of the results • Approach to a field of engineering technique that is not "saturated " • Possibility of agreements with public authorities or private companies that need an advanced know-how for development of its own procedures	• Results distorted by unexpected and/or unforeseen conditions and on-board effects cannot be assessed • Returning of untruthful data about the pavement condition • Mistrust of the Public Administrations in the use of these devices for pavement management

Results, even if preliminary and based on a limited sample size, show that FWD can be effectively used to test and calibrate the new sensor proposed in the paper. It is possible to state that the proposed monitoring solution can be used to detect the occurrence of damages in asphalt concrete pavements using meaningful features extracted from the time (i.e., related to signals' amplitude) and the frequency (i.e., related to signals' Power Spectral Density, PSD) domain of analysis. Future works include: 1) The improvement of the proposed monitoring system (to obtain a self-powered wireless system) in order to address some of the objectives of the ongoing Italian project USR342 (cited above); 3) extension of the test sections including different pavement thickness and crack patterns; 3) The extension of the data analysis (e.g., using machine learning algorithms) by mean of further features manually or automatically extracted (Cafiso et al., 2020).

ACKNOWLEDGEMENT

This work has been financed by the PRIN 2017-2022 within the ongoing project USR342 and by the University of Catania within the project TIMUC (PIACERI).

REFERENCES

Al-Qadi, I., & Lahouar, S., 2005. *Part 4: Portland cement concrete pavement: measuring rebar cover depth in rigid pavements with ground-penetrating radar*. Transportation Research Record: Journal of the Transportation Research Board, (1907), 80–85

ASTM D4748-10, 201. *Standard Test Method for Determining the Thickness of Bound Pavement Layers Using Short-Pulse Radar*, ASTM International, West Conshohocken, PA, 2015, doi: https://doi.org/10.1520/D4748-10R15

Cafiso, S. and Di Graziano A., 2009. *Monitoring and performance of AC pavements reinforced with steel mesh*. International Journal of Pavement Research and Technology, Volume 2 Number 3, ISSN 1996-6814, Chinese Society of Pavement Engineering

Cafiso, S., Capace, B., D'agostino, C., Delfino, E., Di Graziano, A., 2018. *Introduction of new systems for evaluation of ballast bearing capacity*. Proceedings of the 10th International Conference on the Bearing Capacity of Roads, Railways and Airfields, pp. 1993–2001

Cafiso, S., Di Graziano, A., Goulias, D., D'Agostino C., 2019. *Distress and profile data analysis for condition assessment in pavement management systems*. International Journal of Pavement Research and Technology - Volume 12, Issue 5, September 2019, pp 527–536

Cafiso, S., Di Graziano, A., Fedele, R., Marchetta, V., Praticò, F. (2020) *Sensor-based pavement diagnostic using acoustic signature for moduli estimation*. International Journal of Pavement Research and Technology, 13 (6), pp. 573–580.

Ceylan, H., Bayrak, M.B., Gopalakrishnan, K., 2014. *Neural networks applications in pavement engineering: A recent survey*. International Journal of Pavement Research and Technology.

Ceylan, H., Gopalakrishnan, K., Kim, S., Taylor, P.C., Prokudin, M., Buss, A.F., 2013. *Highway infrastructure health monitoring using micro-electromechanical sensors and systems (MEMS)*. Journal of Civil Engineering and Management.

Daniels, D.J., 2004. *Ground Penetrating Radar*, 2nd Edition. IET

Di Graziano A., Marchetta V., Cafiso S. 2020 Structural health monitoring of asphalt *pavements using smart sensor networks: a comprehensive review*. Journal of Traffic and Transportation Engineering (in press).

Fedele, R., 2020. *Smart road infrastructures through vibro-acoustic signature analyses*. Smart Innovation, Systems and Technologies 178, 1481–1490.

Fedele, R., Merenda, M., Praticò, F.G., Carotenuto, R., Della Corte, F.G., 2018. *Energy harvesting for IoT road monitoring systems*. Instrum. Mesure Metrologie 17, 605–623.

Fedele, R., Praticò, F.G., 2019. *Monitoring infrastructure asset through its acoustic signature*. Inter-Noise 2019 Madrid - 48th Int. Congress and Exhibition on Noise Control Eng.

Fernandes, F.M., Pais, J.C., 2017. *Laboratory observation of cracks in road pavements with GPR*. Construction and Building Materials.

Goulias, D.G., Cafiso, S., Di Graziano, A., Saremi, S.G., Currao, V., 2020. *Condition Assessment of Bridge Decks through Ground-Penetrating Radar in Bridge Management Systems*. Journal of Performance of Constructed Facilities, 34 (5) pp 1–13

Grace, R., 2015. *Sensors to support the IoT for infrastructure monitoring: technology and applications for smart transport/smart buildings*. MEPTEC-IoT.

Hasni, H., Alavi, A.H., Chatti, K., Lajnef, N., 2017. *A self-powered surface sensing approach for detection of bottom-up cracking in asphalt concrete pavements: Theoretical/numerical modeling*. Construction and Building Materials 144, 728–746.

Iodice, M., Muggleton, J.M., Rustighi, E., 2020. *The in-situ evaluation of surface-breaking cracks in asphalt using a wave decomposition method*. Nondestructive Testing and Evaluation.

Iuele, T., Praticò, F.G., Vaiana, R., 2019. *Fine aggregate properties vs asphalt mechanical behavior: An experimental investigation*. Proceedings of the World Conference on Pavement and Asset Management, WCPAM 2017.

Lahouar, S., Al-Qadi, I., Loulizi, A., Clark, T., & Lee, D., 2002. *Approach to determining in situ dielectric constant of pavements: development and implementation at interstate 81 in Virginia*. Transportation Research Record: Journal of the TRB, (1806), 81–87

Merenda, M., Carotenuto, R., Della Corte, F.G., Giammaria Pratico, F., Fedele, R., 2020. *Self-powered wireless IoT nodes for emergency management*. 20th IEEE Mediterranean Electrotechnical Conference, MELECON 2020 - Proceedings.

Mounier, D., Di Benedetto, H., Sauzéat, C., 2012. *Determination of bituminous mixtures linear properties using ultrasonic wave propagation*. Construction and Building Materials 36, 638–647.

Odoki J.B., Di Graziano A., Akena R., 2015. *A multicriteria methodology for optimising road investments*. Proceedings of the Institution of Civil Engineers ICE Transport, Volume 168 Issue TR1, pages 34–47

Ouma, Y.O., Hahn, M., 2017. Pothole detection on asphalt pavements from 2D-colour pothole images using fuzzy c-means clustering and morphological reconstruction. Automation in Construction 83, 196–211. https://doi.org/10.1016/j.autcon.2017.08.017

Praticò, F.G., 2001. *Roads and Loudness: A More Comprehensive Approach*. Road Materials and Pavement Design 2, 359–377.

Praticò, F.G., Fedele, R., Naumov, V., Sauer, T., 2020. *Detection and monitoring of bottom-up cracks in road pavement using a machine-learning approach*. Algorithms. Kraków, Pol

Praticò, F.G., Fedele, R., Vizzari, D., 2017. *Significance and reliability of absorption spectra of quiet pavements*. Construction and Building Materials 140, 274–281.

Saarenketo, T., 2006. *Electrical properties of road materials and subgrade soils and the use of ground penetrating radar in traffic infrastructure surveys* (PhD Thesis). University of Oulu

Schubert, E., Wolfe, J., 2006. *Timbral brightness and spectral centroid*. Acta Acustica United with Acustica 92, 820–825.

Yi, C.W., Chuang, Y.T., Nian, C.S., 2015. *Toward Crowdsourcing-Based Road Pavement Monitoring by Mobile Sensing Technologies*. IEEE Transactions on Intelligent Transportation Systems 16, 1905–1917.

Yu, Y., Zhao, X., Shi, Y., Ou, J., 2013. *Design of a real-time overload monitoring system for bridges and roads based on structural response*. Measurement: Journal of the International Measurement Confederation 46, 345–352.

Zhou, Z., Liu, W., Huang, Y., Wang, H., Jianping, H., Huang, M., Jinping, O., 2012. *Optical fiber Bragg grating sensor assembly for 3D strain monitoring and its case study in highway pavement*. Mechanical Systems and Signal Processing 28, 36–49.

Author Index

Abdelmawla, A. 448
Adomako, S. 131
Ahmed, A. 277
Akhtar, N. 131
Alamnie, M. 118
Alenezi, D. 108
Aponte, D. 172
Ashlock, J.C. 139
Ashtiani, R. 355
Ashtiani, R.S. 335
Astolfi, A. 325

Baloochi, H. 172
Baltzer, S. 288
Barbieri, D.M. 131
Barra, M. 172
Bjurström, H. 87

Cafiso, S. 475
Carreño, N. 28
Cepriá, J.J. 172
Cetin, B. 139
Ceylan, H. 139
Chen, Z.G. 78
Chen, Z.N. 78
Cho, S. 299
Crispino, M. 226

Danner, T. 131
Decarlo, C. 387
Di Graziano, A. 475
Dinegdae, Y. 277
Dohnalkova, B. 18
Dutney, H. 367

Elshaer, M. 387
Engelsen, C.J. 131
Eriksson, O. 48
Erlingsson, S. 48, 239, 277, 409

Fedele, R. 325, 475
Feng, D.C. 38
Fladvad, M. 409
Fleischel, O. 28
Flintsch, G.W. 430
Fountain, G. 162
Francois, A. 153, 440

Gaudefroy, V. 87
Georgouli, K. 466
Gewanlal, C. 182
Giancontieri, G. 87
Gkyrtis, K. 466
Gong, S. 48
Graziani, A. 87
Grosek, J. 18
Guarin, A. 48

Hall, F. 3
Haritonovs, V. 96
Hoeller, S. 263
Hoff, I. 59
Hoff, I. 118
Hornych, P. 87

Jamshidi, A. 108, 250
Jansen, D. 288, 419
Jensen, K. 458
Jing, P. 305

Kalantari, M. 288
Kalman, B. 87
Katicha, S.W. 430
Kawalec, J. 182
Ketabdari, M. 226
Kim, S.S. 448
Kolisoja, P. 399

László, P. 299
Levenberg, E. 346

Li, C. 139
Li, J.J. 38
Lo Presti, D. 87
Loizos, A. 68, 466
Lundberg, J. 48

Macan, T. 18
Martínez, A. 172
Mazurowski, P. 182
Mehta, Y. 153, 440
Merijs-Meri, R. 96
Mignini, C. 87
Miró, R. 172
Moharekpour, M. 263
Mollenhauer, K. 87
Morovatdar, A. 335

Nevosad, Z. 18
Nielsen, C.P. 399, 458
Nielsen, J. 346
Norby, M. 131

Oeser, M. 28, 263
Offenbacker, D. 153, 440
Olsen, K. 346
Orejana, R. 172

Pei, Z.S. 38
Pellicano, G. 475
Pettinari, M. 288
Plati, C. 68, 466
Pomoni, M. 68
Praticò, F. 475
Praticò, F.G. 325

Qin, W.J. 78

Radenberg, M. 315
Rahimi Nahouiy, M. 315
Rahman, M.S. 239, 277

Riekstins, A. 96
Rodriguez, E. 355

Saarenketo, T. 399
Satvati, S. 139
Scavone, M. 430
Severino, A. 475
Skar, A. 346

Taddesse, E. 118
Tamrakar, P. 162
Thorstensen, R.T. 131
Toraldo, E. 226

Tóth, C. 299

Varin, P. 399
Vennapusa, P. 162
Vieira, T. 48

Wayne, M.H. 162
Weisser, H. 195
White, D.J. 162
White, G. 3, 108, 195, 210, 250, 367
Winter, M. 87
Wu, Y. 139

Xiao, J. 305
Xue, L. 305
Xue, Z. 139

Yi, J.Y. 38, 78

Zamara, K. 182
Zavrel, T. 18
Zeilinger, M. 28
Zhang, D. 305
Zhang, X. 59
Zhou, W.Y. 78
Zicāns, J. 96